SEMIDEFINITE OPTIMIZATION *and* CONVEX ALGEBRAIC GEOMETRY

MOS-SIAM Series on Optimization

This series is published jointly by the Mathematical Optimization Society and the Society for Industrial and Applied Mathematics. It includes research monographs, books on applications, textbooks at all levels, and tutorials. Besides being of high scientific quality, books in the series must advance the understanding and practice of optimization. They must also be written clearly and at an appropriate level for the intended audience.

Series Volumes

Blekherman, Grigoriy, Parrilo, Pablo A., and Thomas, Rekha R., editors, *Semidefinite Optimization and Convex Algebraic Geometry*

Delfour, M. C., *Introduction to Optimization and Semidifferential Calculus*

Ulbrich, Michael, *Semismooth Newton Methods for Variational Inequalities and Constrained Optimization Problems in Function Spaces*

Biegler, Lorenz T., *Nonlinear Programming: Concepts, Algorithms, and Applications to Chemical Processes*

Shapiro, Alexander, Dentcheva, Darinka, and Ruszczyński, Andrzej, *Lectures on Stochastic Programming: Modeling and Theory*

Conn, Andrew R., Scheinberg, Katya, and Vicente, Luis N., *Introduction to Derivative-Free Optimization*

Ferris, Michael C., Mangasarian, Olvi L., and Wright, Stephen J., *Linear Programming with MATLAB*

Attouch, Hedy, Buttazzo, Giuseppe, and Michaille, Gérard, *Variational Analysis in Sobolev and BV Spaces: Applications to PDEs and Optimization*

Wallace, Stein W. and Ziemba, William T., editors, *Applications of Stochastic Programming*

Grötschel, Martin, editor, *The Sharpest Cut: The Impact of Manfred Padberg and His Work*

Renegar, James, *A Mathematical View of Interior-Point Methods in Convex Optimization*

Ben-Tal, Aharon and Nemirovski, Arkadi, *Lectures on Modern Convex Optimization: Analysis, Algorithms, and Engineering Applications*

Conn, Andrew R., Gould, Nicholas I. M., and Toint, Phillippe L., *Trust-Region Methods*

SEMIDEFINITE OPTIMIZATION *and* CONVEX ALGEBRAIC GEOMETRY

Edited by

Grigoriy Blekherman
Georgia Institute of Technology
Atlanta, Georgia

Pablo A. Parrilo
Massachusetts Institute of Technology
Cambridge, Massachusetts

Rekha R. Thomas
University of Washington
Seattle, Washington

Society for Industrial and Applied Mathematics
Philadelphia

Mathematical Optimization Society
Philadelphia

10 9 8 7 6 5 4 3 2 1

Library of Congress Cataloging-in-Publication Data

Semidefinite optimization and convex algebraic geometry / edited by Grigoriy Blekherman, Georgia Institute of Technology, Atlanta, Georgia, Pablo A. Parrilo, Massachusetts Institute of Technology, Cambridge, Massachusetts, Rekha R. Thomas, University of Washington, Seattle, Washington.

pages cm. – (MOS-SIAM series on optimization)

Includes bibliographical references and index.

ISBN 978-1-611972-28-3

1. Semidefinite programming. 2. Convex geometry. 3. Geometry, Algebraic. I. Blekherman, Grigoriy. II. Parrilo, Pablo A. III. Thomas Rekha R., 1967-

QA402.5.S445 2012

516.3'5–dc23

2012027582

Contents

List of Contributors

Grigoriy Blekherman
Georgia Institute of Technology

João Gouveia
University of Coimbra

William Helton
University of California, San Diego

Igor Klep
The University of Auckland

Scott McCullough
University of Florida

Jiawang Nie
University of California, San Diego

Pablo A. Parrilo
Massachusetts Institute of Technology

Mihai Putinar
University of California, Santa Barbara

Philipp Rostalski
University of Frankfurt and Drägerwerk AG & Co. KGaA, Lübeck

Bernd Sturmfels
University of California, Berkeley

Rekha Thomas
University of Washington

List of Figures

Preface

In the past decade there has been a surge of interest in algebraic approaches to optimization problems defined in terms of multivariate polynomials. Fundamental mathematical challenges that arise in this program include understanding the structure of nonnegative polynomials, the interplay between efficiency and complexity of different representations of algebraic sets, and the development of effective algorithms. Remarkably, and perhaps unexpectedly, convexity provides a new viewpoint and a powerful framework for addressing these questions. This naturally brings us to the intersection of *algebraic geometry*, *optimization*, and *convex geometry*, with an emphasis on algorithms and computation. This emerging area has become known as *convex algebraic geometry*.

Our aim is to provide an accessible and unifying introduction to the many facets of this fast-growing interdisciplinary area. Each chapter addresses a fundamental aspect of convex algebraic geometry, ranging from the well-established core mathematical theory to the forefront of current research and open questions. Throughout we showcase the rich interactions between theory and applications.

This book is suitable as a textbook in a graduate course in mathematics and engineering. The chapters make connections to several areas of pure and applied mathematics and contain exercises at many levels, providing multiple entry points for readers with varied backgrounds.

We thank the National Science Foundation for funding a Focused Research Group grant (2008–2011) awarded to Bill Helton, Jiawang Nie, Pablo A. Parrilo, Bernd Sturmfels, and Rekha R. Thomas. This award enabled a flurry of research activity in semidefinite optimization and convex algebraic geometry. Several workshops and conferences were organized under this grant's support. In particular this book was inspired by the lectures at the workshop LMIPO organized by Bill Helton and Jiawang Nie at the University of California, San Diego in March 2010.

We thank all our contributors for their hard work and perseverance through multiple rounds of edits. We also thank Tom Liebling, Sara Murphy, and Ann Manning Allen at SIAM for their support and patience with the production of this book. Special thanks to our students and colleagues who read versions of this book and sent us comments, in particular Chris Aholt, Hamza Fawzi, Fabiana Ferracina, Alexander Fuchs, Chris Jordan-Squire, Frank Permenter, James Pfeiffer, Stefan

Richter, Richard Robinson, Raman Sanyal, James Saunderson, Rainer Sinn, and Thao Vuong.

Greg Blekherman[1]	Pablo A. Parrilo[2]	Rekha R. Thomas[3]
Atlanta, GA	Cambridge, MA	Seattle, WA

[1]The work of Greg Blekherman was supported by a Sloan Fellowship, NSF grant DMS-0757212, the Mittag-Leffler Institute Sweden, and IPAM UCLA.

[2]The work of Pablo A. Parrilo was supported by NSF grant DMS-0757207 and a Finmeccanica Career Development Chair.

[3]The work of Rekha R. Thomas was supported by NSF grants DMS-0757371 and DMS-1115293 and a Robert R. and Elaine F. Phelps Endowed Professorship.

List of Notation

Basics:	
fields, rings	$\mathbb{R}, \mathbb{C}, \mathbb{P}, \mathbb{Q}, \mathbb{Z}$
nonnegative integers	$\mathbb{N}$
nonnegative orthant	$\mathbb{R}^n_+$
positive orthant	$\mathbb{R}^n_{++}$
standard simplex in $\mathbb{R}^n_+$	$\Delta_n := \{x \in \mathbb{R}^n_+ : \sum x_i = 1\}$
standard basis vectors	e_i
Matrices:	
$m \times n$ matrices	$\mathbb{R}^{m \times n}$
matrix brackets	[ ]
$n \times n$ symmetric matrices	$\mathcal{S}^n$
$n \times n$ positive semidefinite definite matrices	$\mathcal{S}^n_+$
$n \times n$ positive definite matrices	$\mathcal{S}^n_{++}$
inner product in $\mathcal{S}^n$	$\langle A, B \rangle$
matrix multiplication	$A \cdot B$
trace	Tr
matrix transpose	A^T
determinant	$\det M$
rank	$\operatorname{rank} M$
diagonal of a matrix M as a vector	$\operatorname{diag}(M)$
diagonal matrix obtained from a matrix M	$\operatorname{Diag}(M)$
lower triangular matrix from matrix M	$\operatorname{Tril}(M)$
turning a vector v into a diagonal matrix	$\operatorname{Diag}(v)$
block diagonal matrix with blocks A, B etc	$\operatorname{BlockDiag}(A, B, ...)$
positive semidefinite	$\succeq 0$
positive definite	$\succ 0$
max/min eigen/singular value	$\lambda_{\max}, \sigma_{\min}$
Geometry:	
p-norm	$\|u\|_p$
ball with center u, radius r	$B(u, r)$
vector space dual	V^*
orthogonal complement of vector space	$V^\perp$
dimension	$\dim V$

codimension	$\operatorname{codim} V$
cone dual	C^*
polar dual of convex body	P°
dual face to an exposed face	$F^\diamond$
dual variety	X^*
interior of a set	$\mathrm{int}(C)$
boundary of set	∂C
algebraic boundary	$\partial_a C$
closure of set	$\mathrm{cl}(C)$ or $\overline{C}$
convex hull of set C	$\mathrm{conv}(C)$
conical hull of set C	$\mathrm{cone}(C)$
gauge function of a convex body K	$G_K(x)$

Optimization:

optimal solution	$u^\star$
semidefinite program	SDP
kth theta body of ideal I	$\mathrm{TH}_k(I)$
characteristic vector of a set S	χ^S

Algebra:

ideal generated by	$\langle f_1, \dots, f_m \rangle$
variety of ideal	$V_{\mathbb{R}}(I), V_{\mathbb{C}}(I)$
vanishing ideal of a set	$I(S)$
Jacobian	$\mathrm{Jac}(\ )$
gradient	∇
Hessian	∇^2
singular locus	$\mathrm{Sing}(\ )$
smooth points in a variety	X_{reg}
polynomial ring in n variables	$\mathbb{R}[x], \mathbb{C}[x]$
polynomials in n variables, degree at most d	$\mathbb{R}[x]_{n,d}$
if n clear	$\mathbb{R}[x]_d$
monomials of degree at most d	$[x]_d$
$\alpha \in \mathbb{N}^n$ (for exponents of monomials)	$\lvert\alpha\rvert = \sum \alpha_i$
nonnegative polynomials in n variables, degree at most $2d$	$P_{n,2d}$
if n is clear	P_{2d}
sum of squares in n variables of degree at most $2d$	$\Sigma_{n,2d}$
if n is clear	Σ_{2d}
forms in n variables, degree equal to d	$\mathbb{R}[x]_{\mathbf{n,d}}$
if n clear	$\mathbb{R}[x]_{\mathbf{d}}$
monomials of degree d	$[x]_{\mathbf{d}}$
nonnegative forms in n variables, degree $2d$	$P_{\mathbf{n,2d}}$
if n is clear	$P_{\mathbf{2d}}$
sos forms in n variables of degree $2d$	$\Sigma_{\mathbf{n,2d}}$
if n is clear	$\Sigma_{\mathbf{2d}}$

sos polynomials mod an ideal I	$\Sigma(I)$
polynomials in $\mathbb{R}[x]_{n,d}$ that are k-sos mod I	$\Sigma^k_{n,d}(I)$
if n is clear	$\Sigma^k_d(I)$
affine linear polynomials in above	$\Sigma^k_1(I)$
Newton polytope of f	$\mathcal{N}(f)$
linear functionals on $\mathbb{R}[x]$	ℓ
linear functionals that are evaluations at v	ℓ_v
quadratic forms on $\mathbb{R}[x]_{\mathbf{n},\mathbf{d}}$	$S^{n,d}$
nonnegative quadratic forms in $S^{n,d}$	$S^{n,d}_+$
preorder of $g_1, \ldots, g_m$/truncated	$\mathbf{preorder}(g_1, \ldots, g_m)$, $\mathbf{preorder}_k(g_1, \ldots, g_m)$
quadratic module of $g_1, \ldots, g_m$/truncated	$\mathbf{qmodule}(g_1, \ldots, g_m)$, $\mathbf{qmodule}_k(g_1, \ldots, g_m)$

Chapter 1

What is Convex Algebraic Geometry?

Grigoriy Blekherman, Pablo A. Parrilo, and Rekha R. Thomas

Convex algebraic geometry is an evolving subject area arising from a synthesis of ideas and techniques from optimization, convex geometry, and algebraic geometry. The central objects of study in this rapidly developing field are convex sets with algebraic structure. Such sets occur naturally, and have been analyzed independently, in convex geometry, real algebraic geometry, optimization, and analysis, but only recently has a unified perspective that systematically takes advantage of the interactions between algebra and convexity emerged. This viewpoint provides rich connections across the mathematical sciences and novel tools for applied mathematics and engineering. This book presents the foundations of convex algebraic geometry and provides an accessible entry point for students and researchers.

A fundamental class of algebraically defined convex sets arises from intersections of the cone of positive semidefinite matrices with affine subspaces. These sets are called *spectrahedra* and are automatically convex and endowed with rich algebraic structure. The problem of optimizing a linear function over a spectrahedron is called *semidefinite programming*. Such problems admit efficient algorithms, enable many applications, and have been studied extensively in the past few decades. These basic concepts are introduced in Chapter 2.

The structure of nonnegative polynomials is a central theme in polynomial optimization and real algebraic geometry. A classical question is the existence of a representation that makes the nonnegativity of a polynomial apparent. Such representations naturally involve *sums of squares* and provide *certificates* for nonnegativity. In addition to classical existence questions, convex algebraic geometry is concerned with constructive aspects and efficient computation. Semidefinite optimization is the algorithmic engine behind the effective computation of sums of squares certificates. Chapter 3 provides a gentle introduction to these techniques. The underlying geometric aspects of nonnegative and sum of squares polynomials are then analyzed in detail in Chapter 4.

Chapter 5 presents a unified viewpoint of *duality*, a powerful and recurring theme across algebraic geometry, convexity, and optimization. As such, it naturally plays a central role in convex algebraic geometry. The philosophy of this chapter is that the different notions of duality become nearly identical when applied to convex sets with algebraic structure.

A natural question in optimization is to determine what problems can be modeled as semidefinite programs, which translates into the problem of representing or efficiently approximating convex sets as spectrahedra or their projections. These questions are addressed in Chapter 6. A particularly nice yet important and challenging class of sets to represent and approximate are convex hulls of *real varieties.* This is the subject of Chapter 7. Sums of squares provide a universal approach to the above representability questions, although a full picture, particularly with regard to efficiency issues, is still elusive.

Sums of squares are also prominent in *noncommutative* and *Hermitian* contexts. Nonnegativity is a much more rigid property in the noncommutative setting, and thus some parts of the classical commutative theory become more elegant and structured. Chapter 8 offers a friendly tour through noncommutative convexity and nonnegativity. The Hermitian case is motivated by fundamental questions in operator theory and complex analysis, and analytic considerations offer new insights and methods. This is the topic of Chapter 9. Both of these areas have deep roots in classical mathematics and strong connections to engineering applications.

Besides these central themes, convex algebraic geometry offers fertile ground for synergies with other areas such as representation theory, computational complexity, combinatorics, harmonic analysis, and probability theory. These interactions provide exciting opportunities for theoretical developments, computational methods, and practical applications, as can be witnessed by the growing literature.

The different chapters in this book are interwoven by many recurring themes and common ideas. However they can also be read independently by a reader who is interested in a specific topic. Chapters 2 and 3 introduce the reader to the core ideas and techniques in the book. The following chapters delve deeper into their own topics while also presenting applications and links to the rest of the book.

Chapter 2

Semidefinite Optimization

Pablo A. Parrilo

In this chapter we introduce one of the core theoretical and computational techniques in convex algebraic geometry, namely, *semidefinite optimization*. We begin by reviewing linear programming and proceed to define and discuss semidefinite programs from the algebraic, geometric, and computational perspectives. We define *spectrahedra* as the feasible sets of semidefinite programs, study their properties, and discuss numerous examples. Despite the many parallels, the duality theory of semidefinite optimization is more complicated than in the case of linear programming, and we elaborate on the similarities and differences. We also showcase a number of applications of semidefinite optimization in several areas of applied mathematics and engineering and give a short discussion of algorithmic and software aspects. For the convenience of the reader, we present additional background material on convex geometry and optimization in Appendix A.

2.1 From Linear to Semidefinite Optimization

Semidefinite optimization is a branch of convex optimization that is of great theoretical and practical interest. Informally, the main idea is to generalize linear programming and the associated feasible sets (polyhedra) to the case where the decision variables are symmetric matrices, and the inequalities are to be understood as matrices being positive semidefinite. Formal definitions and examples will be presented shortly in Subsection 2.1.2, preceded by a review of the familiar case of linear programming. A few selected standard references for linear programming and their applications are the books [5, 12, 29, 42].

2.1.1 Linear Programming

Linear programming is the problem of minimizing a linear function subject to linear constraints. A linear programming problem (LP) in standard form is usually written as

$$\begin{array}{ll} \text{minimize} & c^T x \\ \text{subject to} & Ax = b, \\ & x \geq 0, \end{array} \tag{LP-P}$$

where $A \in \mathbb{R}^{m \times n}$, $b \in \mathbb{R}^m$, and we are minimizing over the decision variable $x \in \mathbb{R}^n$. The inequality $x \geq 0$ is interpreted componentwise, i.e., $x_i \geq 0$ for $i = 1, \dots, n$.

Geometrically, an LP problem has a nice and natural interpretation. Its feasible set is the intersection of an affine subspace (defined by the equations $Ax = b$), and the nonnegative orthant. Since it is the intersection of two convex sets, the feasible set of (LP-P) is always convex. In general, a set defined by finitely many linear inequalities or equations is called a *polyhedron*, and it is always convex. Thus, linear programming corresponds exactly to the minimization of a linear function over a polyhedron. If a polyhedron is bounded, it is called a *polytope*.

Perhaps one of the most remarkable and useful features of linear programming is that to every LP problem we can associate a corresponding *dual* problem. This is another LP problem ("its dual LP"), which for the case of (LP-P) is

$$\begin{array}{l} \text{maximize } b^T y \\ \text{subject to } \; A^T y \leq c. \end{array} \tag{LP-D}$$

Notice that here we are again optimizing a linear function over a polyhedron. As we will see, there are very natural and direct algebraic relationships between the primal problem (LP-P) and its dual problem (LP-D).

Remark 2.1. *In practice, LP problems may not naturally present themselves in the form* (LP-P)*, where all the decision variables are nonnegative and only equality constraints are present, or the form* (LP-D)*, where there are no sign restrictions on the variables and only inequalities appear. However, they can always be put in either form, by introducing additional slack variables and/or splitting variables if necessary. The details can be found in any textbook on linear programming.*

Example 2.2. Consider the following LP problem:

$$\text{minimize } x_1 - 8x_2 \qquad \text{subject to} \quad \left\{ \begin{array}{rcl} -x_1 + 3x_2 + x_3 & = & 4, \\ 4x_1 - x_2 + x_4 & = & 6, \\ x_1, x_2, x_3, x_4 & \geq & 0. \end{array} \right. \tag{2.1}$$

The feasible region is a two-dimensional polyhedron. Its projection into the (x_1, x_2)-plane is drawn in Figure 2.1. Notice that the optimal solution is achieved at a vertex, namely, $x^\star = (2, 2, 0, 0)$, with optimal cost $p^\star = -14$.

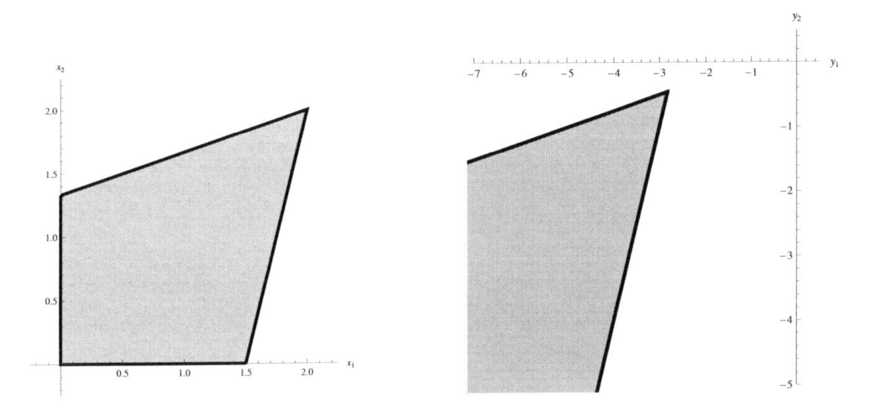

Figure 2.1. *Feasible sets of the primal and dual LP problems* (2.1) *and* (2.2).

The corresponding dual LP is

$$\text{maximize } \ 4y_1 + 6y_2 \qquad \text{subject to} \quad \begin{cases} -y_1 + 4y_2 & \leq & 1, \\ 3y_1 - y_2 & \leq & -8, \\ y_1 & \leq & 0, \\ y_2 & \leq & 0. \end{cases} \tag{2.2}$$

The dual feasible set (y_1, y_2) is presented in the same figure, with optimal solution $y^\star = (-\frac{31}{11}, -\frac{5}{11})$ and optimal cost $d^\star = -14$. For this example we have

$$p^\star = d^\star = -14,$$

and thus the optimal values of the primal and dual problems are the same. ∎

Even in this simple example, we can observe many of the important features of linear programming. The following facts are well known.

Geometry of the feasible set: The feasible sets of linear programs are *polyhedra.* The geometry of polyhedra is quite well understood. In particular, the Minkowski-Weyl theorem (e.g., Appendix A, [5], or [48, Section 1.1]) states that every polyhedron P is finitely generated, i.e., it can be written as

$$P = \operatorname{conv}(u_1, \dots, u_r) + \operatorname{cone}(v_1, \dots, v_s),$$

where u_i, v_i are the *vertices* and *extreme rays* of P, respectively, and the *convex hull* and *conical hull* are defined by

$$\operatorname{conv}(u_1, \dots, u_r) = \left\{ \sum_{i=1}^{r} \lambda_i u_i \;\middle|\; \sum_{i=1}^{r} \lambda_i = 1, \quad \lambda_i \geq 0, \quad i = 1, \dots, r \right\}$$

and

$$\operatorname{cone}(v_1, \dots, v_s) = \left\{ \sum_{i=1}^{s} \lambda_i v_i \;\middle|\; \lambda_i \geq 0, \quad i = 1, \dots, s \right\}.$$

Rational solutions: Unless the problem is unbounded, the optimal solution of a linear programming problem is always achieved at extreme points of the feasible set. Since these correspond to vertices of a polyhedron, the solution can be characterized in terms of a system of linear equations, corresponding to the equations and inequalities that are active at the optimal point. Thus, if the problem description (i.e., the matrices A, b, c) is given by rational numbers, there are always extreme points that are rational and achieve the optimal cost.

Weak duality: For *any* feasible solutions x, y of (LP-P) and (LP-D), respectively, it always holds that

$$c^T x - b^T y = x^T c - (Ax)^T y = x^T (c - A^T y) \geq 0, \tag{2.3}$$

where the last inequality follows from the feasibility conditions $x \geq 0$ and $A^T y \leq c$. Thus, from any feasible dual solution one can obtain a lower bound on the value of the primal. Conversely, primal feasible solutions give upper bounds on the value of the dual.

Strong duality: If both primal and dual problems are feasible, then they achieve exactly the same optimal value, and there exist optimal feasible solutions $x^\star, y^\star$ such that $c^T x^\star = b^T y^\star$. This is a consequence of the separation theorems for convex sets; see, e.g., Section A.3.3 in Appendix A.

Complementary slackness: Strong duality, combined with (2.3), implies that at optimality we must have

$$x_i^\star (c - A^T y^\star)_i = 0, \qquad i = 1, \ldots, n.$$

In other words, there is a correspondence between primal variables and dual inequalities that says that whenever a primal variable is nonzero, the corresponding dual inequality must be tight.

In the linear programming case, these properties are well known and relatively easy to prove. Interestingly, as we will see in the next section, some of these properties will break down as soon as we leave linear programming and go to the more general case of semidefinite programming. These technical aspects will cause some minor difficulties, although with the right assumptions in place, the resulting theory will closely parallel the linear programming case.

Exercise 2.3. Consider a finite set of points $S = \{a_1, a_2, \ldots, a_n\}$ in $\mathbb{R}^d$, where $n > d$. Prove using linear programming duality that exactly one of the following statements must hold:

- The origin is in the convex hull of S.
- There exists a hyperplane passing through the origin, such that all points a_i are strictly on one side of the hyperplane.

Exercise 2.4. Consider the set of $n \times n$ matrices with nonnegative entries that have all row and column sums equal to 1 (i.e., the *doubly stochastic* matrices).

1. Write explicitly the equations and inequalities describing this set for $n = 2, 3, 4$.

2. Compute (using `CDD`, `lrs`, or other software; see Section 2.3.2) all the extreme points of these polytopes.

3. How many extreme points did you find? What is the structure of the extreme points? Can you conjecture what happens for arbitrary values of n?

4. Google "Birkhoff–Von Neumann theorem," and check your guess.

2.1.2 Semidefinite Programming

Semidefinite programming is a broad generalization of linear programming, where the decision variables are symmetric matrices. A semidefinite programming problem (SDP) corresponds to the optimization of a linear function subject to *linear matrix inequality* (LMI) constraints. Semidefinite programs are convex optimization problems and have very appealing numerical properties (e.g., [7, 44, 45]).

Our notation is as follows: the set of real symmetric $n \times n$ matrices is denoted by $\mathcal{S}^n$. A matrix $A \in \mathcal{S}^n$ is *positive semidefinite* if $x^T A x \geq 0$ for all $x \in \mathbb{R}^n$ and is *positive definite* if $x^T A x > 0$ for all nonzero $x \in \mathbb{R}^n$. Equivalently, A is positive semidefinite if its eigenvalues $\lambda_i(A)$ satisfy $\lambda_i(A) \geq 0, i = 1, \dots, n$, and is positive definite if $\lambda_i(A) > 0$, $i = 1, \dots, n$. The set of $n \times n$ positive semidefinite matrices is denoted $\mathcal{S}^n_+$, and the set of positive definite matrices is denoted $\mathcal{S}^n_{++}$. As we will prove soon, $\mathcal{S}^n_+$ is a proper cone (i.e., closed, convex, pointed, and solid). We use the inequality signs "$\preceq$" and "$\succeq$" to denote the partial order induced by $\mathcal{S}^n_+$ (usually called the *Löwner* partial order); i.e., we write $A \succeq B$ if and only if $A - B$ is positive semidefinite. For a square matrix A, its *trace* is defined as $\mathrm{Tr}(A) = \sum_i A_{ii}$. See Section A.1 for further characterizations and general properties of positive semidefinite matrices.

Spectrahedra. Recall that a polyhedron is a set defined by finitely many linear inequalities and that feasible sets of LPs are polyhedra. Similarly, we define *spectrahedra* as sets defined by finitely many LMIs. These sets will correspond exactly to feasible sets of semidefinite programming problems.

Definition 2.5. *A linear matrix inequality (LMI) has the form*

$$A_0 + \sum_{i=1}^{m} A_i x_i \succeq 0,$$

where $A_i \in \mathcal{S}^n$ are given symmetric matrices.

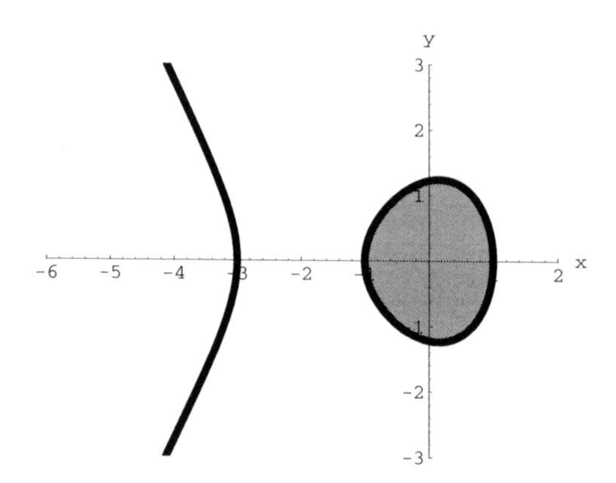

Figure 2.2. *The shaded set is a spectrahedron, with a semidefinite representation given by* (2.4).

Definition 2.6. *A set* $S \subset \mathbb{R}^m$ *is a* spectrahedron *if it has the form*

$$S = \left\{ (x_1, \ldots, x_m) \in \mathbb{R}^m \; : \; A_0 + \sum_{i=1}^{m} A_i x_i \succeq 0 \right\},$$

for some given symmetric matrices $A_0, A_1, \ldots, A_m \in \mathcal{S}^n$.

Geometrically, a spectrahedron is defined by intersecting the positive semidefinite cone and an affine subspace (the span of $A_1, \ldots, A_m$, translated to A_0). Spectrahedra are closed convex sets, since a matrix inequality is equivalent to infinitely many scalar inequalities of the form $v^T(A_0 + \sum_{i=1}^m A_i x_i)v \geq 0$, one for each value of $v \in \mathbb{R}^n$. Since it is always possible to "bundle" several matrix inequalities into a single LMI (by choosing the matrices A_i to be block-diagonal), there is no loss of generality in defining spectrahedra in terms of a single matrix inequality. In particular, this shows that polyhedra are a particular case of spectrahedra, corresponding to all matrices A_i being diagonal.

Recall that the positive semidefiniteness of a matrix can be characterized in terms of scalar inequalities on the coefficients of its characteristic polynomial or its principal minors (see Proposition A.1). Thus, one can obtain an explicit description of a spectrahedron in terms of a finite collection of unquantified scalar polynomial inequalities in the variables x_i. In other words, spectrahedra are *basic semialgebraic sets*, that are convex.

Example 2.7 (elliptic curve). Consider the spectrahedron in $\mathbb{R}^2$ given by

$$\left\{ (x, y) \in \mathbb{R}^2 \; : \; A(x, y) := \begin{bmatrix} x+1 & 0 & y \\ 0 & 2 & -x-1 \\ y & -x-1 & 2 \end{bmatrix} \succeq 0 \right\}. \tag{2.4}$$

This set is shown in Figure 2.2. To obtain scalar inequalities defining the set, let $p_A(t) = \det(tI - A(x, y)) = t^3 + p_2 t^2 + p_1 t + p_0$ be the characteristic polynomial of

$A(x, y)$. Positive semidefiniteness of $A(x, y)$ is then equivalent to the conditions

$$\begin{aligned} -p_2 &= x + 5 \geq 0, \\ p_1 &= -x^2 + 2x - y^2 + 7 \geq 0, \\ -p_0 &= 3 + x - x^3 - 3x^2 - 2y^2 \geq 0. \end{aligned}$$

It can be seen that this spectrahedron corresponds to the "oval" of the elliptic curve $3 + x - x^3 - 3x^2 - 2y^2 = 0$. Notice that the boundary of the set is given by the determinant of the matrix inequality (why?), and the role of the other inequalities is to cut down and isolate the relevant component. ∎

As defined above, a spectrahedron S is a closed convex subset of the affine space $\mathbb{R}^m$. Following standard usage, we will also use "spectrahedron" to denote the set $\{A_0 + \sum_{i=1}^m A_i x_i \mid x \in \mathbb{R}^m\} \cap \mathcal{S}_+^n$. Notice that this is a convex set of *matrices* instead of a subset of $\mathbb{R}^m$, but if the matrices A_i are linearly independent, these two convex sets are affinely equivalent.

Projected spectrahedra. Also of interest are the linear projections of spectrahedra, which we will call *projected spectrahedra*:

Definition 2.8. *A set $S \subset \mathbb{R}^m$ is a* projected spectrahedron *if it has the form*

$$S = \left\{ (x_1, \ldots, x_m) \in \mathbb{R}^m : \exists (y_1, \ldots, y_p) \in \mathbb{R}^p, \quad A_0 + \sum_{i=1}^m A_i x_i + \sum_{j=1}^p B_j y_j \succeq 0 \right\}, \tag{2.5}$$

where $A_0, A_1, \ldots, A_m, B_1, \ldots, B_p$ are given symmetric matrices.

As the name indicates, geometrically this corresponds to a spectrahedron in $\mathbb{R}^{m+p}$ that is projected under the linear map $\pi : \mathbb{R}^{m+p} \to \mathbb{R}^m$, $(x, y) \mapsto x$. Since spectrahedra are semialgebraic sets, by the Tarski–Seidenberg theorem (Section A.4.4 in Appendix A) projected spectrahedra are also semialgebraic. Thus, they can be defined in terms of finite unions of sets defined by polynomial inequalities involving only the variables x_i, although in practice it is not always easy or convenient to do so.

Example 2.9. Consider the projected spectrahedron in $\mathbb{R}^2$ given by

$$\left\{ (x, y) \in \mathbb{R}^2 : \exists z \in \mathbb{R}, \quad \begin{bmatrix} z + y & 2z - x \\ 2z - x & z - y \end{bmatrix} \succeq 0, \quad z \leq 1 \right\}. \tag{2.6}$$

This set is shown in Figure 2.3. It corresponds to the projection on $\mathbb{R}^2$ of the spectrahedron in $\mathbb{R}^3$ defined by the intersection of a quadratic cone and a halfspace (see Figure 2.4).

For any fixed value of z, the set described by the 2×2 matrix inequality is a disk of radius z centered at $(2z, 0)$. Thus, this spectrahedron is the convex hull of the disk of unit radius centered at $(2, 0)$ and the origin. ∎

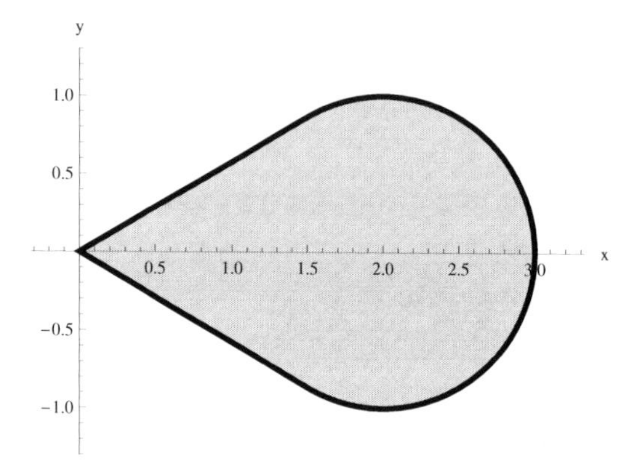

Figure 2.3. *A projected spectrahedron defined by* (2.6).

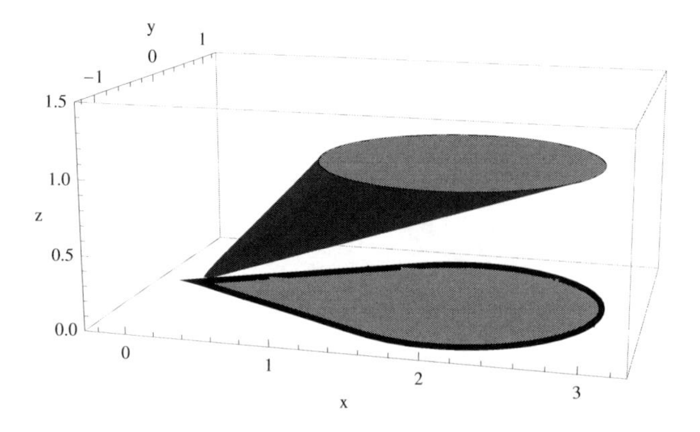

Figure 2.4. *A spectrahedron and its projection.*

As we will see later in much more detail in Chapters 3 and 6, there are simple examples of projected spectrahedra that are *not* spectrahedra (in fact, the set in Example 2.9 is one such case). This is in strong contrast with the case of polyhedra, for which we know (e.g., via Fourier–Motzkin elimination) that the linear projection of a polyhedron is always a polyhedron. Thus, this is a key distinguishing feature of semidefinite programming, since by adding additional *slack* or *lifting* variables, we can significantly expand the expressibility of our class of sets.

Projected spectrahedra are very important for optimization. Indeed, by including the additional "lifting" variables y_i, we will see that it is possible to reduce a linear optimization problem over a projected spectrahedron to the solution of a standard semidefinite program. Furthermore, projected spectrahedra have very high expressive power, in the sense that many convex sets of interest can be represented in this form. Although in general it may be hard to explicitly represent projected spectrahedra in terms of their defining inequalities in their ambient space (see Section 5.6 in Chapter 5), having a representation of the form (2.5) will often be enough for optimization purposes.

Exercise 2.10. Both spectrahedra and projected spectrahedra are convex sets. Show that spectrahedra are always closed sets. What about projected spectrahedra?

Primal SDP formulation. Semidefinite programs are linear optimization problems over spectrahedra. An SDP problem in standard primal form is written as

$$\begin{array}{lll} \text{minimize} & \langle C, X\rangle & \\ \text{subject to} & \langle A_i, X\rangle = b_i, & i = 1, \dots, m, \\ & X \succeq 0, & \end{array} \qquad \text{(SDP-P)}$$

where $C, A_i \in \mathcal{S}^n$, and $\langle X, Y\rangle := \mathrm{Tr}(X^T Y) = \sum_{ij} X_{ij} Y_{ij}$. The matrix $X \in \mathcal{S}^n$ is the variable over which the minimization is performed. The inequality in the third line means that the matrix X must be positive semidefinite. Notice the strong formal similarities to the LP formulation (LP-P). As we will see in Section 2.1.4, this formal analogy can be pushed even further to *conic optimization problems.*

Let us make a few quick comments before presenting examples of semidefinite programs. The set of feasible solutions of (SDP-P), i.e., the set of matrices X that satisfy the constraints, is a spectrahedron, and thus it is always convex. This follows directly from the fact that the feasible set is the intersection of an affine subspace and the positive semidefinite cone $\mathcal{S}_+^n$, both of which are convex sets. However, unlike the linear programming case, in general the set of feasible solutions will not be polyhedral.

Example 2.11. Consider the semidefinite optimization problem

$$\begin{array}{ll} \text{minimize} & 2x_{11} + 2x_{12} \\ \text{subject to} & x_{11} + x_{22} = 1, \\ & \begin{bmatrix} x_{11} & x_{12} \\ x_{12} & x_{22} \end{bmatrix} \succeq 0. \end{array} \qquad (2.7)$$

Clearly, this has the form (SDP-P), with $m = 1$ and

$$C = \begin{bmatrix} 2 & 1 \\ 1 & 0 \end{bmatrix}, \qquad A_1 = \begin{bmatrix} 1 & 0 \\ 0 & 1 \end{bmatrix}, \qquad b_1 = 1.$$

The constraints are satisfied if and only if $x_{11}(1 - x_{11}) \geq x_{12}^2$, and thus the feasible set is a closed disk, which is *not* polyhedral. Figure 2.5 shows the feasible set, parametrized by the variables (x_{11}, x_{12}). The optimal solution is equal to

$$X^\star = \begin{bmatrix} \frac{2-\sqrt{2}}{4} & -\frac{1}{2\sqrt{2}} \\ -\frac{1}{2\sqrt{2}} & \frac{2+\sqrt{2}}{4} \end{bmatrix},$$

with optimal cost $1 - \sqrt{2}$, which is clearly *not* rational. ∎

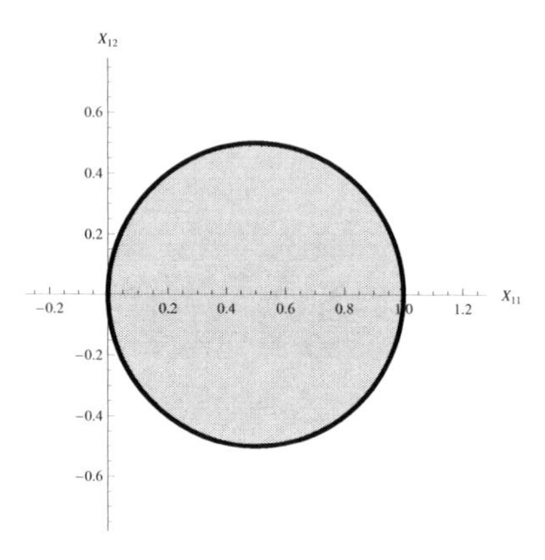

Figure 2.5. *Feasible set of the primal SDP problem* (2.7).

As we have seen from this simple example, SDP problems with rational data do not necessarily have rational optimal solutions. Since the solutions are nevertheless algebraic numbers, a natural question is to analyze their *algebraic degree*, i.e., the minimum degree of a polynomial with integer coefficients needed to specify the solution. The algebraic degree of semidefinite programming is studied in Chapter 5, Section 5.5.

In the particular case when $C = 0$ in (SDP-P), the problem reduces to whether or not the constraints can be satisfied for some matrix X. This is referred to as a *feasibility problem.* As described later, the algebraic nature and convexity of semidefinite programming has made it possible to develop sophisticated and reliable analytical and numerical methods to solve them.

Duality. A very important feature of semidefinite programming, from both the theoretical and applied viewpoints, is the associated *duality theory.* For every semidefinite program of the form (SDP-P) (usually called the *primal problem*), there is another associated SDP, called the *dual problem*, that can be stated as

$$\begin{aligned} &\text{maximize } b^T y \\ &\text{subject to } \sum_{i=1}^{m} A_i y_i \preceq C, \end{aligned} \tag{SDP-D}$$

where $b = (b_1, \ldots, b_m)$, and $y = (y_1, \ldots, y_m)$ are the dual decision variables.

As in the linear programming case, the key relationship between the primal and the dual problems is that feasible solutions of one problem can be used to bound the values of the other. Indeed, let X and y be any two feasible solutions of the primal and dual problems, respectively. We then have the following inequality:

$$\langle C, X\rangle - b^T y = \langle C, X\rangle - \sum_{i=1}^{m} y_i \langle A_i, X\rangle = \left\langle C - \sum_{i=1}^{m} A_i y_i, X \right\rangle \geq 0, \tag{2.8}$$

where the last inequality follows from the fact that the inner product of two positive semidefinite matrices is nonnegative. From (SDP-P) and (SDP-D) we can see that the left-hand side of (2.8) is the difference between the primal and dual objective functions. The inequality in (2.8) tells us that the value of the primal objective function evaluated at any feasible matrix X is always greater than or equal to the dual objective function at any dual feasible y. This is known as *weak duality*. Thus, we can use any X for which (SDP-P) is feasible to compute an upper bound for the value of $b^T y$ in (SDP-D), and we can also use any feasible y of (SDP-D) to compute a lower bound for the value of $\langle C, X\rangle$ in (SDP-P). Furthermore, in the case of feasibility problems (i.e., $C = 0$), the dual problem can be used to certify nonexistence of solutions to the primal problem. This property will be crucial in our later developments.

If X and Y are positive semidefinite matrices, then $\langle X, Y\rangle = 0$ if and only if $XY = YX = 0$ (e.g., Corollary A.24). Thus, the expression (2.8) allows us to give a simple sufficient characterization of optimality.

Lemma 2.12 (optimality conditions for SDP). *Assume* (X, y) *are primal and dual feasible solutions of* (SDP-P) *and* (SDP-D), *respectively, that satisfy the* complementary slackness *condition*

$$\left(C - \sum_{i=1}^{m} A_i y_i\right) X = 0 \tag{2.9}$$

(and thus achieve the same cost $\langle C, X\rangle = b^T y$*). Then,* (X, y) *are primal and dual* optimal *solutions of the SDP problem.*

In general, the converse statement may require some additional assumptions, to be discussed shortly.

Example 2.13. Here we continue Example 2.11. The SDP dual to (2.7) is

$$\begin{aligned} &\text{maximize } y \\ &\text{subject to } \begin{bmatrix} 2-y & 1 \\ 1 & -y \end{bmatrix} \succeq 0. \end{aligned}$$

The optimal solution is $y^\star = 1 - \sqrt{2}$, with optimal cost $1 - \sqrt{2}$. Notice that in this example, the optimal values of the primal and dual problems are equal. Furthermore, complementary slackness holds:

$$\left(C - \sum_{i=1}^{m} A_i y_i^\star\right) X^\star = \begin{bmatrix} 1+\sqrt{2} & 1 \\ 1 & \sqrt{2}-1 \end{bmatrix} \begin{bmatrix} \frac{2-\sqrt{2}}{4} & -\frac{1}{2\sqrt{2}} \\ -\frac{1}{2\sqrt{2}} & \frac{2+\sqrt{2}}{4} \end{bmatrix} = 0. \quad \blacksquare$$

As opposed to the linear programming case, *strong* duality may fail in general semidefinite programming. We present below a simple example (from [36]), for which both the primal and dual problems are feasible, but their optimal values are different (i.e., there is a nonzero finite duality gap). Further examples and a detailed discussion will be presented in Section 2.1.5.

Example 2.14. Let $\alpha \geq 0$, and consider the primal-dual pair

$$\begin{array}{ll} \text{minimize} & \alpha X_{11} \\ \text{subject to} & X_{22} = 0, \\ & X_{11} + 2X_{23} = 1, \\ & X \succeq 0, \end{array} \qquad \begin{array}{ll} \text{maximize} & y_2 \\ \text{subject to} & \begin{bmatrix} y_2 & 0 & 0 \\ 0 & y_1 & y_2 \\ 0 & y_2 & 0 \end{bmatrix} \preceq \begin{bmatrix} \alpha & 0 & 0 \\ 0 & 0 & 0 \\ 0 & 0 & 0 \end{bmatrix}. \end{array}$$

For a primal feasible point, X being positive semidefinite and $X_{22} = 0$ imply $X_{23} = 0$, and thus $X_{11} = 1$. The primal optimal cost $p^\star$ is then equal to α (and is achieved). On the dual side, the vanishing of the $(3,3)$ entry implies that y_2 must be zero, and thus $d^\star = 0$. The duality gap $p^\star - d^\star$ is then equal to α. ∎

The example above (and others like it), are somewhat "pathological." We will see in Section 2.1.5 that under relatively mild conditions, usually called *constraint qualifications*, strong duality will also hold in semidefinite programming. The simplest and most useful case corresponds to the so-called *Slater conditions*, where the primal and/or dual problems are required to be *strictly feasible*. On the primal side, this means that there exists $X \succ 0$ that satisfies the linear constraints, and on the dual side, there exists y such that $C - \sum_i A_i y_i \succ 0$ (notice that the inequalities are strict). In this case, the situation is as nice as in the linear programming case.

Theorem 2.15. *Assume that both the primal* (SDP-P) *and dual* (SDP-D) *semidefinite programs are* strictly *feasible. Then, both problems have optimal solutions, and the corresponding optimal costs are equal; i.e., there is no duality gap.*

This statement will reappear, in a more general setting, in Section 2.1.5. For many problems (for instance, the ones discussed in the next section), these assumptions hold and are relatively straightforward to verify. In full generality, however, they may be restrictive, and thus we investigate in Section 2.1.5 the geometric reasons why strong duality may fail in semidefinite optimization, as well as possible workarounds.

Exercise 2.16. Consider the following SDP problem:

$$\text{minimize } x \qquad \text{subject to} \quad \begin{bmatrix} x & 1 \\ 1 & y \end{bmatrix} \succeq 0.$$

1. Draw the feasible set. Is it convex?

2. Is the primal strictly feasible? Is the dual strictly feasible?

3. What can you say about strong duality? Are the results consistent with Theorem 2.15?

Exercise 2.17. Do the assumptions of Theorem 2.15 hold for Example 2.14?

2.1.3 Spectrahedra and Their Properties

Before proceeding further, we present several interesting examples of sets that are expressible in terms of semidefinite programming. We will revisit several of these throughout the different chapters in this book.

Spectraplex: The *spectraplex* or *free spectrahedron* $\mathcal{O}_n$ is the set of $n \times n$ positive semidefinite matrices of trace one, i.e.,

$$\mathcal{O}_n = \{X \in \mathcal{S}^n \mid \quad X \succeq 0, \quad \operatorname{Tr} X = 1\}.$$

The hyperplane $\operatorname{Tr} X=1$ intersects $\mathcal{S}^n_+$ on a compact set and thus defines a base for this cone. The extreme points of $\mathcal{O}_n$ are exactly the rank one matrices of the form $X = xx^T$, where $x \in \mathbb{R}^n$ and $\|x\| = 1$. The two-dimensional spectraplex $\mathcal{O}_2$ is affinely isomorphic to the unit disk in the plane and has already appeared in Example 2.11.

Elliptope and dual elliptope: Let $\bar{\mathcal{E}}_n$ be the set of positive semidefinite matrices with unit diagonal, i.e.,

$$\bar{\mathcal{E}}_n = \{X \in \mathcal{S}^n \mid \quad X \succeq 0, \quad X_{ii} = 1, \quad i = 1, \ldots, n\}.$$

The convex set $\bar{\mathcal{E}}_n$ is contained in a subspace of $\mathcal{S}^n$ of codimension n, defined by the constraints $X_{ii} = 1$. It is often useful to consider it instead as a full-dimensional convex body in $\mathbb{R}^{\binom{n}{2}}$. For this, define an orthogonal projection $\pi : \mathcal{S}^n \to \mathbb{R}^{\binom{n}{2}}$ that projects a matrix X onto its off-diagonal entries X_{ij} for $i < j$.

The *elliptope* $\mathcal{E}_n$ is defined as $\mathcal{E}_n = \pi(\bar{\mathcal{E}}_n)$ and is a full-dimensional compact convex set in $\mathbb{R}^{\binom{n}{2}}$. As we will see in Section 2.2.2, this set is of great importance when studying semidefinite relaxations of combinatorial problems. Many geometric aspects of elliptopes have been extensively studied, e.g., in [26].

The elliptope $\mathcal{E}_n$ is a convex body containing the origin in its interior. Thus, we can define its polar dual set $\mathcal{E}_n^\circ = \{y \in \mathbb{R}^{\binom{n}{2}} : y^T x \le 1 \quad \forall x \in \mathcal{E}_n\}$, known as the *dual elliptope*. It follows from the expressions above that $\mathcal{E}_n^\circ$ is a (scaled) *projection* of the spectraplex onto the off-diagonal entries:

$$\mathcal{E}_n^\circ = -2\pi(\mathcal{O}_n). \tag{2.10}$$

For nice pictures of the 3×3 elliptope and its dual body, see Figure 5.8 in Chapter 5.

Operator and nuclear norms: Let $A \in \mathbb{R}^{n_1 \times n_2}$ be a matrix. The *spectral* or *operator* norm of A is given by its maximum norm gain, i.e.,

$$\|A\| = \max_{v \in \mathbb{R}^{n_2}, \|v\|=1} \|Av\| = \sigma_1(A),$$

where $\sigma_1(A)$ is the largest singular value of A.

The *nuclear norm* of a matrix is equal to the sum of its singular values, i.e.,

$$\|A\|_* := \sum_{i=1}^{r} \sigma_i(A), \tag{2.11}$$

where r is the rank of A. The nuclear norm is alternatively known by several other names including the Schatten 1-norm, the Ky Fan r-norm, and the trace class norm. As we will see in Section 2.2.6, the nuclear norm is particularly useful in optimization problems involving ranks of matrices.

The operator norm and the nuclear norm are *dual norms* in the sense that their unit balls are convex bodies that are polar duals, i.e.,

$$\{A \in \mathbb{R}^{n_1 \times n_2} \,:\, \|A\| \leq 1\}^\circ = \{B \in \mathbb{R}^{n_1 \times n_2} \,:\, \|B\|_* \leq 1\}.$$

Therefore, any two matrices A and B satisfy

$$\langle A, B \rangle \leq \|A\| \|B\|_*.$$

Furthermore, the following inequalities hold for any matrix A of rank at most r:

$$\|A\| \leq \|A\|_F \leq \|A\|_* \leq \sqrt{r}\|A\|_F \leq r\|A\|, \tag{2.12}$$

where $\|A\|_F$ is the Frobenius norm, defined as $\|A\|_F := (\operatorname{Tr} A^T A)^{\frac{1}{2}} = (\sum_{ij} a_{ij}^2)^{\frac{1}{2}}$.

Both the operator norm and the nuclear norm have nice characterizations in terms of semidefinite programming. In particular, the operator norm $\|A\|$ is the optimal solution of the primal-dual pair of semidefinite programs

$$\begin{array}{ll} \text{maximize} & \operatorname{Tr} 2A^T X_{12} \\ \text{subject to} & \operatorname{Tr} \begin{bmatrix} X_{11} & X_{12} \\ X_{12}^T & X_{22} \end{bmatrix} = 1, \\ & X \succeq 0, \end{array} \qquad \begin{array}{ll} \text{minimize} & t \\ \text{subject to} & \begin{bmatrix} tI_{n_1} & A \\ A^T & tI_{n_2} \end{bmatrix} \succeq 0. \end{array} \tag{2.13}$$

To see the exact correspondence between the standard form (SDP-P)-(SDP-D) and this formulation, notice that we can take $m = 1$, X is a block $(n_1 + n_2) \times (n_1 + n_2)$ matrix, A_1 is the $(n_1 + n_2) \times (n_1 + n_2)$ identity matrix, $b_1 = 1$, and the cost matrix C is the block matrix $\left(\begin{smallmatrix} 0 & -A \\ -A^T & 0 \end{smallmatrix}\right)$. Notice that we have the factor of 2 here because $\operatorname{Tr} CX = \operatorname{Tr} 2A^T X_{12}$, and we have "maximize" in (2.13) instead of "minimize" in (SDP-P) due to change of sign in the objective function.

Similarly (or "dually"), the nuclear norm $\|A\|_*$ corresponds to the optimal value of the primal-dual pair

$$
\begin{array}{llll}
\text{maximize} & \operatorname{Tr} A^T Y & \text{minimize} & \frac{1}{2}(\operatorname{Tr} W_1 + \operatorname{Tr} W_2) \\
\text{subject to} & \begin{bmatrix} I_{n_1} & Y \\ Y^T & I_{n_2} \end{bmatrix} \succeq 0, & \text{subject to} & \begin{bmatrix} W_1 & A \\ A^T & W_2 \end{bmatrix} \succeq 0.
\end{array}
\tag{2.14}
$$

Since the operator norm and the nuclear norm are dual norms, their unit balls are dual polar convex bodies. In Figure 2.6 we illustrate these convex sets for the case of a 2×2 symmetric matrix given by

$$
A = \begin{bmatrix} x & y \\ y & z \end{bmatrix}. \tag{2.15}
$$

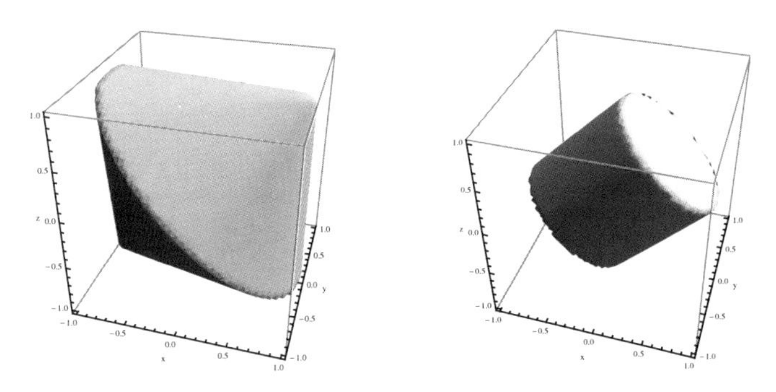

Figure 2.6. *Unit balls of the spectral norm and the nuclear norm, for the space of* 2×2 *symmetric matrices.*

***k*-ellipse:** We consider a class of planar convex sets defined by the algebraic curves known as *k-ellipses* [33]. Recall that the standard ellipse in $\mathbb{R}^2$ is defined as the locus of points with the sum of distances to two fixed points (the *foci*) a fixed constant. Extending this definition to k foci, one can define the k-ellipse as the algebraic curve in $\mathbb{R}^2$ consisting of all points whose sum of distances from k given points is a fixed number. More formally, fix a positive real number d, and fix k distinct points $(u_1, v_1), (u_2, v_2), \ldots, (u_k, v_k)$ in $\mathbb{R}^2$. The *k-ellipse* with *foci* (u_i, v_i) and *radius* d is the following curve in the plane:

$$
\left\{ (x, y) \in \mathbb{R}^2 \;\middle|\; \sum_{i=1}^{k} \sqrt{(x - u_i)^2 + (y - v_i)^2} = d \right\}. \tag{2.16}
$$

In Figure 2.7, we present a few k-ellipses with different numbers of foci. In contrast to the classical circle (corresponding to $k = 1$) and ellipse ($k = 2$), a k-ellipse does not necessarily contain all the foci in its interior. We define the closed convex set $\mathcal{C}_k$ to be the region whose boundary is the k-ellipse, and it is a sublevel set of the

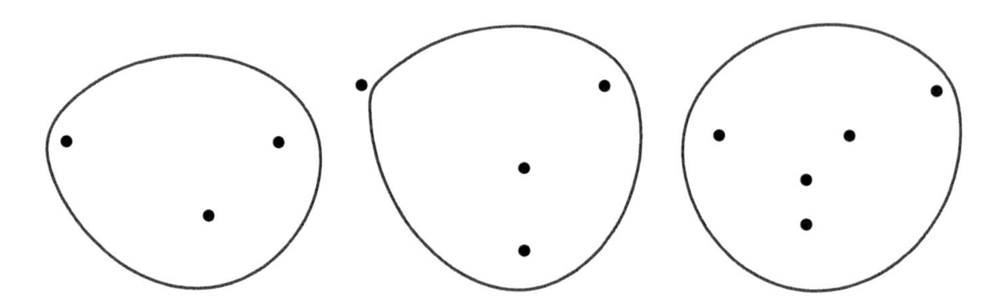

Figure 2.7. *A 3-ellipse, a 4-ellipse, and a 5-ellipse, each with its foci.*

convex function

$$(x, y) \quad \mapsto \quad \sum_{i=1}^{k} \sqrt{(x - u_i)^2 + (y - v_i)^2}. \tag{2.17}$$

In order for $\mathcal{C}_k$ to be nonempty, it is necessary and sufficient that the radius d be greater than or equal to the global minimum $d^\star$ of the convex function (2.17).

The set $\mathcal{C}_k$ is a projected spectrahedron, since it admits a semidefinite representation. This can be easily obtained by adding slack variables d_i and rewriting the function (2.17) in terms of 2×2 matrices. The region $\mathcal{C}_k$ is given by the points (x, y) for which there exist $(d_1, \ldots, d_k)$ satisfying

$$\sum_{i=1}^{k} d_i \leq d, \qquad \begin{bmatrix} d_i + x - u_i & y - v_i \\ y - v_i & d_i - x + u_i \end{bmatrix} \succeq 0, \qquad i = 1, \ldots, k.$$

To see this, notice that the 2×2 matrix above is positive semidefinite if and only if $(x - u_i)^2 + (y - v_i)^2 \leq d_i^2$ and $d_i \geq 0$.

In a less obvious fashion, the k-ellipse can also be represented without additional slack variables, so it is also a spectrahedron. However, in this case the size of the matrices is much bigger. Below we present a concrete statement; see [33] for a sharper result and an explicit construction of this representation.

Theorem 2.18. *The convex set $\mathcal{C}_k$ whose boundary is the k-ellipse of foci (u_i, v_i) and radius d is defined by the LMI*

$$x \cdot A_k + y \cdot B_k + C_k \succeq 0, \tag{2.18}$$

where A_k, B_k, C_k are symmetric $2^k \times 2^k$ matrices. The entries of A_k and B_k are integer numbers, and the entries of C_k are linear forms in the parameters $d, u_1, v_1, \ldots, u_k, v_k$.

For illustration, we present the case $k = 3$ of the theorem. A spectrahedral representation of the 3-ellipse is obtained by requiring the following 8×8 matrix to be positive semidefinite:

$$\begin{bmatrix} d{+}3x{-}u_1{-}u_2{-}u_3 & y{-}v_1 & y{-}v_2 & 0 & y{-}v_3 & 0 & 0 & 0 \\ y{-}v_1 & d{+}x{+}u_1{-}u_2{-}u_3 & 0 & y{-}v_2 & 0 & y{-}v_3 & 0 & 0 \\ y{-}v_2 & 0 & d{+}x{-}u_1{+}u_2{-}u_3 & y{-}v_1 & 0 & 0 & y{-}v_3 & 0 \\ 0 & y{-}v_2 & y{-}v_1 & d{-}x{+}u_1{+}u_2{-}u_3 & 0 & 0 & 0 & y{-}v_3 \\ y{-}v_3 & 0 & 0 & 0 & d{+}x{-}u_1{-}u_2{+}u_3 & y{-}v_1 & y{-}v_2 & 0 \\ 0 & y{-}v_3 & 0 & 0 & y{-}v_1 & d{-}x{+}u_1{-}u_2{+}u_3 & 0 & y{-}v_2 \\ 0 & 0 & y{-}v_3 & 0 & y{-}v_2 & 0 & d{-}x{-}u_1{+}u_2{+}u_3 & y{-}v_1 \\ 0 & 0 & 0 & y{-}v_3 & 0 & y{-}v_2 & y{-}v_1 & d{-}3x{+}u_1{+}u_2{+}u_3 \end{bmatrix}.$$

Exercise 2.19. Prove the relation (2.10) between the elliptope and the spectraplex.

Exercise 2.20. Show that the two semidefinite programs in (2.14) are indeed a primal-dual pair.

Exercise 2.21. Prove the correctness of the semidefinite characterizations of the operator and nuclear norms given in (2.13) and (2.14).

Exercise 2.22. Show that for the symmetric matrix in (2.15), the inequalities that define the boundary of the unit balls of the operator and spectral norms shown in Figure 2.6 are

$$y^2 + (x+z) - xz \le 1, \qquad y^2 - (x+z) - xz \le 1$$

and

$$(x-z)^2 + 4y^2 \le 1, \qquad x+z \le 1, \qquad -(x+z) \le 1,$$

respectively.

Exercise 2.23. Analyze the structure of the convex sets in Figure 2.6. What are the matrices associated with the flat facets (or the vertices)? How can you interpret the rotational symmetries of these convex bodies?

2.1.4 Conic Programming

The strong formal similarities between linear programming and semidefinite programming (equations (LP-P)-(LP-D) vs. (SDP-P)-(SDP-D)) suggest that a more

$$\begin{array}{llcll} \text{minimize} & c^T x & \quad & \text{maximize} & b^T y \\ \text{subject to} & Ax = b & & \text{subject to} & A^T y \leq c \\ & x \geq 0 & & & \end{array} \qquad \text{(LP)}$$

$$\begin{array}{llcll} \text{minimize} & \langle C, X \rangle & \quad & \text{maximize} & b^T y \\ \text{subject to} & \langle A_i, X \rangle = b_i, & & \text{subject to} & \sum_i A_i y_i \preceq C \\ & X \succeq 0 & & & \end{array} \qquad \text{(SDP)}$$

$$\begin{array}{llcll} \text{minimize} & \langle c, x \rangle_S & \quad & \text{maximize} & \langle y, b \rangle_T \\ \text{subject to} & \mathcal{A}x = b, & & \text{subject to} & c - \mathcal{A}^* y \in \mathcal{K}^* \\ & x \in \mathcal{K} & & & \end{array} \qquad \text{(CP)}$$

Table 2.1. *Primal-dual formulations of linear programming (LP), semidefinite programming (SDP), and general conic programming (CP).*

general formulation, encompassing both cases, may be possible. Indeed, a general class of optimization problems that unifies linear and semidefinite optimization (as well as a few other additional cases) is *conic programming*. We describe the conic framework next, explaining first the key idea, followed by the mathematical formulation.

The starting point is the geometric interpretation of linear and semidefinite programming. The feasible set of an LP problem in standard form (LP-P) is the intersection of an affine subspace (described by the equations $Ax = b$) and the nonnegative orthant $\mathbb{R}^n_+$. Similarly, the feasible set of a semidefinite program (SDP-P) is the intersection of an affine subspace (described by $\langle A_i, X \rangle = b_i$) with the set of positive semidefinite matrices $\mathcal{S}^n_+$. Since both $\mathbb{R}^n_+$ and $\mathcal{S}^n_+$ are closed convex cones (in fact, they are *proper* cones—see below), one can define a general class of optimization problems where the feasible set is the intersection of a proper cone and an affine subspace. This is exactly what conic optimization will do!

We present a formal description next. We will be a bit more careful than usual here in the definition of the respective spaces and mappings. It does not make much of a difference if we are working in $\mathbb{R}^n$ (since we can identify a space and its dual through the inner product), but it is "good hygiene" to keep these distinctions in mind and will prove useful when dealing with more complicated spaces. We consider two real vector spaces, S and T, and a linear mapping $\mathcal{A} : S \to T$. Recall that every real vector space has an associated dual space, which is the vector space of real-valued linear functionals. We denote these dual spaces by S^* and T^*, respectively, and the pairing between an element of a vector space and one of the dual as $\langle \cdot, \cdot \rangle$

(i.e., $f(x) = \langle f, x \rangle$). Recall that the *adjoint mapping* of $\mathcal{A}$ is the unique linear map $\mathcal{A}^* : T^* \to S^*$ defined by

$$\langle \mathcal{A}^* y, x \rangle_S = \langle y, \mathcal{A}x \rangle_T \qquad \forall x \in S,\, y \in T^*.$$

Notice here that the brackets on the left-hand side of the equation represent the pairing in S, and those on the right-hand side correspond to the pairing in T.

A cone $\mathcal{K} \subset S$ is *pointed* if $\mathcal{K} \cap (-\mathcal{K}) = \{0\}$ and is *solid* if it is full-dimensional (i.e., $\dim \mathcal{K} = \dim S$). A cone that is convex, closed, pointed, and solid is called a *proper* cone. Given a cone $\mathcal{K}$, its *dual cone* is $\mathcal{K}^* := \{z \in S^* : \langle z, x \rangle_S \geq 0 \quad \forall x \in \mathcal{K}\}$. The dual of a proper cone is also a proper cone; see Exercise 2.24. An element x is in the interior of the proper cone $\mathcal{K}$ if and only if $\langle x, z \rangle > 0\ \forall z \in \mathcal{K}^*, z \neq 0$.

Standard conic programs. Given a linear map $\mathcal{A} : S \to T$ and a proper cone $\mathcal{K} \subset S$, we define the primal-dual pair of (conic) optimization problems

$$\begin{array}{llcll}
\text{minimize} & \langle c, x \rangle_S & \qquad & \text{maximize} & \langle y, b \rangle_T \\
\text{subject to} & \mathcal{A}x = b, & & \text{subject to} & c - \mathcal{A}^* y \in \mathcal{K}^*, \\
 & x \in \mathcal{K}, & & &
\end{array}$$

where $b \in T$, $c \in S^*$. Notice that exactly the same proof presented earlier works here to show weak duality:

$$\begin{aligned}
\langle c, x \rangle_S - \langle y, b \rangle_T &= \langle c, x \rangle_S - \langle y, \mathcal{A}x \rangle_T \\
&= \langle c, x \rangle_S - \langle \mathcal{A}^* y, x \rangle_S \\
&= \langle c - \mathcal{A}^* y, x \rangle_S \\
&\geq 0.
\end{aligned} \tag{2.19}$$

In the usual cases (e.g., LP and SDP), all vector spaces are finite-dimensional and thus isomorphic to their duals. The specific correspondence between these is given through whatever inner product we use.

Among the classes of problems that can be interpreted as particular cases of the general conic formulation we have linear programs, second-order cone programs (SOCP), and semidefinite programs, when we take the cone $\mathcal{K}$ to be the nonnegative orthant $\mathbb{R}^n_+$, the second-order cone $\mathcal{L}^n_+$ (Exercise 2.25), or the positive semidefinite cone $\mathcal{S}^n_+$, respectively. Two other important cases are when $\mathcal{K}$ is the *hyperbolicity cone* associated with a given hyperbolic polynomial [22, 40] and the cone $\Sigma_{n,2d}$ of multivariate polynomials that are *sums of squares*. We discuss this latter example in much more detail in Chapter 3.

Despite the formal similarities, there are a number of differences between linear programming and general conic programming. We have already seen in (2.19) that weak duality always holds for conic programming. However, recall from Example 2.14 that in semidefinite programming (and thus, in general conic programming) there may be a nonzero duality gap. In the next section, we explore the geometric reasons for the possible failure of strong duality in conic programming.

Exercise 2.24. Let $\mathcal{K} \subset S$ be a proper cone. Show that its dual cone $\mathcal{K}^* \subset S^*$ is also a proper cone, and $\mathcal{K}^{**} = \mathcal{K}$.

Exercise 2.25. The *second-order* (or Lorentz) cone is defined as

$$\mathcal{L}_+^n = \left\{ (x_0, x_1, \ldots, x_n) \in \mathbb{R}^{n+1} : \left(\sum_{i=1}^n x_i^2 \right)^{\frac{1}{2}} \leq x_0 \right\}.$$

Show that $\mathcal{L}_+^n$ is a proper cone and is isomorphic to its dual cone.

Exercise 2.26. Classify the following statements as true or false. A proof or counterexample is required.

Let $\mathcal{A} : \mathbb{R}^n \to \mathbb{R}^m$ be a linear mapping and $K \subset \mathbb{R}^n$ a cone.

1. If K is convex, then $\mathcal{A}(K)$ is convex.
2. If K is solid, then $\mathcal{A}(K)$ is solid.
3. If K is pointed, then $\mathcal{A}(K)$ is pointed.
4. If K is closed, then $\mathcal{A}(K)$ is closed.

Do the answers change if $\mathcal{A}$ is injective and/or surjective? How?

2.1.5 Strong Duality

As we have indicated earlier, strong duality in semidefinite programming is a bit more delicate than in the linear programming case. Most of the time (and particularly, in applications) this will not be a source of too many difficulties. However, it is important to understand the geometry behind this, as well as what conditions we can impose to ensure that strong duality will hold.

As we showed in (2.19), weak duality always holds in conic programming (and thus, also for semidefinite programming (2.8)). However, it is possible to have finite duality gaps (as in Example 2.14), or other "anomalies," as the following simple example illustrates.

Example 2.27. Consider the primal-dual SDP pair

$$\begin{array}{ll} \text{minimize} & x_{11} \\ \text{subject to} & 2x_{12} = 1, \\ & \begin{bmatrix} x_{11} & x_{12} \\ x_{12} & x_{22} \end{bmatrix} \succeq 0, \end{array} \qquad \begin{array}{ll} \text{maximize} & y \\ \text{subject to} & \begin{bmatrix} 0 & y \\ y & 0 \end{bmatrix} \preceq \begin{bmatrix} 1 & 0 \\ 0 & 0 \end{bmatrix}. \end{array}$$

For the dual problem, $y = 0$ provides an optimal solution, with optimal value $d^\star = 0$. On the primal side, however, we cannot have $x_{11} = 0$, since this would violate the positive semidefiniteness constraint. However, by choosing $x_{11} = \epsilon$, $x_{22} = 1/\epsilon$, we obtain a cost p^* that is arbitrarily small but always strictly positive. ∎

The example above shows that, in contrast with the case of linear programming, in semidefinite or conic programming optimal solutions *may not be attained*, even if there is zero duality gap.

There are several geometric interpretations of what causes the failure of strong duality for general conic problems. Perhaps the most natural one is based on the fact that the image of a proper cone under a linear map may not be closed, and thus it is *not* necessarily a proper cone. This fact may seem a bit surprising (or perhaps wrong!) the first time one encounters it, but after a while it becomes quite reasonable. (If this is the first time you have heard about this, we strongly encourage you to stop reading and think of a counterexample right now! Or, see Exercise 2.30.)

Strong duality and infeasibility certificates. To better understand strong duality, we begin with a simple geometric interpretation in the conic setting, in terms of the *separating hyperplane theorem.* Recall that this theorem (see Section A.3.3 in Appendix A for several versions of this important result) establishes that if we have two disjoint convex sets, where one of them is closed and the other compact, there always exists a hyperplane that separates the two sets. For simplicity, we concentrate only on the case of conic feasibility, i.e., where we are interested in deciding the existence of a solution x to the equations

$$\mathcal{A}x = b, \qquad x \in \mathcal{K}, \tag{2.20}$$

where as before $\mathcal{K}$ is a proper cone in the vector space S. We want to understand when this problem is feasible and how to certify its infeasibility whenever there are no solutions.

To do this, consider the image $\mathcal{A}(\mathcal{K})$ of the cone under the linear mapping. Notice that feasibility of (2.20) is equivalent to the point b being contained in $\mathcal{A}(\mathcal{K})$. We have now two convex sets in T, namely, $\mathcal{A}(\mathcal{K})$ and the singleton $\{b\}$, and we want to know whether these sets intersect or not. *If* these sets satisfy certain properties (for instance, closedness and compactness), then we could go on to apply the (strict) separating hyperplane theorem and produce a linear functional y that will be positive on one set and negative on the other. In particular, nonnegativity of y on $\mathcal{A}(\mathcal{K})$ implies

$$\langle y, \mathcal{A}x\rangle \geq 0 \;\; \forall x \in \mathcal{K} \quad \Longleftrightarrow \quad \langle \mathcal{A}^* y, x\rangle \geq 0 \;\; \forall x \in \mathcal{K} \quad \Longleftrightarrow \quad \mathcal{A}^* y \in \mathcal{K}^*.$$

Thus, if (2.20) is infeasible, and provided the hypotheses of the separating hyperplane theorem apply, there exists a (suitably normalized) linear functional y which satisfies

$$\langle y, b\rangle = -1, \qquad \mathcal{A}^* y \in \mathcal{K}^*. \tag{2.21}$$

This yields a *certificate* of the infeasibility of the conic system (2.20).

When can we actually do this? The set $\{b\}$ is certainly compact, so a natural condition is that $\mathcal{A}(\mathcal{K})$ be a closed set. However, as we have mentioned, the image of a proper cone is not necessarily closed, so we cannot automatically conclude this. However, under certain conditions, we can ensure that this set will be closed. Well-known sufficient conditions for this are the following.

Theorem 2.28. *Let $\mathcal{K} \subset S$ be a proper cone and $\mathcal{A} : S \to T$ be a linear map. The following two conditions are equivalent:*

(i) $\mathcal{K} \cap \ker \mathcal{A} = \{0\}$.

(ii) *There exists $y \in T^*$ such that $\mathcal{A}^* y \in \operatorname{int}(\mathcal{K}^*)$.*

Furthermore, if these conditions hold, then $\mathcal{A}(\mathcal{K})$ is a closed cone.

The first condition, while intuitive, has the drawback that it is not directly verifiable. The second condition is often more convenient, since it can be certified by exhibiting such a y, and can be interpreted as the range of $\mathcal{A}^*$ properly intersecting $\mathcal{K}^*$.

Proof. The equivalence of (i) and (ii) follows from Exercise 2.32, taking $L = \ker \mathcal{A}$, and thus $L^\perp = \operatorname{range} \mathcal{A}^*$.

Assume now that (ii) holds, and define $C = \{x \in \mathcal{K} : \langle \mathcal{A}^* y, x \rangle = 1\}$. We claim that the set C is compact. Indeed, C is closed (being the intersection of two closed sets), and it is also bounded, since if there is a sequence $x_k \in C$ with $\|x_k\|$ going to infinity, then defining $z = \lim_{k\to\infty} x_k/\|x_k\|$ (passing to a subsequence if necessary) gives an element of $\mathcal{K}$ (by closedness of $\mathcal{K}$), for which $\langle \mathcal{A}^* y, z \rangle = \lim_{k\to\infty} \langle \mathcal{A}^* y, x_k \rangle / \|x_k\| = \lim_{k\to\infty} 1/\|x_k\| = 0$, contradicting $\mathcal{A}^* y \in \operatorname{int}(\mathcal{K}^*)$.

The set $\mathcal{A}(C)$ is also compact (being the linear image of a compact set) and does not include the origin, since for all $x \in C$ we have $\langle y, \mathcal{A}x \rangle = \langle \mathcal{A}^* y, x \rangle = 1$. Thus, since $\mathcal{A}(\mathcal{K}) = \operatorname{cone}(\mathcal{A}(C))$, it follows from Exercise 4.17 in Chapter 4 that $\mathcal{A}(\mathcal{K})$ is closed. $\square$

To recap, having strictly feasible solutions in $(\operatorname{range} \mathcal{A}^*) \cap \operatorname{int} \mathcal{K}^*$ is a natural condition for the existence of infeasibility certificates of the form (2.21).

For the case of a general conic optimization problem (not just feasibility), similar conditions can be used to ensure that there will be no duality gap between the primal and dual conic programs. The basic idea is to reduce the optimization problem to a pure feasibility question by adjoining a new inequality corresponding to the cost function. In this case, imposing a Slater-type condition will guarantee that optimal solutions for both problems are achieved, with no gap (compare with the semidefinite programming case, Theorem 2.15).

Theorem 2.29. *Consider a conic optimization problem* (CP), *where both the primal and dual problems are* strictly *feasible. Then, both problems have nonempty, compact sets of optimal solutions, and there is no duality gap.*

Besides Theorem 2.28, many other conditions are known that ensure the closedness of $\mathcal{A}(\mathcal{K})$. In particular, when $\mathcal{K}$ is polyhedral this image is always closed, with no interior-point requirements needed. This corresponds to the case of linear programming and is the reason why strong duality always holds in the LP case.

In Section 3.4.2 of Chapter 3 we will explore in much more detail general infeasibility certificates for different kinds of systems of equations and inequalities.

Exercise 2.30. Consider the set $\mathcal{K} = \{(x, y, z) : y^2 \leq xz,\ z \geq 0\}$. Show that $\mathcal{K}$ is a proper cone. Show that its projection onto the (x, y) plane is not a proper cone.

Exercise 2.31. Let $\mathcal{K}_1$, $\mathcal{K}_2$ be closed convex cones. Show, via a counterexample, that the Minkowski sum $\mathcal{K}_1 + \mathcal{K}_2$ does not have to be closed.

Exercise 2.32. Let $L \subset S$ be a subspace, and $\mathcal{K} \subset S$ be a proper cone. Show that the following two propositions are equivalent:

(i) $L \cap \mathcal{K} = \{0\}$.

(ii) There exists $z \in L^\perp \cap \operatorname{int}(\mathcal{K}^*)$.

Hint: For the "difficult" direction (i) $\Rightarrow$ (ii), argue by contradiction, and use homogeneity and the separation theorem for convex sets.

Although as we have seen, "standard" duality may fail in semidefinite (or conic) programming, it is nevertheless possible to formulate a more complicated semidefinite dual program (called the "Extended Lagrange–Slater Dual" in [36]) for which strong duality always holds, regardless of interior-point assumptions. For details, as well as a comparison with the more general "minimal cone" approach, we refer the reader to [36, 37].

2.2 Applications of Semidefinite Optimization

There have been *many* applications of semidefinite optimization in a variety of areas of applied mathematics and engineering. We present here just a few, to give a flavor of what is possible; many others will follow in other chapters. The subsections corresponding to the different examples presented here can be read independently and are not essential for the remainder of the developments in the book.

2.2.1 Lyapunov Stability and Control of Dynamical Systems

One of the earliest and most important applications of semidefinite optimization is in the context of dynamical systems and control theory. The main reason is that it is possible to characterize dynamical properties (e.g., stability) in terms of algebraic statements such as the feasibility of specific systems of inequalities. We describe below a relatively simple example of these ideas that captures many of the features of more complicated problems.

Stability of linear systems. Consider a linear difference equation given by

$$x[k+1] = A\,x[k], \qquad x[0] = x_0. \tag{2.22}$$

This kind of linear recurrence equation is a simple example of a discrete-time dynamical system, where the *state* $x[k]$ evolves over time, starting from an initial condition x_0. The difference equation (2.22), or its continuous-time analogue (the linear differential equation $\frac{d}{dt}x(t) = Ax(t)$), is often used to model the time evolution of quantities such as temperature of objects, size of a population, voltage of electrical circuits, and concentration of chemical mixtures.

A natural and important question about (2.22) is the long-term behavior of the state. In particular, as $k \to \infty$, under what conditions can we guarantee that the state $x[k]$ remains bounded, or converges to zero? It is well known (and easy to prove; see Exercise 2.35) that $x[k]$ converges to zero for all initial conditions x_0 if and only if the spectral radius of the matrix A is smaller than one, i.e., all the eigenvalues λ_i satisfy $|\lambda_i(A)| < 1$ for $i = 1, \ldots, n$. In this case we say that the system (2.22), or the matrix A, is *stable* (or *Schur stable*, if the discrete-time aspect is not clear from the context).

While this spectral characterization is very useful, an alternative viewpoint is sometimes even more convenient. The basic idea is to consider a generalization and abstraction of the notion of *energy*, usually known as a *Lyapunov function*. These are functions of the state $x[k]$, with the property that they decrease monotonically along trajectories of the system (2.22). It turns out that for linear systems there is a simple characterization of stability in terms of a *quadratic* Lyapunov function $V(x[k]) = x[k]^T P x[k]$. Notice first that the monotonicity condition $V(x[k+1]) \leq V(x[k])$ (for all states $x[k]$) can be equivalently expressed in terms of the matrix inequality $A^T P A - P \preceq 0$. We then have the following result.

Theorem 2.33. *Given a matrix $A \in \mathbb{R}^{n \times n}$, the following conditions are equivalent:*

1. *All eigenvalues of A are inside the unit circle; i.e., $|\lambda_i(A)| < 1$ for $i = 1, \ldots, n$.*
2. *There exists a matrix $P \in \mathcal{S}^n$ such that*

$$P \succ 0, \qquad A^T P A - P \prec 0.$$

Proof. (2) $\Rightarrow$ (1): Let $Av = \lambda v$, where $v \neq 0$. Then

$$0 > v^*(A^T P A - P)v = (|\lambda|^2 - 1)\underbrace{v^* P v}_{>0},$$

and therefore $|\lambda| < 1$.

(1) $\Rightarrow$ (2): Let $P := \sum_{k=0}^{\infty} (A^k)^T A^k$. The sum converges by the eigenvalue assumption. Then

$$A^T P A - P = \sum_{k=1}^{\infty} (A^k)^T A^k - \sum_{k=0}^{\infty} (A^k)^T A^k = -I \prec 0. \qquad \square$$

Thus, the characterization given above enables the study of the stability properties of the linear difference equation (2.22) in terms of a semidefinite programming problem, whose feasible solutions correspond to Lyapunov functions. In Section 3.6.2 we will explore extensions of these ideas to more complicated dynamics, not necessarily linear.

Control design. Consider now the case of a linear system, where there is a *control input* $u[k]$:

$$x[k+1] = A\,x[k] + B\,u[k], \qquad x[0] = x_0, \tag{2.23}$$

where $B \in \mathbb{R}^{n \times m}$. The idea here is that by properly choosing the control input $u[k] \in \mathbb{R}^m$ at each time instant, we may be able (under certain conditions), to affect or steer the behavior of $x[k]$ toward some desired goal. We are interested in the case where the matrix A is not stable, but we can use linear state feedback to set $u[k] = Kx[k]$ for some fixed matrix K (to be chosen appropriately). It is easy to see that after this substitution, the system is described by (2.22), where the matrix A is replaced by $A(K) = A + BK$. Thus, our goal is "stabilization"; i.e., we want to find a matrix K such that $A + BK$ is stable (all eigenvalues have absolute value smaller than one).

Although this problem seems (and is!) fairly complicated due to the nonlinear dependence of the eigenvalues of $A + BK$ on the unknown matrix K, it turns out that it can be nicely solved using semidefinite optimization and the Lyapunov characterization given earlier. Indeed, we can use Schur complements (see Appendix A) to rewrite the condition

$$(A + BK)^T P(A + BK) - P \prec 0, \qquad P \succ 0,$$

as

$$\begin{bmatrix} P & (A + BK)^T P \\ P(A + BK) & P \end{bmatrix} \succ 0.$$

Although nicer, this condition is not quite an SDP yet, since it is bilinear in (P, K) (and, thus, not jointly convex). However, defining $Q := P^{-1}$, and left- and right-multiplying the equation above with the matrix $\operatorname{BlockDiag}(Q, Q)$, we obtain

$$\begin{bmatrix} Q & Q(A + BK)^T \\ (A + BK)Q & Q \end{bmatrix} \succ 0.$$

Notice that this expression contains both Q and KQ, but there is no single appearance of the variable K. Thus, we can define a new variable $Y := KQ$, to obtain

$$\begin{bmatrix} Q & QA^T + Y^T B^T \\ AQ + BY & Q \end{bmatrix} \succ 0. \tag{2.24}$$

This problem is now linear in the new variables (Q, Y). In fact, it is a semidefinite programming problem! After solving it, we can recover the controller K via $K = Q^{-1}Y$. We summarize our discussion in the following result.

Theorem 2.34. *Given two matrices A and B, there exists a matrix K such that $A + BK$ is stable if and only if the spectrahedron described by* (2.24) *is nonempty, i.e., there exist matrices (Q, Y) satisfying this (strict) linear matrix inequality.*

Hence our control design problem is equivalent to solving a semidefinite programming feasibility problem.

Semidefinite programming techniques have become quite central in the analysis and design of control systems. The example above describes only the tip of the iceberg in terms of the many design problems that can be attacked with these techniques; we refer the reader to the works [6, 47] and the references therein.

We remark that the formulas in this example (e.g., (2.24)) do not explicitly depend on the dimensions of the matrices A, B, K, Y, Q. Hence, these kinds of problems are sometimes called *dimension-free*. This dimension-free feature applies to many classical problems in linear systems and has strong implications. Linear control theory problems can often be reduced to polynomials in matrix variables where the feasible set is defined by these polynomials being positive semidefinite. Analyzing this situation requires a theory of inequalities for free noncommutative polynomials extending classical real geometry for commutative polynomials. The convexity aspects of this new area, noncommutative real algebraic geometry, is the subject of Chapter 8.

Exercise 2.35. Show that for the linear difference equation (2.22), the state $x[k]$ converges to zero for all initial conditions x_0 if and only if $|\lambda_i(A)| < 1$ for $i = 1, \ldots, n$. Hint: show that $x[k] = A^k x_0$, and consider first the case where the matrix A is diagonalizable.

Exercise 2.36. The system (2.23) has a *nonstabilizable mode* if the matrix A has a left eigenvector w such that $w^T A = \lambda w^T$, $w^T B = 0$, and $|\lambda| \geq 1$. Show that if this is the case, then the SDP (2.24) cannot be feasible. Interpret this statement in terms of the eigenvalues of $A + BK$. What does this say about the dual SDP?

2.2.2 Binary Quadratic Optimization

Binary (or Boolean) quadratic optimization is a classical combinatorial optimization problem. In the version we consider, we want to minimize a quadratic function, where the decision variables can take only the values ± 1. In other words, we are minimizing an (indefinite) quadratic form over the vertices of an n-dimensional hypercube. The problem is formally expressed as

$$\begin{array}{ll} \text{minimize} & x^T Q x \\ \text{subject to} & x_i \in \{-1, 1\}, \end{array} \tag{2.25}$$

where $Q \in \mathcal{S}^n$. There are many well-known problems that can be naturally written in the form above. Among these, we mention the maximum cut (MAXCUT) problem, 0-1 knapsack, etc.

Notice that the Boolean constraints can be modeled using quadratic equations, i.e.,

$$x_i \in \{-1, 1\} \quad \Leftrightarrow \quad x_i^2 = 1.$$

These n quadratic equations define a finite set, with an exponential number of elements, namely, all the n-tuples with entries in $\{-1, 1\}$. There are exactly 2^n points in this set, so a direct enumeration approach to (2.25) is computationally prohibitive when n is large (already for $n = 30$ we have $2^n \approx 10^9$).

We write the equivalent polynomial formulation

$$\begin{aligned} \text{minimize} \quad & x^T Q x \\ \text{subject to} \quad & x_i^2 = 1, \end{aligned} \tag{2.26}$$

and we denote the optimal value and optimal solution of this problem as $f_\star$ and $x_\star$, respectively. It is well known that the decision version of this problem is *NP-complete* (e.g., [18]). Notice that this is true even if the objective function is convex (i.e., the matrix Q is positive definite), since we can always assume $Q \succeq 0$ by adding to it a large constant multiple of the identity (this only shifts the objective by a constant).

Computing "good" solutions to the binary optimization problem (2.26) is a quite difficult task, so it is of interest to produce accurate bounds on its optimal value. As in all minimization problems, *upper bounds* can be directly obtained from feasible points. In other words, if $x_0 \in \mathbb{R}^n$ has entries equal to ± 1, it always holds that $f_\star \leq x_0^T Q x_0$ (of course, for a poorly chosen x_0, this upper bound may be very loose).

To prove *lower bounds*, we need a different technique. There are several approaches to doing this, but many of them will turn out to be exactly equivalent in the end. In particular, we can provide a lower bound in terms of the following primal-dual pair of semidefinite programming problems:

$$\begin{aligned} \text{minimize} \quad & \operatorname{Tr} QX & \qquad \text{maximize} \quad & \operatorname{Tr} \Lambda \\ \text{subject to} \quad & X_{ii} = 1, & \qquad \text{subject to} \quad & Q \succeq \Lambda, \\ & X \succeq 0, & & \Lambda \text{ diagonal}. \end{aligned} \tag{2.27}$$

These semidefinite programs can be interpreted in a number of ways. For instance, it is clear that the optimal solution $X^\star$ of the primal formulation in (2.27) yields a lower bound, since for every x in (2.26), the matrix $X = xx^T$ gives a feasible solution of (2.27) with the same cost: $\operatorname{Tr} QX = \operatorname{Tr} Qxx^T = x^T Q x$. Similarly, for every feasible solution $\Lambda = \operatorname{Diag}(\lambda_1, \ldots, \lambda_n)$ of the dual SDP, we have

$$x^T Q x \geq x^T \Lambda x = \sum_{i=1}^{n} \lambda_i x_i^2 = \operatorname{Tr} \Lambda,$$

thus yielding a lower bound on (2.26).

In certain cases, these SDP-based bounds are *provably* good. Well-known cases are when $(-Q)$ is diagonally dominant or positive semidefinite or has a bipartite structure, in which case results due to Goemans–Williamson [20], Nesterov [31], or Grothendieck/Krivine [30, 2, 25], respectively, have shown that there is at most a small constant factor between the "true" solutions and the SDP relaxations. We discuss these bounds next.

Rounding. As described, the optimal value of the SDP relaxation (2.27) provides a lower bound on the optimal value of the binary minimization problem (2.26). Two natural questions arise:

1. Feasible solutions: can we use the SDP relaxations to provide feasible points that yield good (or optimal) values of the objective?

2. Approximation guarantees: is it possible to quantify the quality of the bounds obtained by SDP?

By suitably "rounding" in an appropriate manner the optimal solution of the SDP relaxation, both questions can be answered in the affirmative. The basic idea is to produce a binary vector x from the SDP solution matrix X, using the following "hyperplane rounding" method [20]:

- Factorize the SDP solution X as $X = V^T V$, where $V = [v_1 \dots v_n] \in \mathbb{R}^{r \times n}$ and r is the rank of X.

- Since $X_{ij} = v_i^T v_j$ and $X_{ii} = 1$, this factorization gives n vectors v_i on the unit sphere in $\mathbb{R}^r$. Thus, instead of assigning either 1 or -1 to each variable, so far we have assigned to each x_i a point on the unit sphere in $\mathbb{R}^r$.

- Now, choose a uniformly distributed random hyperplane in $\mathbb{R}^r$ (passing through the origin), and assign to each variable x_i either a $+1$ or a -1, depending on which side of the hyperplane the point v_i lies.

Since the last step involves a random choice, this is a *randomized rounding method.* By a simple geometric argument, it is possible to quantify the expected value of the objective function.

Lemma 2.37. *Let $x = \operatorname{sign}(V^T r)$, where $X = V^T V$ and r is a standard random Gaussian vector. Then, $\mathbf{E}[x_i x_j] = \frac{2}{\pi} \arcsin X_{ij}$.*

By linearity of expectations, we have the following relationship between the lower bound given by the optimal value of the SDP, the "true" optimal value $f_\star$, and the expected value of the rounded solution x:

$$\operatorname{Tr} QX \leq f_\star \leq \mathbf{E}[x^T Q x] = \frac{2}{\pi} \operatorname{Tr} Q \arcsin[X]. \tag{2.28}$$

The notation $\arcsin[\cdot]$ indicates that the arcsine function is applied componentwise, i.e., $(\arcsin[X])_{ij} = \arcsin X_{ij}$.

Exercise 2.38. Prove Lemma 2.37, and verify that it implements the hyperplane rounding scheme.

Approximation ratios. In many problems, we want to understand how far these upper and lower bounds are from each other. Depending on the specific assumptions on the cost function, the hyperplane rounding method (or slight variations) will give solutions with different guaranteed approximation ratios. Since the approximation algorithms literature often considers *maximization* problems (instead of the minimization version (2.26)), in this section we use

$$\begin{aligned} \text{maximize} \quad & x^T A x \\ \text{subject to} \quad & x_i^2 = 1 \end{aligned} \tag{2.29}$$

and state below our assumptions in terms of the matrix A (or, equivalently, the matrix $-Q$ in the minimization formulation (2.25)).

We describe next three well-known cases where *constant* approximation ratios can be obtained.

Diagonally dominant: A symmetric matrix A is *diagonally dominant* if $a_{ii} \geq \sum_{j \neq i} |a_{ij}|$ for all i. This is an important case that corresponds, for instance, to the MAXCUT problem, where the cost function to be maximized is the Laplacian of a graph (V, E), given by $\frac{1}{4}\sum_{(i,j)\in E}(x_i - x_j)^2$. Every diagonally dominant quadratic form can be written as a nonnegative linear combination of terms of the form x_i^2 and $(x_i \pm x_j)^2$ [4]. Thus, to analyze the performance of hyperplane rounding when A is diagonally dominant, it is enough to consider the inequality

$$\mathbf{E}[(x_i \pm x_j)^2/2] = \mathbf{E}[1 \pm x_i x_j] = 1 \pm \frac{2}{\pi} \arcsin X_{ij} \geq \alpha_{GW} \cdot (1 \pm X_{ij}),$$

where $\alpha_{GW} = \min_{t\in[-1,1]}(1 - \frac{2}{\pi}\arcsin t)/(1-t) \approx 0.878$. Combining this with (2.28), and taking into account the change of signs (since $A = -Q$), it follows that

$$\alpha_{GW} \cdot \operatorname{Tr} AX \leq \mathbf{E}[x^T A x] \leq f_\star \leq \operatorname{Tr} AX;$$

i.e., the vector x obtained by randomly rounding the SDP solution matrix X is at most 13% suboptimal in expectation. This analysis is due to Goemans and Williamson [20] and yields the best currently known approximation ratio for the MAXCUT problem.

Positive semidefinite: Nesterov [31] first analyzed the case of maximizing a convex quadratic function, i.e., when the matrix A is positive semidefinite. Notice that here we do not have any information on the sign of the individual entries a_{ij}, and thus a "global" analysis is needed instead of the term-by-term analysis of the previous case. The key idea is to use the following result.

Lemma 2.39. *Let $f : \mathbb{R} \to \mathbb{R}$ be a function whose Taylor expansion has only nonnegative coefficients. Given a symmetric matrix X, define a matrix Y as $Y_{ij} = f(X_{ij})$ (equivalently, $Y = f[X]$). Then $X \succeq 0$ implies $Y \succeq 0$.*

This lemma is a rather direct consequence of the Schur product theorem; see Exercise 2.42. Since the scalar function $f(t) = \arcsin(t) - t$ has only nonnegative Taylor coefficients, if $X \succeq 0$, we have $\arcsin[X] \succeq X$, and thus

$$\mathbf{E}[x^T Ax] = \frac{2}{\pi}\operatorname{Tr} A \arcsin[X] \geq \frac{2}{\pi} \cdot \operatorname{Tr} AX.$$

Thus, in this case we have

$$\frac{2}{\pi} \cdot \operatorname{Tr} AX \leq \mathbf{E}[x^T Ax] \leq f_\star \leq \operatorname{Tr} AX.$$

Notice that $\frac{2}{\pi} \approx 0.636$, so the approximation ratio in this case is slightly worse than for the diagonally dominant case.

Bipartite: This case corresponds to the cost function being bilinear and has been analyzed in [2, 30]. We assume that the matrix A has a structure

$$A = \frac{1}{2}\begin{bmatrix} 0 & S \\ S^T & 0 \end{bmatrix}.$$

Letting $x = [p; q]$, an equivalent formulation is in terms of a *bilinear* optimization problem

$$\text{maximize } p^T Sq,$$

where $S \in \mathbb{R}^{n \times m}$ and p, q are in $\{+1, -1\}^n$ and $\{+1, -1\}^m$, respectively.

This problem has a long history in operator theory and functional analysis and was first analyzed (in a quite different form) by Grothendieck. For this class of problems, it follows from his results that a *constant ratio* approximation is possible. In fact, the worst-case ratio (over all instances) between the values of the semidefinite relaxation and the bilinear binary optimization problem is called the *Grothendieck constant* and is usually denoted K_G,

$$K_G := \sup_A \frac{\operatorname{Tr} AX}{f_\star},$$

where X is, as before, the optimal solution of the SDP relaxation. The exact value is this constant is unknown at this time. The argument below is essentially due to Krivine [25] and provides an upper bound to the Grothendieck constant.

Since there are no assumptions about the sign of the entries of the matrix S, we cannot directly apply the techniques discussed earlier to prove a bound on the quality of hyperplane rounding. The basic strategy in Krivine's approach is the following: instead of using hyperplane rounding directly on the solution X of the SDP relaxation, we will apply first a particular componentwise transformation, to obtain a matrix Y, and then apply hyperplane rounding to Y. The reason is that this will considerably simplify the computation of the expected value of the objective function.

To do this, we use a "block" version of Lemma 2.39.

Lemma 2.40. *Let $f, g : \mathbb{R} \to \mathbb{R}$ be functions such that both $f+g$ and $f-g$ have nonnegative Taylor coefficients. Let*

$$X = \begin{bmatrix} X_{11} & X_{12} \\ X_{12}^T & X_{22} \end{bmatrix}, \quad Y = \begin{bmatrix} f(X_{11}) & g(X_{12}) \\ g(X_{12}^T) & f(X_{22}) \end{bmatrix}. \tag{2.30}$$

Then $X \succeq 0$ implies $Y \succeq 0$.

The result now follows from a clever choice of f and g. Let

$$f(t) = \sinh(c_K \pi t/2), \qquad g(t) = \sin(c_K \pi t/2),$$

where the constant $c_K = \frac{2}{\pi} \sinh^{-1}(1) = \frac{2}{\pi} \log(1+\sqrt{2}) \approx 0.5611$ is chosen so $f(1) = 1$. Since

$$\sinh(t) = \sum_{k=0}^{\infty} \frac{t^{2k+1}}{(2k+1)!}, \qquad \sin(t) = \sum_{k=0}^{\infty} (-1)^k \frac{t^{2k+1}}{(2k+1)!},$$

both $f+g$ and $f-g$ have nonnegative Taylor expansions.

Let X be the optimal solution of the SDP relaxation, and define Y as in (2.30). Notice that the matrix Y satisfies $Y \succeq 0$ and $Y_{ii} = 1$. We can therefore apply hyperplane rounding to it to obtain a vector y. Computing the expected value of this solution, we have

$$\mathbf{E}[y^T A y] = \frac{2}{\pi} \operatorname{Tr} A \arcsin[Y] = \frac{2}{\pi} \cdot \operatorname{Tr} S(c_K \pi X_{12}/2) = c_K \cdot \operatorname{Tr} S X_{12},$$

and therefore this gives us a randomized algorithm with expected value c_K times the value of the SDP relaxation. Notice that no inequalities are used in the analysis, so the expected cost of the solution y for this rounding scheme is *exactly* equal to c_K times the optimal value of the SDP:

$$c_K \cdot \operatorname{Tr} S X_{12} = \mathbf{E}[y^T A y] \leq f_\star \leq \operatorname{Tr} S X_{12}.$$

This analysis gives an upper bound for the Grothendieck constant of $\frac{\operatorname{Tr} S X_{12}}{f_\star} \leq K_G \leq 1/c_K \approx 1.7822$. It has been recently shown that this rounding method (and thus, the value $1/c_K$) is not the best possible one [8], but the exact approximation ratio is not currently known.

Exercise 2.41. Show that the optimal values of the primal and dual semidefinite programs in (2.27) are equal, i.e., there is no duality gap.

Exercise 2.42. The *entrywise product* $A \circ B$ of two matrices is given by $(A \circ B)_{ij} = A_{ij} B_{ij}$. This product is also known as the *Hadamard* or *Schur* product. The Schur product theorem says that if two matrices A, B are positive semidefinite, so is their product $A \circ B$.

1. Prove the Schur product theorem. (Hint: What happens if one of the matrices is rank one?)

2. Prove Lemmas 2.39 and 2.40.

2.2.3 Stable Sets and the Theta Function

Given an undirected graph $G = (V, E)$, a *stable set* (or *independent set*) is a subset of the set of vertices V with the property that the induced subgraph has no edges. In other words, none of the selected vertices are adjacent to each other.

The *stability number* of a graph, usually denoted by $\alpha(G)$, is the cardinality of the largest stable set. Computing the stability number of a graph is NP-hard. There are many interesting applications of the stable set problem. In particular, it can be used to provide upper bounds on the *Shannon capacity of a graph* [28], a problem that appears in coding theory (when computing the zero-error capacity of a noisy channel [43]). In fact, this was one of the first appearances of semidefinite programming.

In many problems, it is of interest to compute upper bounds on $\alpha(G)$. The Lovász theta function of the graph G is denoted by $\vartheta(G)$ and is defined as the solution of the primal-dual SDP pair:

$$
\begin{array}{llll}
\text{maximize} & \operatorname{Tr} JX & & \\
\text{subject to} & \operatorname{Tr} X = 1 & & \\
& X_{ij} = 0, & (i,j) \in E, & \\
& X \succeq 0, & &
\end{array}
\qquad
\begin{array}{llll}
\text{minimize} & t & & \\
\text{subject to} & Y \preceq tI & & \\
& Y_{ii} = 1, & i \in V, & \\
& Y_{ij} = 1, & (i,j) \notin E, &
\end{array}
\tag{2.31}
$$

where J is the matrix with all entries equal to one.

The theta function is an upper bound on the stability number, i.e.,

$$\alpha(G) \leq \vartheta(G).$$

The inequality is easy to prove. Consider the indicator vector $\chi(S)$ of any stable set S, and define the matrix $X := \frac{1}{|S|}\chi(S)\chi(S)^T$. It is easy to see that this X is a feasible solution of the primal SDP in (2.31), and it achieves an objective value equal to $|S|$. As a consequence, the inequality above directly follows.

For a class of graphs known as *perfect graphs*,[1] the upper bound given by the theta function is exact; i.e., it is equal to the stability number. Many classes of graphs, such as bipartite, chordal, and comparability graphs, are perfect. Thus, for these graphs one can compute in polynomial time the size of the largest stable set (and a maximum stable set) by solving the SDPs (2.31). Interestingly, at this time no polynomial-time combinatorial methods (not based on semidefinite programming) are known to compute this quantity for all perfect graphs. Further material

[1]A graph is *perfect* if, for every induced subgraph, the chromatic number is equal to size of the largest clique.

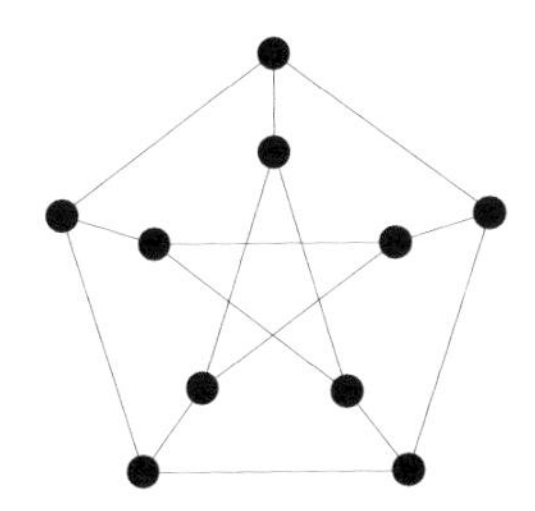

Figure 2.8. *Petersen graph.*

on the theta function of a graph and its applications in combinatorial optimization can be found in Lovász's original paper [28], or the references [19, 21].

Exercise 2.43. Consider the graph in Figure 2.8, known as the Petersen graph. Compute the semidefinite programming upper bound on the size of its largest stable subset (i.e., the Lovász theta function). Is this bound tight? Can you find a stable set that achieves this value?

Exercise 2.44. The *chromatic number* $\chi(G)$ of a graph G is the minimum number of colors needed to color all vertices, in such a way that adjacent vertices receive distinct colors. Show that the inequality

$$\vartheta(G) \leq \chi(\bar{G})$$

holds, where $\bar{G}$ is the complement of the graph G.

Hint: Given a coloring of $\bar{G}$, construct a feasible solution of the dual SDP in (2.31).

2.2.4 Bounded analytic interpolation

In many applications, one tries to find a function in a given function class, that takes specific values at prescribed points. These kinds of questions are known as *interpolation* problems. A classical and important class of interpolation problems involves *bounded analytic functions.* The mathematical background for these problems is reviewed and developed further in Chapter 9. Good general references include [3] for the theoretical aspects, and [24, 47] for specific applications of interpolation in systems and control theory.

We discuss here two specific problems related to this area. The first is the computation of the $\mathcal{H}_\infty$-norm of an analytic function, and the second is the classical Nevanlinna–Pick interpolation problem. Additional connections between analytic interpolation and convex optimization can be found in [6].

Norms of rational analytic functions. Let $\mathbb{D}$ be the complex open unit disk $\mathbb{D} = \{z \in \mathbb{C} : |z| < 1\}$. Consider a scalar rational function of a complex variable z given by

$$f(z) = c^T(z^{-1}I - A)^{-1}b + d, \tag{2.32}$$

where $A \in \mathbb{R}^{n\times n}$, $b, c \in \mathbb{R}^{n\times 1}$, and $d \in \mathbb{R}$. We assume that all the eigenvalues of A are in $\mathbb{D}$: $|\lambda_i(A)| < 1$ (i.e., A is *Schur stable*). It follows that $z^{-1}I - A$ is nonsingular on $|z| \leq 1$, and thus $f(z)$ is analytic[2] on the domain $\mathbb{D}$.

The question of interest is to compute the $\mathcal{H}_\infty$-norm of the function $f(z)$, i.e., its maximum absolute value on the unit disk:

$$\|f\|_\infty = \sup_{z\in\mathbb{D}} |f(z)|. \tag{2.33}$$

It can be shown, by using the maximum principle in complex analysis, that it is enough to compute the supremum of $f(z)$ on the boundary of the domain, i.e., the unit circle $|z| = 1$. A fairly complete characterization of this question is available. It is known in the literature under several names, such as the *Kalman–Yakubovich–Popov* lemma [38], or the *bounded real lemma*, or (as a special case of) the *structured singular value* theory [34], among others. The statement, presented below, characterizes this norm in terms of the solution of a semidefinite programming problem.

Theorem 2.45. *Consider a function $f(z)$ as in (2.32), with $|\lambda_i(A)| < 1$. Then, $\|f\|_\infty < \gamma$ if and only if the semidefinite program*

$$\begin{bmatrix} A & b \\ c^T & d \end{bmatrix}^T \begin{bmatrix} P & 0 \\ 0 & 1 \end{bmatrix} \begin{bmatrix} A & b \\ c^T & d \end{bmatrix} \prec \begin{bmatrix} P & 0 \\ 0 & \gamma^2 \end{bmatrix}, \qquad P \succ 0, \tag{2.34}$$

is feasible, where the decision variable is the matrix $P \in \mathcal{S}^n$.

A full proof can be found, for instance, in [3, 47]. We present here only the "easy" direction, i.e., showing that if (2.34) holds, then we have $\|f(z)\|_\infty < \gamma$. For this, let $v = (z^{-1}I - A)^{-1}b$, and multiply the first inequality in (2.34) left and right by $[v^* \; 1]$ and its conjugate transpose, respectively. From the identity

$$\begin{bmatrix} A & b \\ c^T & d \end{bmatrix} \begin{bmatrix} v \\ 1 \end{bmatrix} = \begin{bmatrix} z^{-1}v \\ f(z) \end{bmatrix},$$

we have that

$$(|z^{-1}|^2 - 1)(v^* P v) + (|f(z)|^2 - \gamma^2) < 0,$$

and thus the conclusion directly follows. The converse direction takes a bit more work; see Chapter 9. There are extensions of this result to the matrix case, i.e., where $f(z)$ is matrix-valued.

Exercise 2.46. Use the given formulation to compute the $\mathcal{H}_\infty$-norm of the analytic function $f(z) = \frac{z-2}{z^3+z^2-z+3}$. How can you compute, from the semidefinite formulation, a value of z at which the maximum is achieved?

[2] We remark that the notation used here is slightly different from the usual notation in systems and control theory, where z is used instead of z^{-1} in (2.32). The reason is that for interpolation, it is more natural to use functions that are analytic on $\mathbb{D}$ (poles outside the unit circle) than functions that are analytic outside $\mathbb{D}$. To avoid distracting technical issues of controllability and/or observability, we use strict inequalities throughout.

Exercise 2.47. Formulate a similar statement for the matrix case. Do the same formulas work?

Nevanlinna–Pick interpolation. Consider now the following problem. We want to find an analytic function on $\mathbb{D}$ satisfying the interpolation constraints:

$$f(a_k) = c_k \qquad \text{for} \quad k = 1, \dots, m, \tag{2.35}$$

where $a_k \in \mathbb{D}$. When does there exist an analytic function, satisfying the interpolation conditions, whose absolute value is bounded by 1 on the unit disk?

Clearly, a necessary condition is that the interpolated values c_k must satisfy $|c_k| \leq 1$ for all k. However, due to the analyticity constraint, this is not sufficient. Consider, for instance, the case $m = 2$ and the constraints $f(0) = 0$ and $f(1/2) = c$. In this case, a necessary condition is $|c| \leq 1/2$, which is stronger than the "obvious" condition $|c| \leq 1$. To see this, notice that, due to the first interpolation constraint, $f(z)$ must have the form $f(z) = zg(z)$, where $g(z) = f(z)/z$ is also analytic on $\mathbb{D}$ and bounded by one (by the maximum modulus theorem, since $|f(z)| = |g(z)|$ on the unit circle). Thus, we must have $1 \geq |g(1/2)| = 2|c|$, and thus $|c| \leq 1/2$.

Necessary and sufficient conditions for the interpolation problem to be feasible are given by the Nevanlinna–Pick theorem; see Chapter 9. The formulation below is convenient from the optimization viewpoint.

Theorem 2.48. *There exists a function $f(z)$ analytic on $\mathbb{D}$, satisfying the norm bound $\|f(z)\|_\infty \leq \gamma$ and the interpolation constraints* (2.35) *if and only if*

$$\begin{bmatrix} \gamma Z & C^* \\ C & \gamma Z^{-1} \end{bmatrix} \succeq 0, \tag{2.36}$$

where $Z_{jk} = \frac{1}{1-a_j^* a_k}$ and $C = \mathrm{Diag}(c_1, \dots, c_m)$.

Using Schur complements, it can be easily seen that this formulation is equivalent to the more usual characterization where the $m \times m$ Pick matrix P given by

$$P_{jk} = \frac{\gamma^2 - c_j^* c_k}{1 - a_j^* a_k}$$

is required to be positive semidefinite (e.g., Section 9.8). The advantage of condition (2.36) is that it is *linear* in the interpolation values c_k. This allows its use in a variety of system identification problems; see, for instance, [11, 35]. The Nevanlinna–Pick interpolation problem has many important applications in systems and control theory; see, for instance, [14] and [47] and the references therein.

2.2.5 Euclidean Distance Matrices

Assume we are given a list of pairwise distances between a finite number of points. Under what conditions can the points be embedded in some finite-dimensional space and those distances be realized as the *Euclidean* metric between the embedded points? This problem appears in a large number of applications, including distance geometry, computational chemistry, sensor network localization, and machine learning.

Concretely, assume we have a list of distances d_{ij} for $1 \leq i < j \leq n$. We would like to find points $x_i \in \mathbb{R}^k$ (for some value of k) such that $\|x_i - x_j\| = d_{ij}$ for all i, j. What are necessary and sufficient conditions for such an embedding to exist? In 1935, Schoenberg [41] gave an exact characterization in terms of the semidefiniteness of the matrix of squared distances.

Theorem 2.49. *The distances d_{ij} can be embedded in a Euclidean space if and only if the $n \times n$ matrix*

$$D := \begin{bmatrix} 0 & d_{12}^2 & d_{13}^2 & \dots & d_{1n}^2 \\ d_{12}^2 & 0 & d_{23}^2 & \dots & d_{2n}^2 \\ d_{13}^2 & d_{23}^2 & 0 & \dots & d_{3n}^2 \\ \vdots & \vdots & \vdots & \ddots & \vdots \\ d_{1n}^2 & d_{2n}^2 & d_{3n}^2 & \dots & 0 \end{bmatrix}$$

is negative semidefinite on the subspace orthogonal to the vector $e := (1, 1, \dots, 1)$.

Proof. We show only the necessity of the condition. Assume an embedding exists, i.e., there are points $x_i \in \mathbb{R}^k$ such that $d_{ij} = \|x_i - x_j\|$. Consider now the Gram matrix G of inner products

$$G := \begin{bmatrix} \langle x_1, x_1 \rangle & \langle x_1, x_2 \rangle & \dots & \langle x_1, x_n \rangle \\ \langle x_2, x_1 \rangle & \langle x_2, x_2 \rangle & \dots & \langle x_2, x_n \rangle \\ \vdots & \vdots & \ddots & \vdots \\ \langle x_n, x_1 \rangle & \langle x_n, x_2 \rangle & \dots & \langle x_n, x_n \rangle \end{bmatrix} = [x_1, \dots, x_n]^T [x_1, \dots, x_n],$$

which is positive semidefinite by construction. Since $D_{ij} = \|x_i - x_j\|^2 = \langle x_i, x_i \rangle + \langle x_j, x_j \rangle - 2\langle x_i, x_j \rangle$, we have

$$D = \text{Diag}(G) \cdot e^T + e \cdot \text{Diag}(G)^T - 2G,$$

from which the result directly follows. □

Notice that the dimension of the embedding is given by the rank k of the Gram matrix G.

For more on this and related embedding problems, good starting points are Schoenberg's original paper [41] as well as the book [15].

Exercise 2.50. Consider the Euclidean distance matrix characterization in Theorem 2.49. Show that it implies the triangle inequality $d_{ik} \leq d_{ij} + d_{jk}$ for all triples (x_i, x_j, x_k) of points. Is the converse true?

2.2.6 Rank Minimization and Nuclear Norm

An interesting class of optimization problems appearing in many application domains is *rank minimization* problems. These have the form

$$\begin{array}{ll} \text{minimize} & \operatorname{rank} X \\ \text{subject to} & X \in \mathcal{C}, \end{array} \tag{2.37}$$

where the matrix $X \in \mathbb{R}^{m \times n}$ is the decision variable, and $\mathcal{C}$ is a given convex constraint set. Notice that the cost function is integer-valued, and thus (unless the problem is trivial) these optimization problems are not convex.

Rank minimization questions arise in many different areas, since notions such as order, complexity, and dimensionality can often be expressed by means of the rank of an appropriate matrix. For example, a low-rank matrix could correspond to a low-degree statistical model for a random process (e.g., factor analysis), a low-order realization of a linear dynamical system, or a low-dimensional embedding of data in Euclidean space (as in Section 2.2.5). If the set of models that satisfy the desired constraints is convex, then choosing the simplest one in a given family can be formulated as a rank minimization problem of the form (2.37).

In general, rank minimization problems can be quite difficult to solve, both in theory and practice. However, several researchers have proposed heuristic techniques to obtain good approximate solutions. A particularly interesting method is the *nuclear norm* heuristic, originally proposed in [17, 16]. In this method, instead of directly solving the problem (2.37), one solves instead

$$\begin{array}{ll} \text{minimize} & \|X\|_* \\ \text{subject to} & X \in \mathcal{C}, \end{array} \tag{2.38}$$

where $\|\cdot\|_*$ is the *nuclear norm* defined earlier in (2.11). In other words, the "difficult" objective function (rank) is replaced by a "nicer" cost function (nuclear norm) which is convex, and thus the resulting problem is convex.

Under certain conditions on the set $\mathcal{C}$, it has been shown that the solution of the problem (2.38) coincides with the lowest-rank solution, i.e., the "true" solution of (2.37). For example, a typical formulation (see, e.g., [39] for a specific statement) would establish that if the set $\mathcal{C}$ is a subspace of dimension $O(n \log n)$, uniformly chosen according to a natural rotation-invariant probability measure, then the nuclear norm heuristic succeeds with high probability.

Atomic norms. An interesting generalization of these methods is obtained by considering more general *atomic norms* [10]. Consider a set $\mathcal{A}$ of *atoms* v_i in some vector space V (the set $\mathcal{A}$ can be finite or infinite). Given an element $a \in V$, we are interested in the "smallest" decomposition of a in terms of the elements v_i, i.e., the one that satisfies

$$\begin{array}{ll} \text{minimize} & \sum_i |\alpha_i| \\ \text{subject to} & a = \sum_i \alpha_i v_i. \end{array} \tag{2.39}$$

We can then define the *atomic norm* $\|a\|_{\mathcal{A}}$ as the optimal value of this optimization problem. If the set of atoms is finite, this is a linear programming problem. In most situations of interest, however, the set $\mathcal{A}$ is either infinite or exponentially large, in which case an LP formulation is impractical. In certain cases, however, we can still compute this norm efficiently. For instance, in the case where the set of atoms $\mathcal{A}$ corresponds to the rank one matrices uv^T, where $\|u\| = \|v\| = 1$, then this norm corresponds exactly to the matrix nuclear norm defined earlier.

For many problems, however, we would like to consider more general sets of atoms. A particularly interesting case is when the atoms are the rank one matrices with ± 1 entries. In other words, the atoms are given by $\mathcal{A} = \{vw^T \in \mathbb{R}^{m \times n} : v \in \mathbb{R}^m, v_i^2 = 1, w \in \mathbb{R}^n, w_i^2 = 1\}$. In this case, the norm (2.39) is in general NP-hard to compute. However, a nice computable approximation is available, known as the γ_2 or max-norm. This norm is defined as $\|A\|_{\gamma_2} := \max_{\|u\|=1, \|v\|=1} \|A \circ uv^T\|_*$, where $\circ$ is the entrywise product, and can be computed as the optimal value of the primal-dual pair of semidefinite programs:

$$\begin{array}{ll} \text{maximize} & \operatorname{Tr} A^T Y \\ \text{subject to} & \begin{bmatrix} \operatorname{Diag}(p) & Y \\ Y^T & \operatorname{Diag}(q) \end{bmatrix} \succeq 0, \\ & \sum_{i=1}^{m} p_i + \sum_{i=1}^{n} q_i = 2, \end{array} \qquad \begin{array}{ll} \text{minimize} & t \\ \text{subject to} & \begin{bmatrix} V & A \\ A^T & W \end{bmatrix} \succeq 0, \\ & V_{ii} = t, \\ & W_{ii} = t. \end{array} \tag{2.40}$$

It can be easily seen that (2.40) gives a lower bound on the optimal value of (2.39), i.e., $\|A\|_{\gamma_2} \leq \|A\|_{\mathcal{A}}$. Indeed, if $A = \sum_i \alpha_i v_i w_i^T$, where the v_i and w_i are ± 1 vectors, then choosing $V = \sum_i |\alpha_i| v_i v_i^T$, $W = \sum_i |\alpha_i| w_i w_i^T$, and $t = \sum_i |\alpha_i|$ gives a feasible solution for the right-hand side of (2.40). As discussed in Exercise 2.53, the γ_2-norm actually yields a constant approximation ratio to the atomic norm for this specific set of atoms. The γ_2-norm is of great importance in a number of applications, including communication complexity; see, e.g., [27].

Exercise 2.51. Check that the expression (2.39) correctly defines a matrix norm by verifying homogeneity and the triangle inequality. What properties are needed on the atom set $\mathcal{A}$ to ensure that the norm is well defined and nonzero at every nontrivial point?

Exercise 2.52. Let the set of atoms $\mathcal{A}$ be the rank one matrices of the form vw^T, where $\|v\| = \|w\| = 1$. Show that the corresponding atomic norm is the "standard" nuclear norm (sum of singular values).

Exercise 2.53. Using the results in Section 2.2.2, show that in the case where the atoms are the rank one matrices with ± 1 entries, the following inequality holds:

$$\|A\|_{\gamma_2} \leq \|A\|_{\mathcal{A}} \leq K_G \|A\|_{\gamma_2},$$

where K_G is the Grothendieck constant.

Exercise 2.54. Based on the previous exercise, explain the geometric relationship between the unit ball of the γ_2-norm in $\mathbb{R}^{m\times n}$ and the elliptope $\mathcal{E}_{m+n}$ defined earlier in Section 2.1.3.

2.3 Algorithms and Software

2.3.1 Algorithms

In this section we describe a few algorithmic and complexity aspects of the numerical solution of semidefinite optimization problems. For a complete treatment, we refer the reader to articles and monographs such as [13, 32, 44, 45].

Semidefinite programs are convex optimization problems and, as such, can be solved using general convex optimization techniques. Under "natural" assumptions (e.g., to rule out doubly exponentially small solutions), semidefinite optimization is solvable in polynomial time, in the sense that ϵ-suboptimal, weakly feasible solutions can be computed in time polynomial in $\log \frac{1}{\epsilon}$. This follows, for instance, from general results about the ellipsoid method [21].

Despite these nice theoretical results, the ellipsoid method is often too slow in practice. Since SDP is a generalization of linear programming, it is natural that some of the most effective practical methods for SDP have been inspired by state-of-the-art techniques from LP. This has led to the development of interior-point methods [1, 32] for SDP. The basic idea of interior-point methods is to consider the optimality conditions of Lemma 2.12 and to perturb the complementarity slackness condition to $(C - \sum_i A_i y_i)X = \mu I$. As μ varies, these equations implicitly define a curve (X_μ, y_μ) called the *central path*, and to solve the original problem we need to compute (X_μ, y_μ) as $\mu \to 0$. These equations are relatively easy to solve for large μ, and by carefully decreasing the value of μ, it is possible to use Newton's method to efficiently track solutions as μ decreases to zero. There are several different versions of these methods (depending on the exact form of the equations to which Newton's method is applied), although they all share fairly similar features. In particular, primal-dual interior-point methods of this kind are among the most efficient known methods for small- and medium-scale SDP problems.

Besides interior-point methods, there are several alternative techniques for solving SDPs that are sometimes preferable to "pure" primal-dual methods due to speed or memory efficiency issues. Examples of these are techniques based on low-rank factorizations [9], spectral bundle methods [23], or augmented Lagrangian methods for large-scale problems [46], among others.

2.3.2 Software

There are a number of useful software packages for polyhedral computations, linear and semidefinite programming, and algebraic visualization. We present below a partial annotated selection. A few good up-to-date web resources for general information about semidefinite programming include Christoph Helmberg's SDP page www-user.tu-chemnitz.de/~helmberg/semidef.html and the SDPA website sdpa.sourceforge.net.

Polyhedral computations. The first class of software packages we discuss is polyhedral manipulation codes and libraries. Almost all of them allow us to convert an inequality representation of a polyhedron (usually called an H-representation) into vertices/extreme rays (V-representation), and vice versa, as well as much more complicated operations between polyhedra.

- `cdd`, by Komei Fukuda.
  www.ifor.math.ethz.ch/~fukuda/cdd_home.

- `lrs`, by David Avis.
  cgm.cs.mcgill.ca/~avis/C/lrs.html.

- `polymake`, by Ewgenij Gawrilow and Michael Joswig (main authors).
  polymake.org.

- `PORTA`, by Thomas Christof and Andreas Löbel.
  typo.zib.de/opt-long_projects/Software/Porta.

Linear programming. For formulating and solving linear programs, many codes are available, ranging from academic implementations suitable for relatively small problems to industrial-scale solvers. The following is a necessarily partial list:

- `GLPK` – GNU Linear Programming Kit
  www.gnu.org/s/glpk. This is an open-source package for solving large-scale linear programming problems, using either simplex or interior-point methods. GLPK can also solve integer programming problems and can be used as a callable C library.

- `CLP` – LP solver, part of the COIN-OR (COmputational INfrastructure for Operations Research) suite of open source software. www.coin-or.org

- `CPLEX` – Perhaps the best-known commercial solver, now being developed and marketed by IBM.

Semidefinite programming. Although SDP is much more recent than linear programming, fortunately many good software packages are already available. Among the most well-known are the following:

- CSDP, originally by Brian Borchers, now a COIN-OR project:
  projects.coin-or.org/Csdp

- SDPA, by the research group of Masakazu Kojima, sdpa.sourceforge.net. Several versions of the SDPA solver are available, including parallel and variable-precision floating-point arithmetic, in MATLAB and C++ versions.

- SDPT3, by Kim-Chuan Toh, Reha Tütüncü, and Michael Todd.
  www.math.nus.edu.sg/~mattohkc/sdpt3.html. SDPT3 is a MATLAB package for linear, quadratic, and semidefinite programming. It can also handle determinant maximization problems, as well as problems with complex data.

- SeDuMi, originally by Jos Sturm, currently being maintained by the optimization group at Lehigh University (sedumi.ie.lehigh.edu), is a widely used MATLAB package for linear, quadratic, second order conic, and semidefinite optimization, and any combination of these.

An easy and convenient way to "try out" many of these packages, without installing them in a local machine, is through the NEOS Optimization server (neos-server.org), currently hosted by the University of Wisconsin-Madison.

Parsers. In practice, specifying a semidefinite programming problem by explicitly defining matrices A_i, C, and b in (SDP-P) can be cumbersome and error-prone. A much more convenient and reliable way is to use a "natural" description of the variables and inequalities and to automatically translate these into standard form using a parser or modeling language. Two well-known and convenient modeling environments for semidefinite programming are the following:

- CVX, by Michael Grant and Stephen Boyd.
  cvxr.com/cvx. CVX is a MATLAB-based "disciplined convex programming" software. It is particularly well suited to conic optimization, including semidefinite and geometric programming.

- YALMIP, by Johan Löfberg.
  yalmip.org. YALMIP is a MATLAB-based parser and solver for the modeling and solution of convex and nonconvex optimization problems.

Bibliography

[1] F. Alizadeh. Interior point methods in semidefinite programming with applications to combinatorial optimization. *SIAM J. Optim.*, 5(1):13–51, 1995.

[2] N. Alon and A. Naor. Approximating the cut-norm via Grothendieck's inequality. In *Proceedings of the Thirty-Sixth Annual ACM Symposium on Theory of Computing*, ACM, New York, 2004, pp. 72–80.

[3] J.A. Ball, I. Gohberg, and L. Rodman. *Interpolation of Rational Matrix Functions.* Birkhäuser, Basel, 1990.

[4] G.P. Barker and D. Carlson. Cones of diagonally dominant matrices. *Pacific J. Math.*, 57(1):15–32, 1975.

[5] D. Bertsimas and J. N. Tsitsiklis. *Introduction to Linear Optimization.* Athena Scientific, Cambridge, MA, 1997.

[6] S. Boyd, L. El Ghaoui, E. Feron, and V. Balakrishnan. *Linear Matrix Inequalities in System and Control Theory,* Studies in Applied Mathematics 15. SIAM, Philadelphia, 1994.

[7] S. Boyd and L. Vandenberghe. *Convex Optimization.* Cambridge University Press, Cambridge, UK, 2004.

[8] M. Braverman, K. Makarychev, Y. Makarychev, and A. Naor. The Grothendieck constant is strictly smaller than Krivine's bound. In the *IEEE 52nd Annual Symposium on Foundations of Computer Science (FOCS)*, IEEE, Washington, DC, 2011, pp. 453–462.

[9] S. Burer and R. D.C. Monteiro. A nonlinear programming algorithm for solving semidefinite programs via low-rank factorization. *Mathematical Programming*, 95(2):329–357, 2003.

[10] V. Chandrasekaran, B. Recht, P. A. Parrilo, and A.S. Willsky. The convex geometry of linear inverse problems. *Foundations of Computational Mathematics*, 12:805–849, 2012.

[11] J. Chen, C.N. Nett, and M.K.H. Fan. Worst case system identification in $\mathcal{H}_\infty$: Validation of a priori information, essentially optimal algorithms, and error bounds. *IEEE Transactions on Automatic Control*, 40(7):1260–1265, 1995.

[12] V. Chvátal. *Linear Programming*. W.H. Freeman, New York, 1983.

[13] E. de Klerk. *Aspects of Semidefinite Programming: Interior Point Algorithms and Selected Applications*, Applied Optimization 65. Kluwer Academic Publishers, Dordrecht, The Netherlands, 2002.

[14] P. Delsarte, Y. Genin, and Y. Kamp. On the role of the Nevanlinna–Pick problem in circuit and system theory. *International Journal of Circuit Theory and Applications*, 9(2):177–187, 1981.

[15] M. M. Deza and M. Laurent. *Geometry of Cuts and Metrics*, Algorithms and Combinatorics 15. Springer-Verlag, Berlin, 1997.

[16] M. Fazel. *Matrix Rank Minimization with Applications*. Ph.D. thesis, Stanford University, Stanford, CA, 2002.

[17] M. Fazel, H. Hindi, and S.P. Boyd. A rank minimization heuristic with application to minimum order system approximation. In *Proceedings of the American Control Conference*, volume 6, IEEE, Washington, DC, 2001, pp. 4734–4739.

[18] M. R. Garey and D. S. Johnson. *Computers and Intractability: A Guide to the Theory of NP-Completeness*. W. H. Freeman, New York, 1979.

[19] M. X. Goemans. Semidefinite programming in combinatorial optimization. *Math. Programming*, 79(1–3):143–161, 1997.

[20] M. X. Goemans and D. P. Williamson. Improved approximation algorithms for maximum cut and satisfiability problems using semidefinite programming. *Journal of the ACM*, 42(6):1115–1145, 1995.

[21] M. Grötschel, L. Lovász, and A. Schrijver. *Geometric Algorithms and Combinatorial Optimization*, 2nd ed., Algorithms and Combinatorics 2. Springer-Verlag, Berlin, 1993.

[22] O. Güler. Hyperbolic polynomials and interior point methods for convex programming. *Math. Oper. Res.*, 22(2):350–377, 1997.

[23] C. Helmberg and F. Rendl. A spectral bundle method for semidefinite programming. *SIAM Journal on Optimization*, 10(3):673–696, 2000.

[24] J.W. Helton. *Operator Theory, Analytic Functions, Matrices, and Electrical Engineering*. CBMS Regional Conference Series in Mathematics 68. AMS, Providence, RI, 1987.

[25] J.L. Krivine. Constantes de Grothendieck et fonctions de type positif sur les spheres. *Adv. Math*, 31:16–30, 1979.

[26] M. Laurent and S. Poljak. On a positive semidefinite relaxation of the cut polytope. *Linear Algebra and Its Applications*, 223:439–461, 1995.

[27] T. Lee and A. Shraibman. Lower bounds in communication complexity. *Foundations and Trends in Theoretical Computer Science*, 3(4), 2009.

[28] L. Lovász. On the Shannon capacity of a graph. *IEEE Transactions on Information Theory*, 25(1):1–7, 1979.

[29] J. Matoušek and B. Gärtner. *Understanding and Using Linear Programming*. Springer-Verlag, New York, 2007.

[30] A. Megretski. Relaxations of quadratic programs in operator theory and system analysis. In *Systems, Approximation, Singular Integral Operators, and Related Topics (Bordeaux,* 2000*)*, Oper. Theory Adv. Appl. 129. Birkhäuser, Basel, 2001, pp. 365–392.

[31] Y. Nesterov. Semidefinite relaxation and nonconvex quadratic optimization. *Optimization Methods and Software*, 9:141–160, 1998.

[32] Y. E. Nesterov and A. Nemirovski. *Interior Point Polynomial Methods in Convex Programming*, Studies in Applied Mathematics 13. SIAM, Philadelphia, 1994.

[33] J. Nie, P. A. Parrilo, and B. Sturmfels. Semidefinite representation of the k-ellipse. *IMA Volumes in Mathematics and Its Applications*, 146:117–132, 2008.

[34] A. Packard and J. C. Doyle. The complex structured singular value. *Automatica J. IFAC*, 29(1):71–109, 1993.

[35] P. A. Parrilo, M. Sznaier, R.S. Sánchez Peña, and T. Inanc. Mixed time/frequency-domain based robust identification. *Automatica J. IFAC*, 34(11):1375–1389, 1998.

[36] M. V. Ramana. An exact duality theory for semidefinite programming and its complexity implications. *Math. Programming*, 77(2, Ser. B):129–162, 1997.

[37] M. V. Ramana, L. Tunçel, and H. Wolkowicz. Strong duality for semidefinite programming. *SIAM J. Optim.*, 7(3):641–662, 1997.

[38] A. Rantzer. On the Kalman-Yakubovich-Popov lemma. *Systems & Control Letters*, 28:7–10, 1996.

[39] B. Recht, M. Fazel, and P. A. Parrilo. Guaranteed minimum-rank solutions of linear matrix equations via nuclear norm minimization. *SIAM Review*, 52(3):471–501, 2010.

[40] J. Renegar. Hyperbolic programs, and their derivative relaxations. *Found. Comput. Math.*, 6(1):59–79, 2006.

[41] I. J. Schoenberg. Remarks to Maurice Fréchet's article "Sur la définition axiomatique d'une classe d'espace distanciés vectoriellement applicable sur l'espace de Hilbert." *Ann. of Math.* (2), 36(3):724–732, 1935.

[42] A. Schrijver. *Theory of Linear and Integer Programming.* Wiley, New York, 1986.

[43] C. Shannon. The zero error capacity of a noisy channel. *IRE Transactions on Information Theory*, 2(3):8–19, 1956.

[44] M. Todd. Semidefinite optimization. *Acta Numerica*, 10:515–560, 2001.

[45] L. Vandenberghe and S. Boyd. Semidefinite programming. *SIAM Review*, 38(1):49–95, 1996.

[46] X.Y. Zhao, D. Sun, and K.C. Toh. A Newton-CG augmented Lagrangian method for semidefinite programming. *SIAM Journal on Optimization*, 20:1737–1765, 2010.

[47] K. Zhou, K. Glover, and J. C. Doyle. *Robust and Optimal Control.* Prentice Hall, Englewood Cliffs, NJ, 1995.

[48] G. M. Ziegler. *Lectures on Polytopes*, Graduate Texts in Mathematics 152. Springer-Verlag, New York, 1995.

Chapter 3

Polynomial Optimization, Sums of Squares, and Applications

Pablo A. Parrilo

We begin the study of one of the main themes of the book, namely, the relationships between *nonnegative polynomials*, *sums of squares*, and *semidefinite programming*. The two key ideas around which this chapter is structured are

> sum of squares decompositions of polynomials can be computed using semidefinite programming,

and

> the search for infeasibility certificates for real polynomial systems is a convex problem. Given an upper bound on the degree of the certificates, they can be found by solving a sum of squares program.

In the rest of this chapter, we define and explain the basic concepts needed to make these assertions precise. For this, in Section 3.1 we introduce *nonnegative polynomials*, *sum of squares decompositions*, and the notion of *sum of squares programs*, followed by a few simple but important applications in Section 3.2. In Section 3.3 we explore how the presence of additional algebraic structure, such as symmetries or sparsity, enables more efficient computations. We then explain how these results can be used to provide *infeasibility certificates* for systems of polynomial inequalities and the important implications for polynomial optimization (Section 3.4). Section 3.5 explores the dual side, including geometric and probabilistic interpretations. Finally, in Section 3.6, we present additional applications of the methods in diverse areas of applied mathematics and engineering, concluding with a short discussion of current software implementations.

3.1 Nonnegative Polynomials and Sums of Squares

3.1.1 Nonnegative Polynomials

We consider polynomials in n variables, with real coefficients. A multivariate polynomial $p(x_1, \ldots, x_n)$ is *nonnegative* if it takes only nonnegative values, i.e.,

$$p(x_1, \ldots, x_n) \geq 0 \quad \text{for all } (x_1, \ldots, x_n) \in \mathbb{R}^n. \tag{3.1}$$

The characterization of nonnegativity of multivariate polynomials is a ubiquitous question throughout mathematics, with many rich and surprising connections.

From the algorithmic and computational viewpoints, perhaps the immediate first questions that one can ask include the following:

Decision question. Given a polynomial $p(x)$, how do we decide if it is nonnegative?

Certification. Is it possible to *certify* nonnegativity efficiently? In other words, imagine you are trying to convince a friend that $p(x)$ is actually nonnegative, or that it is not. Is there a more efficient way of doing this than having them run an algorithm themselves?

Complexity. What computational resources are needed to decide polynomial nonnegativity?

Structural questions. What is the structure of the set of nonnegative polynomials?

Before proceeding to answer these questions in the general case, it makes sense to consider first a few simple special cases.

Univariate polynomials. A good starting point is the case of polynomials in a single variable, i.e., when $n = 1$:

$$p(x) = p_d x^d + p_{d-1} x^{d-1} + \cdots + p_1 x + p_0. \tag{3.2}$$

We normally assume that the leading coefficient p_d is not zero, and occasionally we will normalize it to $p_d = 1$, in which case we say that $p(x)$ is *monic*. The *roots* are the values of x at which $p(x)$ vanishes. By the fundamental theorem of algebra, there is a unique factorization

$$p(x) = p_d \cdot \prod_{i=1}^{d} (x - x_i), \tag{3.3}$$

where the (complex) roots x_i may have multiplicities, i.e., they are not necessarily all distinct.

How do we decide if p is nonnegative? Clearly, an obvious necessary condition is that the degree of $p(x)$ be even. Otherwise, if the degree is odd, then either as $x \to \infty$ or as $x \to -\infty$, the polynomial $p(x)$ will become negative.

In some simple cases, it is possible to give direct characterizations.

Example 3.1. Let $p(x) = x^2 + p_1 x + p_0$ be a monic quadratic polynomial. What conditions must p_1, p_0 satisfy for $p(x)$ to be nonnegative? Since $p(x)$ defines a convex function that achieves its minimum, it is enough to verify the nonnegativity condition only for its minimum value. Solving for the minimizer of $p(x)$ by setting its derivative to zero (i.e., $2x + p_1 = 0$), we obtain $x_\star = -p_1/2$, $p(x_\star) = p_0 - p_1^2/4$, and thus we have

$$\{(p_0, p_1) : p(x) \geq 0 \quad \forall x \in \mathbb{R}\} = \{(p_0, p_1) : 4p_0 - p_1^2 \geq 0\}. \quad \blacksquare$$

Thus, in the special case of polynomials of degree 2, we were able to write an explicit inequality condition in the coefficients of $p(x)$ to ensure its nonnegativity.

What can we say in the general (univariate) case? Reasoning directly in terms of coefficients does not seem too promising. However, it can be easily seen that nonnegativity imposes strong restrictions on the *roots* of $p(x)$. Assume the leading coefficient of $p(x)$ is positive. If $p(x) \geq 0$, then either $p(x)$ has no real roots, or, if it has real roots, they must have even multiplicity (why?). However, since in general the roots are nonelementary functions of the coefficients of the polynomial, this approach does not directly yield a good characterization (we will, however, use this insight later in Section 3.1.3).

There are several explicit algorithms for deciding nonnegativity of univariate polynomials. These methods will not require the computation of the roots and may in fact be implemented in exact rational arithmetic. A classical formulation is based on *Sturm sequences*; see, e.g., [19]. We describe an alternative technique instead, known as the *Hermite* or *trace form* method; its justification is developed in Exercise 3.7. Consider a monic univariate polynomial (3.2) and define its associated *Hermite matrix* as the following $d \times d$ symmetric Hankel matrix:

$$H_1(p) = \begin{bmatrix} s_0 & s_1 & \cdots & s_{d-1} \\ s_1 & s_2 & \cdots & s_d \\ \vdots & \vdots & \ddots & \vdots \\ s_{d-1} & s_d & \cdots & s_{2d-2} \end{bmatrix}, \qquad s_k = \sum_{j=1}^{d} x_j^k, \tag{3.4}$$

where, as before, x_j are the roots of $p(x)$. The quantities s_k are known as the *power sums* and, remarkably, can be obtained directly from the coefficients of $p(x)$ using the *Newton identities*, with no root computation needed; see Exercise 3.5. When $p(x)$ is monic, the s_k are polynomials of degree k in the coefficients of $p(x)$.

It turns out that we can count the real roots of $p(x)$ by analyzing the *inertia* of its Hermite matrix (see Appendix A for background material on matrix inertia). The following theorems make this connection precise.

Theorem 3.2. *The rank of the Hermite matrix $H_1(p)$ is equal to the number of distinct (complex) roots. Its signature is equal to the number of distinct real roots.*

Theorem 3.3. *Let $p(x)$ be a monic univariate polynomial of degree $2d$. Then, the following are equivalent:*

1. *The polynomial $p(x)$ is strictly positive.*

2. *The polynomial $p(x)$ has no real roots.*

3. *The inertia of the Hermite matrix is $\mathcal{I}(H_1(p)) = (k, 2d-k, k)$ for some $1 \leq k \leq d$.*

Recall that the inertia of a matrix can be computed efficiently, in polynomial time, by diagonalization with a congruence transformation (e.g, via the LDL^T decomposition; see Appendix A), so a decision method for strict positivity based on this theorem can be effectively implemented.

Example 3.4. Consider again the quadratic univariate polynomial $p(x) = x^2 + p_1 x + p_0$. The power sums are $s_0 = 2$, $s_1 = -p_1$, and $s_2 = p_1^2 - 2p_0$. The Hermite matrix is then

$$H_1(p) = \begin{bmatrix} 2 & -p_1 \\ -p_1 & p_1^2 - 2p_0 \end{bmatrix}.$$

Let $\Delta = \det H_1(p) = p_1^2 - 4p_0$. The inertia of the Hermite matrix is

$$\mathcal{I}(H_1(p)) = \begin{cases} (0,0,2) & \text{if } \Delta > 0, \\ (0,1,1) & \text{if } \Delta = 0, \\ (1,0,1) & \text{if } \Delta < 0, \end{cases}$$

and thus p is strictly positive if and only if $p_1^2 - 4p_0 < 0$. ∎

Exercise 3.5. Let $p(x)$ be a monic univariate polynomial as in (3.2). Show that the power sums s_k satisfy the recursive equations:

$$s_0 = d, \qquad s_k = \sum_{j=1}^{k} (-1)^{j-1} p_j s_{k-j}, \quad k = 1, 2, \ldots.$$

These equations are known as the *Newton identities*.

Exercise 3.6. Show that the determinant of the matrix $H_1(p)$ is (up to a constant) equal to the *discriminant* [32] of $p(x)$. Hint: Express $\det H_1(p)$ in terms of the roots of $p(x)$.

Exercise 3.7. Given a univariate polynomial $p(x)$ of degree d, define the *Hermite quadratic form* or *trace form* $H_1(p) : \mathbb{R}[x]_d \mapsto \mathbb{R}$ as

$$H_1(p)[f] = \sum_{i=1}^{d} f(x_i)^2,$$

where $x_1, \ldots, x_d$ are the roots of $p(x)$.

1. Find a matrix representation of the quadratic form $H_1(p)$.

2. When is $H_1(p)$ singular?

3. Find a factorization of the Hermite matrix in terms of the Vandermonde matrix of the roots. If necessary, assume that roots x_i are all distinct, and describe the required modifications for the general case.

4. Prove Theorem 3.2.

Exercise 3.8. Can you find a criterion for polynomial nonnegativity (not strict positivity) based *solely* on the inertia of the Hermite matrix? Describe your proposed criterion in detail, or explain why additional information may be necessary. Hint: Consider the polynomials $(x+1)x^2(x-1)^3$ and $(x+1)^2x^2(x-1)^2$.

Exercise 3.9. Find necessary and sufficient conditions for the quartic polynomial $p(x) = x^4 + p_1 x + p_0$ to be positive for all real values of x. Plot the number of real roots as a function of the parameters (p_0, p_1).

Multivariate polynomials. Now we move on to the multivariate case. Let $P_{n,2d}$ be the set of nonnegative polynomials in n variables of degree less than or equal to $2d$, i.e.,

$$P_{n,2d} = \{p \in \mathbb{R}[x]_{n,2d} \, : \, p(x) \geq 0 \quad \forall x \in \mathbb{R}^n\}.$$

By identifying a polynomial with its $N := \binom{n+d}{d}$ coefficients, and noticing that the constraints $p(x) \geq 0$ are affine in the coefficients of p for every fixed x, it follows directly that $P_{n,2d}$ is a *convex* set in $\mathbb{R}[x]_{n,2d} \sim \mathbb{R}^N$. Furthermore, the following is true.

Theorem 3.10. *The set of nonnegative polynomials $P_{n,2d}$ is a proper cone (i.e., closed, convex, pointed, and solid) in $\mathbb{R}[x]_{n,2d} \sim \mathbb{R}^N$.*

Example 3.11. Consider the case of polynomials of degree $2d = 2$, i.e., *quadratic* polynomials in n variables. Every such polynomial can be represented as

$$p(x) = \frac{1}{2} x^T A x + 2b^T x + c,$$

where $A \in \mathcal{S}^n$ is a symmetric matrix. It can be shown (Exercise 3.16) that $p(x) \geq 0$ for all $x \in \mathbb{R}^n$ if and only if

$$\begin{bmatrix} A & b \\ b^T & c \end{bmatrix} \succeq 0.$$

Thus, in this case, the set $P_{n,2}$ is isomorphic to the positive semidefinite cone $\mathcal{S}_+^{n+1}$.

Notice that for the particular case of a univariate quadratic polynomial $p(x) = p_2 x^2 + p_1 x + p_0$, this reduces to the condition

$$\begin{bmatrix} p_2 & p_1/2 \\ p_1/2 & p_0 \end{bmatrix} \succeq 0.$$

This agrees with Example 3.1, which corresponds to the monic case where $p_2 = 1$. ∎

As we will shortly see, although always convex, the cone of nonnegative polynomials has a fairly complicated geometry in the general case. In Chapter 4, further features of this set will be studied in detail.

Exercise 3.12. Prove Theorem 3.10.

Except for special situations like the quadratic case of Example 3.11, it will not be easy to efficiently obtain explicit descriptions of $P_{n,2d}$. The reason is that the algebraic and combinatorial structure of the set of nonnegative polynomials can be extremely complicated, even though it is a convex set. As a consequence, obtaining general explicit inequalities (e.g., on the coefficients) that define when a polynomial is nonnegative can be a very complex, or even hopeless, task.

To understand this situation in more detail, we discuss the algebraic and geometric situation with the help of a few examples, followed by a discussion of the computational complexity aspects.

$P_{n,2d}$ is semialgebraic but is not basic semialgebraic. Recall that in Example 3.1 we provided explicit inequalities for the set $P_{1,2}$ of univariate quadratics. Since this description did not include quantifiers or logical operations (e.g., set unions, implications), we obtained a *basic semialgebraic set* (see Section A.4.4 in Appendix A). As we will see, such convenient descriptions are not possible in general, since the set of nonnegative polynomials is *not* basic semialgebraic for $2d \geq 4$.

To see why this is the case, consider the following example, describing a particular affine section of $P_{1,4}$.

Example 3.13. Let $p(x)$ be the quartic univariate polynomial $p(x) = x^4 + 2ax^2 + b$. For what values of a, b is $p(x)$ nonnegative? Since the leading term x^4 has even degree and is strictly positive, $p(x)$ is strictly positive if and only if it has no real roots. The discriminant[1] of $p(x)$ is equal to $\mathrm{Dis}_x(p) = 256\, b\, (a^2 - b)^2$. For the number of real roots to change, the discriminant must vanish, and thus the zero set of the discriminant partitions the set of parameters (a, b) into regions where the number of real roots is constant. The subset of $(a, b) \in \mathbb{R}^2$ for which $p(x)$ is positive corresponds to the case of no real roots, with its closure being the region of nonnegativity. Notice that (as expected) this subset is convex and is shown in Figure 3.1. ∎

As the example illustrates, in the univariate case it is easy to see that if $p(x)$ lies on the boundary of the set $P_{1,2d}$, then it must have a real root, of multiplicity at least two. Indeed, if there is no real root, then $p(x)$ is in the strict interior of $P_{1,2d}$ (small enough perturbations will not create a root), and if it has a simple real root it clearly cannot be nonnegative. Thus, on the boundary of $P_{1,2d}$, the discriminant

[1] The discriminant $\mathrm{Dis}_x(p)$ of a univariate polynomial $p(x)$ is a polynomial in the coefficients of p that vanishes if and only if p has a multiple root. It is defined as the resultant between $p(x)$ and its derivative $p'(x)$; see [32] or [120] for an introduction to polynomial resultants and discriminants.

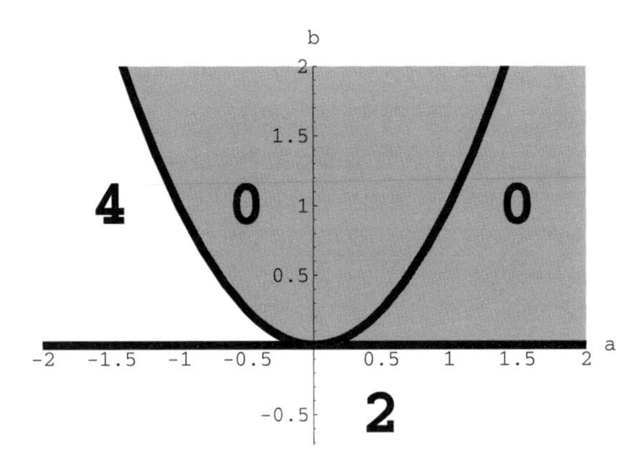

Figure 3.1. *The discriminant* $\mathrm{Dis}_x(p)$ *partitions the parameter space* (a, b) *into regions where the number of real roots is constant. The numbers indicate how many real roots the polynomial* $x^4 + 2ax^2 + b$ *has whenever* (a, b) *are in the corresponding region. The shaded set corresponds to the polynomial being nonnegative.*

$\mathrm{Dis}_x(p)$ must necessarily vanish. However, it turns out that the discriminant does not vanish *only* on the boundary, but it may also vanish at points *inside* the set; see Figure 3.1. The algebraic reason is that pairs of *complex* roots may coincide, which will cause the discriminant to vanish, even though this does not directly affect nonnegativity of p.

This situation can create some serious difficulties. For instance, even though we have a perfectly valid analytic expression for the boundary of the set, we cannot get a good sense of "how far we are" from the boundary by looking at the absolute value of the discriminant (this would be very useful for numerical optimization over $P_{n,2d}$). A more algebraic way of describing the situation is that $P_{n,2d}$ is a convex set with the complicating feature that the *Zariski closure* of the boundary intersects the interior of the set.

In general, these sets are not very convenient to work with since we cannot describe them in terms of *unquantified* inequalities.

Lemma 3.14. *The set discussed in Example* 3.13 *and presented in Figure* 3.1 *is* not *basic semialgebraic.*

The fact that $P_{n,2d}$ is not basic semialgebraic (for $2d \geq 4$) means that there is *no* description of $P_{n,2d}$ in terms of a finite collection of polynomial inequalities $\{g_1(p_\alpha) \geq 0, \ldots, g_m(p_\alpha) \geq 0\}$ in the coefficients p_α. In other words, any characterization of the set $P_{n,2d}$ using polynomial inequalities must necessarily include logical operations between sets (e.g., unions, complements) or other similar complications.

Things can be even more complicated than what Figure 3.1 suggests in the sense that (as opposed to what may be inferred from this figure) in higher dimensions it is impossible to remove the "undesired" component (i.e., the discriminant does not factor, as it did in this example). Consider the case of a quartic polynomial of

Figure 3.2. *The zero set of the discriminant of the polynomial* $x^4 + 4ax^3 + 6bx^2 + 4cx + 1$. *The convex set inside the "bowl" corresponds to the region of nonnegativity. There is an additional one-dimensional component inside the set.*

the form $p(x) = x^4 + 4ax^3 + 6bx^2 + 4cx + 1$. Its discriminant (up to a nonessential numerical factor) is the irreducible polynomial

$$1 - 27a^4 - 64c^3a^3 + 108bca^3 - 54b^3a^2 + 36b^2c^2a^2 - 6c^2a^2 + 54ba^2 \\ + 108bc^3a - 180b^2ca - 12ca + 81b^4 - 27c^4 - 18b^2 - 54b^3c^2 + 54bc^2.$$

The zero set of this discriminant, shown in Figure 3.2, is an algebraic surface that defines the boundary of a three-dimensional convex set, corresponding to the values of (a, b, c) for which $p(x)$ is nonnegative. It can be shown that this convex set is the convex hull of two parabolas, defined parametrically as

$$t \mapsto \left(t, \frac{2t^2+1}{3}, t\right), \qquad t \mapsto \left(t, \frac{2t^2-1}{3}, -t\right),$$

respectively, and that the surface is singular along these parabolas (these correspond to the cases when the polynomial factors as $p(x) = (x^2 + 2tx \pm 1)^2$).

From the numerical optimization viewpoint, the presence of "extraneous" components of the discriminant in the interior of the feasible set is also an important roadblock for the availability of "easily computable" barrier functions for these sets (even in the univariate case). Indeed, every polynomial that vanishes on the boundary of the set $P_{1,2d}$ must necessarily contain the discriminant as a factor. This is a striking difference from the case of the nonnegative orthant or the positive semidefinite cone, where the standard barriers are given (up to a logarithm) by products of the linear constraints or a determinant (which are polynomials). A possible solution to this problem is to produce nonpolynomial barrier functions, either by partial minimization from a higher-dimensional barrier (i.e., projection) or other constructions such as the "universal" barrier function introduced by Nesterov and Nemirovski [84].

Remark 3.15. *In principle, explicit conditions describing the set $P_{n,2d}$ can be obtained via quantifier elimination techniques, such as Tarski-Seidenberg, cylindrical algebraic decomposition (CAD), and related algorithms; see, e.g.,* [13, 26]. *To do this, consider the quantified formula*

$$\forall x_1 \forall x_2 \cdots \forall x_n \;\; p(x_1, \ldots, x_n) \geq 0,$$

and eliminate the quantified variables $(x_1, \ldots, x_n)$ to obtain a description of $P_{n,2d}$ as a semialgebraic set in terms of the coefficients of p only. Notice that this shows that the nonnegativity problem is decidable. Although extremely powerful from the theoretical viewpoint, these methods often run into serious practical difficulties, given their doubly exponential dependence on the number of variables (modern versions reduce this to doubly exponential in the number of quantifier alternations). In practice, they can only be used for problems of fairly small size. A high-quality implementation of these methods is the software QEPCAD [23].

Exercise 3.16. Prove the characterization of nonnegativity of quadratic polynomials given in Example 3.11.

Exercise 3.17. In this exercise, we consider sets that are semialgebraic but not basic semialgebraic. For much more about this, see [7, 8].

1. Consider the set $S = \mathbb{R}^2 \setminus T$, where T is the open nonnegative orthant $T = \{(x, y) \in \mathbb{R}^2 : x > 0,\; y > 0\}$. Write S as a union of basic semialgebraic sets. Show that S is *not* basic semialgebraic.

2. Prove Lemma 3.14.

Exercise 3.18. Recall that an extreme point v of a convex set S is *exposed* if there exists a supporting hyperplane H of S such that $\{v\} = S \cap H$. Show that the closed convex set in Figure 3.1 has an extreme point that is *not* exposed.

Exercise 3.19. Explore the geometry of the convex set in Figure 3.2. In particular, analyze the "swallowtail" singularities of the discriminant variety at the points $(1, 1, 1)$ or $(-1, 1, 1)$ and the one-dimensional component that joins them.

Computational complexity. A different but related viewpoint on why the set of nonnegative polynomials is difficult to characterize is based on computational complexity arguments; see, e.g., [47] for an introduction to computational complexity. The goal here is to quantify the computational resources (e.g., time, memory) required to decide membership in $P_{n,2d}$ and, in particular, to understand how these resources scale as a function of the problem input size.

Recall the situation of quadratic polynomials discussed in Example 3.11, where nonnegativity of a quadratic polynomial was shown to be equivalent to the positive semidefiniteness of a symmetric matrix. Thus, for this case, polynomial nonnegativity (equivalently, membership in $P_{n,2}$) can be decided in polynomial time using, for instance, Gaussian elimination, or LDL^T, or Cholesky matrix decompositions. Similarly, in the univariate case, there are algorithms that will decide, in time polynomial in the input size (i.e., the bit-length of the coefficients), whether a univariate polynomial is nonnegative. This can be done, for instance, with minor variations of the Hermite matrix method described earlier (which, as described, applied only to strict positivity); see [19] for a complete treatment and other related methods.

Unfortunately, the situation is drastically different for multivariate polynomials of degree four or higher. When $2d \geq 4$ it is known that deciding polynomial nonnegativity is an NP-hard problem (for fixed degree, as a function of the number of variables). Essentially, this means that unless the complexity-theoretic statement P = NP holds (which is generally considered very unlikely), there cannot be a polynomial-time algorithm that can decide whether a polynomial is nonnegative. This includes, of course, the possibility of writing a "small" list of explicit conditions on the coefficients.

Exercise 3.20. Give a reduction from any known NP-hard problem (e.g., satisfiability, independent set, binary integer programming, etc.) to nonnegativity of multivariate quartic polynomials.

A way out: Describing sets as projections. As we have seen, even in the univariate case, the set of nonnegative polynomials $P_{n,2d}$ has fairly complicated features, such as not being basic semialgebraic. However, it turns out that at least in some cases one can provide nicer representations.

To do this, we will represent (or approximate) these sets as a *projection* from a higher dimensional space, where the object "upstairs" will have nicer properties, and all complicating features will be a consequence of the projection. As an example, recall the set discussed in Example 3.13 and Figure 3.1. This is a two-dimensional set, describing a particular section of the set of univariate nonnegative quartics. Although, as we showed, this set is *not* basic semialgebraic, it is however the *projection* of the convex basic semialgebraic set

$$\{(a, b, t) \in \mathbb{R}^3 : \quad b \geq (a-t)^2, \quad t \geq 0\}.$$

In Figure 3.3 we present a plot of this three-dimensional convex set and its projection onto the plane (a, b) that gives exactly the set of Figure 3.1.

As we shall see in detail in the next section, this idea will allow us to exactly represent the set $P_{1,2d}$ of univariate nonnegative polynomials as the projection of a "nice" spectrahedral set. Furthermore, the same techniques will make it possible to obtain good approximations for the set $P_{n,2d}$ of multivariate nonnegative polynomials. The techniques will be based on the connection between *sums of squares* polynomials and semidefinite programming.

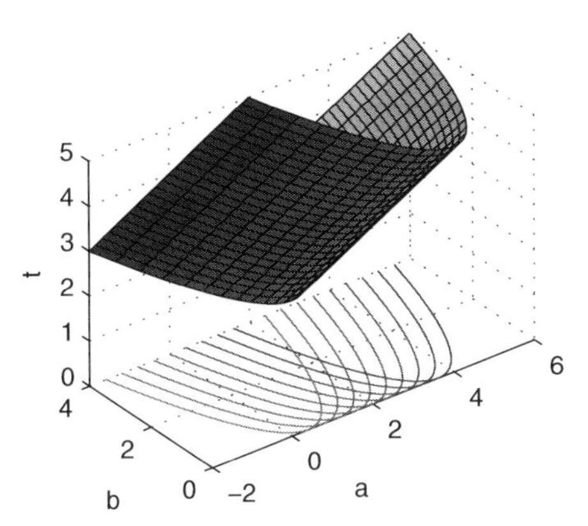

Figure 3.3. *A three-dimensional convex set, described by a quadratic and a linear inequality, whose projection on the (a, b) plane is equal to the set in Figure 3.1.*

Exercise 3.21. Recall that the *Minkowski sum* of two sets $S_1, S_2 \subset \mathbb{R}^n$ is the set $S_1 + S_2 := \{s_1 + s_2 \,:\, s_1 \in S_1,\ s_2 \in S_2\}$. Consider a set $S \subset \mathbb{R}^2$, given by the Minkowski sum of a disk and a line segment. Show that S is not basic semialgebraic. Give a representation of S as a projection of a convex semialgebraic set in $\mathbb{R}^3$.

Exercise 3.22. Consider the set

$$\{(x, y, z) \in \mathbb{R}^3 \,:\, xyz \geq 1, \quad x \geq 0, \quad y \geq 0, \quad z \geq 0\}.$$

1. Is it convex?
2. Is it a spectrahedron?
3. Is it a projected spectrahedron?

Hint: If you need help with item 2, try the "real zero" condition in Chapter 6.

Exercise 3.23. Prove the validity of the set containment relationships described in Figure 3.4, and give counterexamples for all noninclusions.

3.1.2 Sums of Squares

A multivariate polynomial $p(x_1, \ldots, x_n)$ is a *sum of squares* (sos) if it can be written as the sum of squares of some other polynomials. Formally, we have the following.

Definition 3.24. *A polynomial $p(x) \in \mathbb{R}[x]_{n,2d}$ is a* sum of squares *(sos) if there exist $q_1, \ldots, q_m \in \mathbb{R}[x]_{n,d}$ such that*

$$p(x) = \sum_{k=1}^{m} q_k^2(x). \tag{3.5}$$

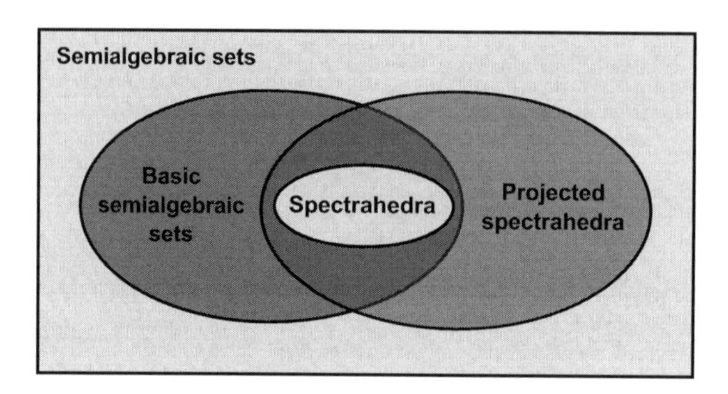

Figure 3.4. *Relationships between set classes.*

We will use $\Sigma_{n,2d}$ for the set of sos polynomials in n variables of degree less than or equal to $2d$. If a polynomial $p(x)$ is a sum of squares, then it obviously satisfies $p(x) \geq 0$ for all $x \in \mathbb{R}^n$. Thus, an sos condition is a sufficient condition for global nonnegativity, i.e., $\Sigma_{n,2d} \subseteq P_{n,2d}$.

In general, sos decompositions are not unique.

Example 3.25. The polynomial $p(x_1, x_2) = x_1^2 - x_1 x_2^2 + x_2^4 + 1$ is a sum of squares. Among infinitely many others, it has the decompositions

$$\begin{aligned} p(x_1, x_2) &= \frac{3}{4}(x_1 - x_2^2)^2 + \frac{1}{4}(x_1 + x_2^2)^2 + 1^2 \\ &= \frac{1}{9}(3 - x_2^2)^2 + \frac{2}{3}x_2^2 + \frac{1}{288}(9x_1 - 16x_2^2)^2 + \frac{23}{32}x_1^2. \qquad \blacksquare \end{aligned}$$

It quickly follows from its definition that the set $\Sigma_{n,2d}$ of sos polynomials is invariant under nonnegative scalings and convex combinations; i.e., it is a convex cone. In fact, more is true, as follows.

Theorem 3.26. *The set of sos polynomials $\Sigma_{n,2d}$ is a proper cone (i.e., closed, convex, pointed, and solid) in $\mathbb{R}[x]_{n,2d} \sim \mathbb{R}^N$.*

One of the central questions in convex algebraic geometry is to understand the relationships between the two cones $P_{n,2d}$ and $\Sigma_{n,2d}$. In the remainder of this chapter, as well as in Chapter 4, we analyze this problem from the algebraic, geometric, and computational viewpoints.

Exercise 3.27. Consider the sum of squares representation (3.5). Show that if $p(x)$ has degree $2d$, then the polynomials q_i necessarily have degree less than or equal to d, by considering the coefficients corresponding to the highest order terms.

Exercise 3.28. Using finitely many squares in Definition 3.24 may seem restrictive at first. Show using Caratheodory's theorem (Theorem A.10 in Appendix A) that in Definition 3.24 we can always take $m \leq \binom{n+d}{d}$.

When is nonnegativity equal to sum of squares? Since sum of squares implies nonnegativity (i.e., $\Sigma_{n,2d} \subseteq P_{n,2d}$), a natural question is to understand under what conditions the converse holds, i.e., when a nonnegative polynomial can be expressed as a sum of squares. We will study many aspects of this question extensively in this book, particularly in Chapter 4.

More than a century ago, David Hilbert showed that equality between the set of nonnegative polynomials $P_{n,2d}$ and sos polynomials $\Sigma_{n,2d}$ holds only in the following three cases:

- Univariate polynomials (i.e., $n = 1$).
- Quadratic polynomials ($2d = 2$).
- Bivariate quartics ($n = 2$, $2d = 4$).

For all other cases, there always exist nonnegative polynomials that are *not* sums of squares. Perhaps the most celebrated example is the bivariate sextic ($n = 2$, $2d = 6$) due to Motzkin, given by (in dehomogenized form)

$$M(x,y) = x^4y^2 + x^2y^4 + 1 - 3x^2y^2. \tag{3.6}$$

This polynomial is nonnegative but is not a sum of squares. Nonnegativity of $M(x,y)$ follows from the arithmetic-geometric inequality applied to $(x^4y^2, x^2y^4, 1)$ (or, alternatively, from the identity (3.19)) and the fact that it is not a sum of squares from Exercise 3.97.

The first two cases (univariate and quadratic) of Hilbert's classification are relatively straightforward and are discussed in Exercises 3.30 and 3.32, respectively. In Chapter 4 the more subtle remaining case will be proved, along with an in-depth study of the structure of these sets.

Another immediate question is related to the algorithmic aspects of sos polynomials. Given a polynomial, how can we decide if it is a sum of squares? Equivalently, how can we decide membership in the cone $\Sigma_{n,2d}$? We answer these questions in the next section, where we describe the connections between sos conditions on polynomials and semidefinite programming.

Exercise 3.29. Let $p(x)$, $q(x)$ be sos polynomials.

1. Show that the sum $p(x) + q(x)$ and the product $p(x)q(x)$ are also sums of squares.
2. Furthermore, show that if both $p(x)$ and $q(x)$ are each the sum of *two* squares, then so is their product $p(x)q(x)$.
   Hint: Recall complex multiplication. For $w, z \in \mathbb{C}$, $|w|^2|z|^2 = |wz|^2$ holds. Consider the real and imaginary parts of this expression.

Exercise 3.30. In this exercise, we show that univariate nonnegative polynomials are sums of squares, and, in fact, *two* squares suffice.

1. Show that if $p(x) = p_{2d}x^{2d} + \cdots + p_1 x + p_0$ is nonnegative, it has a factorization of the form

$$p(x) = p_{2d} \cdot \prod_j (x - r_j)^{n_j} \cdot \prod_k [(x - z_k)(x - z_k^*)]^{m_k},$$

where r_j and (z_k, z_k^*) are the real and complex roots of $p(x)$, $p_{2d} > 0$, and the multiplicities n_j of the real roots are even.

2. Show that if z is a complex number, the quadratic polynomial $(x - z)(x - z^*)$ is a sum of two squares.

3. Use Exercise 3.29 to conclude that $p(x)$ is itself a sum of two squares.

Exercise 3.31. Using the previous exercise, compute a decomposition of $p(x) = x^4 + 2x^3 + 6x^2 - 22x + 13$ as a sum of two squares.

Exercise 3.32. Let $p(x_1, \ldots, x_n)$ be a quadratic polynomial (i.e., $2d = 2$). Show that if $p(x_1, \ldots, x_n)$ is nonnegative, then it is a sum of squares.
Hint: Recall Example 3.11 and matrix factorizations.

3.1.3 Univariate Polynomials

In this section we explain in detail the computation of sos decompositions of univariate polynomials, with a full discussion of the multivariate case in the next section. The main reason for starting with the univariate case is that it is notationally simpler, and it is fairly similar to the general case.

Consider a univariate polynomial $p(x)$ of degree $2d$:

$$p(x) = p_{2d}x^{2d} + p_{2d-1}x^{2d-1} + \cdots + p_1 x + p_0. \tag{3.7}$$

Assume that $p(x)$ is a sum of squares; i.e., it can be written as in (3.5). Notice that the degree of the polynomials q_k must be at most equal to d, since the coefficient of the highest term of each q_k^2 is positive, and thus there cannot be any cancellation in the highest power of x (cf. Exercise 3.27). Then, we can write

$$\begin{bmatrix} q_1(x) \\ q_2(x) \\ \vdots \\ q_m(x) \end{bmatrix} = V \begin{bmatrix} 1 \\ x \\ \vdots \\ x^d \end{bmatrix}, \tag{3.8}$$

where $V \in \mathbb{R}^{m \times (d+1)}$, and its kth row contains the coefficients of the polynomial q_k. For future reference, let $[x]_d$ be the vector of monomials on the right-hand side of (3.8), and define the matrix $Q := V^T V$. We then have

$$p(x) = \sum_{k=1}^{m} q_k^2(x) = (V[x]_d)^T (V[x]_d) = [x]_d^T V^T V [x]_d = [x]_d^T Q [x]_d.$$

This immediately suggests the following characterization of sos polynomials.

Lemma 3.33. *Let $p(x)$ be a univariate polynomial of degree $2d$. Then, $p(x)$ is a sum of squares if and only if there exists a symmetric matrix $Q \in \mathcal{S}^{d+1}$ that satisfies*

$$p(x) = [x]_d^T Q [x]_d, \qquad Q \succeq 0. \tag{3.9}$$

The matrix Q is usually called the *Gram matrix* of the sos representation. One direction of the lemma follows directly from noticing that the matrix $Q = V^T V$ constructed above is positive semidefinite. For the other direction, assume there exists a positive semidefinite matrix Q for which (3.9) holds. Then, by factorizing $Q = V^T V$ (e.g., via Cholesky or square root factorization), we obtain an sos decomposition of $p(x)$.

Although perhaps not immediately obvious at first, the condition in (3.9) is a semidefinite program! Indeed, notice that the constraint $p(x) = [x]_d^T Q [x]_d$ is *affine* in the matrix Q, and thus the set of possible Gram matrices Q is given exactly by the intersection of an affine subspace and the cone of positive semidefinite matrices.

To obtain explicit equations for this semidefinite program, we index the rows and columns of Q by $\{0, \dots, d\}$ as

$$[x]_d^T Q [x]_d = \sum_{i=0}^{d} \sum_{j=0}^{d} Q_{ij} x^{i+j} = \sum_{k=0}^{2d} \left(\sum_{i+j=k} Q_{ij} \right) x^k.$$

Thus, for this expression to be equal to $p(x)$, it must be the case that

$$p_k = \sum_{i+j=k} Q_{ij}, \qquad k = 0, \dots, 2d. \tag{3.10}$$

This is a system of $2d+1$ linear equations between the entries of Q and the coefficients of $p(x)$. Thus, since Q is simultaneously constrained to be positive semidefinite, and to belong to the affine subspace defined by these equations, an sos condition is exactly equivalent to a semidefinite programming problem. We have shown, then, the following.

Lemma 3.34. *A univariate polynomial $p(x) = \sum_{k=0}^{2d} p_k x^k$ is a sum of squares if and only if there exists a positive semidefinite matrix $Q \in \mathcal{S}^{d+1}$ satisfying (3.10). This is a semidefinite programming problem.*

Recall that in the univariate case, nonnegativity and sum of squares are equivalent conditions. Thus, Lemma 3.34 completely characterizes the set of univariate nonnegative polynomials and shows that the set $P_{1,2d} = \Sigma_{1,2d}$ is a *projected spectrahedron.*

Example 3.35. Consider the univariate polynomial

$$p(x) = x^4 + 4x^3 + 6x^2 + 4x + 5,$$

for which we want to find an sos decomposition. Proceeding as described earlier, we consider the expression

$$\begin{aligned} p(x) &= \begin{bmatrix} 1 \\ x \\ x^2 \end{bmatrix}^T \begin{bmatrix} q_{00} & q_{01} & q_{02} \\ q_{01} & q_{11} & q_{12} \\ q_{02} & q_{12} & q_{22} \end{bmatrix} \begin{bmatrix} 1 \\ x \\ x^2 \end{bmatrix} \\ &= q_{22}x^4 + 2q_{12}x^3 + (q_{11} + 2q_{02})x^2 + 2q_{01}x + q_{00}. \end{aligned}$$

Matching coefficients, we obtain the following linear equality constraints:

$$\begin{aligned} x^4 &: \quad 1 = q_{22}, \\ x^3 &: \quad 4 = 2q_{12}, \\ x^2 &: \quad 6 = q_{11} + 2q_{02}, \\ x &: \quad 4 = 2q_{01}, \\ 1 &: \quad 5 = q_{00}. \end{aligned}$$

We need to find a positive semidefinite matrix Q that satisfies these linear equations (i.e., solve a semidefinite program). In this case, the semidefinite program is feasible, and we can obtain a solution given by

$$Q = \begin{bmatrix} 5 & 2 & 0 \\ 2 & 6 & 2 \\ 0 & 2 & 1 \end{bmatrix} = V^T V, \qquad V = \begin{bmatrix} 0 & 2 & 1 \\ \sqrt{2} & \sqrt{2} & 0 \\ \sqrt{3} & 0 & 0 \end{bmatrix},$$

which yields the sos decomposition

$$p(x) = (x^2 + 2x)^2 + 2(1 + x)^2 + 3. \qquad \blacksquare$$

In certain special cases, it may be possible to construct sos representations of a fixed polynomial without necessarily having to solve semidefinite programs. In Exercise 3.36 we explore the case of univariate polynomials; see Exercise 3.84 for an extension of these results to a more complicated situation. However, as we discuss in Section 3.1.7, the reason why the SDP reformulation is crucial is because it will allow us to *search* for sos polynomials, even in the presence of additional convex constraints.

Exercise 3.36. Consider the following algorithm, which computes an sos decomposition of a monic univariate polynomial, using linear algebra techniques, in a numerically stable way.

Algorithm 3.1. SOS decomposition of a univariate polynomial.

Input: A monic univariate polynomial $p(x) = x^{2d} + \cdots + p_1 x + p_0$.
Output: An sos decomposition of $p(x)$.

1: Form the companion matrix $\mathcal{C}_p$, defined by

$$\mathcal{C}_p := \begin{bmatrix} 0 & 0 & \cdots & 0 & -p_0 \\ 1 & 0 & \cdots & 0 & -p_1 \\ 0 & 1 & \cdots & 0 & -p_2 \\ \vdots & \vdots & \ddots & \vdots & \vdots \\ 0 & 0 & \cdots & 1 & -p_{2d-1} \end{bmatrix}.$$

2: Find a complex Schur decomposition of the companion matrix, i.e.,

$$C_p = U\Lambda U^* = \begin{bmatrix} U_{11} & U_{12} \\ U_{21} & U_{22} \end{bmatrix} \begin{bmatrix} \Lambda_{11} & \Lambda_{12} \\ 0 & \Lambda_{22} \end{bmatrix} \begin{bmatrix} U_{11} & U_{12} \\ U_{21} & U_{22} \end{bmatrix}^*,$$

where U is unitary, Λ is upper triangular, and the spectra of Λ_{11}, Λ_{22} are complex conjugates of each other.

3: Let $q := vU_{12}^{-1}$, where v is the first row of U_{22}. Let q_r and q_i be the real and imaginary parts of q, respectively.

4: Define

$$\begin{bmatrix} q_1(x) \\ q_2(x) \end{bmatrix} = \begin{bmatrix} -q_r & 1 \\ -q_i & 0 \end{bmatrix} \begin{bmatrix} 1 \\ x \\ \vdots \\ x^d \end{bmatrix}.$$

5: **return** sos decomposition $p(x) = q_1^2(x) + q_2^2(x)$.

1. Implement this algorithm, and test it in a few examples.
2. If $p(x)$ is not nonnegative, where does the algorithm fail?
3. Prove that the algorithm is correct, i.e., it always produces a valid sos decomposition.

Hint: What properties does the complex polynomial $q(x) = q_1(x) + iq_2(x)$ have?

Exercise 3.37. The results presented in this section for "standard" univariate polynomials can be easily extended to real *trigonometric* polynomials, i.e., expressions of the form

$$p(\theta) = a_0 + \sum_{k=1}^{d} (a_k \cos k\theta + b_k \sin k\theta).$$

This is a trigonometric polynomial of degree d.

1. Show that if $p(\theta) \geq 0$ for all $\theta \in [-\pi, \pi]$ and d is even, then there is an sos decomposition $p(\theta) = q_1^2(\theta) + q_2^2(\theta)$, where q_1, q_2 are trigonometric polynomials. What is the corresponding statement for the case when d is odd?

2. Give a semidefinite programming formulation to decide if a trigonometric polynomial is nonnegative. The formulation should be in terms of a $(d+1) \times (d+1)$ real symmetric matrix, where d is the degree of the polynomial. It may be helpful to consider separately the case where d is odd or even.

3. Find an sos decomposition of the polynomial

$$p(\theta) = 4 - \sin\theta + \sin 2\theta - 3\cos 2\theta.$$

4. Find an sos decomposition of the polynomial

$$p(\theta) = 5 - \sin\theta + \sin 2\theta - 3\cos 3\theta.$$

3.1.4 Multivariate Polynomials

The general multivariate case is quite similar to the univariate case discussed in the previous section. The main differences are the need of multi-index notation for monomials, and the fact that sos will only be a sufficient condition for nonnegativity.

Consider a polynomial $p(x_1, \ldots, x_n)$ of degree $2d$ in n variables. The number of coefficients of p is equal to $\binom{n+2d}{2d}$. We let $p(x) = \sum_\alpha p_\alpha x^\alpha$, where α are tuples of exponents $\alpha \in \{(\alpha_1, \ldots, \alpha_n) \ : \ \alpha_1 + \cdots + \alpha_n \leq 2d, \ \alpha_i \geq 0 \ \forall i = 1, \ldots, n\}$.

Let $[x]_d := [1, x_1, \ldots, x_n, x_1^2, x_1 x_2, \ldots, x_n^d]^T$ be the vector of all $\binom{n+d}{d}$ monomials in $x_1, \ldots, x_n$ of degree less than or equal to d, and consider the equation

$$p(x) = [x]_d^T \, Q \, [x]_d, \tag{3.11}$$

where Q is an $\binom{n+d}{d} \times \binom{n+d}{d}$ symmetric matrix. Proceeding exactly as in the previous section, and indexing the matrix Q by the $\binom{n+d}{d}$ monomials in n variables of degree d (or, more precisely, the associated exponent tuples), we obtain the following conditions:

$$p_\alpha = \sum_{\beta+\gamma=\alpha} Q_{\beta\gamma}, \qquad Q \succeq 0. \tag{3.12}$$

This is a system of $\binom{n+2d}{2d}$ linear equations, one for each coefficient of $p(x)$. As before, these equations are affine conditions relating the entries of Q and the coefficients of $p(x)$. Thus, we can decide membership in, or optimize over, the set of sos polynomials by solving an SDP problem.

Example 3.38. We want to determine whether the bivariate quartic polynomial

$$p(x, y) = 2x^4 + 5y^4 - x^2y^2 + 2x^3y + 2x + 2$$

is a sum of squares. Since this polynomial has degree $2d = 4$, the vector $[x]_d$ contains all monomials of degree less than or equal to 2, i.e., $[x]_d = [1, x, y, x^2, xy, y^2]^T$. Writing the expression (3.11) for a generic matrix Q (which, for consistency with (3.12), though perhaps at the expense of clarity, we index with exponent tuples), we have

$$p(x,y) = \begin{bmatrix} 1 \\ x \\ y \\ x^2 \\ xy \\ y^2 \end{bmatrix}^T \begin{bmatrix} q_{00,00} & q_{00,10} & q_{00,01} & q_{00,20} & q_{00,11} & q_{00,02} \\ q_{00,10} & q_{10,10} & q_{10,01} & q_{10,20} & q_{10,11} & q_{10,02} \\ q_{00,01} & q_{10,01} & q_{01,01} & q_{01,20} & q_{01,11} & q_{01,02} \\ q_{00,20} & q_{10,20} & q_{01,20} & q_{20,20} & q_{20,11} & q_{20,02} \\ q_{00,11} & q_{10,11} & q_{01,11} & q_{20,11} & q_{11,11} & q_{11,02} \\ q_{00,02} & q_{10,02} & q_{01,02} & q_{20,02} & q_{11,02} & q_{02,02} \end{bmatrix} \begin{bmatrix} 1 \\ x \\ y \\ x^2 \\ xy \\ y^2 \end{bmatrix}.$$

Expanding the right-hand side, and matching coefficients, we obtain $\binom{2+4}{4} = 15$ linear equations, one per each possible coefficient of $p(x,y)$. For instance, the equations corresponding to the monomials x^4, x^2y^2, and y^2 are

$$\begin{aligned} x^4: &\quad 2 = q_{20,20}, \\ x^2y^2: &\quad -1 = q_{00,22} + 2\,q_{01,21} + q_{11,11}, \\ y^2: &\quad 0 = 2\,q_{00,02} + q_{01,01}. \end{aligned}$$

Again, finding a positive semidefinite matrix Q subject to these 15 linear equations is an SDP problem. Solving it, we obtain a feasible solution:

$$Q = \frac{1}{3}\begin{bmatrix} 6 & 3 & 0 & -2 & 0 & -2 \\ 3 & 4 & 0 & 0 & 0 & 0 \\ 0 & 0 & 4 & 0 & 0 & 0 \\ -2 & 0 & 0 & 6 & 3 & -4 \\ 0 & 0 & 0 & 3 & 5 & 0 \\ -2 & 0 & 0 & -4 & 0 & 15 \end{bmatrix}.$$

Any factorization of this positive semidefinite matrix will give an explicit sos decomposition of $p(x,y)$, for instance,

$$\begin{aligned} p(x,y) = \frac{4}{3}y^2 + \frac{1349}{705}y^4 + \frac{1}{12}(4x+3)^2 + \frac{1}{15}(3x^2+5xy)^2 + \\ + \frac{1}{315}(-21x^2+20y^2+10)^2 \\ + \frac{1}{59220}(328y^2-235)^2. \quad \blacksquare \end{aligned}$$

We summarize the contents of this section in the following theorem, describing the direct relation between positive semidefinite matrices and an sos condition.

Theorem 3.39. *A multivariate polynomial $p(x) = \sum_\alpha p_\alpha x^\alpha$ in n variables and degree $2d$ is a sum of squares if and only if there exists $Q \in \mathcal{S}_+^{\binom{n+d}{d}}$ satisfying (3.12). As a consequence, membership in $\Sigma_{n,2d}$ can be decided via semidefinite programming.*

The matrix size of the semidefinite program appearing in Theorem 3.39 is $\binom{n+d}{d}$, which grows polynomially in the number of variables n for fixed degree d. Since $\binom{n+d}{d} = \binom{n+d}{n}$, it also grows polynomially in d for fixed n.

Corollary 3.40. *The cone $\Sigma_{n,2d}$ of sos polynomials is a projected spectrahedron of dimension $\binom{n+2d}{2d}$.*

The connections between sos conditions, the Gram matrix representation, and convexity can be traced back to the work of Shor [113], as well as Reznick and collaborators [106, 30]. The links with semidefinite programming were made explicit in [89, 91] and were also explored independently by Nesterov [83] and Lasserre [72].

These results will be of crucial importance in the remainder of the chapter. Notice in particular the striking constrast with the case of nonnegative polynomials: while membership in $P_{n,2d}$ is an NP-hard problem for $2d \geq 4$ (and thus, practically infeasible for most problems of interest), membership in $\Sigma_{n,2d}$ can be reduced to a polynomially sized SDP problem.

3.1.5 Computational Formulations

A nice and useful coordinate-free interpretation of our earlier discussion (and in particular, of (3.11)) is that writing a polynomial of degree $2d$ as a sum of squares is equivalent to expressing it as a quadratic form on the vector space of polynomials $\mathbb{R}[x]_{n,d}$. Although this coordinate-free viewpoint is very advantageous for theoretical work, when solving these problems in practice it is necessary to express the corresponding semidefinite programs in a specific set of coordinates. The choice of basis, although irrelevant from the mathematical (or exact arithmetic) viewpoint, may have significant consequences for the numerical conditioning of the resulting optimization problem.

When writing down the semidefinite programs associated to the sos decomposition of a polynomial, as we did in the previous section, there is an implicit choice of bases for two vector spaces: one for the space of polynomials $\mathbb{R}[x]_{n,d}$, and one for the dual space $\mathbb{R}[x]^*_{n,2d}$. Indeed, in our formulation, the polynomial $p(x)$ was expressed as a quadratic form on the vector space $\mathbb{R}[x]_{n,d}$, represented by the matrix Q with respect to the monomial basis $[x]_{n,d}$; see (3.11). Similarly, the constraints (3.12) correspond to the coefficients of $p(x) = [x]_d^T Q [x]_d$ with respect to the monomial basis $[x]_{n,2d}$ of $\mathbb{R}[x]_{n,2d}$. While these choices are perhaps "canonical," there are several alternative bases that can be used instead, and these can have very different algebraic and numerical properties.

Based on this discussion, we can write the following more general SDP formulation for sums of squares.

Theorem 3.41. *Let $p(x)$ be a polynomial in n variables and degree $2d$. Choose bases $\{v_1, \dots, v_s\}$ and $\{w_1, \dots, w_t\}$ of $\mathbb{R}[x]_{n,d}$ and $\mathbb{R}[x]^*_{n,2d}$, respectively (and thus, $s = \binom{n+d}{d}$ and $t = \binom{n+2d}{2d}$). Then, $p(x)$ is a sum of squares if and only if there exists a positive semidefinite matrix $Q \in \mathcal{S}^{\binom{n+d}{d}}$ satisfying the affine constraints:*

$$\langle p(x), w_k \rangle = \sum_{i,j=1}^{s} Q_{ij} \langle v_i v_j, w_k \rangle, \qquad k = 1, \dots, t.$$

From the exact arithmetic viewpoint, this statement is of course completely equivalent to Theorem 3.39 and simply corresponds to a change of basis in both primal and dual variables of the corresponding semidefinite program. However, when solving the corresponding semidefinite programs in floating-point arithmetic, there may be very significant differences in the numerical stability of the corresponding formulations. When choosing a particular basis $\{v_i\}$ for the space of polynomials, it will often be convenient to pick $\{w_j\}$ as the corresponding *dual basis*, i.e., so it satisfies $\langle v_i, w_j \rangle = \delta_{ij}$. However, this need not always be the case, and there may be advantages (numerical or otherwise) in not doing so.

In what follows, we discuss four specific bases for the space of polynomials $\mathbb{R}[x]_{n,d}$, briefly mentioning some of their relative advantages and disadvantages.

Monomial basis. This basis is given by the monomials

$$\mathcal{B}_m = \{x^\alpha\},$$

where $\alpha = (\alpha_1, \ldots, \alpha_n)$, with $|\alpha| \leq d$. This is perhaps the most usual choice, and as we did in Section 3.1.4, much of the literature in this area implicitly or explicitly uses this basis. While convenient from the notational and implementation viewpoints, it can have very poor numerical properties.

Scaled monomials. A small modification of the monomial basis is given by the *scaled monomial* basis. The main motivation for this is to achieve certain natural and appealing invariance properties, as explained below. This basis is defined as

$$\mathcal{B}_s = \left\{ \binom{d}{\alpha}^{\frac{1}{2}} x^\alpha \right\},$$

where $\binom{d}{\alpha}$ denotes the multinomial coefficient $\binom{d}{\alpha_1, \ldots, \alpha_n} = \frac{d!}{\alpha_1! \alpha_2! \ldots \alpha_n!}$.

The rationale behind this choice is the following: consider the inner product between polynomials given by

$$\langle p(x), q(x) \rangle := \left\langle \sum_\alpha p_\alpha x^\alpha, \sum_\alpha q_\alpha x^\alpha \right\rangle = \sum_\alpha \binom{d}{\alpha}^{-1} p_\alpha q_\alpha.$$

This inner product is known under many different names, such as the *apolar*, *Fischer*, *Calderón*, or *Bombieri* inner product. Its defining property is the direct relationship between powers of linear forms and point evaluations. Indeed, if p is a homogeneous polynomial of degree d, we have

$$\langle p(x), (v^T x)^d \rangle = \sum_\alpha \binom{d}{\alpha}^{-1} p_\alpha \binom{d}{\alpha} v^\alpha = p(v).$$

As a consequence, this inner product satisfies the invariance property

$$\langle p(Ax), q(x) \rangle = \langle p(x), q(A^T x) \rangle,$$

where A is an $n \times n$ matrix.

The scaled monomial basis is simply an orthonormal basis with respect to this invariant inner product.

Orthogonal polynomials. Similar to the previous case, assume that there is a naturally defined inner product in the space of polynomials. In this case, a natural choice is to pick an orthonormal basis with respect to this inner product. Many well-known families of polynomials (e.g., Chebyshev, Lagrange, Gegenbauer, etc.) fall into this class.

As a concrete illustration, consider the case of an inner product that is induced by integration against a strictly positive measure. For instance, in the case of univariate polynomials, for certain problems it may be natural to have an inner product defined by the Gaussian measure, i.e.,

$$\langle p(x), q(x) \rangle = \frac{1}{\sqrt{2\pi}} \int_{-\infty}^{\infty} p(x)q(x)e^{-\frac{x^2}{2}}\,dx.$$

For this example, such an orthonormal family would be the well-known Hermite polynomials.

Orthogonal polynomials generally enjoy much nicer numerical stability properties than the monomial basis. This is particularly true whenever the underlying measure is chosen in an appropriate way, consistent with the problem to be solved.

Lagrange interpolation. Yet another choice is given by Lagrange interpolating polynomials with respect to a given fixed set of nodes. For simplicity, we discuss here the univariate case only, although the discussion extends naturally to the multivariate case.

Fix $d+1$ distinct points $x_0, \ldots, x_d$ in $\mathbb{R}$. It is well known that the Lagrange interpolating polynomials

$$\ell_i(x) := \prod_{k \neq i} \frac{x - x_k}{x_i - x_k}, \qquad i = 0, \ldots, d,$$

form a basis of $\mathbb{R}[x]_{1,d}$. Also of interest is that the corresponding dual basis of the dual space $\mathbb{R}[x]_{1,d}^*$ is then given by the point evaluations ℓ_{x_i} that satisfy $\ell_{x_i}(p) = p(x_i)$.

This choice is particularly appealing in the case where the polynomial is presented in terms of its values at a given set of points, instead of an explicit description in terms of coefficients. This approach also has some convenient numerical properties related to the use of interior-point methods in the solution of the corresponding semidefinite programs; see [75] for more details.

Exercise 3.42. Consider a univariate cubic polynomial $p(x)$ on the interval $[a,b]$, for which we want to describe the convex hull of its graph, i.e., the set

$$S = \operatorname{conv}\left(\{(t, p(t)) \in \mathbb{R}^2 \,:\, t \in [a,b]\}\right).$$

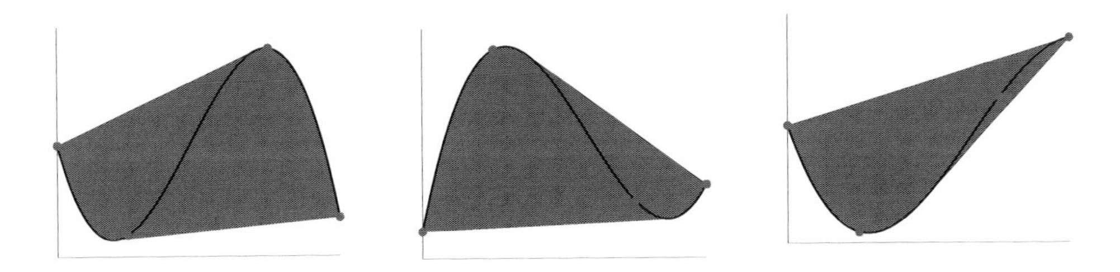

Figure 3.5. *Convex hulls of the graphs of cubic polynomials on an interval.*

See Figure 3.5 for a few examples. We provide below a description of the set S as a projected spectrahedron. Define the interpolation points

$$x_1 = a, \quad x_2 = a + \frac{1}{4}(b-a), \quad x_3 = a + \frac{3}{4}(b-a), \quad x_4 = b.$$

Consider the spectrahedron in the variables $(x, y, \alpha_1, \alpha_2, \alpha_3, \alpha_4) \in \mathbb{R}^6$ defined by

$$\begin{bmatrix} 3\alpha_2 & 0 \\ 0 & 12\alpha_4 \end{bmatrix} + \alpha_3 \begin{bmatrix} 1 & 2 \\ 2 & 4 \end{bmatrix} \succeq 0,$$
$$\begin{bmatrix} 3\alpha_3 & 0 \\ 0 & 12\alpha_1 \end{bmatrix} + \alpha_2 \begin{bmatrix} 1 & 2 \\ 2 & 4 \end{bmatrix} \succeq 0,$$
$$\sum_{i=1}^{4} \alpha_i = 1, \qquad \sum_{i=1}^{4} \alpha_i x_i = x, \qquad \sum_{i=1}^{4} \alpha_i p(x_i) = y.$$

The set S is then given by the projection of this spectrahedron onto the variables (x, y). Notice that in this description, the explicit expression of the polynomial $p(x)$ is never used, but instead only the interpolation values $p(x_i)$ appear.

1. Prove the validity of this description using an sos formulation based on Lagrange interpolation.
2. Generalize this representation to univariate polynomials of any degree.

3.1.6 Rational Sos Decompositions

We have seen in previous sections how to compute sos decompositions using semidefinite programming. These convex optimization problems are usually solved numerically, using floating-point arithmetic. Although floating-point techniques in principle allow for numerical approximations of arbitrary precision, the computed solutions will typically not be exact. This may mean, for instance, that the equation $p(x) = [x]_d^T Q [x]_d$ is only approximately satisfied, or that the matrix Q may have very small negative eigenvalues.

In many applications, particularly those arising from problems in pure mathematics, it is desirable or necessary to obtain *exact* solutions. Examples of this are the use of sos methods for geometric theorem proving (e.g., Section 3.6.5) for establishing the validity of certain algebraic inequalities between matrices [68], or a case of the monotone column permanent (MCP) conjecture [64]. A remarkable recent application is the work in [10], where sos methods were used to prove new upper bounds on kissing numbers, a well-known problem in sphere packings. A common element in all these works is the use of exact algebraic identities obtained from inspection of a numerically computed solution as the basic ingredients in a rigorous proof.

In this section, we show that under a *strict feasibility* assumption, we can obtain a *rational* sos representation from an *approximate solution* to the semidefinite program of Theorem 3.39. The basic idea is to round and project the numerically obtained Gram matrix onto the feasible subspace. We quantify the relation between the numerical error in the subspace and semidefinite constraints, versus the rounding tolerance, that will guarantee that the rounded and projected solution remains feasible. For a full exposition of these ideas, as well as alternative approaches and improvements, we refer the reader to [98], [60], [65], and the references therein.

To obtain rational sos decompositions, it is enough to focus on rational Gram matrices. This follows from the LDL^T decomposition; see Exercise 3.46.

Theorem 3.43. *There exists a rational sos decomposition, i.e.,* $p(x) = \sum_i p_i(x)^2$*, where* $p_i(x) \in \mathbb{Q}[x]$*, if and only if there is a Gram matrix with rational entries.*

The approach we will use to obtain rational sums of squares is to take advantage of interior point solvers' computational efficiency: we first compute an approximate numerical solution, and in a second step we round this numerical solution to an exact rational one. We have the following standing assumption.

Assumption. There exists a positive definite Gram matrix Q for $p(x)$.

This assumption is equivalent to the polynomial $p(x)$ being in the *interior* of the cone of sums of squares. The method described here could fail in general for sums of squares that are not strictly positive: if there is an $x^\star$ such that $p(x^\star) = 0$, it follows from the identity $p(x^\star) = [x^\star]_d^T Q [x^\star]_d$ that the monomial vector $[x^\star]_d$ is in the kernel of Q. Hence Q cannot be positive definite. Nevertheless, this assumption is reasonable for many problems of interest. Furthermore, very recent work of Scheiderer [108] shows that this assumption (or a similar one) is required by giving a construction of sos polynomials with rational coefficients for which no rational decompositions exist.

We assume the sos problem is posed as a semidefinite problem in primal form, as described in Section 3.1.4. After solving the SDP problem in general the numerical solution Q will not exactly satisfy (3.11). For an exact representation of the original polynomial $p(x)$, we have to find a rational approximation to Q which satisfies the equality constraints. The simplest procedure is to compute a rational approximation $\tilde{Q}$, either by naive rounding or more sophisticated techniques like continued fractions. This rational approximation $\tilde{Q}$ is then projected onto the subspace defined by the equations. Since this subspace is defined by rational data

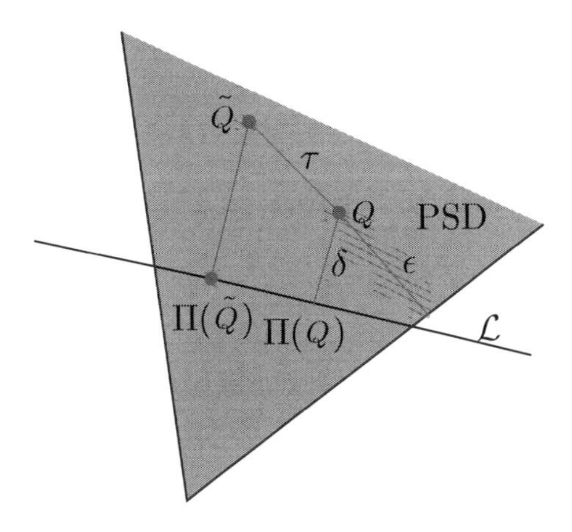

Figure 3.6. *Projection of a rounded solution. The matrix Q is the numerical solution of the SDP problem, and the orthogonal projections of the matrices Q and $\tilde{Q}$ onto the subspace $\mathcal{L}$ are denoted by $\Pi(Q)$ and $\Pi(\tilde{Q})$, respectively. The shaded cone PSD represents the cone of positive semidefinite matrices.*

(the coefficients of $p(x)$), an orthogonal projection Π onto this subspace will yield a rational matrix $\Pi(\tilde{Q})$; see Exercise 3.47.

Now we obtain conditions to ensure that the truncated and projected matrix $\Pi(\tilde{Q})$ remains positive semidefinite. For this, we will estimate the rounding tolerance needed. Assuming strict feasibility of the numerical solution Q returned by the SDP solver, we quantify how far it is from the boundary of the positive semidefinite cone and the affine subspace through two parameters ϵ and δ. The parameter $\epsilon > 0$ will satisfy $Q \succeq \epsilon I$ and is a lower bound on the minimum eigenvalue of Q. The parameter δ quantifies the distance of Q to the subspace, and thus $\|Q - \Pi(Q)\|_F \leq \delta$, where $\|\cdot\|_F$ denotes the Frobenius norm. The matrix Q will be approximated by a rational matrix $\tilde{Q}$ such that $\|Q - \tilde{Q}\|_F \leq \tau$, where τ is the desired tolerance. Figure 3.6 depicts the whole situation.

Theorem 3.44. *Let ϵ, δ, and τ be defined as above. Assume $\delta < \epsilon$, and choose $\tau \leq \sqrt{\epsilon^2 - \delta^2}$. Then, the orthogonal projection $\Pi(\tilde{Q})$ of the rounded matrix $\tilde{Q}$ onto the affine subspace $\mathcal{L}$ is rational and positive semidefinite, and thus it is a valid rational sos decomposition.*

Hence if the SDP problem is strictly feasible, and the numerical solution Q satisfies $\delta < \epsilon$, it is in principle always possible to compute a valid rational solution by using sufficiently many digits for the approximated solution. The allowed rounding tolerance τ depends on the minimum eigenvalue of the positive definite matrix Q and its distance from the affine space $\mathcal{L}$. Under the strict feasibility assumption, there always exists a solution with δ sufficiently small such that the inequality above can be fulfilled (in particular, we can just take $\delta = 0$, although using larger values of δ, if possible, will yield solutions with smaller denominators).

As described in [96], these ideas have been implemented in the software package SOS.m2 for the computer algebra system Macaulay 2 [54]. This package can be used to compute rational sos decompositions and is available for download at [97]. Similar concepts have been recently implemented by Harrison in the open source theorem prover HOL Light [60].

In SOS.m2, the main function is getSOS, which tries to compute a rational sos decomposition for a given polynomial. In the following example we demonstrate how to use the getSOS command for computing an sos decomposition of a polynomial of degree 4 with 4 variables.

Example 3.45. Consider the polynomial

$$p(x,y,z,w) = 2x^4 + x^2y^2 + y^4 - 4x^2z - 4xyz - 2y^2w + y^2 - 2yz + 8z^2 - 2zw + 2w^2.$$

We first load the SOS package and define $p(x,y,z,w)$:

```
i1 : loadPackage "SOS";
i2 : P = QQ[x,y,z,w];
i3 : p = 2*x^4 + x^2*y^2 + y^4 - 4*x^2*z - 4*x*y*z - 2*y^2*w +
          y^2 - 2*y*z + 8*z^2 - 2*z*w + 2*w^2;
```

If successful, the function getSOS returns a weighted sos representation such that $p(x,y,z,w) = \sum_i d_i g_i(x,y,z,w)^2$. Otherwise an error message is displayed.

```
i4 : (g,d) = getSOS p

... omitted output ...

        1  2   1       1        1         2  2    2         8  2    1
o8 = ({- -*x  - -*x*y - -*y + z - -*w,  - --*x  - --*x*y - --*y  - --*y + w,
        4      4       8        8         15      15       15      15
     ------------------------------------------------------------------------
      2   4        4  2     2          18  2   20        81  2       2
     x  - --*x*y - --*y  - --*y, x*y - --*y  - --*y, - ---*y  + y, y },
          11       11      11          59      59      205
     ------------------------------------------------------------------------
         15  22  59  41   66
     {8, --, --, --, --, ----})
          8  15  55  59  1025
```

Hence $p(x,y,z,w)$ may be written as

$$\begin{aligned} p(x,y,z,w) =& 8\left(-\frac{1}{4}x^2 - \frac{1}{4}xy - \frac{1}{8}y + z - \frac{1}{8}w\right)^2 + \frac{15}{8}\left(-\frac{2}{15}x^2 - \frac{2}{15}xy - \frac{8}{15}y^2 - \frac{1}{15}y + w\right)^2 \\ &+ \frac{22}{15}\left(x^2 - \frac{4}{11}xy - \frac{4}{11}y^2 - \frac{2}{11}y\right)^2 + \frac{59}{55}\left(xy - \frac{18}{59}y^2 - \frac{20}{59}y\right)^2 \\ &+ \frac{41}{59}\left(-\frac{81}{205}y^2 + y\right)^2 + \frac{66}{1025}y^4. \end{aligned}$$

Correctness of the obtained decomposition may be verified with the function sumSOS, which expands a weighted sum of squares decomposition:

```
i5 : sumSOS (g,d) - p

o5 = 0
```
■

Exercise 3.46. Prove Theorem 3.43. Use the LDL^T decomposition (see Appendix A, Section A.1.2).

Exercise 3.47. Consider the affine subspace in $\mathbb{R}^n$ defined by the equations $Ax = b$, and a point $x_0 \in \mathbb{R}^n$. Show that the orthogonal projection of x_0 onto the subspace is given by

$$\Pi(x_0) = A^+ b + (I - A^+ A)x_0,$$

where A^+ is the Moore–Penrose pseudoinverse of A. If the rows of A are linearly independent, we have $A^+ = A^T(AA^T)^{-1}$, and thus this formula can be written as

$$\Pi(x_0) = x_0 - A^T(AA^T)^{-1}(Ax_0 - b).$$

Show that if the matrices A and b are rational, and x_0 is a rational point, then so is $\Pi(x_0)$. Prove these facts, and show how to use them to convert an approximate Gram matrix into a rational Gram matrix.

Exercise 3.48. Prove Theorem 3.44.

3.1.7 Sum of Squares Programs

We have described in previous sections how to check whether a given, fixed multivariate polynomial is a sum of squares. These results can be nicely generalized to define a natural class of convex optimization problems which we will call *sum of squares (sos) programs.*

Recall that the main objects of interest in semidefinite programming are

quadratic forms that are **positive semidefinite**.

When attempting to generalize this to homogeneous polynomials of higher degree, a difficulty appears: deciding nonnegativity for quartic or higher degree forms is NP-hard. Therefore, a computationally tractable replacement is the following:

even degree polynomials that are **sums of squares**.

Sum of squares programs can then be defined as conic optimization problems, where the feasible set is given by the intersection of an affine family of polynomials and the proper cone $\Sigma_{n,2d}$ of sos polynomials. As in the case of "pure" semidefinite programming, there are several possible equivalent descriptions. We choose below a free variables formulation to highlight the analogy with the standard SDP dual form (SDP-D) discussed in Chapter 2.

Definition 3.49. *An* sos optimization problem *or* sos program *is a convex optimization problem of the form*

$$\begin{array}{lll} \text{maximize}_y & b_1 y_1 + \cdots + b_m y_m & \\ \text{subject to} & p_i(x; y) \text{ are sos in } \mathbb{R}[x], & i = 1, \ldots, k, \end{array} \tag{3.13}$$

where $p_i(x;y) := c_i(x) + a_{i1}(x)y_1 + \cdots + a_{im}(x)y_m$*, and the* c_i, a_{ij} *are given multivariate polynomials in* $\mathbb{R}[x]$.

Notice that the $p_i(x;y)$ are arbitrary polynomial expressions that are affine in the "parameters" $y_1, \ldots, y_m$ (the decision variables). Also, note that the variables x are "dummy variables," in the sense that we are *not* optimizing over them, but they are the indeterminates of the underlying polynomials. Sum of squares programs are very useful, since they directly operate with polynomials as their basic "data type," thus providing a quite natural modelling formulation for many problems. We will discuss several examples later in this chapter, including Lyapunov functions for nonlinear systems [89, 87], probability inequalities [16], and convex relaxations for nonconvex optimization [89, 72].

Example 3.50. Consider the following simple sos program:

$$\begin{array}{lll} \text{maximize}_y & y_1 + y_2 & \\ \text{subject to} & x^4 + y_1 x + (2 + y_2) & \text{is sos,} \\ & (y_1 - y_2 + 1)\, x^2 + y_2\, x + 1 & \text{is sos.} \end{array}$$

The constraints involve two univariate polynomials (in x), whose coefficients are affine functions of the parameters (or decision variables) (y_1, y_2). Notice that the feasible set (i.e., the set of y_1, y_2 for which both polynomials are sos) is necessarily convex, since it is defined by the intersection of an affine subspace and the sos cone. ■

Interestingly enough, despite their apparently greater generality, sos programs are in fact *equivalent* to SDPs. To see this, notice that, on the one hand, by choosing the polynomials $c_i(x), a_{ij}(x)$ to be quadratic forms, we recover the standard SDP formulation. On the other hand, it is possible to exactly embed every sos program into a larger semidefinite program. Indeed, the constraints requiring $p_i(x;y)$ to be sos in $\mathbb{R}[x]$ are equivalent to the existence of matrices $Q_i \succeq 0$ satisfying

$$p_i(x;y) = [x]_d^T Q_i [x]_d, \qquad i = 1, \ldots, k.$$

Expanding and matching coefficients as before, we obtain linear equations between the coefficients of $p_i(x;y)$ and the entries of Q_i. Since the coefficients of $p_i(x;y)$ are affine in y, the equations above reduce to linear equations between the decision variables y_i and the entries of the matrices Q_i. Thus, the sos program (3.13) is equivalent to a (larger) SDP in the variables $(y_1, \ldots, y_m, Q_1, \ldots, Q_k)$.

Example 3.51. Consider again the sos program of Example 3.50. Using the Gram matrix reformulation described in earlier sections, the sos constraints are equivalent to

$$x^4 + y_1 x + (2 + y_2) = \begin{bmatrix} 1 \\ x \\ x^2 \end{bmatrix}^T \begin{bmatrix} q_{00} & q_{01} & q_{02} \\ q_{01} & q_{11} & q_{12} \\ q_{02} & q_{12} & q_{22} \end{bmatrix} \begin{bmatrix} 1 \\ x \\ x^2 \end{bmatrix},$$

$$(y_1 - y_2 + 1)x^2 + y_2 x + 1 = \begin{bmatrix} 1 \\ x \end{bmatrix}^T \begin{bmatrix} r_{00} & r_{01} \\ r_{01} & r_{11} \end{bmatrix} \begin{bmatrix} 1 \\ x \end{bmatrix},$$

where the matrices Q, R are positive semidefinite. Expanding and equating the left- and right-hand sides, we obtain affine equations between the decision variables y_1, y_2 and the entries of the matrices Q, R. For instance, for the first constraint we obtain

$$\begin{aligned} x^4: &\quad 1 = q_{22}, \\ x^3: &\quad 0 = 2q_{12}, \\ x^2: &\quad 0 = q_{11} + 2q_{02}, \\ x: &\quad y_1 = 2q_{01}, \\ 1: &\quad 2 + y_2 = q_{00}, \end{aligned}$$

while for the second we obtain

$$\begin{aligned} x^2: &\quad y_1 - y_2 + 1 = r_{11}, \\ x: &\quad y_2 = 2r_{01}, \\ 1: &\quad 1 = r_{00}. \end{aligned}$$

Putting together these linear equations with the conditions $Q \succeq 0$ and $R \succeq 0$ yields a standard semidefinite program. ■

As we see, the conversion process from an sos program to a standard semidefinite program is fully algorithmic (and somewhat messy and cumbersome if done by hand!). For these reasons, it has been implemented in several parsers/solvers such as SOSTOOLS [101], YALMIP [74], and SPOT [78]. Furthermore, it is quite useful from both theoretical and practical viewpoints to "abstract out" the fact that (under the hood) sos programs are solved via semidefinite programming and instead just think of them as a tractable class of convex optimization problems that we can freely use for modeling and implementation. In fact, from the next chapter on, we will rarely mention semidefinite programming, and all our formulations will be given directly in terms of sos programs.

Although sos programs and semidefinite programming are "equivalent" in the sense described earlier, the rich algebraic structure of sos programs makes possible a much deeper understanding of their special properties. This also enables customized, more efficient algorithms for their numerical solution [50, 75, 107]. As illustrated in later sections, there are numerous questions in a number of application domains, as well as foundational issues in nonconvex optimization that have simple and natural formulations as sos programs.

Exercise 3.52. Plot the feasible set of the sos program of Example 3.50. Find the corresponding optimal solution $(y_1^\star, y_2^\star)$ as well as explicit sos decompositions of the constraint polynomials at optimality.

Exercise 3.53. Show that sos programs can be written as conic optimization problems in terms of the cone $\Sigma_{n,2d}$ of sos polynomials. Write the corresponding dual conic program.

3.2 Applications of Sum of Squares Programs

In this section we elaborate on several natural extensions of the basic sos methods discussed so far. In combination with the more advanced techniques presented later, these will serve as building blocks for more complex, domain-specific applications developed in Section 3.6.

3.2.1 Unconstrained Polynomial Optimization

Our first application is the global optimization of a univariate polynomial $p(x)$. Although this is a relatively simple task that could be handled with a variety of alternative methods, it nicely illustrates many of the features of much more complicated problems. In this section, we consider only the unconstrained case (i.e., minimization over the whole real line); the constrained case will be considered later.

Rather than directly computing a minimizer $x_\star$ for which $p(x_\star)$ is as small as possible, we instead focus on the alternative viewpoint of obtaining a good (or the best possible) lower bound on its optimal value. It is easy to see that a number γ is a global lower bound of a polynomial $p(x)$ if and only if the polynomial $p(x) - \gamma$ is nonnegative, i.e.,

$$p(x) \geq \gamma \quad \forall x \in \mathbb{R} \qquad \Longleftrightarrow \qquad p(x) - \gamma \geq 0 \quad \forall x \in \mathbb{R}.$$

Notice that the polynomial $p(x) - \gamma$ has coefficients that depend affinely on γ. This suggests considering the optimization problem

$$\underset{\gamma}{\text{maximize}}\ \gamma \qquad \text{subject to} \qquad p(x) - \gamma \ \text{is nonnegative.} \tag{OPT-NN}$$

Clearly, this is a *convex* problem, since the feasible set is defined by an infinite number of linear inequalities (one for each value of x). Its optimal solution $p_\star$ is equal to the global minimum of the polynomial, $p(x_\star)$.

Consider now instead the following optimization problem, where the nonnegativity condition has been replaced by an sos constraint:

$$\underset{\gamma}{\text{maximize}}\ \gamma \qquad \text{subject to} \qquad p(x) - \gamma \ \text{is sos.} \tag{OPT-SOS}$$

The key distinction between the problems (OPT-NN) and (OPT-SOS) is the replacement of *nonnegativity* by an sos condition. However, since in the univariate case nonnegativity is equivalent to sum of squares, these two optimization problems are, in fact, equivalent. Furthermore, (OPT-SOS) has exactly the form of an sos program, and it is thus equivalent to a standard semidefinite program; see Exercise 3.54 for its explicit formulation.

As a consequence, we can obtain the value of the *global* minimum of a univariate polynomial by solving an sos program. Notice also that at optimality we have $0 = p(x_\star) - p_\star = \sum_{k=1}^{m} q_k^2(x_\star)$ and thus all the q_k simultaneously vanish at $x_\star$, which in principle gives a way of computing the minimizer $x_\star$. As we shall see later, a better alternative is to obtain the solution $x_\star$ directly from the dual SDP problem by using complementary slackness.

Even though $p(x)$ may be highly nonconvex, the proposed convex formulation nevertheless effectively computes its global minimum. This will extend, with suitable modifications, to the general multivariate case.

Exercise 3.54. Let $p(x) = \sum_{k=0}^{2d} c_k x^k$. Give an explicit SDP formulation to compute the value of the global minimum of $p(x)$. Apply your formulation to the polynomial $p(x) = x^4 - 20x^2 + x$.

3.2.2 Rational Functions

What happens if we want to minimize a univariate *rational function* instead of a polynomial? Consider a rational function given as a ratio of polynomials $p(x)/q(x)$, where $q(x)$ is strictly positive. From the equivalence

$$\frac{p(x)}{q(x)} \geq \gamma \qquad \Leftrightarrow \qquad p(x) - \gamma\, q(x) \geq 0,$$

it follows that one can find the global minimum of the rational function by solving

$$\text{maximize } \gamma \qquad \text{subject to} \quad p(x) - \gamma\, q(x) \text{ is sos.}$$

The constrained case (i.e., minimization over a finite or semi-infinite interval) is very similar and can be formulated using the results in Section 3.3.1. The details are left to the exercises.

Exercise 3.55. Compute numerically the global minimum and the global maximum of the rational function $(x^3 - 8x + 1)/(x^4 + x^2 + 12)$.

Exercise 3.56. Why did we assume that the denominator $q(x)$ is strictly positive? Is this restriction necessary?

3.2.3 Multivariate Optimization

Consider now the case of unconstrained polynomial optimization of a multivariate polynomial $p(x_1, \ldots, x_n)$. As in the univariate case discussed in Section 3.2.1, we can write the following formulation for the global minimum of $p(x_1, \ldots, x_n)$:

$$\underset{\gamma}{\text{maximize}}\ \gamma \qquad \text{subject to} \qquad p(x_1, \ldots, x_n) - \gamma \text{ is nonnegative.} \qquad \text{(MOPT-NN)}$$

Despite being convex (why?), this formulation is in general intractable, since the constraint set involves the set of nonnegative polynomials. As in the univariate case, this suggests considering its sos alternative:

$$\underset{\gamma}{\text{maximize}}\ \gamma \qquad \text{subject to} \qquad p(x_1, \ldots, x_n) - \gamma \text{ is sos.} \qquad \text{(MOPT-SOS)}$$

Let $p_\star$ be the optimal value of (MOPT-NN) (i.e., the global minimum[2] of the polynomial $p(x_1, \ldots, x_n)$) and p_{sos} be the optimal value of (MOPT-SOS). It should

[2] Unlike in the univariate case, a multivariate polynomial that is bounded below need not achieve its global minimum (as an example, consider the polynomial $x^2 + (1 - xy)^2$). Therefore, to make things fully rigorous one should consider here the supremum rather than the maximum.

be clear that one can compute p_{sos} efficiently by solving the corresponding sos program (e.g., using an SDP solver).

Recall that for the general multivariate case, nonnegativity and sum of squares are no longer equivalent. Thus, since the feasible set of the second problem is a (possibly strict) subset of the feasible set of the first problem, we have the inequality

$$p_{sos} \leq p_\star,$$

and thus the sos technique is (in principle) only guaranteed to produce a lower bound on the value of the global minimum of p. Notice that, on computational complexity grounds, this is to be expected, since multivariate polynomial optimization is NP-hard, while semidefinite programming is polynomial-time (to any given accuracy).

Interestingly, there is strong experimental evidence that shows that, at least for relatively small problems, we very often have $p_\star = p_{sos}$; see, e.g., [94]. The reasons for this phenomenon are not yet completely understood, except in particular cases. As explained in Chapter 4, perhaps the opposite trend should be expected for large enough dimension. Nevertheless, as we shall see shortly in Section 3.2.6, even in those situations where $p_{sos} < p_\star$, we will be able to produce "stronger" sos conditions that will improve upon the "plain" sos lower bound p_{sos}.

Exercise 3.57. Find the value of p_{sos} for the trivariate polynomial

$$p(x, y, z) = x^4 + y^4 + z^4 - 4xyz + 2x + 3y + 4z.$$

Is the computed value of p_{sos} equal to the global minimum $p_\star$?

Exercise 3.58. Find a bivariate polynomial $p(x, y)$ for which $p_{sos} < p_\star$.

Exercise 3.59. Assume that $p(x)$ is bounded below. Is p_{sos} necessarily finite? Prove or disprove with a counterexample.

3.2.4 Nonnegativity on Sets and Constrained Optimization

An sos representation is an "obvious" certificate of the nonnegativity of a polynomial $p(x_1, \ldots, x_n)$ over the whole space $\mathbb{R}^n$. What if we only care about $p(x)$ being nonnegative on a given subset $S \subseteq \mathbb{R}^n$, as in the case of constrained optimization? Are there similarly simple and natural sufficient conditions for nonnegativity that we can write in this case? We present below an answer to these questions. We remark up-front, however, that in this section we are concerned only with the sufficiency of our conditions, and we postpone all possible concerns about the converse direction to Section 3.4.

The set S could be specified in very different forms (e.g., using only equations, or only inequalities, or a combination of both). As a consequence, the proposed conditions for nonnegativity of $p(x)$ on S that we discuss below will naturally depend on how the set S is presented.

Equations. For simplicity, let us assume first that the set S is described by a set of polynomial equations, i.e., that it is a real algebraic variety of the form

$$S = \{x \in \mathbb{R}^n \,:\, f_1(x) = 0, \ldots, f_m(x) = 0\}.$$

Recalling the formal similarity with weak duality and Lagrange multipliers, it is natural to write a condition of the following type:

$$p(x) + \sum_{i=1}^{m} \lambda_i(x) f_i(x) \qquad \text{is sos,} \tag{3.14}$$

where $\lambda_i(x)$ are arbitrary polynomials. Notice that this condition does what we want, since it "obviously" implies that $p(x)$ is nonnegative on the set S. Indeed, if (3.14) holds, by evaluating this expression at any point $x_0 \in S$, we immediately conclude that $p(x_0) \geq 0$. Notice also that the expression (3.14) is *affine* in the unknown polynomials $\lambda_i(x)$, and once the set of allowable multipliers $\lambda_i(x)$ has been fixed (e.g., by restricting their degrees), this condition has the form of an sos program.

In more algebraic terms, condition (3.14) considers the *polynomial ideal* I generated by the constraints $f_i(x)$. If $p(x)$ is congruent with a sum of squares modulo the ideal I, then this "obviously" certifies nonnegativity of $p(x)$. We elaborate more on this algebraic viewpoint in Section 3.3.5 and Chapter 7.

Inequalities. If the set S is described using polynomial inequalities (as opposed to equations), we can do something very similar. Assume the set S has a description:

$$S = \{x \in \mathbb{R}^n \,:\, g_1(x) \geq 0, \ldots, g_m(x) \geq 0\}.$$

Similar to the previous subsection, and again inspired by weak duality, one can now consider expressions of the type

$$p(x) = s_0(x) + \sum_{i=1}^{m} s_i(x) g_i(x), \tag{3.15}$$

where $s_0(x)$ and $s_i(x)$ are sos polynomials. Indeed, this serves as a "self-evident" certificate of nonnegativity of $p(x)$ on the set S, since evaluating such a representation at any point $x_0 \in S$ will directly prove $p(x_0) \geq 0$. In addition, notice that we can consider more powerful expressions by allowing finite products of constraints of the form

$$p(x) = s_0(x) + \sum_{i=1}^{m} s_i(x) g_i(x) + \sum_{ij}^{m} s_{ij}(x) g_i(x) g_j(x) + \cdots, \tag{3.16}$$

where as before the polynomials $s_0(x)$, $s_i(x)$, $s_{ij}(x), \ldots$ are sums of squares. Again, once the structure of these polynomials has been fixed (e.g., by restricting their degrees), the conditions boil down to sos programs. Any representation of the type (3.16) serves as an obvious certificate of nonnegativity of $p(x)$ on S.

Remark 3.60. *In principle, one could perhaps think of using nonnegative polynomials instead of sum of squares for the $s_i(x)$ in the previous expressions, since evaluating them at candidate points x_0 would certainly show nonnegativity of $p(x)$ on the set S. Notice, however, that in this case one would have to rely on a "promise" that the polynomials s_i indeed have the stated property. The reason why sums of squares are of relevance is that their (unconstrained) positivity is certified by the sos decomposition itself, and thus they serve as a* bona fide *mathematical proof of nonnegativity of $p(x)$ on S.*

Under certain assumptions, *converse* results or *representation theorems* will ensure that whenever $p(x)$ is nonnegative on a given set S, a certificate of a specified form must exist. We emphasize, however, that in most practical applications of sos techniques only the "easy" direction is actually used, in the sense that once an sos certificate has effectively been computed, it transparently proves the desired property (e.g., polynomial nonnegativity, etc.).

S-procedure. In the particular case when the $g_i(x)$ are quadratic forms, and the $s_i(x)$ are nonnegative scalars, the sufficient condition (3.15) is known as the *S-procedure* in the mathematical optimization and control literature. Under suitable assumptions, this condition is *lossless*; i.e., it exactly characterizes nonnegativity of a quadratic form on a quadratically constrained set.

Lemma 3.61 (S-lemma). *Let $p(x)$ and $g_1(x)$ be quadratic forms, and assume that the set S has an interior point (i.e., there exists an $x_0 \in \mathbb{R}^n$ such that $g_1(x_0) > 0$). In this case, if $p(x)$ is nonnegative on S, it has a representation as in* (3.16), *i.e.,*

$$p(x) = s_0(x) + s_1\, g_1(x),$$

where $s_0(x)$ is a positive semidefinite quadratic form, and s_1 is a nonnegative constant.

For more about the S-procedure, the S-lemma, and their many applications, see the books [21, 15] or the survey [99].

Exercise 3.62. Let $p(x) = x^4 - 3x^2 + 1$. Give an sos certificate of the nonnegativity of $p(x)$ on the set $S = \{x \in \mathbb{R} : x^3 - 4x = 1\}$.

Exercise 3.63. Allowing products of constraints (as in (3.16) as opposed to (3.15)) sometimes makes possible the existence of much more concise nonnegativity certificates (or even makes possible their existence). Consider, for instance, the polynomial $p(x, y) = xy$, which is obviously nonnegative on the compact set $S = \{(x, y) \in \mathbb{R}^2 : x \geq 0, y \geq 0, x + y \leq 1\}$.

1. Show that no nonnegativity certificate of the form (3.15) exists.

2. Give a nonnegativity certificate of the form (3.16).

Exercise 3.64. Assume that the set S is described using both equations and inequalities; i.e., it has the form

$$S = \{x \in \mathbb{R}^n : f_1(x) = 0, \ldots, f_k(x) = 0, g_1(x) \geq 0, \ldots, g_m(x) \geq 0\}.$$

What conditions would you propose to use to certify nonnegativity of a polynomial $p(x)$ on S?

3.2.5 Bounding the Distance to a Variety

The following problem is of interest in many applications: given a real algebraic variety V and a point x_0 that is *not* on V, we want to lower bound the distance from x_0 to V. This distance can be measured according to different metrics, but for simplicity we consider here only the case of the squared Euclidean norm $\|\cdot\|^2$. A common engineering motivation for this problem occurs, for instance, when the point x_0 represents the nominal behavior of a system, while the variety V corresponds to an "undesired" operating region. In this situation, we want to quantify how large the perturbations to x_0 can be, while guaranteeing that the undesired region described by V cannot be reached.

There are numerous important instances of this situation that appear mostly in robust optimization [14] and robust control [125] problems. For instance, a typical formulation in the robust control literature is the case where the point x_0 represents the parameter values of a feedback control system (given, e.g., by differential or difference equations), and the variety V is described by a determinantal condition that ensures that the system is stable. More complicated situations may require the undesirable set to be a semialgebraic set (instead of an algebraic variety), but the underlying techniques are essentially the same.

Let the real variety V be defined by polynomials $f_1(x), \ldots, f_m(x)$, i.e., $V = \{x \in \mathbb{R}^n : f_1(x) = 0, \ldots, f_m(x) = 0\}$. As we will see, such "safe" regions can be computed by considering the constrained polynomial optimization problem:

$$\text{minimize } \|x - x_0\|^2 \qquad \text{subject to} \quad f_i(x) = 0, \;\; i = 1, \ldots, m.$$

The true minimum value $d_\star$ of this problem yields the distance from x_0 to the variety V, and thus any valid lower bound on $d_\star$ will give a guaranteed neighborhood of x_0 that does not intersect the variety. Based on the same arguments as in the previous section, it should be clear that one can compute lower bounds on $d_\star$ and "safe neighborhoods" by considering sos problems of the form

$$\text{maximize } \gamma \qquad \text{subject to} \quad (\|x - x_0\|^2 - \gamma) + \sum_{j=1}^{m} \lambda_i(x) f_i(x) \quad \text{is sos.} \tag{3.17}$$

Any feasible solution γ_{sos} of this problem gives a ball $B = \{x \in \mathbb{R}^n : \|x - x_0\|^2 < \gamma_{sos}\}$ and a certificate that B that does not intersect the variety V. Indeed, evaluating the constraint in (3.17) at any point $x \in V$, we directly obtain $\|x - x_0\|^2 \geq \gamma_{sos}$.

Example 3.65. Consider a linear difference equation

$$x[k+1] = Ax[k].$$

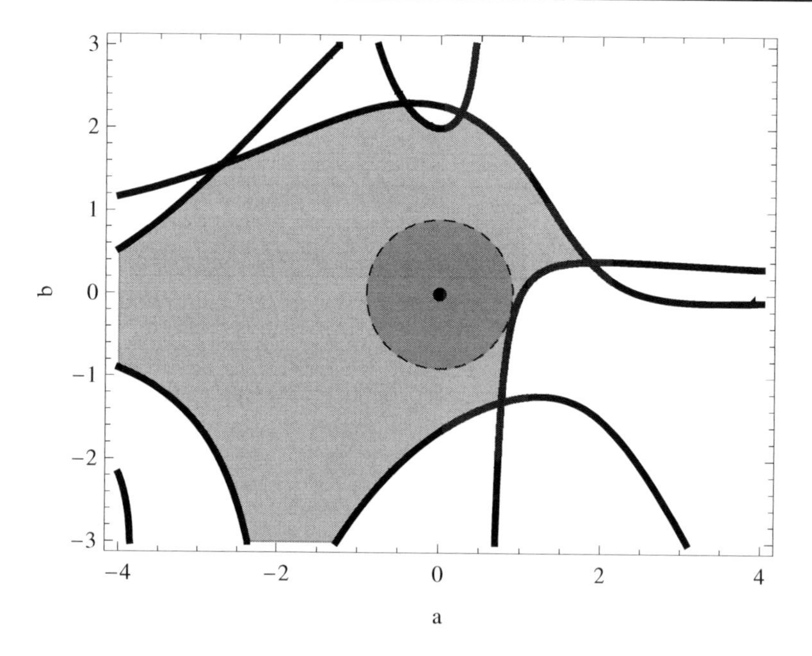

Figure 3.7. *The boundary of the domain of stability is defined by* $\bar{f}(a,b) = 0$*. Also shown is the computed certified stable region of the form* $a^2 + b^2 < \gamma_1$*.*

Recall (e.g., from Section 2.2.1) that this linear difference equation is *stable* (i.e., solutions converge to the origin as $k \to \infty$ for all initial conditions $x[0]$) if and only if all eigenvalues of A are inside the unit disk.

Now let A be the matrix

$$A = \frac{1}{3} \begin{bmatrix} 1+b & 0 & a \\ a & 2-b+a & -1 \\ 0 & -b & 2 \end{bmatrix},$$

whose characteristic polynomial is

$$\det(tI - A) = [27t^3 + (-45 - 9a)t^2 + (24 + 9a + 3ab - 3b^2)t \\ + (-4 - 2a - b - 2ab + a^2 b + 3b^2)]/27.$$

When the parameters (a,b) vanish, i.e., for $(a,b) = (0,0)$, the eigenvalues of A are $(1/3, 2/3, 2/3)$, and thus the difference equation is clearly stable. We want to determine how large a perturbation in (a,b) can be (measured in the Euclidean norm) for the difference equation to remain stable.

To apply the methods described in this section, we can consider the algebraic variety defined by the Zariski closure of the boundary of the region of stability. Clearly, A is on the boundary of stability if and only if some eigenvalue λ_i lies on the unit circle, i.e., satisfies $\lambda_i^* \lambda_i = 1$. We can easily characterize this condition algebraically. For instance, one can consider the polynomial

$$f(a,b) := \det(A \otimes A - I),$$

since the eigenvalues of the Kronecker product $A \otimes A$ are the products $\lambda_i \lambda_j$, and because A is real its eigenvalues appear in complex conjugate pairs. For our example, after removing constants and multiplicities from the factors, this yields the polynomial

$$\begin{aligned}\bar{f}(a,b) &= (2 - 2a - b + ab + a^2 b)(-100 - 20a - b - 5ab + a^2 b + 6b^2) \\ &\cdot (-245 + 133a - 14a^2 - 37b + 2ab + 27a^2 b + 5a^3 b + 31b^2 + 19ab^2 \\ &\quad + 2a^2 b^2 - 4a^3 b^2 + a^4 b^2 - 6b^3 - 12ab^3 + 6a^2 b^3 + 9b^4). \end{aligned} \tag{3.18}$$

This polynomial defines the variety of interest, and it can be seen that it factors into three components. This factorization is structural and corresponds to the conditions of the matrix A having eigenvalues at 1, at -1, or on the remainder of the unit circle. (As an aside, a more efficient alternative is to directly compute a factorized representation using the *bialternate matrix product* instead of the Kronecker product, since this removes multiplicities associated with the pairs $\lambda_i \lambda_j$ and $\lambda_j \lambda_i$; see, e.g., [57].)

We can now compute, using (3.17), the size γ of a neighborhood of (a, b) that is guaranteed not to intersect this variety. Notice that, for our example, since the variety is defined by a single polynomial that factors, it is possible (and more efficient) to consider each factor separately. In this case, for each of the three factors in (3.18), we obtain values

$$\gamma_1 \approx 0.8875, \qquad \gamma_2 \approx 9.0696, \qquad \gamma_3 \approx 2.1974.$$

Of these three, γ_1 defines the smallest neighborhood, and thus it yields a region $a^2 + b^2 < 0.8875$ where the linear difference equation is certified to be stable. This neighborhood and the corresponding varieties are presented in Figure 3.7. ∎

Remark. In the robust control literature, there are several methods that can partially exploit the determinantal structure of these kinds of problems. The notion of *structured singular value* and associated convex bounds are particularly relevant; see e.g. [18, 43, 125] and the references therein.

Remark 3.66. *Notice that in the optimization problem* (3.17) *the unknown multipliers* $\lambda_i(x)$ *are otherwise unconstrained. We will see in Section 3.3.5 and Chapter 7 that it is possible to exploit this structure for more efficient computation by computing sums of squares on the quotient ring* $\mathbb{R}[x]/\mathcal{I}(V)$.

3.2.6 What If "Simple" Sums of Squares Are Not Enough?

In many of the applications described earlier, we replaced the set of nonnegative polynomials $P_{n,2d}$, which is computationally intractable, with its tractable equivalent, the sos polynomials $\Sigma_{n,2d}$. In certain cases (e.g., univariate, quadratic) these two sets coincide, but in general $\Sigma_{n,2d}$ is a strictly smaller subset (quantitative estimates of the difference between these sets will be presented in Chapter 4).

What do we do in the cases where the set of nonnegative polynomials is no longer equal to sum of squares, and a simple sos approximation is not powerful enough to obtain the desired results? As outlined below, it is possible to produce stronger, more refined approximations to the set of nonnegative polynomials that strictly improve over what is achievable by "simple" sums of squares.

The power of multipliers. As a preview, and a hint at the techniques that will be considered later, let us show how to prove nonnegativity of a particular polynomial which is *not* a sum of squares. Recall that the Motzkin polynomial was defined as

$$M(x,y) = x^4y^2 + x^2y^4 + 1 - 3x^2y^2$$

and is a nonnegative polynomial that is not a sum of squares.

Despite $M(x,y)$ not being a sum of squares, we can try *multiplying* it by another polynomial which is *known to be positive* and then check whether the resulting product is a sum of squares. Clearly, if this is the case, we have succeeded in proving nonnegativity of the original polynomial (why?). For instance, for our example, consider multiplying $M(x,y)$ by the obviously positive factor $q(x) := (x^2 + y^2)$. In this case, the product will be a sum of squares, and in fact we have the explicit sos decomposition

$$(x^2+y^2) \cdot M(x,y) = y^2(1-x^2)^2 + x^2(1-y^2)^2 + x^2y^2(x^2+y^2-2)^2, \qquad (3.19)$$

which clearly certifies that $M(x,y) \geq 0$, despite the fact that $M(x,y)$ itself is not a sum of squares.

We will discuss a far-reaching generalization of this basic idea in Section 3.4, where we explain how to approximate *any* semialgebraic problem (including of course the simple case of a single polynomial being nonnegative) by sos techniques. However, let us elaborate at this point on a number of interesting connections.

Sums of squares of rational functions. A simple explanation of why a multiplier $q(x)$ makes possible more powerful nonnegativity certificates can be obtained by considering the case where $q(x) = \sum_i q_i(x)^2$ is a sum of squares. In this case, we can reinterpret an sos certificate for the product as

$$q(x) \cdot p(x) = \sum_j s_j^2(x) \qquad \Longleftrightarrow \qquad p(x) = \sum_i \sum_j \left(\frac{s_j(x)q_i(x)}{q(x)} \right)^2.$$

In other words, we now obtain a representation of the polynomial $p(x)$ as a *sum of squares of rational functions* (instead of a sum of squares of polynomials). It was conjectured by Hilbert (in fact, this is exactly the statement of the celebrated Hilbert's 17th problem; see, e.g., [106]) and later proved by Artin that every nonnegative polynomial has a representation as a sum of squares of rational functions.

Searching over multipliers. In the Motzkin example presented earlier, we produced the multiplier $q(x) = x^2 + y^2$ in an ad hoc fashion. Notice, however, that if $p(x)$ is a fixed polynomial for which we are trying to prove nonnegativity, we can systematically *search* for a multiplier $q(x)$ by solving a modified convex optimization problem (assuming a fixed bound on the degree of $q(x)$). Indeed, the problem of finding a polynomial $q(x)$ such that

$$q(x) \text{ is sos}, \qquad q(x) \cdot p(x) \text{ is sos}$$

is clearly affine in the unknown polynomial $q(x)$ and thus can be reduced to an sos program (and solved via semidefinite programming).

Uniform denominators and Pólya's theorem. Artin's solution to Hilbert's 17th problem ensures that for every nonnegative polynomial there is a decomposition as a sum of squares of rational functions, or alternatively, a suitable multiplier always exists. In many situations, it is convenient or necessary to restrict the structure of the possible multipliers (we will see examples of this later when discussing copositive matrices in Section 3.6.1). Recall that a *form* is a homogeneous polynomial, i.e., one for which all monomials have the same degree. A well-known theorem by Pólya about forms that are positive on the nonnegative orthant states precisely a case where this situation holds.

Theorem 3.67 ([59, Section 2.24]). *Given a form $f(x_1, x_2, \ldots, x_n)$ strictly positive for $x_i \geq 0$, $\sum_i x_i > 0$, then f can be expressed as*

$$f = \frac{g}{h},$$

where g and h are forms with positive coefficients. In particular, we can choose

$$h = (x_1 + x_2 + \cdots + x_n)^r$$

for a suitable r.

As we see, a representation of this kind gives an "obvious" certificate of the nonnegativity of f on the nonnegative orthant. To see the relationship with sums of squares, notice that if f is positive on the nonnegative orthant, then we can write $\widehat{f}(x_1, \ldots, x_n) := f(x_1^2, \ldots, x_n^2) = g(x_1^2, \ldots, x_n^2)/(x_1^2 + \cdots + x_n^2)^r$, and thus Pólya's theorem yields a representation of the positive even form $\widehat{f}$ as a sum of squares of rational functions, with a denominator of a fixed form. Pólya's theorem was generalized by Reznick [105], who showed that for *any* strictly positive form (not necessarily even), after multiplying by a suitable factor $(\sum_i x_i^2)^r$ it becomes a sum of squares (for r large enough). Furthermore, he also provided quantitative estimates for the exponent r.

Exercise 3.68. Let $q(x)p(x)$ and $q(x)$ be sums of squares, where the multiplier $q(x)$ is not the zero polynomial. Show that $p(x)$ is nonnegative.

Exercise 3.69. Consider the quartic form in four variables

$$p(w, x, y, z) := w^4 + x^2y^2 + x^2z^2 + y^2z^2 - 4wxyz.$$

1. Show that $p(w, x, y, z)$ is not a sum of squares.
2. Find a multiplier $q(w, x, y, z)$ such that $q(w, x, y, z) \cdot p(w, x, y, z)$ is a sum of squares.

Exercise 3.70. The conditions for a Pólya-type nonnegativity certificate can be fairly stringent. Consider the quadratic form $f(x,y) = (x-y)^2 + \epsilon xy$, which is obviously positive on the nonnegative orthant for all $\epsilon > 0$. Estimate how large the exponent r must be, as a function of ϵ, for the polynomial $(x+y)^r f(x,y)$ to have only positive coefficients.

3.3 Special Cases and Structure Exploitation

In Section 3.1.4 we introduced a general characterization of sums of squares in terms of its "standard" SDP formulation. In many applications, the polynomials under consideration have further structure that can be characterized algebraically in a variety of ways. In this section we analyze different situations that often appear in practice and the consequent theoretical and computational simplifications.

3.3.1 Univariate Intervals

For univariate polynomials, we have seen how to exactly characterize global nonnegativity (i.e., for $x \in (-\infty, \infty)$) in terms of semidefinite programming. But what if we are interested in polynomials that are nonnegative only on an interval (either finite or semi-infinite)? As explained below, we can use very similar ideas and two classical characterizations usually associated to the names Pólya–Szegö, Fekete, or Markov–Lukacs. The basic results are the following.

Theorem 3.71. *A univariate polynomial $p(x)$ is nonnegative on $[0,\infty)$ if and only if it can be written as*

$$p(x) = s(x) + x \cdot t(x),$$

where $s(x), t(x)$ are sums of squares. If $\deg(p) = 2d$, then we have $\deg(s) \leq 2d$, $\deg(t) \leq 2d-2$, while if $\deg(p) = 2d+1$, then $\deg(s) \leq 2d$, $\deg(t) \leq 2d$.

A similar result holds for closed finite intervals.

Theorem 3.72. *Let $a < b$. Then the univariate polynomial $p(x)$ is nonnegative on $[a,b]$ if and only if it can be written as*

$$\begin{cases} p(x) = s(x) + (x-a)\cdot(b-x)\cdot t(x) & \text{if } \deg(p) \text{ is even,} \\ p(x) = (x-a)\cdot s(x) + (b-x)\cdot t(x) & \text{if } \deg(p) \text{ is odd,} \end{cases}$$

where $s(x), t(x)$ are sums of squares. In the first case, we have $\deg(p) = 2d$, and $\deg(s) \leq 2d$, $\deg(t) \leq 2d-2$. In the second, $\deg(p) = 2d+1$, and $\deg(s) \leq 2d$, $\deg(t) \leq 2d$.

Notice the similarity to the conditions discussed in Section 3.2.4 and the fact that these representations "obviously" certify that $p(x) \geq 0$ on the corresponding set. From the existence of these sos representations, it also follows directly that nonnegative polynomials on any interval (finite or semi-infinite) can be exactly characterized using "small" sos programs.

As we will see later, these sos characterizations, suitably dualized, can be used to give a complete characterization of the set of valid moments of probability measures with support on univariate intervals. We will discuss the details in Section 3.5.3, followed by an application to game theory in Section 3.6.6.

Exercise 3.73. Prove Theorem 3.71. Hint: $p(x)$ is nonnegative on $[0, \infty)$ if and only if $q(t) := p(t^2)$ is a nonnegative polynomial.

Exercise 3.74. Let $p(x)$ be a univariate polynomial of degree d that satisfies $|p(x)| \leq 1$ for $x \in [-1, 1]$. How large can its leading coefficient be?

1. Give an sos formulation for this problem, and solve it numerically for $d = 2, 3, 4, 5$.
2. What is the largest value of d for which you can numerically solve this problem (using the monomial basis) in a reliable way? Experiment using different polynomial bases, as explained in Section 3.1.5.
3. Can you guess what the general solution is as a function of d? Can you give an exact characterization of the optimal polynomial?

Exercise 3.75. Give an sos formulation to the problem of minimizing a univariate rational function $p(x)/q(x)$ on the interval $[a, b]$. What condition is needed on the denominator $q(x)$, if any?

3.3.2 Sum of Squares Matrices

The notions of positive semidefiniteness and sums of squares of scalar polynomials can be naturally extended to polynomial matrices, i.e., matrices with entries in $\mathbb{R}[x_1, \ldots, x_n]$. Sum of squares matrices are of interest in many situations, including the characterization of sos convexity (Section 3.3.3) and representations for symmetry-invariant polynomials (Section 3.3.6).

We say that a symmetric polynomial matrix $P(x) \in \mathbb{R}[x]^{m \times m}$ is positive semidefinite if $P(x) \succeq 0$ for all $x \in \mathbb{R}^n$ (i.e., it is pointwise positive semidefinite). The definition of an sos matrix is as follows [69, 48, 109].

Definition 3.76. *A symmetric polynomial matrix $P(x) \in \mathbb{R}[x]^{m \times m}$, $x \in \mathbb{R}^n$, is an* sos matrix *if there exists a polynomial matrix $M(x) \in \mathbb{R}[x]^{s \times m}$ for some $s \in \mathbb{N}$, such that $P(x) = M^T(x)M(x)$.*

When $m = 1$, i.e., for scalar polynomials, this corresponds to the standard sos notion. Also, when P is a constant matrix, then the condition simply states that P is positive semidefinite. Thus, sos matrices are a common generalization of positive semidefinite (constant) matrices and sos polynomials.

Example 3.77. Consider the polynomial matrix

$$P(x) = \begin{bmatrix} x^2 - 2x + 2 & x \\ x & x^2 \end{bmatrix}.$$

This is an sos matrix since it admits the factorization

$$P(x) = \begin{bmatrix} 1 & x \\ x-1 & 0 \end{bmatrix}^T \begin{bmatrix} 1 & x \\ x-1 & 0 \end{bmatrix}. \quad \blacksquare$$

Since an $m \times m$ matrix is simply a representation of an m-variate quadratic form, we can always interpret an sos matrix in terms of a polynomial with m additional variables. The following result makes this precise.

Lemma 3.78. *Let $P(x) \in \mathbb{R}[x]^{m\times m}$ be a symmetric polynomial matrix, with $x \in \mathbb{R}^n$. Let $p(x,y) := y^T P(x) y$ be the associated scalar polynomial in $m+n$ variables $[x; y]$, where $y = [y_1, \dots, y_m]$.*

1. *The matrix $P(x)$ is positive semidefinite if and only if $p(x,y)$ is nonnegative.*
2. *The matrix $P(x)$ is an sos matrix if and only if $p(x,y)$ is a sum of squares (in $\mathbb{R}[x;y]$).*

Example 3.79. Here we continue Example 3.77. The matrix $P(x)$ is an sos matrix since the scalar polynomial $y^T P(x) y$ has the sos decomposition

$$y^T P(x) y = (y_1 + x y_2)^2 + (x-1)^2 y_1^2. \quad \blacksquare$$

Notice that Lemma 3.78 allows us to easily decide whether a given polynomial matrix is an sos matrix using the same semidefinite programming techniques already described in Section 3.1.4. While these results establish that sos matrices are not a completely new concept (since they are fully equivalent to scalar sos polynomials), the main advantage is that they allow for a more concise notation, since they appear naturally in many contexts (e.g., sos-convexity in Section 3.3.3, or symmetry reduction in Section 3.3.6).

When are positive semidefinite matrices sums of squares? A celebrated result about sos matrices that has been rediscovered many times is the fact that in the univariate case, the sos condition is also necessary.

Theorem 3.80. *Let $P(x) \in \mathbb{R}[x]^{m\times m}$ be a symmetric polynomial matrix, where the variable x is scalar (i.e., $x \in \mathbb{R}$). Then the matrix $P(x)$ is positive semidefinite if and only if it is an sos matrix.*

For a proof and historical details, see, e.g., [28], [9], and the references therein. Notice that this is a simultaneous generalization of two of the classical Hilbert cases where nonnegativity is equal to sum of squares (scalar polynomials and quadratic forms). For more details about univariate polynomial matrices, references to the literature, as well as an efficient eigenvalue-based method for finding their sos decomposition, we refer the reader to [9].

In the multivariate case, however, not all positive polynomial matrices are sums of squares. A well-known counterexample is due to Choi [27], who constructed a positive semidefinite biquadratic form that is *not* a sum of squares of bilinear forms. His counterexample can be rewritten as the polynomial matrix

$$C(x) = \begin{bmatrix} x_1^2 + 2x_2^2 & -x_1x_2 & -x_1x_3 \\ -x_1x_2 & x_2^2 + 2x_3^2 & -x_2x_3 \\ -x_1x_3 & -x_2x_3 & x_3^2 + 2x_1^2 \end{bmatrix}, \tag{3.20}$$

which is positive semidefinite for all $(x_1, x_2, x_3) \in \mathbb{R}^3$ but is not an sos matrix.

Exercise 3.81. Prove Lemma 3.78.

Exercise 3.82. Let $P(x)$ be an sos matrix. Show that all principal minors of $P(x)$ are scalar sos polynomials. (Hint: Use the Cauchy–Binet matrix identity.)

Exercise 3.83. Show that the Choi matrix (3.20) is positive semidefinite for all real values of (x_1, x_2, x_3) but is not an sos matrix.

Exercise 3.84. Modify the algorithm given in Exercise 3.36 so that it will compute a decomposition of a univariate sos matrix $P(x)$.

Exercise 3.85. Certain optimization problems include constraints that are naturally expressed in matrix form. For instance, a set S could be defined as

$$S = \left\{ (x_1, x_2, x_3) \in \mathbb{R}^3 \ : \ G(x) := \begin{bmatrix} 1 & x_1 & x_2 \\ x_1 & 1 & x_3 \\ x_2 & x_3 & 1 \end{bmatrix} \succeq 0 \right\}$$

(notice that this corresponds to the 3-dimensional elliptope discussed in Section 2.1.3). While these descriptions could be "scalarized" and rewritten in terms of scalar polynomial inequalities (e.g., by considering minors, or coefficients of the characteristic polynomial of $G(x)$), it is often much more convenient to preserve their structure and keep them in matrix form.

Consider a scalar polynomial $p(x)$, for which we want to show that it is nonnegative on the set S.

1. Show that a sufficient condition for nonnegativity of p on the set S is the existence of a scalar sos polynomial $s_0(x)$ and an sos matrix $S_1(x)$, such that

$$p(x) = s_0(x) + \langle S_1(x), G(x) \rangle.$$

2. Explain how to compute $s_0(x)$ and $S_1(x)$ via sos programs and semidefinite programming.

3. Give an sos certificate of nonnegativity of $p(x) := 4 - (x_1^4 + x_2^4 + x_3^4)$ on the set S.

3.3.3 Sum of Squares Convexity

The notion of *sos-convexity* is a tractable algebraic replacement for convexity of a polynomial function. Informally, the (difficult to verify) requirement of positive semidefiniteness of the Hessian matrix is replaced with a tractable condition, the existence of an sos decomposition. Besides its computational implications, sos-convexity is an appealing concept since it bridges the *geometric* and *algebraic* aspects of convexity. Indeed, while the usual definition of convexity is concerned only with the geometry of the epigraph, in sos-convexity this geometric property (or the nonnegativity of the Hessian) must be *certified* through a "simple" algebraic identity, namely, an sos factorization of the Hessian.

Recall that a multivariate polynomial $p(x) := p(x_1, \ldots, x_n)$ is convex if and only if its Hessian is positive semidefinite for all $x \in \mathbb{R}^n$. This is a pointwise condition that the Hessian must satisfy at every point x. The notion of sos-convexity requires instead a global algebraic certificate for this property.

Definition 3.86. *A polynomial $p(x)$ is* sos-convex *if its Hessian $H(x)$ is an sos matrix, i.e., if it factors as $H(x) = M(x)^T M(x)$, where $M(x)$ is a polynomial matrix.*

Clearly, an sos-convex polynomial is convex, since the Hessian being an sos matrix implies it is positive semidefinite everywhere. Is the converse true? In other words, is every convex polynomial necessarily sos-convex?

Recall (e.g., from the Choi example in the previous section) that not every positive semidefinite polynomial matrix is an sos matrix. However, this does not necessarily serve as a counterexample, since due to the fact that partial derivatives commute, the Hessian matrix of a polynomial has strong affine dependencies among the different entries, of the form $\frac{\partial H_{ij}(x)}{\partial x_k} = \frac{\partial H_{ik}(x)}{\partial x_j}$. As a consequence, the set of valid Hessians is a lower-dimensional subspace of the space of symmetric polynomial matrices. Thus, due to this special structure, it is perhaps conceivable that convexity and sos-convexity of polynomials could still be equivalent.

The following counterexample from [5] shows that this is not the case.

Theorem 3.87. *The trivariate form of degree 8 given by*

$$\begin{aligned} p(x) &= 32x_1^8 + 118x_1^6x_2^2 + 40x_1^6x_3^2 + 25x_1^4x_2^4 - 43x_1^4x_2^2x_3^2 - 35x_1^4x_3^4 + 3x_1^2x_2^4x_3^2 \\ &\quad -16x_1^2x_2^2x_3^4 + 24x_1^2x_3^6 + 16x_2^8 + 44x_2^6x_3^2 + 70x_2^4x_3^4 + 60x_2^2x_3^6 + 30x_3^8 \end{aligned}$$

is convex but is not sos-convex.

The work [4] presents a complete classification of the cases for which convexity and sos-convexity coincide. This description is in a certain sense the analogue to Hilbert's classification of nonnegativity described in Section 3.1.2.

Another motivation and justification for studying sos-convexity is its computational tractability. Deciding convexity of a multivariate polynomial is an NP-hard problem [3], while it follows from our earlier discussions that sos-convexity can be checked using semidefinite programming. Sos-convexity will appear prominently in the characterization of semidefinite representability of convex sets; see Section 6.4.3 in Chapter 6 for details. For more results and background material on sos-convexity, we refer the reader to [5, 4].

Exercise 3.88. Show that the Choi matrix (3.20) is not the Hessian of any polynomial.

Exercise 3.89. Prove Theorem 3.87. Hint: To show that $p(x)$ is *not* sos-convex, analyze the $(1,1)$ entry of the Hessian.

Exercise 3.90. In this exercise, we explore the use of sos-convexity for the problem of fitting a polynomial to data, under a convexity restriction (e.g., [76]).

Consider a finite set of data $\{\mathbf{x}_i, f_i\}$ for $i = 1, \ldots, N$, where $\mathbf{x}_i \in D \subseteq \mathbb{R}^n$ and $f_i \in \mathbb{R}$. We want to fit these data points with a polynomial function $p(x)$ of degree d, making the least-squares fitting error $\sum_{i=1}^{N}(p(\mathbf{x}_i) - f_i)^2$ as small as possible.

1. Give an sos formulation for this problem, in the case where $p(x)$ is required to be a globally convex polynomial. Explain whether the formulation solves this problem exactly.

2. How would you modify your formulation if we only require that $p(x)$ be convex on the domain D of interest?

3. Generate data points where $\mathbf{x}_i \in D := [-1,1] \times [-1,1]$, and numerically solve your formulation for those two cases ($p(x)$ is convex everywhere, or is only convex on the domain D).

3.3.4 Sparsity and Newton Polytopes

Many of the polynomial systems that appear in practice are far from being "generic" but rather present a number of structural features that, when properly exploited, allow for much more efficient computational techniques. This is quite similar to the situation in numerical linear algebra, where there is a big difference in performance between algorithms that take into account matrix sparsity and those that do not. For matrices, the notion of sparsity is often relatively straightforward and relates mostly to the number of nonzero coefficients. In computational algebra, however, there exists a much more refined notion of sparsity that refers not only to the number of zero coefficients of a polynomial, but also to the underlying combinatorial structure of the nonzero coefficients.

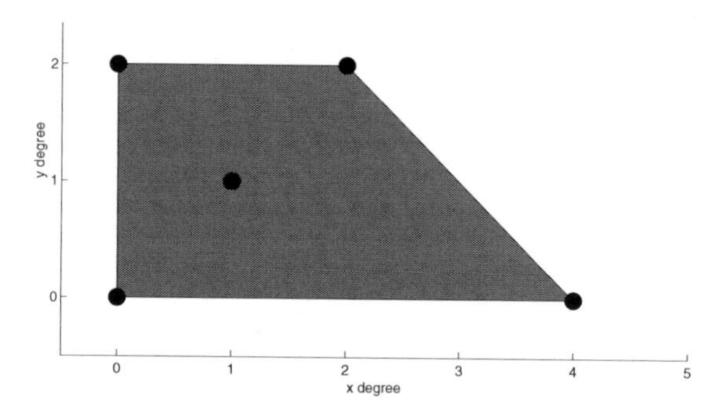

Figure 3.8. *Newton polytope of the polynomial* $5 - xy - x^2y^2 + 3y^2 + x^4$.

Sparsity for multivariate polynomials is usually characterized in terms of their *Newton polytope*, defined below.

Definition 3.91. *Consider a multivariate polynomial* $p(x_1, \ldots, x_n) = \sum_\alpha c_\alpha x^\alpha$. *The* Newton polytope *of* p, *denoted by* $\mathcal{N}(p)$, *is the convex hull of the set of exponents* α, *considered as vectors in* $\mathbb{R}^n$.

Thus, the Newton polytope of a polynomial always has integer extreme points, given by a subset of the exponents of the polynomial.

Example 3.92. Consider the polynomial $p(x, y) = 5 - xy - x^2y^2 + 3y^2 + x^4$. Its Newton polytope $\mathcal{N}(p)$, displayed in Figure 3.8, is the convex hull of the points $(0, 0), (1, 1), (2, 2), (0, 2), (4, 0)$. ∎

Example 3.93. Consider the polynomial $p(x, y) = 1 - x^2 + xy + 4y^4$. Its Newton polytope $\mathcal{N}(p)$ is the triangle in $\mathbb{R}^2$ with vertices $\{(0, 0), (2, 0), (0, 4)\}$. ∎

Newton polytopes are an essential tool when considering polynomial arithmetic because of the following fundamental identity:

$$\mathcal{N}(g \cdot h) = \mathcal{N}(g) + \mathcal{N}(h), \tag{3.21}$$

where + denotes the Minkowski addition of polytopes.

The Newton polytope allows us to introduce a notion of sparsity for a polynomial, related to the size of its Newton polytope. Sparsity (in this algebraic sense) allows a notable reduction in the computational cost of checking sum of squares conditions of multivariate polynomials. The reason is the following theorem due to Reznick.

Theorem 3.94 ([104, Theorem 1]). *If* $p(x) = \sum_i q_i(x)^2$, *then* $\mathcal{N}(q_i) \subseteq \frac{1}{2}\mathcal{N}(p)$.

This theorem allows us, without loss of generality, to restrict the set of monomials appearing in the sos representation (3.12) to those in the Newton polytope of p, scaled by a factor of $\frac{1}{2}$. This reduces the size of the corresponding matrix Q, thus simplifying the semidefinite program to be solved.

Example 3.95. Consider the following polynomial:

$$p(w,x,y,z) := (w^4+1)(x^4+1)(y^4+1)(z^4+1) + 2w + 3x + 4y + 5z,$$

for which we want to compute an sos decomposition. The polynomial p has degree $2d = 16$ and four independent variables ($n = 4$). A naive direct approach, along the lines described in Section 3.1.4, would require a matrix Q indexed by all monomials in (w,x,y,z) of degree less than or equal to $d = 8$, i.e., of size $\binom{n+d}{d} = 495$.

However, its Newton polytope $\mathcal{N}(p)$ is easily seen to be the four-dimensional hypercube with opposite vertices at $(0,0,0,0)$ and $(4,4,4,4)$. Therefore, by Theorem 3.94, the polynomials q_i in the sos decomposition of p must have support in $\frac{1}{2}\mathcal{N}(p)$, which is the hypercube with vertices at $(0,0,0,0)$ and $(2,2,2,2)$. This scaled polytope contains $3^4 = 81$ distinct monomials, and as a consequence a full sos decomposition can be computed by solving a much smaller semidefinite program. ∎

For a discussion of additional techniques for exploiting sparsity in the context of sum of squares, we refer the reader to [70, 124] and the references therein.

Exercise 3.96. Prove identity (3.21).

Exercise 3.97. Consider the Motzkin polynomial (3.6), and compute its Newton polytope. Which monomials could appear in a (hypothetical) sos decomposition of $M(x,y)$? Show, by considering the coefficient of x^2y^2, that this leads to a contradiction, and thus that $M(x,y)$ is not a sum of squares.

Exercise 3.98. *Facial reduction* [20] is a technique by which a conic programming feasibility problem $x \in \mathcal{K} \cap L$ that is feasible, but not strictly feasible, is replaced with a simpler problem that satisfies strict feasibility. The key idea is that if the subspace L does not properly intersect the cone $\mathcal{K}$, one may restrict attention to a smaller face of $\mathcal{K}$ instead (ideally, of minimal possible dimension). For the positive semidefinite cone, faces are themselves isomorphic to smaller dimensional positive semidefinite cones, and thus this procedure yields smaller, but equivalent, semidefinite programs.

Explain how to interpret the Newton polytope technique described above in terms of facial reduction.

3.3.5 Equations, Ideals, and Quotient Rings

Sum of squares decompositions give sufficient conditions for global nonnegativity of a polynomial. However, as discussed in Section 3.2.4, often we are interested in deciding or proving nonnegativity only on certain regions of $\mathbb{R}^n$. In this section we consider the case where the set of interest is defined using equality constraints only; i.e., it is an algebraic variety. The more general case of polynomial inequalities (i.e., basic semialgebraic sets) will be discussed in Section 3.4. As we will see, when explicit equality constraints are present in the problem, notable simplifications in the formulation of the corresponding semidefinite programs are possible.

For concreteness, consider the problem of verifying the nonnegativity of a polynomial $p(x)$ on a set defined by equality constraints: $\{x \in \mathbb{R}^n : f_i(x) = 0, i = 1, \ldots, m\}$ (i.e., an algebraic variety). Let $I = \langle f_1, \ldots, f_m \rangle$ be the ideal generated by the equality constraints, and define the quotient ring $\mathbb{R}[x]/I$ as the set of equivalence classes for congruence modulo the ideal I. Then, provided computations can be effectively performed in this quotient ring, very compact SDP formulations will be possible. This will be usually the case when Gröbner bases for the ideal are either available or easy to compute. The first case usually occurs in combinatorial optimization problems, and the latter when the ideal is generated by a few constraints.

We explain the details next. We want to write sos-like sufficient conditions for the polynomial $p(x)$ to be nonnegative on the variety $V(I)$. As mentioned earlier, the condition

$$p(x) + \sum_i \lambda_i(x) f_i(x) \quad \text{is a sum of squares in } \mathbb{R}[x] \tag{3.22}$$

is a self-evident certificate of nonnegativity that clearly guarantees this. To see this, notice that evaluating this expression on any point x_0 of $V(I)$ gives $p(x_0)$ (since $f_i(x_0) = 0$), and this is a nonnegative value (since the expression is a sum of squares). By passing to the quotient ring (equivalently, considering (3.22) modulo the ideal I), we can rewrite this as

$$f(x) \quad \text{is a sum of squares in } \mathbb{R}[x]/I. \tag{3.23}$$

Both expressions are sufficient conditions for the nonnegativity of p on the variety defined by $f_i(x) = 0$. As we will see, we can use this to give a more efficient version of the SDP formulation of sum of squares.

Sum of squares on quotient rings. We describe next a natural modification of the standard sos methods that will allow us to compute sos decompositions on quotient rings. This can be done by using essentially the same SDP techniques as in the standard case. Since we will need to do effective computations on the quotient, we assume that a *Gröbner basis* $\mathcal{G} = \{b_1, \ldots, b_k\}$ of the polynomial ideal I is available; see Appendix A and [32] for an introduction to computational algebra and Gröbner basis methods.

The method will be basically the same as in the standard case explained in Section 3.1.4 (expressing the polynomial as a quadratic form on a vector of monomials and writing linear equations to obtain a semidefinite program), but with two main differences:

- Instead of indexing the rows and columns of the matrix Q in the semidefinite program by the usual monomials, we use *standard* monomials corresponding to the Gröbner basis $\mathcal{G}$ of the ideal I. These are the monomials that are not divisible by any leading term of the polynomials b_i in the Gröbner basis.
- When equating the left- and right-hand sides to form linear equations defining the subspace of valid Gram matrices, all operations are performed in the quotient ring; i.e., we rewrite the terms in *normal form* after multiplication.

Rather than giving a formal description, it is more transparent to explain the methodology via a simple example.

Example 3.99. Consider the problem of deciding if the polynomial $p := 10 - x^2 - y$ is nonnegative on the variety defined by $f := x^2 + y^2 - 1 = 0$ (the unit circle). We will check whether p is a sum of squares in $\mathbb{R}[x, y]/I$, where I is the ideal $I = \langle f \rangle$. Since the ideal I is principal (generated by a single polynomial), we already have a Gröbner basis, which is simply $\mathcal{G} = \{f\}$. We use a graded lexicographic monomial ordering, where $x \prec y$. The corresponding set of standard monomials is then $\mathcal{B} = \{1, x, y, x^2, xy, x^3, x^2 y, \ldots\}$.

To formulate the corresponding semidefinite program, we pick a partial basis of the quotient ring (i.e., a subset of monomials in $\mathcal{B}$). In this example, we take only $\{1, x, y\}$, and, as before, we write p as a quadratic form in these monomials:

$$\begin{aligned} 10 - x^2 - y &= \begin{bmatrix} 1 \\ x \\ y \end{bmatrix}^T \begin{bmatrix} q_{11} & q_{12} & q_{13} \\ q_{12} & q_{22} & q_{23} \\ q_{13} & q_{23} & q_{33} \end{bmatrix} \begin{bmatrix} 1 \\ x \\ y \end{bmatrix} \\ &= q_{11} + q_{22} x^2 + q_{33} y^2 + 2 q_{12} x + 2 q_{13} y + 2 q_{23} xy \\ &\equiv (q_{11} + q_{33}) + (q_{22} - q_{33}) x^2 + 2 q_{12} x + 2 q_{13} y + 2 q_{23} xy \qquad \bmod I, \end{aligned}$$

where, in the last line, we used reduction modulo the ideal to rewrite some terms as linear combinations of standard monomials only (e.g., the term $q_{33} y^2$ is replaced by $q_{33} - q_{33} x^2$). Matching coefficients between left and right, we obtain the linear equations

$$\begin{aligned} 1: &\quad 10 = q_{11} + q_{33}, \\ x: &\quad 0 = 2 q_{12}, \\ y: &\quad -1 = 2 q_{13}, \\ x^2: &\quad -1 = q_{22} - q_{33}, \\ xy: &\quad 0 = 2 q_{23} \end{aligned}$$

that define the subspace. Thus, we obtain again a simple semidefinite program. Solving it, we have

$$Q = \begin{bmatrix} 9 & 0 & -\frac{1}{2} \\ 0 & 0 & 0 \\ -\frac{1}{2} & 0 & 1 \end{bmatrix} = L^T L, \qquad L = \frac{1}{\sqrt{2}} \begin{bmatrix} 3 & 0 & -\frac{1}{6} \\ 0 & 0 & \frac{\sqrt{35}}{6} \end{bmatrix},$$

and therefore

$$10 - x^2 - y \equiv \left(3 - \frac{y}{6}\right)^2 + \frac{35}{36} y^2 \qquad \bmod I,$$

which shows that p is indeed a sum of squares on $\mathbb{R}[x, y]/I$. A simple geometric interpretation is shown in Figure 3.9. As expected, by the condition above, p coincides with an sos polynomial on the variety, and thus it is obviously nonnegative on that set. ∎

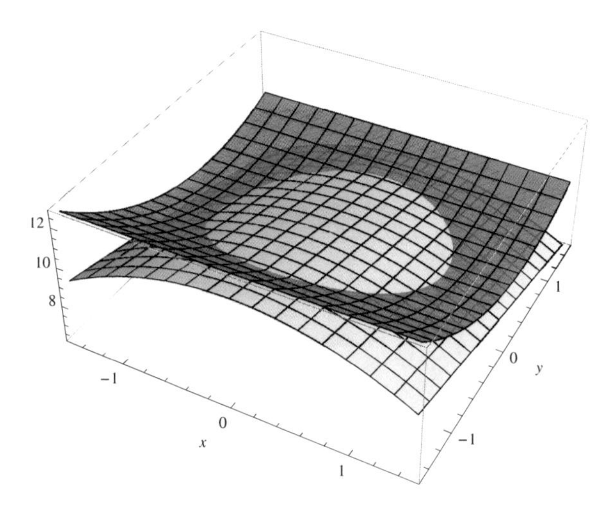

Figure 3.9. *The polynomials* $p = 10 - x^2 - y$ *and* $(3 - \frac{y}{6})^2 + \frac{35}{36}y^2$ *take exactly the same values on the unit circle* $x^2 + y^2 = 1$. *Thus,* p *is nonnegative on the circle.*

Remark 3.100. *Despite the similarities between the "standard" case of sum of squares on the polynomial ring* $\mathbb{R}[x]$ *versus the quotient ring* $\mathbb{R}[x]/I$, *there are a few important differences. A key distinction is related to computational complexity issues. Consider an sos decomposition* $p(x) = \sum_i q_i(x)^2$. *When working on* $\mathbb{R}[x]$, *we can always bound a priori the degree of the polynomials* q_i *in terms of the degree of* p *(namely,* $\deg(q_i) \leq \frac{1}{2}\deg(p)$*). This is* not *true when working on a quotient ring, since monomials can "wrap around" when computing normal forms. This is the reason why when working on* $\mathbb{R}[x]/I$ *we typically have some freedom in choosing a finite set of standard monomials to index the matrix* Q *(unless it is feasible to include all of them).*

In fact, since for the ideal $I = \langle x_1^2 - 1, \ldots, x_n^2 - 1 \rangle$ *every polynomial nonnegative on* $V(I)$ *is a sum of squares on* $\mathbb{R}[x]/I$ *(Exercise* 3.105*), it directly follows that, in the general case, deciding whether a polynomial is sum of squares modulo* I *is NP-hard.*

Even though in the worst case computing a Gröbner basis for I may be troublesome, for many practical problems they are often directly available or relatively easy to compute. A typical example is the case of combinatorial optimization problems, where the equations defining the Boolean ideal $\langle x_1^2 - 1, \ldots, x_n^2 - 1 \rangle$ are already a Gröbner basis. Another frequent situation is when the ideal is defined by a single constraint, in which case the defining equation is again obviously a Gröbner basis of the corresponding ideal.

SDP dimensions and Hilbert series. Another advantage of the ideal-theoretic formulation is the ease with which structural results can be obtained through basic algebraic notions. For instance, consider the following question: what are the matrix dimensions of the semidefinite programs for sum of squares modulo an ideal? Recall that in the "standard" sos case (over $\mathbb{R}[x]$, for a polynomial of degree $2d$), the matrices are indexed by all monomials of degree less than or equal to d and thus have size $\binom{n+d}{d}$. This can be rewritten as $\binom{n+d}{d} = \sum_{k=0}^{d} \binom{n+k-1}{k}$, where each term in the sum corresponds to the number of monomials of total degree k. How can we generalize this?

For quotient rings, there is a nice way of counting the dimensions of the different homogeneous components, known as the *Hilbert series*; see, e.g., [33]. The Hilbert series $H(I,t)$ is the generating function (a formal power series) of the Hilbert function $H_I(k)$, which gives the dimension of the degree k the homogeneous part of the quotient ring, i.e.,

$$H(I,t) = \sum_{k=0}^{\infty} H_I(k)\, t^k,$$

where $H_I(k) = \dim(\mathbb{R}[x]/I \cap \mathbb{R}[x]_{\mathbf{k}})$. If I is a monomial ideal, $H_I(k)$ counts the number of standard monomials of total degree k. If I is an ideal, and $in_{\succ}(I)$ is its initial ideal with respect to a graded monomial ordering, then both have the same Hilbert series. The Hilbert series can be computed from a Gröbner basis of the ideal I, and, as a consequence, this allows us to determine the size of the corresponding semidefinite program, given a bound on the total degree of the standard monomials we will be considering.

For instance, the "standard" case we just discussed corresponds to the trivial ideal $I = \{0\}$. The Hilbert series for $\mathbb{R}[x]/I \cong \mathbb{R}[x]$ is $H(I,t) = 1/(1-t)^n = \sum_{k=0} \binom{n+k-1}{k} t^k$, which corresponds exactly to the dimensions computer earlier.

Example 3.101. Consider the ideal $I = \langle x^2 + y^2 - 1 \rangle$ of Example 3.99. Its Hilbert series is

$$H(I,t) = \frac{1+t}{1-t} = 1 + 2t + 2t^2 + 2t^3 + 2t^4 \cdots,$$

which counts the number of standard monomials of each degree. The terms of the series allow us to determine, given a bound on the total degree of the monomials to be considered, what size the corresponding semidefinite program will be. For instance, since in Example 3.99 we used only monomials of degree less than or equal to 1, the size of the corresponding semidefinite program is $1 + 2 = 3$. ■

In Exercise 3.106 we discuss another natural and important example, namely, the Boolean ideal $\langle x_1^2 - 1, \ldots, x_n^2 - 1 \rangle$. These ideas will appear again in Chapter 7, when computing semidefinite representations of convex hulls of algebraic varieties.

Exercise 3.102. Prove formally that the expressions (3.22) and (3.23) are equivalent.

Exercise 3.103. Consider the polynomial $f(x,y,z) := 1 + xy + yz + xz$, and the variety $V(I)$, where $I = \langle x^2 - 1, y^2 - 1, z^2 - 1 \rangle$. Notice that $V(I)$ is finite.

1. Show, by explicit enumeration, that f is nonnegative on $V(I)$.

2. Write f as a sum of squares on $\mathbb{R}[x]/I$.

Exercise 3.104. Consider the *butterfly* curve in $\mathbb{R}^2$, defined by the equation

$$x^6 + y^6 = x^2.$$

Give an sos certificate that the real locus of this curve is contained in a disk of radius 5/4. Is this the best possible constant?

Exercise 3.105. Consider $\mathbb{R}[x_1, \ldots, x_n]$ and the ideal $I = \langle x_1^2 - 1, \ldots, x_n^2 - 1\rangle$. We will show that every polynomial that is nonnegative on $V(I) \subset \mathbb{R}^n$ is a sum of squares modulo $\mathbb{R}[x]/I$.

1. Show that $V(I)$ corresponds to all the points $\{-1, 1\}^n$ (i.e., the 2^n vertices of the unit hypercube). Thus, a polynomial $p(x)$ is nonnegative on $V(I)$ if and only if evaluates to a nonnegative number on all these vertices.

2. Let $v = (v_1, \ldots, v_n) \in V(I)$. Define the polynomial $\ell_v(x) = \prod_{i=1}^n \left(\frac{v_i + x_i}{2v_i}\right)$. Show that $\ell_v(v) = 1$, and $\ell_v(w) = 0$ for all $w \in V(I)$, with $w \neq v$.

3. Assume that $p(x)$ is nonnegative on $V(I)$. Find an explicit sos decomposition for $p(x)$ on $\mathbb{R}[x]/I$ using the fact that, for all $x \in V(I)$, we have
$$p(x) = \sum_{v \in V(I)} p(v)\ell_v(x).$$

4. Extend this result to all radical zero-dimensional ideals [90].

Exercise 3.106. Consider $\mathbb{R}[x_1, \ldots, x_n]$ and the ideal $I = \langle x_1^2 - 1, \ldots, x_n^2 - 1\rangle$. Show that the standard monomials are the *square-free* monomials and thus are in bijection with the 2^n subsets of $\{1, \ldots, n\}$. Show that the Hilbert series (actually a polynomial in this case) is $H(I, t) = (1 + t)^n = \sum_{k=0}^n \binom{n}{k} t^k$. What does this say about the sizes of the corresponding semidefinite programs when looking at sums of squares modulo I?

3.3.6 Symmetries

Another useful property that can be exploited in the sos context is *symmetry*. Symmetric problems arise very frequently in applications for a variety of reasons. Sometimes symmetry reflects the underlying structure of existing physical systems (e.g., time-invariance, conservation laws), while in some other cases it arises as a result of the chosen mathematical abstraction. Symmetry reduction techniques have been explored in many contexts, with areas such as crystallography, dynamical systems [53], and geometric mechanics [77] being prominent examples.

In optimization, as we shall see, symmetry interacts in a very interesting way with convexity, particularly in the case of semidefinite programming. In general, there are many potential advantages in exploiting symmetries:

Problem size. The first immediate advantage is the reduction in problem size, as the new instance can have a significantly smaller number of variables and constraints.

Degeneracy removal. In symmetric SDP problems, there are repeated eigenvalues of high multiplicity that are difficult to handle numerically. These can be removed by a proper handling of the symmetry.

Conditioning and reliability. Symmetry-aware methodologies have in general much better numerical conditioning, and the resulting smaller size instances are usually less prone to numerical errors.

An in-depth discussion of symmetries in sum of squares and semidefinite programming requires some elements of group representation theory and invariant theory. In this section, we present and isolate the key ideas, referring to the literature for the full technical details; see, e.g., [48, 123]. We consider the simple situation where we want to compute an sos decomposition of a single polynomial, and the underlying symmetry group is finite; the extensions to more general cases are relatively straightforward. The main message is that the presence of symmetry in sos problems can be exploited at three levels of increasing sophistication: (a) convexity, (b) semidefinite programming, and (c) sum of squares.

The set-up is as follows: we consider a polynomial $p(x_1, \dots, x_n)$ that is *invariant* under the action of a finite group G. A formal definition is given below in (3.24), but the idea is that the polynomial in unchanged under certain transformations of the variables. We will use the following as a running example.

Example 3.107. Consider the (nonconvex) quartic trivariate polynomial

$$p(x, y, z) = x^4 + y^4 + z^4 - 4xyz + x + y + z.$$

This polynomial is invariant under all permutations of $\{x, y, z\}$ (the full symmetric group S_3). The global minimum of p is $p_\star \approx -2.1129$ and is achieved at the orbit of global minimizers:

$$(0.988, -1.102, -1.102), \ (-1.102, 0.988, -1.102), \ (-1.102, -1.102, 0.988).$$

For this polynomial, it holds that $p_{sos} = p_\star$. ∎

Recall that a *linear representation* of a group G is a homomorphism $\rho : G \to GL(\mathbb{R}^n)$ (i.e., $\rho(st) = \rho(s)\rho(t)\ \forall s, t \in G$), where $GL(\mathbb{R}^n)$ is the group of invertible $n \times n$ real matrices. The assumption that p is invariant under the group action means that

$$p(\rho(g)x) = p(x) \qquad \forall g \in G. \tag{3.24}$$

Convexity. In general, when minimizing a symmetric function, one cannot always expect that minimizers will also be symmetric (Example 3.107 is a case where this clearly fails). There is, however, an important situation where optimization problems invariant under the action of a group are *guaranteed* to have solutions that are themselves invariant. As we show below, this is the case for *convex* problems, where there is no loss of generality in restricting to symmetric solutions.

Consider the problem of minimizing a convex function $f(x)$ over a convex set S, where both the objective function f and the constraint set S are invariant under the group action. This means that

$$f(\rho(g)x) = f(x) \qquad \forall g \in G$$

and

$$x \in S \Rightarrow \rho(g)x \in S \qquad \forall g \in G,$$

respectively. When these properties hold (symmetry + convexity), then we can always restrict the solution to the *fixed-point subspace* (or subspace of symmetric solutions) defined by

$$\mathcal{F} := \{x \in \mathbb{R}^n \, : \, \rho(g)x = x, \quad \forall g \in G\}.$$

To see why the statement is true, consider any feasible solution $x_0 \in S$, and define the "group average"

$$\widehat{x}_0 = \frac{1}{|G|} \sum_{g \in G} \rho(g) x_0$$

that expresses $\widehat{x}_0$ as a convex combination of the images of x_0 under the group action. By construction, $\widehat{x}_0 \in \mathcal{F}$. Since S is convex and invariant, we have $\widehat{x}_0 \in S$, and convexity and invariance of f yield $f(\widehat{x}_0) \le \frac{1}{|G|} \sum_{g \in G} f(\rho(g)x_0) = f(x_0)$.

Thus, without loss of generality, for invariant convex problems we can restrict the search for optimal solutions to a potentially much smaller subset $S \cap \mathcal{F}$ (of course, this is most useful whenever the dimension of the subspace $\mathcal{F}$ is small). In other words, for convex problems, no "symmetry-breaking" is ever necessary.

Example 3.108. The *entropy* of a probability vector $(p_1, \dots, p_n)$ with $\sum_{i=1}^n p_i = 1$, $p_i \ge 0$, is defined as

$$H(p) := -\sum_{i=1}^{n} p_i \log p_i,$$

where (by continuity) $0 \log 0$ is defined as 0. The (negative) entropy $-H(p)$ is a convex function of p that is clearly symmetric with respect to arbitrary permutations of the p_i. Consider the problem of finding the vector p with largest possible entropy; i.e., we want to minimize the convex symmetric function $-H(p)$ over the convex symmetric set $S = \{p : \sum_{i=1}^n p_i = 1, p_i \ge 0\}$. For this problem, the fixed-point subspace $\mathcal{F}$ is one-dimensional, of the form $(t, t, \dots, t)$, and thus it follows with no calculation that the entropy maximizing vector is given by the uniform distribution $(\frac{1}{n}, \dots, \frac{1}{n})$. ∎

Semidefinite programs, being convex optimization problems, naturally fit into the class discussed above, and thus for invariant SDP problems we will always be able to restrict solutions to their fixed-point subspace. Furthermore, as we shall see next, there is often additional structure to be exploited.

Semidefinite programming. An *invariant semidefinite program* is a semidefinite program whose objective and feasible sets are invariant under the action of a group. As we have just seen, in this case we can always restrict solutions to the *fixed-point subspace* $\mathcal{F}$ of the group action. Remarkably, this subspace will have a very convenient description.

For most semidefinite programs (in particular, those arising from sos decompositions), the group acts on the decision variables in a specific way, where group elements g act on a symmetric matrix by conjugation, i.e., $X \mapsto \rho(g)^T X \rho(g)$. Writing the equations for $\mathcal{F}$, and using the fact that $\rho(g)$ is an orthogonal matrix, we obtain

$$\mathcal{F} = \{X \,:\, X\rho(g) = \rho(g)X \quad \forall g \in G\}; \tag{3.25}$$

i.e., X must *commute* with all matrices in the representation of G. In this case, using Schur's lemma of representation theory, one can show that in the appropriate symmetry-adapted basis, the fixed-point subspace will have a *block-diagonal* structure.

Example 3.109. Consider an invariant semidefinite program where the matrices in the fixed-point subspace have the structure

$$X = \begin{bmatrix} a & b & b \\ b & c & d \\ b & d & c \end{bmatrix}.$$

Notice that these matrices are invariant under simultaneous permutation of the last two rows and columns. We now show that these matrices can be put into a more convenient form. By pre- and postmultiplying by the orthogonal matrix

$$T = \begin{bmatrix} 1 & 0 & 0 \\ 0 & \alpha & \alpha \\ 0 & \alpha & -\alpha \end{bmatrix}, \qquad \alpha = \frac{1}{\sqrt{2}},$$

we obtain

$$T^T X T = \begin{bmatrix} a & \sqrt{2}b & 0 \\ \sqrt{2}b & c+d & 0 \\ 0 & 0 & c-d \end{bmatrix},$$

and the matrix becomes *block diagonal.* ∎

The calculation of a symmetry-adapted basis (i.e., the matrix T in the example above) is fully algorithmic; the details are representation-theoretic (and thus omitted here) but can be found in the literature in [111, 45, 48]. What is important is that this step simplifies the description of $\mathcal{F}$ by replacing a big matrix with a collection of smaller ones (the specific dimensions will of course depend on the problem data). As a consequence, the original SDP problem is reduced to a collection of smaller coupled matrix constraints, with each block corresponding to an "isotypic component," and cardinality equal to the number of irreducible representations of the group that appear nontrivially. This allows for a notable reduction in both the number of decision variables and the size of the semidefinite programs to be solved.

Example 3.110. Consider our running example, Example 3.107. Since $p(x, y, z)$ has $n = 3$ variables, degree $2d = 4$, and a full Newton polytope, its standard sos formulation is indexed by all $\binom{n+d}{d} = \binom{5}{2} = 10$ monomials of degree 2, i.e.,

$$p(x,y,z)-\gamma = \begin{bmatrix} 1 \\ x \\ y \\ z \\ x^2 \\ y^2 \\ z^2 \\ yz \\ xz \\ xy \end{bmatrix}^T \begin{bmatrix} q_{00} & q_{01} & q_{02} & q_{03} & q_{04} & q_{05} & q_{06} & q_{07} & q_{08} & q_{09} \\ q_{01} & q_{11} & q_{12} & q_{13} & q_{14} & q_{15} & q_{16} & q_{17} & q_{18} & q_{19} \\ q_{02} & q_{12} & q_{22} & q_{23} & q_{24} & q_{25} & q_{26} & q_{27} & q_{28} & q_{29} \\ q_{03} & q_{13} & q_{23} & q_{33} & q_{34} & q_{35} & q_{36} & q_{37} & q_{38} & q_{39} \\ q_{04} & q_{14} & q_{24} & q_{34} & q_{44} & q_{45} & q_{46} & q_{47} & q_{48} & q_{49} \\ q_{05} & q_{15} & q_{25} & q_{35} & q_{45} & q_{55} & q_{56} & q_{57} & q_{58} & q_{59} \\ q_{06} & q_{16} & q_{26} & q_{36} & q_{46} & q_{56} & q_{66} & q_{67} & q_{68} & q_{69} \\ q_{07} & q_{17} & q_{27} & q_{37} & q_{47} & q_{57} & q_{67} & q_{77} & q_{78} & q_{79} \\ q_{08} & q_{18} & q_{28} & q_{38} & q_{48} & q_{58} & q_{68} & q_{78} & q_{88} & q_{89} \\ q_{09} & q_{19} & q_{28} & q_{39} & q_{49} & q_{59} & q_{69} & q_{79} & q_{89} & q_{99} \end{bmatrix} \begin{bmatrix} 1 \\ x \\ y \\ z \\ x^2 \\ y^2 \\ z^2 \\ yz \\ xz \\ xy \end{bmatrix},$$

where the matrix Q above will be constrained to be positive semidefinite. Recall that p is invariant under all permutations of the variables (the full symmetric group S_3). Thus, we can constrain the matrix Q to be in the fixed-point subspace, i.e., it should satisfy $Q = \rho(g)^T Q \rho(g)$, where $g \in G$ and $\rho : G \to GL(\mathbb{R}^{10})$ is the induced representation on the vector of monomials that arises from permuting the variables (x, y, z). Solving the equations (3.25) that define the fixed-point subspace, we find that the matrices there have the structure

$$\widehat{Q} = \begin{bmatrix} r_0 & r_1 & r_1 & r_1 & r_2 & r_2 & r_2 & r_3 & r_3 & r_3 \\ r_1 & r_4 & r_5 & r_5 & r_6 & r_7 & r_7 & r_8 & r_9 & r_9 \\ r_1 & r_5 & r_4 & r_5 & r_7 & r_6 & r_7 & r_9 & r_8 & r_9 \\ r_1 & r_5 & r_5 & r_4 & r_7 & r_7 & r_6 & r_9 & r_9 & r_8 \\ r_2 & r_6 & r_7 & r_7 & r_{10} & r_{11} & r_{11} & r_{12} & r_{13} & r_{13} \\ r_2 & r_7 & r_6 & r_7 & r_{11} & r_{10} & r_{11} & r_{13} & r_{12} & r_{13} \\ r_2 & r_7 & r_7 & r_6 & r_{11} & r_{11} & r_{10} & r_{13} & r_{13} & r_{12} \\ r_3 & r_8 & r_9 & r_9 & r_{12} & r_{13} & r_{13} & r_{14} & r_{15} & r_{15} \\ r_3 & r_9 & r_8 & r_9 & r_{13} & r_{12} & r_{13} & r_{15} & r_{14} & r_{15} \\ r_3 & r_9 & r_9 & r_8 & r_{13} & r_{13} & r_{12} & r_{15} & r_{15} & r_{14} \end{bmatrix}. \tag{3.26}$$

Notice that the fixed-point subspace is 16-dimensional, as opposed to the $\binom{11}{2} = 55$ degrees of freedom in the original matrix.

We can now, however, give a nicer description of this subspace. Consider the coordinate transformation (a symmetry-adapted basis) of the form $X \mapsto T^T X T$, where the orthogonal matrix T is given by

$$T = \operatorname{BlockDiag}(1, R, R, R) \cdot \Pi, \qquad R = \begin{bmatrix} \alpha & \alpha & \alpha \\ \alpha & \beta & \gamma \\ \alpha & \gamma & \beta \end{bmatrix},$$

where $\alpha = 1/\sqrt{3}$, $\beta = (3-\sqrt{3})/6$, $\gamma = -(3+\sqrt{3})/6$, and Π is the permutation matrix satisfying $\Pi^T [x_0, x_1, x_2, x_3, x_4, x_5, x_6, x_7, x_8, x_9] = [x_0, x_1, x_4, x_7, x_2, x_5, x_8, x_3, x_6, x_9]$.

It can be verified that under this tranformation, the matrix in (3.26) now takes the form

$$T^T\widehat{Q}T = \text{BlockDiag}(Q_1, Q_2, Q_2),$$

where

$$Q_1 = \begin{bmatrix} r_0 & \sqrt{3}r_1 & \sqrt{3}r_2 & \sqrt{3}r_3 \\ \sqrt{3}r_1 & r_4 + 2r_5 & r_6 + 2r_7 & r_8 + 2r_9 \\ \sqrt{3}r_2 & r_6 + 2r_7 & r_{10} + 2r_{11} & r_{12} + 2r_{13} \\ \sqrt{3}r_3 & r_8 + 2r_9 & r_{12} + 2r_{13} & r_{14} + 2r_{15} \end{bmatrix},$$

$$Q_2 = \begin{bmatrix} r_4 - r_5 & r_6 - r_7 & r_8 - r_9 \\ r_6 - r_7 & r_{10} - r_{11} & r_{12} - r_{13} \\ r_8 - r_9 & r_{12} - r_{13} & r_{14} - r_{15} \end{bmatrix}.$$

Notice that the 10×10 matrix has split into three blocks, one of size 4×4 and two identical blocks of size 3×3. Also, all entries are otherwise linearly independent (in fact, we have the dimension count $\binom{5}{2} + \binom{4}{2} = 10 + 6 = 16$, the number of free parameters in (3.26)).

Since $\widehat{Q} \succeq 0$ if and only if $T^T\widehat{Q}T \succeq 0$, this implies that instead of solving an SDP problem with a positivity constraint on a 10×10 matrix, we have now a 4×4 matrix and a 3×3 matrix instead (clearly, we need only one copy of the two identical 3×3 blocks), which is a lot simpler. ∎

As we can see, exploiting symmetry can allow for a significant reduction in the computational cost. Depending on how much symmetry the problem has, the gains can be very significant and may enable the solution of problems that are otherwise practically impossible to solve.

Sums of squares. We showed in the previous section how to simplify and decompose a *specific* semidefinite program, corresponding to the sos decomposition of a *given polynomial.* We can use similar techniques to simultaneously decompose the semidefinite programs associated to sos decompositions of *all polynomials* invariant under a given symmetry group. In other words, if before we were using a symmetry-adapted basis to split a fixed vector of monomials into isotypic components, now we will instead simultaneously decompose the whole polynomial ring.

The results we present can be expressed in a very appealing form using a few basic concepts of invariant theory. Given a finite group G acting on $(x_1, \ldots, x_n)$, recall that the *invariant ring* is the set of invariant polynomials $\mathbb{R}[x]^G := \{p \in \mathbb{R}[x] : p(\rho(g)x) = p(x)\ \forall g \in G\}$, with the natural operations. For simplicity, we will restrict ourselves to the simple situation where the invariant ring $\mathbb{R}[x]^G$ is isomorphic to a polynomial ring.[3] In this case, we have $\mathbb{R}[x]^G \approx \langle \theta_1, \ldots, \theta_n \rangle$, where $\theta_1, \ldots, \theta_n$ are algebraically independent invariant polynomials.

[3]In general, the invariant ring is a finitely generated algebra but is not necessarily isomorphic to a polynomial ring; i.e., there may not exist a set of algebraically independent generators; see, e.g., [119, 38]. A simple example of this situation is the cyclic group C_3 acting on $\mathbb{R}[x, y, z]$ by cyclically permuting the indeterminates. In this case, a minimal set of generators for the invariant ring $\mathbb{R}[x]^G$ is $\{s_1, s_2, s_3, s_4\} := \{x+y+z, xy+yz+zx, xyz, x^2y+y^2z+z^2y\}$. However, these are algebraically dependent since they satisfy the relation $9s_3^2 + 3s_3s_4 + s_4^2 - 6s_1s_2s_3 - s_1s_2s_4 + s_2^3 + s_1^3s_3 = 0$.

Example 3.111. Consider $\mathbb{R}[x_1, \ldots, x_n]$ and the symmetric group S_n acting by permutation of the variables in the natural way. It is well known that in this case the invariant ring $\mathbb{R}[x]^G$ is isomorphic to a polynomial ring. There are several natural sets of generators for the invariant ring of symmetric polynomials, including the *elementary symmetric functions*

$$\begin{aligned} e_1 &= x_1 + x_2 + \cdots + x_n, \\ e_2 &= x_1x_2 + x_1x_3 + \cdots + x_{n-1}x_n \\ &\vdots \\ e_n &= x_1x_2 \cdots x_n \end{aligned}$$

and the *power sums*

$$\begin{aligned} p_1 &= x_1 + x_2 + \cdots + x_n, \\ p_2 &= x_1^2 + x_2^2 + \cdots + x_n^2 \\ &\vdots \\ p_n &= x_1^n + x_2^n + \cdots + x_n^n. \end{aligned} \quad \blacksquare$$

Because the invariant ring is generated by $\{\theta_1, \ldots, \theta_n\}$, it is possible to rewrite every invariant polynomial $f(x)$ in terms of the generators θ_i to yield a new polynomial $\tilde{f}(\theta)$. This can be done algorithmically, e.g., using Gröbner bases, although more efficient techniques like SAGBI bases can also be used [119].

Example 3.112. Consider the trivariate polynomial of our running example, Example 3.107. We can rewrite $p(x, y, z)$ in terms of the elementary symmetric functions $e_1 = x + y + z$, $e_2 = xy + yz + xz$, and $e_3 = xyz$ as

$$\tilde{p}(e_1, e_2, e_3) = e_1^4 - 4e_1^2e_2 + 2e_2^2 + 4e_1e_3 - 4e_3 + e_1. \quad \blacksquare$$

Rewriting an invariant polynomial $f(x)$ in terms of invariants to obtain $\tilde{f}(\theta)$ is very convenient, since it usually leads to simpler representations. But how does this help us in deciding if $f(x)$ is a sum of squares? In general, if an invariant polynomial is a sum of squares, it may not be a sum of squares of invariant polynomials (see Exercise 3.115), so requiring $\tilde{f}(\theta)$ to be a sum of squares in $\mathbb{R}[\theta]$ would be a very weak condition. The answer is given in the next theorem.

Theorem 3.113. *Let $f(x_1, \ldots, x_n)$ be an sos polynomial that is invariant under the action of a finite group G, and let $\{\theta_1, \ldots, \theta_n\}$ be generators of the corresponding invariant ring. Then $\tilde{f}(\theta) = f(x)$ has a representation of the form*

$$\tilde{f}(\theta) = \sum_i \langle S_i(\theta), \Pi_i(\theta) \rangle,$$

where $\Pi_i \in \mathbb{R}[\theta]^{r_i \times r_i}$ are symmetric matrices that depend only on the group action and $S_i \in \mathbb{R}[\theta]^{r_i \times r_i}$ are sos matrices.

The structure of this representation is very appealing. Given a group G, the matrices Π_i can be precomputed, since they depend only on how the group acts on the polynomial ring. Then, every invariant sos polynomial can be written as a sum of pairings between "coefficients" $S_i(\theta)$ (which are sos matrices) and the matrices Π_i. Since the $S_i(\theta)$ are sos matrices that are subject to affine constraints (equality in the expression above), this is easily reducible to semidefinite programming (which should not be surprising, since this is just the symmetry-reduced version of the original formulation).

The sizes r_i of the matrices Π_i in Theorem 3.113 correspond to the rank of the ith module of equivariants as a free module over the ring of invariants, and the number of terms corresponds to the number of irreducible representations of the group that appear nontrivially in the isotypic decomposition of the polynomial ring. The dimensions of the corresponding semidefinite programs can be determined explicitly using the generating functions known as the *Molien* (or *Hilbert–Poincaré*) series in a similar way as the Hilbert series for ideals discussed in Section 3.3.5. The details are omitted here but can be found in [48].

Example 3.114. For the symmetric group S_3, the invariant ring $\mathbb{R}[x]^G$ is generated by the elementary symmetric functions e_1, e_2, e_3. The corresponding matrices Π_i can be computed to be

$$\begin{aligned}
\Pi_1 &= 1,\\
\Pi_2 &= e_1^2e_2^2 - 4e_2^3 - 4e_1^3e_3 + 18e_1e_2e_3 - 27e_3^2,\\
\Pi_3 &= \begin{bmatrix} 2e_1^2 - 6e_2 & -e_1e_2 + 9e_3 \\ -e_1e_2 + 9e_3 & 2e_2^2 - 6e_1e_3 \end{bmatrix}.
\end{aligned}$$

Thus, every S_3-invariant sos polynomial can be written in the form

$$\tilde{f}(e_1, e_2, e_3) = s_1 \cdot \Pi_1 + s_2 \cdot \Pi_2 + \langle S_3, \Pi_3 \rangle,$$

where s_1, s_2 are scalar sos polynomials and S_3 is a 2×2 sos matrix.

Recall that the global minimum of our polynomial $\tilde{p}(e_1, e_2, e_3)$ is $p_\star = p_{sos} \approx -2.112913$ (an algebraic number of degree 6). We use the representation above to provide a rational certificate that $p_{sos} \geq -\frac{2113}{1000}$ by choosing

$$\begin{aligned}
s_1 &= \tfrac{2113}{1000} + e_1 + \tfrac{79}{47}e_2 - \tfrac{79}{141}e_1^2 - \tfrac{1120}{11511}e_1e_2 - \tfrac{148}{1279}e_1^3 + \tfrac{1439}{2454}e_2^2 - \tfrac{85469}{188958}e_1^2e_2 + \tfrac{85}{693}e_1^4,\\
s_2 &= 0,\\
S_3 &= \begin{bmatrix} \frac{79}{282} + \frac{74}{1279}e_1 + \frac{304}{693}e_1^2 & -\frac{2}{9} + \frac{749}{1636}e_1 \\ -\frac{2}{9} + \frac{749}{1636}e_1 & \frac{3469}{4908} \end{bmatrix}.
\end{aligned}$$

It is easy to check that s_1, s_2, and S_3 are indeed sums of squares and that they satisfy $\tilde{p} + \frac{2113}{1000} = s_1 + \langle S_3, \Pi_3 \rangle$ and therefore serve as a valid algebraic certificate for the lower bound -2.113. ∎

General case	Equality constraints	Symmetries
polynomial ring $\mathbb{R}[x]$	quotient ring $\mathbb{R}[x]/I$	invariant ring $\mathbb{R}[x]^G$
monomials (deg $\leq k$)	standard monomials	isotypic components
$\frac{1}{(1-t)^n} = \sum_{k=0}^{\infty} \binom{n+k-1}{k} \cdot t^k$	Hilbert series	Molien series
	Finite convergence on zero dimensional ideals	Block diagonalization

Table 3.1. *Algebraic structures and sos properties.*

In Table 3.1 we present a summary and comparison of the different techniques to exploit algebraic structure in sos programs.

Exercise 3.115. Let $p(x)$ be an sos polynomial that is invariant under the action of a group. Show that, in general, there may not exist an sos decomposition $p(x) = \sum_i q_i(x)^2$, where all the $q_i(x)$ are invariant polynomials.

Exercise 3.116. An undirected graph $G = (V, E)$ is *vertex transitive* if its automorphism group $\text{Aut}(G)$ acts transitively on the set of vertices V. Consider the standard semidefinite relaxation for MAXCUT, discussed in Section 2.2.2.

1. Explain how to simplify the MAXCUT semidefinite relaxation in the case of vertex-transitive graphs.

2. Apply your results to the k-cycle graph. What are the values of the optimal cut and the corresponding SDP upper bound?

Exercise 3.117. Consider the following sextic form, known as the *Robinson form*:

$$R(x, y, z) = x^6 + y^6 + z^6 - x^4y^2 - y^4x^2 - x^4z^2 - y^4z^2 - x^2z^4 - y^2z^4 + 3x^2y^2z^2.$$

1. Show that $R(x, y, z)$ is invariant under S_3 but is not a sum of squares.

2. Rewrite $(x^2+y^2+z^2)\cdot R(x, y, z)$ in terms of the elementary symmetric functions e_1, e_2, e_3, and give an sos representation as in Theorem 3.113.

3.4 Infeasibility Certificates

At several points in this chapter, we have given sos-based sufficient conditions for different problems (e.g., nonnegativity of polynomials over sets in Section 3.2.4). We now study in more detail the structure of these certificates, as well as the question of when converse results hold, i.e., how to use sos techniques to *certify* properties of systems of equations and inequalities over the real numbers. As we shall see, sos techniques are very powerful in the sense that they can always provide proofs of infeasibility for general basic semialgebraic sets. The key role of sum of squares in these infeasibility certificates is developed in Section 3.4.2, where we introduce the Positivstellensatz, highlighting the similarities to and differences from other well-known algebraic infeasibility certificates.

3.4.1 Valid Constraints: Ideals and Preorders

The feasible set S of an optimization problem is usually described by a finite number of polynomial equations and/or inequalities. However, at least in principle, one could write many other constraints that are equally valid on the set S. For instance, for a linear programming problem, we could consider nonnegative linear combinations of the given inequalities. Recall that this issue appeared already in Section 3.2.4, when considering polynomial nonnegativity over a set, and we described there two techniques (for equations and inequalities, respectively) of producing further valid constraints. We would like to understand the set of *all* possible valid constraints and, in particular, how to algorithmically generate them. To do so, we revisit those constructions next and formalize their properties in terms of two important algebraic objects: *ideals* and *preorders.*

For the case of a set described by equations $f_i(x) = 0$, we were able to produce further polynomials vanishing on the set S by considering linear combinations with polynomial coefficients. The set of all polynomials generated this way is a polynomial ideal. We restate the familiar definition here, for easy comparison with the new concepts introduced later.

Definition 3.118. *Given multivariate polynomials* $\{f_1, \ldots, f_m\}$, *the* ideal *generated by the* f_i *is*

$$\langle f_1, \ldots, f_m \rangle := \{f \,:\, f = t_1 f_1 + \cdots + t_m f_m, \quad t_i \in \mathbb{R}[x]\}\,.$$

Similarly, for a set described by inequalities $g_i(x) \geq 0$, one can generate new valid inequality constraints by multiplying the $g_i(x)$ against sos polynomials, or by taking conic combinations of valid constraints. This is formalized through the notion of *quadratic module.*

Definition 3.119. *Given multivariate polynomials* $\{g_1, \ldots, g_m\}$, *the* quadratic module *generated by the* g_i *is the set*

$$\textbf{qmodule}(g_1, \ldots, g_m) := \{g \,:\, g = s_0 + s_1 g_1 + \cdots + s_m g_m\},$$

where $s_0, s_1, \ldots, s_m \in \mathbb{R}[x]$ *are sums of squares.*

However, as noted earlier, we can also generate further valid constraints by taking *products* of existing valid constraints, which suggests considering the *preorder* generated by the polynomials $g_i(x)$.

Definition 3.120. *Given multivariate polynomials* $\{g_1, \dots, g_m\}$, *the* preorder *generated by the* g_i *is the set*

$$\mathbf{preorder}(g_1, \dots, g_m) := \left\{ g \,:\, g = s_0 + \sum_{\{i\}} s_i g_i + \sum_{\{i,j\}} s_{ij} g_i g_j + \sum_{\{i,j,k\}} s_{ijk} g_i g_j g_k + \cdots \right\},$$

where each term in the sum is a square-free product of the polynomials g_i, *with a coefficient* $s_\alpha \in \mathbb{R}[x]$ *that is a sum of squares. The sum is finite, with a total of* 2^m *terms, corresponding to all subsets of* $\{g_1, \dots, g_m\}$.

Clearly $\mathbf{qmodule}(g_1, \dots, g_m) \subseteq \mathbf{preorder}(g_1, \dots, g_m)$, so, in principle, the latter yields a possibly larger set of valid constraints. By construction, ideals, quadratic modules, and preorders contain only *valid constraints*, which are logical consequences of the given equations and inequalities. Indeed, every polynomial in the ideal $\langle f_1, \dots, f_m \rangle$ vanishes on the solution set of $f_i(x) = 0$. Similarly, every element of $\mathbf{preorder}(g_1, \dots, g_m)$ is clearly nonnegative on the feasible set of $g_i(x) \geq 0$.

A natural question arises: Can *all* valid constraints be generated this way? Unless further assumptions are made, ideals and preorders (and thus, quadratic modules) may not necessarily contain *all* valid constraints; see Exercise 3.121. Remarkably, however, they will be powerful enough to always detect and certify the possible infeasibility (i.e., emptiness) of the corresponding feasible set; the Positivstellensatz (Theorem 3.127) formalizes this statement.

The notions of *ideal*, *preorder*, and *quadratic module* as used above are standard in real algebraic geometry; see, for instance, [19] (the preorders are sometimes also referred to as a *cones*). Notice that, as geometric objects, ideals are affine sets, and quadratic modules and preorders are closed under convex combinations and nonnegative scalings (i.e., they are actually cones in the convex geometry sense). These convexity properties, coupled with the relationships between semidefinite programming and sums of squares, will be key for our developments in the next section.

Exercise 3.121. In general, ideals and preorders may not contain all valid constraints. In this exercise, we illustrate a few cases where things may go wrong.

1. Let $S = \{x \in \mathbb{R} \,:\, x^2 = 0\}$. Show that the polynomial x vanishes on the feasible set but is not in the ideal $\langle x^2 \rangle$.

2. Let $S = \{(x,y) \in \mathbb{R}^2 : x^2 + y^2 = 0\}$. Show that the polynomial x vanishes on the feasible set but is not in the ideal $\langle x^2 + y^2 \rangle$.

3. Let $S = \{x \in \mathbb{R} : x^3 \geq 0\}$. Show that the polynomial x is nonnegative on the feasible set but is not in **preorder**(x) (and thus, is not in **qmodule**(x)).

4. Let $S = \{x \in \mathbb{R} : x \geq 0, y \geq 0\}$. Show that the polynomial xy is nonnegative on the feasible set but is not in **qmodule**(x,y) (but it is in **preorder**(x,y)).

These examples "fail" for a variety of reasons that are related to either multiplicities, real versus complex solutions, or impossibility of degree cancellations. As we shall see, using suitable modifications to take into account the differences between $\mathbb{C}$ and $\mathbb{R}$, and/or additional assumptions, all these difficulties can be avoided.

3.4.2 Certificates of Infeasibility

A central theme throughout convex optimization is the concept of *infeasibility certificates*, or, equivalently, *theorems of the alternative.* The key links relating algebraic techniques and optimization will be the facts that infeasibility of a given polynomial system can *always* be certified through a particular algebraic identity, and that this identity itself can be found via convex optimization.

Let us start by considering the following question: If a system of equations does not have solutions, how can we *prove* this fact? In particular, what kind of evidence could we show to a third party to convince them that the given equations are indeed unsolvable?

Remark 3.122. *Notice the asymmetry between this question (proving or certifying nonexistence of solutions) versus providing evidence that the equations truly have solutions. The latter could be certified (at least in principle) by producing a candidate point x_0 that satisfies all equations (finding such a point x_0 could be very hard, but that is not the issue here). In complexity-theoretic terms, this is essentially the distinction between the NP and co-NP complexity classes (over either the Turing or the real computation model).*

Fortunately, for problems with algebraic structure, there are quite natural ways of providing *infeasibility certificates.* These are formal algebraic identities that give irrefutable evidence about the inexistence of solutions. We briefly recall and illustrate several well-known special cases before proceeding to the general case of polynomial systems over the reals. Table 3.2 contains a summary of the infeasibility certificates to be discussed and the associated computational techniques.

Linear equations. We consider first linear systems of equations over either the real or the complex numbers (in fact, any field will do). It is a well-known result from linear algebra that if a set of linear equations $Ax = b$ is infeasible, there exists a linear combination of the given equations such that the left-hand side is identically zero, but the right-hand side does not vanish (and thus, infeasibility is evident). Such a linear combination can be found, for instance, by Gaussian elimination. This result is also known as the *Fredholm alternative.*

Degree \ field	Complex	Real
Linear	*Range/kernel* Linear algebra	*Farkas' lemma* Linear programming
Polynomial	*Nullstellensatz* Bounded degree: Linear algebra Gröbner bases	*Positivstellensatz* Bounded degree: SDP

Table 3.2. *Infeasibility certificates and associated computational techniques.*

Theorem 3.123 (Range/kernel). *Consider the linear system $Ax = b$. Then,*

$$Ax = b \quad \textit{is infeasible}$$
$$\Updownarrow$$
$$\exists\, \mu \ \textit{s.t.}\ A^T\mu = 0,\ b^T\mu = -1.$$

Notice that one direction of the theorem (existence of a suitable μ implies infeasibility) is obvious: premultiply the equations with μ^T to obtain

$$Ax = b \quad \Rightarrow \quad \mu^T Ax = \mu^T b \quad \Rightarrow \quad 0 = -1,$$

which is clearly a contradiction. Thus, if a vector μ satisfies the conditions in the second half of the theorem, it provides an easily checkable certificate of the infeasibility of the system $Ax = b$. Notice that in this particular case, not only is it easy to *verify* that a given vector μ is a valid certificate, but one can also efficiently *find* such a μ (e.g., by Gaussian elimination).

Polynomial systems over $\mathbb{C}$. For systems of polynomial equations over an algebraically closed field, infeasibility is characterized through one of the central results in algebraic geometry.

Theorem 3.124 (Hilbert's Nullstellensatz). *Let $f_i(z), \dots, f_m(z)$ be polynomials in complex variables $z_1, \dots, z_n$. Then,*

$$f_i(z) = 0 \quad (i = 1, \dots, m) \quad \textit{is infeasible in } \mathbb{C}^n$$
$$\Updownarrow$$
$$-1 \in \langle f_1, \dots, f_m \rangle.$$

Again, the "easy" direction is almost trivial. If -1 is in the ideal generated by the f_i, there exist polynomials $t_1(z), \dots, t_m(z)$ such that

$$t_1(z)f_1(z) + \cdots + t_m(z)f_m(z) = -1.$$

Evaluating this expression at any candidate solution of the polynomial system, we obtain a contradiction, since the left-hand side vanishes, while the right-hand side does not. The polynomials t_i prove infeasibility of the given equations and constitute a *Nullstellensatz refutation* for the polynomial system. Their effective computation can be accomplished in a variety of ways. This could be done, for instance, via

Gröbner basis techniques, or, if a bound on the degree of the polynomials t_i is assumed a priori, via straightforward (but possibly inefficient) linear algebra.

At this point, we should mention an important complexity-theoretic distinction between this case and the simpler case of linear equations discussed earlier. Since deciding feasibility of polynomial equations includes propositional satisfiability (which is NP-hard) as a special case, it would be unreasonable to expect that "short" certificates of infeasibility always exist. Thus, in general one should not expect to always be able to produce certificates $t_i(z)$ of small degree for every infeasible system. In fact, explicit systems of equations are known whose Nullstellensatz refutations necessarily have large degree; see Exercise 3.135, as well as [24, 55, 36] and the references therein.

Remark 3.125. *The two results discussed above deal only with* equations *(either linear equations over any field, or polynomial equations over the complex numbers). Working with* inequalities, *or trying to distinguish between* real versus complex *solutions, will bring additional algebraic challenges. As we will see, to do this one needs to take into account special properties of the* reals *(mainly, the fact that $\mathbb{R}$ is an* ordered field*) that are not true for the complex numbers.*

Linear inequalities. For systems of linear inequalities, strong LP duality provides efficient certificates of infeasibility. These are essentially an algebraic interpretation of the separation theorem for polyhedral sets and are usually presented in terms of theorems of the alternative such as the celebrated Farkas lemma.

Theorem 3.126 (Farkas' lemma).

$$\left\{\begin{array}{rcl} Ax+b & = & 0, \\ Cx+d & \geq & 0 \end{array}\right. \quad \textit{is infeasible}$$

$$\Updownarrow$$

$$\exists \lambda \geq 0, \, \mu \;\; s.t. \; \left\{\begin{array}{rcl} A^T\mu + C^T\lambda & = & 0, \\ b^T\mu + d^T\lambda & = & -1. \end{array}\right.$$

As in the previous cases, the "easy" direction is straightforward. It is equivalent to the *weak duality* of linear programming and follows from direct syntactic manipulations (premultiply the first equation by μ^T and the second equation by λ^T, and add to obtain a contradiction). The "difficult" converse direction is equivalent to *strong duality*, which always holds for linear programming problems. A suitable certificate pair (λ, μ) can be obtained by solving the corresponding LP, which can be done in polynomial time using the ellipsoid algorithm or interior-point methods.

These classical results can be generalized and unified to handle the case of systems of polynomial equations and inequalities over the real numbers. This will yield a simultaneous generalization of Farkas' lemma (to allow for polynomial inequalities), as well as the possibility of distinguishing between real and complex solutions (unlike the Nullstellensatz).

3.4.3 The Positivstellensatz

Consider a general system of polynomial equations and inequalities for which one wants to show that it has no solutions over the real numbers. How do we certify its infeasibility? As we describe next, a very natural class of algebraic certificates exists for this case, under no assumptions whatsoever. This result is known as the *Positivstellensatz* and is one of the cornerstones of real algebraic geometry. It essentially appears in this form in [19] and is due to Stengle [114].

Theorem 3.127 (Positivstellensatz).

$$\left\{\begin{array}{rcl} f_i(x) & = & 0 \quad (i=1,\dots,m), \\ g_i(x) & \geq & 0 \quad (i=1,\dots,p) \end{array}\right. \quad \textit{is infeasible in } \mathbb{R}^n$$

$$\Updownarrow \tag{3.27}$$

$$\exists\, F(x), G(x) \in \mathbb{R}[x] \ \textit{s.t.} \ \left\{\begin{array}{l} F(x) + G(x) = -1, \\ F(x) \in \langle f_1, \dots, f_m \rangle, \\ G(x) \in \mathbf{preorder}(g_1, \dots, g_p). \end{array}\right.$$

The theorem states that for every infeasible system of polynomial equations and inequalities, there exists a simple algebraic identity that directly certifies the inexistence of real solutions. The certificate has a very simple form: a polynomial $F(x)$ from the ideal generated by the equality constraints and a polynomial $G(x)$ from the preorder generated by the equations that add up to the polynomial -1. The "easy" direction is immediate: by construction, evaluating $F(x) + G(x)$ at any feasible point should produce a nonnegative number. However, since this expression is identically equal to the polynomial -1, we arrive at a contradiction. Remarkably, the Positivstellensatz holds under *no assumptions* whatsoever on the polynomials.

Naturally, we are concerned with the effective computation of these certificates. Recall that for the cases of Theorems 3.123–3.126, the corresponding refutations can be obtained using either linear algebra, linear programming, or Gröbner bases techniques. For the Positivstellensatz, we have established that ideals and preorders are *convex cones* in the space of polynomials. As a consequence, the conditions in Theorem 3.127 for a certificate to exist are *convex*, regardless of any convexity property of the original problem. Furthermore, the same property holds if we consider only bounded-degree sections, i.e., the intersection with the subspace of polynomials of degree less than or equal to a given number D. In this case, the conditions in the Positivstellensatz have *exactly* the form of an sos program. This implies that we can find bounded-degree certificates by solving semidefinite programs.

Theorem 3.128. *Consider a system of polynomial equations and inequalities that has no real solutions. The search for bounded-degree Positivstellensatz infeasibility certificates is an sos program and thus is solvable via semidefinite programming. If the degree bound is sufficiently large, infeasibility certificates $F(x), G(x)$ for the original system will be obtained from the corresponding sos program.*

Since infeasibility certificates are naturally ordered by their degree, this gives rise to a natural *hierarchy of semidefinite relaxations* for semialgebraic problems, indexed by certificate degree [89, 91]. The Positivstellensatz guarantees that this hierarchy is complete in the sense that, for every infeasible system, a suitable refutation will eventually be found.

Example 3.129. Consider the following polynomial system:

$$\begin{aligned} f_1 &:= x_1^2 + x_2^2 - 1 = 0, \\ g_1 &:= 3x_2 - x_1^3 - 2 \geq 0, \\ g_2 &:= x_1 - 8x_2^3 \geq 0. \end{aligned}$$

We will prove that it has no solutions $(x_1, x_2) \in \mathbb{R}^2$. By the Positivstellensatz, the system is infeasible if and only if there exist polynomials $t_1, s_0, s_1, s_2, s_{12} \in \mathbb{R}[x_1, x_2]$ that satisfy

$$\underbrace{f_1 \cdot t_1}_{\text{ideal } \langle f_1 \rangle} + \underbrace{s_0 + s_1 \cdot g_1 + s_2 \cdot g_2 + s_{12} \cdot g_1 \cdot g_2}_{\textbf{preorder}(g_1, g_2)} = -1, \tag{3.28}$$

where s_0, s_1, s_2, and s_{12} are sos polynomials.

We will look for solutions where all the terms on the left-hand side have degree bounded by D. For each degree bound D, this is an sos program and thus is solvable via semidefinite programming. For instance, for $D = 4$ we find the certificate (written in fully explicit sos form)

$$\begin{aligned} t_1 &= -3x_1^2 + x_1 - 3x_2^2 + 6x_2 - 2, \\ s_0 &= \frac{5}{43}x_1^2 + \frac{387}{44}\left(x_1x_2 - \frac{52}{129}x_1\right)^2 + \frac{11}{5}\left(-x_1^2 - \frac{1}{22}x_1x_2 - \frac{5}{11}x_1 + x_2^2\right)^2 \\ &\quad + \frac{1}{20}\left(-x_1^2 + 2x_1x_2 + x_2^2 + 5x_2\right)^2 + \frac{3}{4}\left(2 - x_1^2 - x_2^2 - x_2\right)^2, \\ s_1 &= 3, \qquad s_2 = 1, \qquad s_{12} = 0. \end{aligned}$$

The resulting identity (3.28) thus certifies the inconsistency of the system $\{f_1 = 0, g_1 \geq 0, g_2 \geq 0\}$. ∎

In the worst case, of course, the degree of the infeasibility certificates $F(x)$, $G(x)$ could be large (this is to be expected due to the NP-hardness of polynomial infeasibility). In fact, as in the Nullstellensatz case, there are explicit counterexamples where large degree refutations are necessary [55]. Nevertheless, for many problems of practical interest, it is often possible to prove infeasibility using relatively low-degree certificates. There is significant numerical evidence that this is the case, as indicated by the large number of practical applications where sos techniques have provided solutions of very high quality. An outstanding open research question is to understand classes of polynomial systems that can be solved, either in an exact or approximate fashion, using certificates of low degree.

To summarize our discussions, there is a direct path connecting general polynomial optimization problems to semidefinite programming, via Positivstellensatz infeasibility certificates. Pictorially, we have the following:

$$\begin{array}{c}\text{Polynomial systems}\\ \downarrow \\ \text{Positivstellensatz certificates}\\ \downarrow \\ \text{Sum of squares programs}\\ \downarrow \\ \text{Semidefinite programming.}\end{array}$$

Even though so far we have discussed only feasibility problems, there are obvious straightforward connections with optimization questions, which we make more concrete in the next section. As we did earlier in the case of unconstrained optimization, by considering the emptiness of the sublevel sets of the objective function, sequences of converging bounds indexed by certificate degree can be directly constructed.

Exercise 3.130. Consider a single quadratic polynomial equation $ax^2 + bx + c = 0$. What conditions must (a, b, c) satisfy for this equation to have no real solutions? Assuming this condition holds, give a Positivstellensatz certificate of the nonexistence of real solutions.

Exercise 3.131. Explain how Theorem 3.127 simplifies in the following cases:

1. There are no equality constraints.

2. There are no inequality constraints. Is this case equivalent to Hilbert's Nullstellensatz? Explain why or why not.

Exercise 3.132. Consider the polynomial system $\{x + y^3 = 2,\ x^2 + y^2 = 1\}$.

1. Is it feasible over $\mathbb{C}$? How many solutions are there?

2. Is it feasible over $\mathbb{R}$? If not, give a Positivstellensatz-based infeasibility certificate of this fact.

Exercise 3.133. Assume that in the statement of the Positivstellensatz, we replace $\textbf{preorder}(g_1, \dots, g_p)$ with the (potentially smaller) set $\textbf{qmodule}(g_1, \dots, g_p)$. Is the result still true? Prove, or disprove via a counterexample.

Exercise 3.134. Prove, using the Positivstellensatz, that every nonnegative polynomial is a sum of squares of rational functions. (Hint: A polynomial $f(x)$ satisfies $f(x) \geq 0$ for all $x \in \mathbb{R}^n$ if and only if the set $\{(x, y) \in \mathbb{R}^n \times \mathbb{R} \,:\, f(x) \leq 0,\ y \cdot f(x) = 1\}$ is empty.)

Exercise 3.135. In this exercise we compare the relative power of Nullstellensatz and Positivstellensatz based proofs in the context of a specific example. Consider the set of equations $\{\sum_{i=1}^{n} x_i = 1,\ x_i^2 = 0 \text{ for } i = 1, \ldots, n\}$.

1. Show that the given equations are infeasible (either over $\mathbb{C}$ or $\mathbb{R}$).
2. Give a short Positivstellensatz proof of infeasibility (degree 2 should be enough).
3. Show that every Nullstellensatz proof of infeasibility must have degree greater than or equal to n.

3.4.4 Positivity on Compact Sets

In many problems, such as constrained optimization, it is of interest to obtain explicit certificates of positivity of a polynomial over a set. In what follows, S is a basic closed semialgebraic set defined as

$$S = \{x \in \mathbb{R}^n \,:\, g_1(x) \geq 0, \ldots, g_m(x) \geq 0\}. \tag{3.29}$$

Using the Positivstellensatz, it can be easily shown (Exercise 3.139) that if a polynomial $p(x)$ is *strictly positive* on the set S, then it has a representation of the form

$$p(x) = \frac{1 + q_1(x)}{q_2(x)}, \qquad q_1(x), q_2(x) \in \mathbf{preorder}(g_1, \ldots, g_m), \tag{3.30}$$

which obviously certifies its strict positivity on S.

Under further assumptions on the set S, this representation can be simplified. The following result, due to Schmüdgen, provides a denominator-free representation for positive polynomials on *compact* sets.

Theorem 3.136 ([110]). *Let S be a* compact *set, defined as in* (3.29). *If a polynomial $p(x)$ is strictly positive on S, then $p(x)$ is in* $\mathbf{preorder}(g_1, \ldots, g_m)$.

Adding an additional assumption (not just compactness of the set S, but an algebraic certificate of its compactness), even more is true. It is convenient to introduce the following *Archimedean property.*

Definition 3.137. *A quadratic module is* Archimedean *if there exists $N \in \mathbb{N}$ such that the polynomial $N - \sum_i x_i^2$ is in the quadratic module.*

Notice that if $\mathbf{qmodule}(g_1, \ldots, g_m)$ is Archimedean, then the set S is contained in the ball $\sum_i x_i^2 \leq N$, and thus it is necessarily compact. The following theorem by Putinar gives a representation of positive polynomials for the Archimedean case.

Theorem 3.138 ([102]). *Let S be a* compact *set, defined as in* (3.29). *Furthermore, assume that* $\mathbf{qmodule}(g_1, \ldots, g_m)$ *is Archimedean. If a polynomial $p(x)$ is strictly positive on S, then $p(x)$ is in* $\mathbf{qmodule}(g_1, \ldots, g_m)$.

As we can see, these representations are "simpler" in the sense that the conditions involve fewer sos multipliers (recall that the preorder contains terms corresponding to squarefree products between inequalities). Notice, however, that these results say nothing about the *degrees* of the corresponding sos polynomials. It may be possible, at least in certain cases, that the degrees appearing in "simpler" representations are much larger than those of more complicated ones; see, e.g., [115]. We explore some of these issues in the exercises.

Hierarchies of relaxations. All the sos conditions that we have discussed, including Positivstellensatz certificates (Theorem 3.127) and the representation theorems of Schmüdgen (Theorem 3.136) and Putinar (Theorem 3.138), depend on the degree of the sos multipliers. Thus, each of these theorems gives rise to a corresponding hierarchy of sos relaxations, obtained by increasing the corresponding certificate degree. For instance, when minimizing a polynomial $p(x)$ over a set S of the form (3.29), we can consider as before Positivstellensatz certificates of the form

$$p(x) - \gamma = \frac{1 + q_1(x)}{q_2(x)},$$

where $q_1, q_2 \in \mathbf{preorder}(g_1, \ldots, g_m)$, or Schmüdgen/Putinar representations

$$p(x) - \gamma \in \mathbf{preorder}(g_1, \ldots, g_m),$$

$$p(x) - \gamma \in \mathbf{qmodule}(g_1, \ldots, g_m),$$

respectively, depending on what form of certificate is desired (or what assumptions the set S satisfies). For any given maximum degree of the sos polynomials appearing on the right-hand side, one can maximize over γ, which can be done via sos programs (possibly combined with bisection). Each of these alternatives will thus produce a monotone sequence of lower bounds converging to the optimal value (provided the assumptions are satisfied, for the case of Schmüdgen and Putinar representations). For the Positivstellensatz, this was presented in [89, 91], and the case of Putinar-type certificates was analyzed by Lasserre in [72] from the dual viewpoint of moment sequences.

Exercise 3.139. Consider a set S as in (3.29). Show, using the Positivstellensatz, that a polynomial $p(x)$ is strictly positive on S if and only if it has a representation of the form (3.30).

Exercise 3.140. Consider the problem of finding a representation certifying the nonnegativity of $p(x) := 1 - x^2$ over the set $S = \{x : (1 - x^2)^3 \geq 0\}$. Notice that the feasible set S is the interval $[-1, 1]$ and that for this example the preorder and the quadratic module coincide. Let $\gamma \geq 0$. Stengle proved in [115] that no representation of the form

$$p(x) + \gamma = s_0(x) + s_1(x)(1 - x^2)^3 \tag{3.31}$$

exists when $\gamma = 0$, where $s_0(x), s_1(x)$ are sums of squares. He also showed that as $\gamma \to 0$, the degrees of s_0, s_1 necessarily have to go to infinity, and provided the bounds $c_1\gamma^{-\frac{1}{2}} \leq \deg(s_0) \leq c_2\gamma^{-\frac{1}{2}} \log\frac{1}{\gamma}$ for some constants c_1, c_2.

1. Give a Positivstellensatz certificate of the form (3.30) for strict positivity of $p(x) + \gamma$ on S. Does the certificate degree depend on γ?
2. Verify that the expressions below give the "best" representation of the form (3.31). Let the degree of $s_0(x)$ be equal to $4N$. Then, the optimal solution that minimizes γ is

$$\gamma_N^\star = \frac{1}{(2N+2)^2 - 1}, \qquad s_0(x) = q_0(x)^2, \quad s_1(x) = q_1(x)^2,$$

where

$$q_0(x) = 2(N+1)\ {}_2F_1\left(-N, N+2\,;\ \tfrac{1}{2}; x^2\right),$$
$$q_1(x) = \frac{1}{\gamma_N^\star}\, x\ {}_2F_1\left(-N-1, N+1\,;\ \tfrac{3}{2}; x^2\right),$$

and ${}_2F_1(a, b; c, x)$ is the standard Gauss hypergeometric function [1, Chapter 15].

Exercise 3.141. Recall the set S from Exercise 3.63:

$$S = \{(x, y) \in \mathbb{R}^2 \,:\, x \geq 0,\ y \geq 0,\ x + y \leq 1\}.$$

The polynomial $p(x, y) = xy + \epsilon$ (for $\epsilon > 0$) is strictly positive on S. Analyze experimentally the smallest values of ϵ, provable using the positivity certificates of Theorems 3.136 and 3.138, as a function of certificate degree. Compare this against the Positivstellensatz certificates (3.30).

3.5 Duality and Sums of Squares

The sets of nonnegative and sos polynomials, being convex cones, have a rich duality structure. In this section we introduce their duals $P_{n,2d}^*$ and $\Sigma_{n,2d}^*$ and explain their natural interpretations. We do this from both a coordinate-free viewpoint that emphasizes the geometric aspects as well as a probabilistic interpretation with strong links to the classical truncated moment problem and applications.

3.5.1 Dual Cones of Polynomials

Recall that the sets of nonnegative polynomials $P_{n,2d}$ and sums of squares $\Sigma_{n,2d}$ are proper cones in $\mathbb{R}[x]_{n,2d}$. It then follows that the corresponding duals $P_{n,2d}^*$ and $\Sigma_{n,2d}^*$ are also proper cones (in the vector space $\mathbb{R}[x]_{n,2d}^*$) and that the reverse containment holds:

$$\Sigma_{n,2d} \subseteq P_{n,2d} \qquad \Longleftrightarrow \qquad \Sigma_{n,2d}^* \supseteq P_{n,2d}^*.$$

What is the interpretation of these dual cones? Are there "natural" objects associated with them?

The dual space. Let us consider first the dual space to polynomials $\mathbb{R}[x]^*_{n,2d}$. The elements of this vector space are linear functionals on polynomials, i.e., linear maps of the form $\ell : \mathbb{R}[x]_{n,2d} \to \mathbb{R}$, that take a polynomial and return a real number. There are many such functionals, and they can superficially look quite different. For instance, some examples of such linear maps are

- evaluation of p at a point $x_0 \in \mathbb{R}^n$: $p \mapsto p(x_0)$,
- integration of p over a subset $S \subset \mathbb{R}^n$: $p \mapsto \int_S p(x)dx$,
- evaluation of derivatives of p at a point $x_0 \in \mathbb{R}^n$: $p \mapsto \frac{\partial p}{\partial x_i \dots \partial x_k}(x_0)$,
- extraction of coefficients: $p \mapsto \text{coeff}(p, x^\alpha)$,
- contraction with a differential operator $q \in \mathbb{R}[\partial_1, \dots, \partial_n]_{n,2d}$: $p \mapsto q \bullet p$.

A distinguished class of linear functionals are the *point evaluations* (our first example above): to any $v \in \mathbb{R}^n$, we can associate $\ell_v \in \mathbb{R}[x]^*_{n,2d}$, with $\ell_v : p \mapsto p(v)$. Naturally, we can generate additional linear functionals by taking linear combinations of point evaluations, i.e., maps of the form $p \mapsto \sum_i \lambda_i \ell_{v_i}(p) = \sum_i \lambda_i p(v_i)$ for $\lambda_i \in \mathbb{R}$ and $v_i \in \mathbb{R}^n$. It turns out that *all* linear functionals can be obtained this way; this is equivalent to the existence of dense multivariate polynomial interpolation schemes (Exercise 3.142).

Dual cone of nonnegative polynomials. What about the dual cone $P^*_{n,2d} = \{\ell \in \mathbb{R}[x]^*_{n,2d} : \ell(p) \geq 0 \ \forall p \in P_{n,2d}\}$? Clearly, it contains all the point evaluations ℓ_v (since for any nonnegative polynomial p, we have $\ell_v(p) = p(v) \geq 0$), as well as their *conic* combinations $\sum_i \lambda_i \ell_{v_i}$, with $\lambda_i \geq 0$ and $v_i \in \mathbb{R}^n$. It can be shown that almost all elements of $P^*_{n,2d}$ have this form in the sense that this dual cone is the *closure* of the convex hull of the point evaluations. The need for a closure condition arises because we are working in an *affine* setting, i.e., with polynomials instead of forms; see Exercise 3.143 for an illustration of why the closure is required. In the homogeneous case, as will be explained in Chapter 4, or when working on a compact set, the situation is nicer, and the convex hull of point evaluations is automatically closed. We discuss a probabilistic interpretation in Section 3.5.2 and revisit this geometric characterization in Section 3.5.4.

Dual cone of sums of squares. For the cone $\Sigma^*_{n,2d}$ (dual of sums of squares), the situation is a bit simpler. Since the cone $\Sigma_{n,2d}$ is generated by the squares, we have almost by definition the description $\Sigma^*_{n,2d} = \{\ell \in \mathbb{R}[x]^*_{n,2d} : \ell(q^2) \geq 0 \ \forall q \in \mathbb{R}[x]_{n,d}\}$. This directly gives a characterization of $\Sigma^*_{n,2d}$ as a spectrahedron; see Exercise 3.144. However, in this case the geometric interpretation is less clear, since in general $\Sigma^*_{n,2d} \supsetneq P^*_{n,2d}$, and thus this cone has extreme rays that do not necessarily correspond to point evaluations.

We remark that from Hilbert's classification of the cases when $P_{n,2d}$ and $\Sigma_{n,2d}$ coincide (Section 3.1.2), one directly obtains the corresponding equalities between $P^*_{n,2d}$ and $\Sigma^*_{n,2d}$ for the same values of n and d.

The coordinate-free viewpoint described above is mathematically natural and notationally simple, and it is analyzed in more detail in Chapter 4. It is also of relevance when doing numerical computations, since, as we have discussed already in Section 3.1.5, it is often essential to use vector space bases with good numerical properties. Nevertheless, given its many applications, it is also important to understand the alternative viewpoint where one identifies the dual space $\mathbb{R}^*_{n,2d}$ with truncated moment sequences of probability distributions. This corresponds to a specific choice of coordinates for the space of polynomials (namely, the monomial basis), and moments constitute the associated dual basis for the dual space $\mathbb{R}^*_{n,2d}$. This viewpoint is further explored in the remainder of the section.

Exercise 3.142. As described earlier, every linear functional on $\mathbb{R}[x]_{n,2d}$ is a linear combination of point evaluations, i.e., for every $\ell \in \mathbb{R}[x]^*_{n,2d}$, there exist $\lambda_1, \dots, \lambda_k \in \mathbb{R}$ and $v_1, \dots, v_k \in \mathbb{R}^n$, such that $\ell(p) = \sum_{i=1}^k \lambda_i p(v_i)$.

1. Prove this statement for the univariate case ($n = 1$). Hint: Use the nonsingularity of the Vandermonde matrix for suitably chosen points.

2. Extend your proof to the general multivariate case.

Exercise 3.143. Consider the vector space of univariate quadratic polynomials $\mathbb{R}[x]_{1,2} \approx \mathbb{R}^3$.

1. Express the linear functional $(p_2 x^2 + p_1 x + p_0) \mapsto \int_2^3 p(x)dx$ as a (finite) linear combination of point evaluations.

2. Express the linear functional $(p_2 x^2 + p_1 x + p_0) \mapsto p_2$ as a linear combination of point evaluations.

3. Show that this linear functional is in $P^*_{1,2}$ but cannot be written as a conic combination of point evaluations.

4. Give a geometric interpretation of the statement above.

Exercise 3.144.

1. Show that $\Sigma^*_{n,2d}$ is a spectrahedron.

2. Show that $\Sigma_{n,2d}$ is a projected spectrahedron but is *not* a spectrahedron.

3. Is $P^*_{n,2d}$ or $\Sigma^*_{n,2d}$ basic semialgebraic?

Exercise 3.145. Find an extreme point of $\Sigma^*_{2,4}$ that is *not* a conic combination of point evaluations. Hint: Think about the Motzkin polynomial. How would you prove that it is not a sum of squares?

3.5.2 Probability and Moments

A particular, but important, interpretation of the dual cone $P^*_{n,2d}$ is in terms of *truncated moment sequences* of probability distributions. The basic idea, discussed below in more detail, is the following: consider the standard monomial basis for $P_{n,2d}$, and let $p = \sum_{|\alpha|\leq 2d} c_\alpha x^\alpha$ be a nonnegative polynomial and μ be a nonnegative measure. Then $\int p d\mu = \sum_\alpha c_\alpha \mu_\alpha \geq 0$, where $\mu_\alpha := \int x^\alpha d\mu$. Conversely, given a set of numbers μ_α, if $\sum_\alpha c_\alpha \mu_\alpha \geq 0$ for all nonnegative p, then the linear functional $\ell_\mu(p) := \sum_\alpha c_\alpha \mu_\alpha$ is in $P^*_{n,2d}$, and thus it is (up to closure) a conic combination of point evaluations. We can interpret this as a nonnegative measure μ, which will satisfy $\mu_\alpha = \ell_\mu(x^\alpha) = \int x^\alpha d\mu$. Thus, we can identify (again, up to closure) the duals space $P^*_{n,2d}$ with the set of *moments* μ_α for which a nonnegative measure matching those moments exists. The following geometric interpretation may be helpful: on compact sets (or in the homogeneous case), by the Riesz representation theorem the duals of the nonnegative continuous functions are the nonnegative measures. Since the set of polynomials is a subspace, $P_{n,2d}$ is a *section* of the cone of nonnegative continuous functions, and thus its dual $P^*_{n,2d}$ must be a *projection* of the cone of measures. In the chosen basis, this projection is the moment map $\mu \mapsto \int x^\alpha d\mu$ that takes a measure into its moments.

In what follows, we explain and elaborate upon this interpretation. For simplicity, we start with the univariate case.

Valid sequences of moments. Consider a real-valued random variable X, or, equivalently, a nonnegative measure μ supported on $\mathbb{R}$, where $\mathbf{P}(X \in E) = \mu(E)$ for all events E. The *moments* of X (or of μ) are defined as the expectation of the pure powers, i.e.,

$$\mu_k := \mathbf{E}[X^k] = \int x^k d\mu(x). \tag{3.32}$$

In particular, for a random variable X we have $\mu_0 = 1$ (normalization) and $\mu_1 = \mathbf{E}[X]$ (mean or expected value).

A natural question to consider is the following: what constraints, if any, should the moments μ_k satisfy? In particular, is it true that for any set of numbers $(\mu_0, \mu_1, \ldots, \mu_k)$ there always exists a nonnegative measure having exactly these moments? This is the classical *truncated moment problem*; see, e.g., [6, 112].

It should be apparent that this is not always the case and that some conditions on the μ_k are required. For instance, consider (3.32) for an *even* value of k. Since the measure μ is nonnegative, it is clear that in this case we must have $\mu_k \geq 0$. However, this condition is clearly not enough, and further restrictions should hold. A simple one can be derived by recalling the relationship between the variance of a random variable and its first and second moments, i.e., $\text{var}(X) = \mathbf{E}[(X - \mathbf{E}[X])^2] = \mathbf{E}[X^2] - \mathbf{E}[X]^2 = \mu_2 - \mu_1^2$. Since the variance is always nonnegative, the inequality $\mu_2 - \mu_1^2 \geq 0$ must always hold.

How to systematically derive conditions of this kind? The previous inequality can be obtained by noticing that for all a_0, a_1,

$$0 \leq \mathbf{E}[(a_0 + a_1 X)^2] = a_0^2 + 2a_0 a_1 \mathbf{E}[X] + a_1^2 \mathbf{E}[X^2] = \begin{bmatrix} a_0 \\ a_1 \end{bmatrix}^T \begin{bmatrix} \mu_0 & \mu_1 \\ \mu_1 & \mu_2 \end{bmatrix} \begin{bmatrix} a_0 \\ a_1 \end{bmatrix},$$

which implies that the 2×2 matrix above must be positive semidefinite. Interestingly, this is equivalent to the inequality obtained earlier.

The same procedure can be repeated for higher-order moments. Let $\boldsymbol{\mu} = (\mu_0, \mu_1, \ldots, \mu_{2d})$ be given. By considering the expectation of the square of a generic polynomial

$$0 \leq \mathbf{E}[(a_0 + a_1 X + \cdots + a_d X^d)^2],$$

we have that the higher order moments of a random variable must satisfy

$$\mathcal{H}(\boldsymbol{\mu}) := \begin{bmatrix} \mu_0 & \mu_1 & \mu_2 & \cdots & \mu_d \\ \mu_1 & \mu_2 & \mu_3 & \cdots & \mu_{d+1} \\ \mu_2 & \mu_3 & \mu_4 & \cdots & \mu_{d+2} \\ \vdots & \vdots & \vdots & \ddots & \vdots \\ \mu_d & \mu_{d+1} & \mu_{d+2} & \cdots & \mu_{2d} \end{bmatrix} \succeq 0. \tag{3.33}$$

Notice that $\mathcal{H}(\boldsymbol{\mu})$ is a Hankel matrix, and the diagonal elements correspond to the even-order moments, which should obviously be nonnegative.

As we will see below, this condition is "almost" necessary and sufficient in the univariate case in the sense that it characterizes the set of valid moments up to closure.

Theorem 3.146. *Let $\boldsymbol{\mu} = (\mu_0, \mu_1, \ldots, \mu_{2d})$ be given, where $\mu_0 = 1$. If $\boldsymbol{\mu}$ is a valid set of moments, then the associated Hankel matrix $\mathcal{H}(\boldsymbol{\mu})$ is positive semidefinite. Conversely, if $\mathcal{H}(\boldsymbol{\mu})$ is (strictly) positive definite, then $\boldsymbol{\mu}$ is valid; i.e., there exists a nonnegative random variable with this set of moments.*

The derivation given earlier shows the necessity of the semidefiniteness condition. Sufficiency will follow from the explicit construction of Section 3.5.5.

Remark 3.147. *For the case of measures supported on the real line, the semidefinite condition in* (3.33) *characterizes the* closure *of the set of moments, but not necessarily the whole set. As an example, consider $\boldsymbol{\mu} = (1, 0, 0, 0, 1)$, corresponding to the Hankel matrix*

$$\mathcal{H}(\boldsymbol{\mu}) = \begin{bmatrix} 1 & 0 & 0 \\ 0 & 0 & 0 \\ 0 & 0 & 1 \end{bmatrix}.$$

Although this matrix is positive semidefinite, there is no nonnegative measure corresponding to those moments (notice that $\mu_2 = 0$). However, the parametrized atomic measure given by

$$\mu_\varepsilon = \frac{\varepsilon^4}{2} \cdot \delta\left(x + \frac{1}{\varepsilon}\right) + (1 - \varepsilon^4) \cdot \delta(x) + \frac{\varepsilon^4}{2} \cdot \delta\left(x - \frac{1}{\varepsilon}\right)$$

has as first five moments $(1, 0, \varepsilon^2, 0, 1)$, *and thus as* $\varepsilon \to 0$ *they converge to those given above.*

As the remark above illustrates, the fact that the semidefinite description is correct only "up to closure" is a consequence of considering measures supported on the whole real line, which is not compact. For the case of compact intervals, the situation will be nicer, as we will see in the next section.

As we move on to the general multivariate case, however, much more serious difficulties will appear (essentially, once again, the difference between polynomial nonnegativity versus sums of squares). We will discuss this situation in Section 3.5.6.

3.5.3 Nonnegative Measures on Intervals

We are interested now in deriving conditions for $\boldsymbol{\mu} = (\mu_0, \mu_1, \ldots, \mu_d)$ to be valid moments of the distribution of a random variable supported on a compact interval of the real line. For simplicity, we concentrate in the case of the interval $[-1, 1]$.

Clearly, the necessary condition described in the previous section (positive semidefiniteness of the Hankel matrix $\mathcal{H}(\boldsymbol{\mu})$) should hold. However, additional conditions may be required to ensure the measure is supported in $[-1, 1]$. Recall how the necessity of the condition $\mathcal{H}(\boldsymbol{\mu}) \succeq 0$ was derived: by considering a nonnegative polynomial $p(x)$, and computing $\mathbf{E}[p(X)]$, which gives a linear condition on the moments. Thus, in order to generate additional valid inequalities that $\boldsymbol{\mu}$ must satisfy, we need to have access to nonnegative polynomials on the domain of interest (the support set of the measure).

Fortunately, we have already discussed in Section 3.3.1 a full sos characterization of the set of polynomials nonnegative on intervals. As shown below, dualizing these conditions, we can similarly obtain a complete characterization for valid moments of a $[-1, 1]$ measure. As in the case of polynomial nonnegativity, depending on whether the index of the largest moment is even or odd, we can write two slightly different (but equivalent) characterizations.

Odd case. Consider the polynomials

$$(1 + x)\left(\sum_{i=0}^{d} a_i x^i\right)^2, \qquad (1 - x)\left(\sum_{i=0}^{d} a_i x^i\right)^2, \tag{3.34}$$

which are obviously nonnegative for $x \in [-1, 1]$. As before, by computing the expectation of these polynomials, we obtain necessary conditions in terms of the quadratic form (in the coefficients a_i):

$$0 \leq \mathbf{E}\left[(1 \pm X)\left(\sum_{i=0}^{d} a_i X^i\right)^2\right] = \sum_{j=0}^{d}\sum_{k=0}^{d}(\mu_{j+k} \pm \mu_{j+k+1}) a_j a_k.$$

Since the polynomials of the form (3.34) generate *all* nonnegative polynomials on $[-1, 1]$, and this interval is compact, these conditions give a full characterization. We formalize this in the next result.

Lemma 3.148. *There exists a nonnegative finite measure supported in $[-1,1]$ with moments $(\mu_0, \mu_1, \ldots, \mu_{2d+1})$ if and only if*

$$\begin{bmatrix} \mu_0 & \mu_1 & \mu_2 & \cdots & \mu_d \\ \mu_1 & \mu_2 & \mu_3 & \cdots & \mu_{d+1} \\ \mu_2 & \mu_3 & \mu_4 & \cdots & \mu_{d+2} \\ \vdots & \vdots & \vdots & \ddots & \vdots \\ \mu_d & \mu_{d+1} & \mu_{d+2} & \cdots & \mu_{2d} \end{bmatrix} \pm \begin{bmatrix} \mu_1 & \mu_2 & \mu_3 & \cdots & \mu_{d+1} \\ \mu_2 & \mu_3 & \mu_4 & \cdots & \mu_{d+2} \\ \mu_3 & \mu_4 & \mu_5 & \cdots & \mu_{d+3} \\ \vdots & \vdots & \vdots & \ddots & \vdots \\ \mu_{d+1} & \mu_{d+2} & \mu_{d+3} & \cdots & \mu_{2d+1} \end{bmatrix} \succeq 0. \quad (3.35)$$

Even case. Consider now instead

$$\left(\sum_{i=0}^{d} a_i x^i\right)^2, \qquad (1-x^2)\left(\sum_{i=0}^{d-1} a_i x^i\right)^2,$$

which are again obviously nonnegative in $[-1,1]$. This yields the following lemma.

Lemma 3.149. *There exists a nonnegative finite measure supported in $[-1,1]$ with moments $(\mu_0, \mu_1, \ldots, \mu_{2d})$ if and only if*

$$\begin{bmatrix} \mu_0 & \mu_1 & \mu_2 & \cdots & \mu_d \\ \mu_1 & \mu_2 & \mu_3 & \cdots & \mu_{d+1} \\ \mu_2 & \mu_3 & \mu_4 & \cdots & \mu_{d+2} \\ \vdots & \vdots & \vdots & \ddots & \vdots \\ \mu_d & \mu_{d+1} & \mu_{d+2} & \cdots & \mu_{2d} \end{bmatrix} \succeq 0,$$

$$\begin{bmatrix} \mu_0 & \mu_1 & \mu_2 & \cdots & \mu_{d-1} \\ \mu_1 & \mu_2 & \mu_3 & \cdots & \mu_d \\ \mu_2 & \mu_3 & \mu_4 & \cdots & \mu_{d+1} \\ \vdots & \vdots & \vdots & \ddots & \vdots \\ \mu_{d-1} & \mu_d & \mu_{d+1} & \cdots & \mu_{2d-2} \end{bmatrix} - \begin{bmatrix} \mu_2 & \mu_3 & \mu_4 & \cdots & \mu_{d+1} \\ \mu_3 & \mu_4 & \mu_5 & \cdots & \mu_{d+2} \\ \mu_4 & \mu_5 & \mu_6 & \cdots & \mu_{d+3} \\ \vdots & \vdots & \vdots & \ddots & \vdots \\ \mu_{d+1} & \mu_{d+2} & \mu_{d+3} & \cdots & \mu_{2d} \end{bmatrix} \succeq 0. \quad (3.36)$$

In both cases, if the measure is normalized (i.e., if it is a probability measure), then additionally the zeroth moment must satisfy $\mu_0 = 1$.

Exercise 3.150. Show that the condition (3.35) implies positive semidefiniteness of the Hankel matrix $\mathcal{H}(\mu_0, \mu_1, \ldots, \mu_{2d})$.

Exercise 3.151. Show that the two given descriptions (odd and even cases) are equivalent in the sense that if the highest-order moment is otherwise unconstrained, the projection of the feasible set of one description is exactly given by the other.

3.5.4 Moment Spaces and the Moment Curve

An appealing geometric interpretation of the set of valid moments described in the previous section is in terms of the so-called *moment curve*. This is the parametric

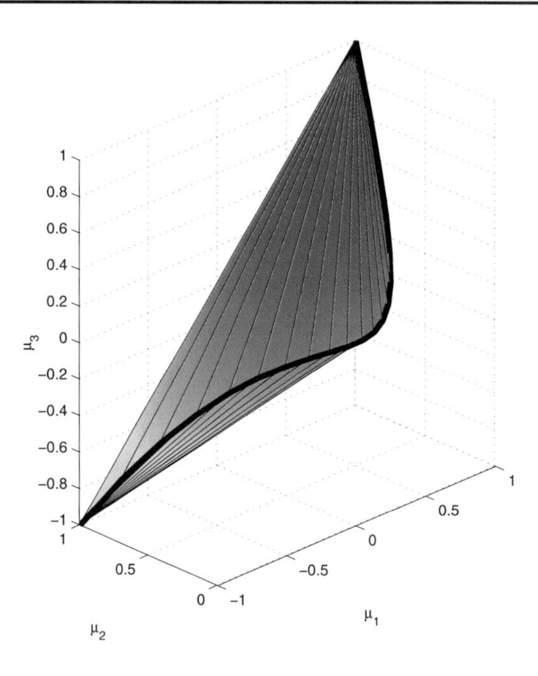

Figure 3.10. *Set of valid moments* (μ_1, μ_2, μ_3) *of a probability measure supported on* $[-1, 1]$*. This is the convex hull of the moment curve* (t, t^2, t^3) *for* $-1 \leq t \leq 1$*. An explicit semidefinite representation is given in* (3.37).

curve in $\mathbb{R}^{d+1}$ given by $t \mapsto (1, t, t^2, \ldots, t^d)$. The convex hull of this curve is known as the *moment space* and corresponds exactly to the set of valid moments; see [66] for background and many more details on this geometric viewpoint.

The reason for this correspondence is simple to understand. Every point on the curve can be associated to a Dirac measure (i.e., one where all the probability is concentrated on a single point). Indeed, for a measure of the form $\delta(x - c)$ (all mass is concentrated at $x = c$), we have $\mu_k = \mathbf{E}[X^k] = \int \delta(x - c) x^k dx = c^k$, and thus the corresponding set of moments is $(1, c, c^2, \ldots, c^d)$. Any other measure can be interpreted as a nonnegative combination of these Dirac measures. Since the moment map that takes a measure into its set of moments is linear, these "probabilistic" nonnegative linear combinations can be interpreted geometrically as convex combinations of points, yielding the convex hull of the curve. Thus, every finite measure on the interval gives a point in the convex hull.

In Figure 3.10 we present an illustration of the moment space for the case of support $[-1, 1]$ and $d = 3$. Notice that, in this case, by Lemma 3.148, we have the semidefinite characterization

$$\begin{bmatrix} \mu_0 & \mu_1 \\ \mu_1 & \mu_2 \end{bmatrix} \pm \begin{bmatrix} \mu_1 & \mu_2 \\ \mu_2 & \mu_3 \end{bmatrix} \succeq 0, \qquad \mu_0 = 1. \tag{3.37}$$

Since both semidefinite constraints are given by 2×2 matrices, the moment space is the intersection of two circular cones.

Exercise 3.152. Explain Remark 3.147 from this geometric perspective. What can you say about the closedness of the convex hull of the moment curve in $\mathbb{R}^d$? Show that if we consider closed intervals (i.e., $t \in [a,b]$), then the corresponding convex hull is compact. What happens in the unconstrained case, i.e., when $t \in (-\infty, \infty)$?

3.5.5 Constructing a Measure

We have given necessary conditions for the existence of a univariate nonnegative measure with given moments. Under the right assumptions (e.g., compactness of support, or strict positivity of the Hankel matrix), these conditions were also sufficient. We describe next a classical algorithm to effectively obtain this measure.

In general, given a set of moments, there may be many measures that exactly match these moments (equivalently, the moment map that takes a measure into a finite set of its moments is not injective). Over the years, researchers have developed a number of techniques to produce specific choices of measures matching a given set of moments (e.g., those that are "simple" according to specific criteria, or that have large entropy, etc.). We review next a classical method for producing an *atomic* measure matching a given set of moments; see, e.g., [112, 39]. This technique (or essentially similar ones) is known under a variety of names, such as *Prony's method*, or the *Vandermonde decomposition of a Hankel matrix.* Other variations of this method are commonly used in areas such as signal processing, e.g., Pisarenko's harmonic decomposition method, where one is interested in producing a superposition of sinusoids with a given covariance matrix.

Consider the set of moments $\boldsymbol{\mu} = (\mu_0, \mu_1, \ldots, \mu_{2d-1})$ for which we want to find an associated nonnegative measure supported on the real line. We assume that the associated Hankel matrix $\mathcal{H}(\boldsymbol{\mu})$ is positive definite. In this method, the resulting measure will be discrete (a sum of d atoms) and will have the form $\sum_{i=1}^{d} w_i \delta(x - x_i)$. To obtain the weights w_i and atom locations x_i, consider the linear system

$$\begin{bmatrix} \mu_0 & \mu_1 & \cdots & \mu_{d-1} \\ \mu_1 & \mu_2 & \cdots & \mu_d \\ \vdots & \vdots & \ddots & \vdots \\ \mu_{d-1} & \mu_d & \cdots & \mu_{2d-2} \end{bmatrix} \begin{bmatrix} c_0 \\ c_1 \\ \vdots \\ c_{d-1} \end{bmatrix} = - \begin{bmatrix} \mu_d \\ \mu_{d+1} \\ \vdots \\ \mu_{2d-1} \end{bmatrix}. \tag{3.38}$$

The Hankel matrix on the left-hand side of this equation is $\mathcal{H}(\boldsymbol{\mu})$, and thus the linear system in (3.38) has a unique solution if the matrix is positive definite. In this case, we let x_i be the roots of the univariate polynomial

$$x^n + c_{n-1}x^{n-1} + \cdots + c_1 x + c_0 = 0,$$

which are all real and distinct (why?). We can then obtain the corresponding weights w_i by solving the nonsingular Vandermonde system given by

$$\sum_{i=1}^{n} w_i x_i^j = \mu_j \qquad (0 \leq j \leq n-1).$$

In Exercise 3.155 we will prove that this method actually works (i.e., the atoms x_i are real and distinct, the weights w_i are nonnegative, and the moments are the correct ones).

Example 3.153. Consider the problem of finding a nonnegative measure whose first six moments are given by $(1, 1, 2, 1, 6, 1)$. The solution of the linear system (3.38) yields the polynomial

$$x^3 - 4x^2 - 9x + 16 = 0,$$

whose roots are -2.4265, 1.2816, and 5.1449. The corresponding weights are 0.0772, 0.9216, and 0.0012, respectively. It can be easily verified that the found measure indeed satisfies the desired constraints. ∎

Remark 3.154. *The measure recovery method described above always works correctly, provided the computations are done in exact arithmetic. In most practical applications, it is necessary or convenient to use floating-point computations. Furthermore, in many settings the moment information may be noisy, and therefore the matrices may contain some (hopefully small) perturbations from their nominal values. For these reasons, it is of interest to understand sensitivity issues at the level of what is intrinsic about both the problem (conditioning) and the specific algorithm used (numerical stability).*

When using floating-point arithmetic, this technique may run into numerical difficulties. On the conditioning side, it is well known that from the numerical viewpoint, the monomial basis (with respect to which we are taking moments) is a "bad" basis for the space of polynomials. On the numerical stability side, the algorithm above does a number of inefficient calculations, such as explicitly computing the coefficients c_i of the polynomial corresponding to the support of the measure. Better approaches involve, for instance, directly computing the nodes x_i as the generalized eigenvalues of a matrix pencil; see, e.g., [51, 52].

Exercise 3.155. Prove that the algorithm described above always produces a valid measure, provided the initial matrix of moments is positive definite. Hint: Show that if $p(x)$ is a polynomial that vanishes at the points x_i then $\mathbf{E}[p(x)^2] = 0$. From this, using the assumed positive definiteness of the Hankel matrix, determine what equations $p(x)$ must satisfy. What is the relation between this matrix and the Hermite form?

Exercise 3.156.

1. Find a discrete measure having the same first eight moments as a standard Gaussian distribution of zero mean and unit variance.

2. What does the previous result imply if we are interested in computing integrals of the type

$$\frac{1}{\sqrt{2\pi}} \int_{-\infty}^{\infty} p(x) e^{-\frac{x^2}{2}} \, dx,$$

where $p(x)$ is a polynomial of degree less than eight? What would you do if $p(x)$ is an arbitrary (smooth) function?

3. Use these ideas to give an approximate numerical value of the definite integral

$$\int_{-\infty}^{\infty} \cos(2x+1)\, e^{-2x^2}\, dx.$$

How does your approximation compare with the exact value $\sqrt{\frac{\pi}{2e}}\cos(1)$?

Note. In the general case where we are matching $2d$ moments of a standard Gaussian, it can be shown that the support of these discrete measures will be given by the d zeros of $H_d(x/\sqrt{2})$, where H_d is the standard Hermite polynomial of degree d. These numerical techniques are called *Gaussian quadrature*; see, e.g., [116, 49] for details.

Exercise 3.157. What is the geometric interpretation of the atomic measure produced by the algorithm described in this section? Explain your answer in terms of Figure 3.10 and the set of moments $\boldsymbol{\mu} = \left(1, \frac{1}{5}, \frac{1}{2}, \frac{1}{7}\right)$.

3.5.6 Moments in Several Variables

The same questions we have considered so far in this section for the univariate case can be formulated for nonnegative measures in several variables. Concretely, given a set of numbers μ_α, with $\alpha \in \mathbb{N}^n$ and $|\alpha| \leq 2d$, does there exist a nonnegative measure in $\mathbb{R}^n$ matching these moments? By our earlier discussions, this is essentially the membership problem for the cone $P^*_{n,2d}$.

Unfortunately, except for a few special situations (e.g., the univariate case and the others that follow from Hilbert's classification) there is no easy answer or an efficient polynomial-time algorithm for this question. This mirrors (in fact, dualizes) the case of polynomial nonnegativity. Recall that the cone of valid moments of nonnegative measures is (up to closure) the dual of the cone of nonnegative polynomials $P_{n,2d}$. It is known that the complexity of the weak membership problem for a convex cone and its dual are equivalent [56], and, as a consequence, deciding membership in $P^*_{n,2d}$ will also be NP-hard. Thus, the computational intractability of nonnegative polynomials implies (and is equivalent to) the intractability of valid multivariate moment sequences.

Remark 3.158. *As in the case of polynomial nonnegativity noted in Remark* 3.15, *the characterization of truncated moment sequences can be reformulated (and, in principle, solved) using decision algebra methods such as quantifier elimination. Indeed, both polynomial nonnegativity and conic convex duality are expressible in first-order logic, and thus (for any fixed dimension and degree) elimination of quantifiers will yield a semialgebraic description of the valid moment sequences, in terms of the variables* μ_α *only. While theoretically useful (since, for instance, this shows decidability of the problem), this approach is practically infeasible except for very small instances.*

Fortunately, we can use the sos methods developed in earlier sections; recall that these yield the (dual) sos outer bound $\Sigma^*_{n,2d} \supseteq P^*_{n,2d}$. Furthermore, we can produce tighter outer approximations to the set $P^*_{n,2d}$ that improve upon the straightforward outer bound $\Sigma^*_{n,2d}$ while still being computationally tractable. To do this, we simply dualize the hierarchies of inner approximations to the set of nonnegative polynomials that we obtained via sos methods. Each variation of the sos methods that we have seen (Positivstellensatz, Pólya/Reznick theorem, Schmüdgen, and Putinar representations) can be used to produce a matching sequence of dual approximations to the corresponding dual cone. For concreteness, we illustrate this discussion with two specific examples.

Polynomial multipliers and rational moments. Recall from Section 3.2.6 that a way of producing stronger sos conditions in the multivariate case was to multiply the given polynomial $p(x)$ by a fixed sos factor $q(x)$. What does this construction correspond to on the dual side?

A dual interpretation of this method is in terms of *rational moments*, i.e., expectations of rational functions

$$\eta_\alpha = \mathbf{E}\left[X^\alpha / q(X)\right].$$

Indeed, one can easily write necessary conditions that these rational moments should satisfy, of the form

$$\mathbf{E}\left[p(X)^2/q(X)\right] \geq 0, \tag{3.39}$$

which, as before (after parametrizing polynomials $p(x)$ up to a given degree), give spectrahedral conditions on the rational moments η_α. Furthermore, the "standard" moments $\mu_\alpha = \mathbf{E}[X^\alpha]$ are given by a linear transformation of the rational moments η_α, since if $q(x) = \sum_\beta c_\beta x^\beta$, then

$$\mu_\alpha = \mathbf{E}[X^\alpha] = \mathbf{E}[q(X)(X^\alpha/q(X))] = \sum_\beta c_\beta \mathbf{E}[X^{\alpha+\beta}/q(X)] = \sum_\beta c_\beta \eta_{\alpha+\beta}.$$

Notice that this yields the normalization condition $\mathbf{E}[1] = \sum_\beta c_\beta \eta_\beta = 1$. These expressions give a refined outer approximation to the set of valid moments as an affine projection of a spectrahedral set (i.e., we are approximating the set of moments with projected spectrahedra). Under suitable conditions on the polynomial $q(x)$ (e.g., those in Pólya's theorem), this method will produce a complete hierarchy of spectrahedral approximations to the set of valid moments.

Example 3.159. In this example we compute a particular projection of the set of moments $P^*_{n,2d}$. We consider bivariate probability distributions ($n = 2$) and moments up to sixth order ($2d = 6$). We are interested in the projection of the set of valid moments onto the two-dimensional plane (α, β) given by $\alpha = \mu_{42} + \mu_{24} = \mathbf{E}[x^4y^2] + \mathbf{E}[x^2y^4]$ and $\beta = \mu_{22} = \mathbf{E}[x^2y^2]$. The simple sos approximation $\Sigma^*_{2,6} \supseteq P^*_{2,6}$ in this case yields the trivial orthant outer bound $\alpha \geq 0$, $\beta \geq 0$.

We can produce tighter bounds by considering the multiplier-based relaxations described earlier. Let us describe the geometry first. For this, define the Motzkin-like family of polynomials $M_t(x,y) = t^3x^4y^2 + t^3x^2y^4 + 1 - 3t^2x^2y^2$ (for $t = 1$, this

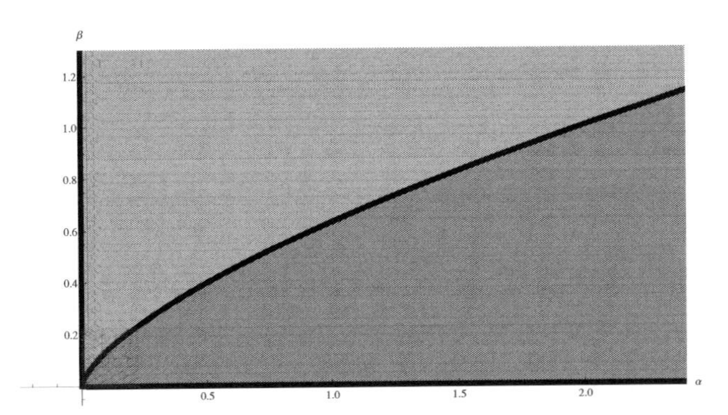

Figure 3.11. *Projection of the set $P^*_{n,2d}$ of valid moments onto $(\alpha, \beta) = (\mu_{42} + \mu_{24}, \mu_{22})$. The outer approximation $\alpha \geq 0$, $\beta \geq 0$ corresponds to the "plain" sos bound $\Sigma^*_{n,2d}$. The inner region is obtained using a polynomial multiplier $q(x, y) = x^2 + y^2$ and gives the exact projection.*

is the standard Motzkin polynomial). It can be shown (e.g., via the arithmetic-geometric inequality or Exercise 3.160) that $M_t(x, y)$ is nonnegative for $t \geq 0$. Therefore, we have the parametrized family of linear inequalities

$$0 \leq \mathbf{E}[M_t(X, Y)] = t^3\alpha + 1 - 3t^2\beta$$

for all $t \geq 0$. Simplifying this expression, we obtain $\alpha \geq 2\beta^{\frac{3}{2}}$, $\beta \geq 0$. These inequalities exactly define the projection of the set of valid moments onto (α, β); see Figure 3.11 and Exercise 3.160.

Let us see how the rational moments interpretation described earlier gives a description of this set as a projected spectrahedron. We choose $q(x, y) = x^2 + y^2$ (as the Pólya–Reznick theorem would suggest) and define rational moments $\eta_{jk} = \mathbf{E}[X^jY^k/(X^2 + Y^2)]$. Parametrizing a generic polynomial $p(x, y) = a_{10}x + a_{01}y + \cdots + a_{13}xy^3 + a_{04}y^4$, we write the inequality (3.39), i.e.,

$$\mathbf{E}[p(X, Y)^2/(X^2 + Y^2)] \geq 0 \qquad \forall p,$$

which is a quadratic form in the coefficients a_i. Expressing this in matrix form, one obtains a 14×14 matrix[4] whose entries are the rational moments η_{jk}. We also have the normalization condition $\mathbf{E}[1] = \eta_{20} + \eta_{02} = 1$. Since $(\alpha, \beta) = (\mu_{24} + \mu_{42}, \mu_{22})$, the desired projection is then given by $\eta_{jk} \mapsto (\eta_{62} + 2\eta_{44} + \eta_{26}, \eta_{42} + \eta_{24})$. ∎

Moments on compact sets. Consider a basic semialgebraic set set $S = \{x \in \mathbb{R}^n : g_1(x) \geq 0, \ldots, g_m(x) \geq 0\}$. We want to describe (or approximate) the set of valid moments of nonnegative measures supported on S.

As before, we can easily write necessary conditions that the moments should satisfy by computing expectations of polynomials that are "obviously" nonnegative

[4]In this specific case, the problem can be much simplified by exploiting the sparsity and symmetry present in the problem. For simplicity, the details are omitted.

on S. Since squares are certainly nonnegative, and so are the products of squares with the defining polynomials g_i, we can consider the expressions

$$\mathbf{E}[p(X)^2] \geq 0, \qquad \mathbf{E}[g_1(X)p(X)^2] \geq 0, \quad \ldots \quad \mathbf{E}[g_m(X)p(X)^2] \geq 0, \tag{3.40}$$

where $p(x)$ are arbitrary polynomials. Exactly as in the univariate case, imposing this condition for all $p(x)$ up to a certain degree, these yield quadratic forms in the coefficients of $p(x)$ that depend linearly on the moments μ_α. Thus, the conditions (3.40) give a family of spectrahedral approximations of the set of moments of S-supported nonnegative measures. By increasing the degree of the polynomial $p(X)$, tighter approximations are obtained. Under the "right" assumptions (essentially, if we can approximate the set of nonnegative polynomials), this dual hierarchy will approximate the set of moments arbitrarily well. For instance, recall from Section 3.4.4 that this will be the case if $\mathbf{qmodule}(q_1, \ldots, q_m)$ satisfies the Archimedean property of Definition 3.137 (and thus, S is compact), as was done in [72]. Notice that these approximations can be strengthened by including products of the form $\mathbf{E}[g_i(X) \cdots g_k(X)p(X)^2] \geq 0$, which correspond to the distinction between preorders and quadratic modules, or, equivalently, Schmüdgen versus Putinar representations.

Constructing multivariate measures. In the univariate case, we have discussed in Section 3.5.5 how to produce an atomic measure matching a given finite set of moments using Prony's method. This is possible because in that case there is a full characterization of the moment space. In the multivariate case, as we have seen, even the decision question ("Are these valid moments?") is NP-hard, and thus, in general, unless further assumptions are satisfied, no such efficient procedure is available.

Given a truncated moment sequence (or, equivalently, a functional $\ell_\mu \in \mathbb{R}^*_{n,2d}$), the positivity condition $\ell_\mu(p^2) \geq 0$ is of course necessary for the existence of a nonnegative measure. A well-known case where it is possible to construct such a measure is whenever the *flat extension* property [34] holds. This is a condition on the given moment sequence that requires the rank of the quadratic form $p \mapsto \ell_\mu(p^2)$ to remain the same when considering polynomials p of degree d or $d+1$ for some value of d. Whenever this condition holds, a natural generalization of the method described in the univariate case can be applied to obtain an atomic measure matching the given moment sequence. The basic idea of this construction is sketched below and appears in a number of related forms in the literature (e.g., Gelfand–Neimark–Segal construction, Stickelberg/Stetter-Möller/eigenvalue method for polynomial equations [32, 121], etc.). Under the flat extension assumption, one can define finite-dimensional commuting multiplication operators (i.e., matrices) associated to each of the variables x_i. To do this, one considers the linear maps $M_{x_i} : f \mapsto x_i f$, where $M_{x_i} : \mathbb{R}[x]_{n,d}/S \to \mathbb{R}[x]_{n,d}/S$ and S is the subspace $\{p \in \mathbb{R}[x]_{n,d} : \ell_\mu(p^2) = 0\}$. By construction, these matrices pairwise commute, and they can be simultaneously diagonalized. From their diagonal representation, one can directly read the components of the support of the measure and then obtain the corresponding weights. For a full exposition of the procedure, we refer the reader to [63, 73].

Exercise 3.160. Consider again Example 3.159.

1. Show that $(x^2 + y^2) \cdot M_t(x, y)$ is a sum of squares when $t \geq 0$.
2. Show, by producing a family of suitable probability distributions, that the inequalities $\alpha \geq 2\beta^{\frac{3}{2}}$, $\beta \geq 0$ fully characterize the projection of the set of valid moments $P^*_{n,2d}$ onto the plane (α, β).
3. Write the explicit form of the corresponding semidefinite program, and verify that it indeed gives the projection of $P^*_{n,2d}$ onto the plane (α, β).

Exercise 3.161. Show that, by allowing the degree of $p(X)$ to grow, the conditions (3.40) can approximate arbitrarily closely the set of valid moments of nonnegative measures supported on S. Notice that this statement is essentially the dual of Putinar's representation theorem (Theorem 3.138).

3.6 Further Sum of Squares Applications

In this section we present several applications from different domains of applied mathematics and engineering where sos techniques have provided new solutions and insights. In each case we present the core mathematical ideas, attempting to reduce as much as possible the use of domain-specific jargon. The main point we want to illustrate is the power and versatility of polynomial optimization and convex optimization in addressing many apparently diverse questions, using virtually the same mathematical and computational machinery. We refer the reader to the cited literature for in-depth discussions of each specific topic.

3.6.1 Copositive Matrices

A symmetric matrix $M \in \mathcal{S}^n$ is *copositive* if, for all $x \in \mathbb{R}^n$,

$$x \geq 0 \quad \Longrightarrow \quad x^T M x \geq 0.$$

Equivalently, the associated quadratic form is nonnegative on the closed nonnegative orthant. If $x^T M x$ takes only positive values on the closed orthant (except the origin), then M will be *strictly copositive*. We will denote the set of $n \times n$ copositive matrices as $\mathcal{C}_n$.

Copositive matrices are of importance in a number of applications. We briefly describe two of them.

Example 3.162. Consider the problem of obtaining a lower bound on the optimal solution of a linearly constrained quadratic optimization problem [103]:

$$f^* = \min_{Ax \geq 0,\, x^T x = 1} x^T Q x.$$

If there exists a feasible solution C to the linear matrix inequality

$$Q - A^T C A \succeq \gamma I$$

where the matrix C is copositive, then by multiplying the inequality above by x^T on the left and x on the right, it immediately follows that $f^* \geq \gamma$. Thus, having "good" convex conditions for copositivity would allow for enhanced bounds for this type of problem. ■

Example 3.163 ([35]). This is an important special case of the problem just described. It corresponds to the computation of the stability number α of a graph G (recall Section 2.2.3 in Chapter 3). From a result of Motzkin and Straus [80], it is known that $\alpha(G)$ can be obtained as

$$\frac{1}{\alpha(G)} = \min_{x_i \geq 0, \sum_i x_i = 1} x^T(A+I)x,$$

where A is the adjacency matrix of the graph G. This result implies that given a graph G with adjacency matrix A, the matrix $\alpha(G)(I+A) - ee^T$ is copositive. In [35], de Klerk and Pasechnik show how to use this result and the semidefinite approximations presented in this section to obtain guaranteed approximations to the stability number that can improve upon the bound provided by the Lovász theta function. ■

The set of copositive matrices $\mathcal{C}_n$ is a closed convex cone (in fact, it is a proper cone; see Exercise 3.167). However, in general it is very difficult to decide if a given matrix belongs to this cone. In the literature there are explicit necessary and sufficient conditions for a given matrix to be copositive, usually expressed in terms of principal minors; see, e.g., [122, 31] and the references therein. Unfortunately, it has been shown that checking copositivity of a matrix is a co-NP-complete problem [81], so this implies that in the worst case, these tests can take an exponential number of operations (unless P = NP). This motivates the need of developing efficient sufficient conditions to guarantee copositivity.

It should be clear that the situation looks very similar to the case of nonnegative polynomials studied in earlier sections. In fact, it is *exactly* the same, since, as we will see, we can identify the set of copositive matrices with a particular *section* of the cone of nonnegative polynomials. Not surprisingly, we will be able to use sos and SDP techniques to provide tractable approximations of the cone $\mathcal{C}_n$.

An apparent distinction between the copositivity question and the problems studied earlier is the presence of nonnegativity constraints on the variables x_i. Thus, to establish the links with sos techniques we will need a way of handling the nonnegativity constraints. There are different ways of doing this, but a straightforward one is to define new variables z_i and to let $x_i = z_i^2$. Then, to decide copositivity of M, we can equivalently study the global nonnegativity of the quartic form given by

$$P(\mathbf{z}) := \mathbf{x}^T M \mathbf{x} = \sum_{i,j} m_{ij} z_i^2 z_j^2.$$

It is easy to verify that M is copositive if and only if the form $P(\mathbf{z})$ is nonnegative, i.e., $P(\mathbf{z}) \geq 0$ for all $\mathbf{z} \in \mathbb{R}^n$. This shows that we can indeed identify the cone

$\mathcal{C}_n$ of copositive matrices with a particular slice of the cone of nonnegative quartic forms $P_{n,4}$.

How to produce "good" approximations to $\mathcal{C}_n$? Based on the characterization given earlier, it should be clear that an obvious sufficient condition for M to be copositive is that $P(\mathbf{z})$ be a sum of squares. Due to the special structure of the polynomial, this condition can be interpreted directly in terms of the matrix M.

Lemma 3.164. *The form $P(\mathbf{z})$ is a sum of squares if and only if M can be written as the sum of a positive semidefinite matrix and a nonnegative matrix, i.e.,*

$$M = P + N, \qquad P \succeq 0, \qquad N_{ij} \geq 0 \qquad \textit{for } i \neq j$$

(without loss of generality, we can take $N_{ii} = 0$). If this holds, then M is copositive.

The condition in Lemma 3.164 is only sufficient for copositivity. A well-known example showing this is the matrix

$$H = \begin{bmatrix} 1 & -1 & 1 & 1 & -1 \\ -1 & 1 & -1 & 1 & 1 \\ 1 & -1 & 1 & -1 & 1 \\ 1 & 1 & -1 & 1 & -1 \\ -1 & 1 & 1 & -1 & 1 \end{bmatrix}. \tag{3.41}$$

This matrix, originally introduced by A. Horn, is copositive even though *it does not* satisfy the $P + N$ condition of Lemma 3.164.

This motivates the definition of a natural hierarchy of approximations to the copositive cone [89, 35]. Consider the family of $2(r + 2)$-forms given by

$$P_r(\mathbf{z}) = \left(\sum_{i=1}^{n} z_i^2\right)^r P(\mathbf{z}), \tag{3.42}$$

and define the cones $\mathcal{K}_r = \{M \in \mathcal{S}^n : P_r(\mathbf{z}) \text{ is sos}\}$ (for simplicity, we omit the dependence on n). It is easy to see that if P_r is a sum of squares, then P_{r+1} is also a sum of squares. The converse proposition, however, does not necessarily hold; i.e., P_{r+1} could be a sum of squares even if P_r is not. Additionally, if $P_r(\mathbf{z})$ is nonnegative, then so is $P(\mathbf{z})$. Thus, by testing whether $P_r(\mathbf{z})$ is a sum of squares, we can guarantee the nonnegativity of $P(\mathbf{z})$ and, as a consequence, the copositivity of M. This yields the hierarchy of inclusions

$$\mathcal{S}^n_+ + \mathbb{R}_+^{\binom{n}{2}} \approxeq \mathcal{K}_0 \subseteq \mathcal{K}_1 \subseteq \cdots \subseteq \mathcal{K}_r \subseteq \cdots \subseteq \mathcal{C}_n, \tag{3.43}$$

where (abusing notation) the first equality expresses the statement of Lemma 3.164. The containment between these cones is in general strict. For instance, the Horn matrix presented in (3.41) is not in $\mathcal{K}_0$, but it is in $\mathcal{K}_1$; see Exercise 3.170.

Clearly, this hierarchy gives computable conditions that are at least as powerful as the $P+N$ test of Lemma 3.164. But how conservative is this procedure? Does it approximate the copositive cone $\mathcal{C}_n$ to arbitrary precision? It follows from our discussion of Pólya's theorem in Section 3.2.6 that for any strictly copositive matrix M, there is a finite r for which $M \in \mathcal{K}_r$. However, the minimum r cannot be chosen as a constant (uniformly over all strictly copositive matrices). In general, the known lower bounds for r usually involve a "condition number" for the form $P(\mathbf{z})$: the minimum r grows as the form tends to degeneracy (nontrivial solutions). This is consistent with the computational complexity results mentioned earlier: if the value of r were uniformly bounded above, then we could always produce a polynomial-time certificate for copositivity (namely, an sos decomposition of $P_r(\mathbf{z})$), contradicting NP $\neq$ co-NP.

Circulant copositive matrices. In general, particularly in high dimensions, the geometry of the copositive cone is very complicated. As such, it is often useful to consider low-dimensional sections, where we can gain some intuition and understanding. A nice case, which we analyze next, is the case of circulant (or cyclic) matrices.

An $n \times n$ matrix is *circulant* if its (i, j) entry depends only on $|i-j| \bmod n$. We denote the subspace of $n \times n$ circulant matrices by $\mathcal{O}_n$. For the case of $n = 5$, we provide below a complete characterization of the circulant copositive matrices and the associated relaxations. A general 5×5 circulant matrix has the form

$$M(a,b,c) = \begin{bmatrix} a & b & c & c & b \\ b & a & b & c & c \\ c & b & a & b & c \\ c & c & b & a & b \\ b & c & c & b & a \end{bmatrix}. \tag{3.44}$$

For circulant matrices, the second relaxation $\mathcal{K}_1$ will be enough to recognize copositivity, i.e., $\mathcal{C}_5 \cap \mathcal{O}_5 = \mathcal{K}_1 \cap \mathcal{O}_5$. Notice that if $a = 0$, then all the other elements must be nonnegative. For later reference, we define the constant

$$\xi = (1+\sqrt{5})/4 \approx 0.809.$$

Theorem 3.165. *Consider a circulant matrix $M = M(a,b,c)$ as in (3.44). Then the following hold.*

1. *The matrix M is in $\mathcal{K}_0$ if and only if*

$$a \geq 0, \qquad \xi a + b \geq 0, \qquad \xi a + c \geq 0, \qquad a + 2b + 2c \geq 0.$$

2. *The matrix M is in $\mathcal{K}_1$ if and only if*

$$a \geq 0, \qquad a + b \geq 0, \qquad a + c \geq 0, \qquad a + 2b + 2c \geq 0,$$

$$\text{if } b < 0, \text{ then } ac \geq 2b^2 - a^2, \qquad \text{if } c < 0, \text{ then } ab \geq 2c^2 - a^2.$$

3. *Furthermore, if M is copositive, then it is in $\mathcal{K}_1$.*

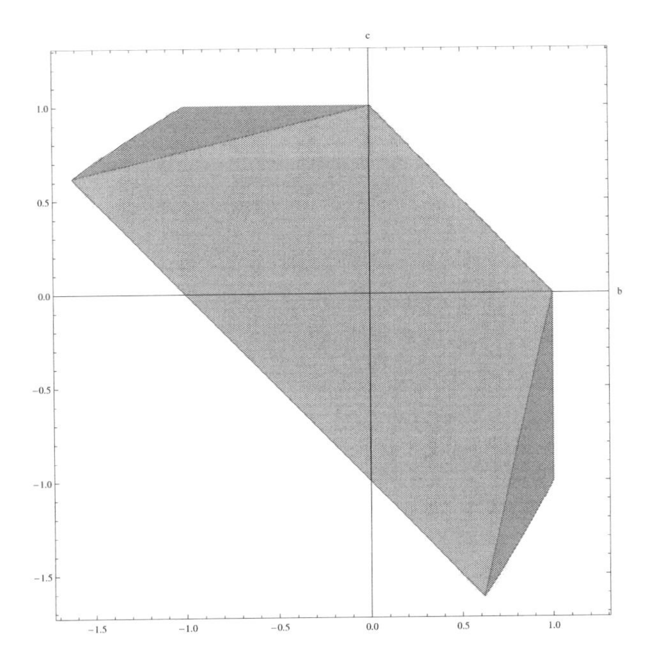

Figure 3.12. *The convex cone of* 5×5 *circulant copositive matrices* (3.44) *and its inner sos approximation* $\mathcal{K}_0$. *This plot corresponds to a compact section of the cone where* $a + b + c = 1$. *This cone is* not *polyhedral, as parts of the boundary are described by quadratic inequalities; see Theorem* 3.165.

Notice that, for this example, the set $\mathcal{K}_0 \cap \mathcal{O}_5$ is basic semialgebraic (in fact, polyhedral), but $\mathcal{K}_1 \cap \mathcal{O}_5 = \mathcal{C}_5 \cap \mathcal{O}_5$ is *not* basic semialgebraic. These sets are presented in Figure 3.12. Notice that the Horn matrix $H = M(1, -1, 1)$ presented in (3.41) corresponds to the extreme point at $b = -1$, $c = 1$.

For general matrices (even 5×5!), the situation is not nearly as nice as the slice described above may lead us to believe. In fact, the following is true.

Theorem 3.166. *Consider the set* $\mathcal{C}_5$ *of copositive* 5×5 *matrices. There is no finite value of* r *for which* $\mathcal{C}_5 = \mathcal{K}_r$.

In fact, it is not yet known whether the set of 5×5 copositive matrices $\mathcal{C}_5$ is a projected spectrahedron.

Exercise 3.167. Show that the set of copositive matrices $\mathcal{C}_n$ is a proper cone (i.e., closed, convex, pointed, and solid).

Exercise 3.168. A matrix $A \in \mathcal{S}^n$ is *completely positive* if $A = VV^T$ for some nonnegative matrix $V \in \mathbb{R}_+^{n \times k}$, i.e.,

$$A = \sum_{i=1}^{k} v_i v_i^T,$$

where $v_i \in \mathbb{R}^n_+$ are the columns of V, and hence nonnegative vectors.

1. Show that the set $\mathcal{B}_n$ of completely positive matrices is a proper cone.

2. Show that $\mathcal{B}_n$ and $\mathcal{C}_n$ are dual cones.

3. Give explicit examples of matrices in the *interior* of the cones $\mathcal{C}_n$ and $\mathcal{B}_n$.

Exercise 3.169. Prove Lemma 3.164.

Exercise 3.170. Prove that the Horn matrix (3.41) is copositive by finding an sos certificate of the nonnegativity of $x^T Hx$ on the nonnegative orthant.

Exercise 3.171. An alternative (and perhaps more natural) interpretation of the approximations (3.43) can be obtained by rewriting the sos certificates directly in terms of the variables x_i. In this case, we have that $M \in \mathcal{K}_0$ if and only if

$$x^T Mx = x^T Px + \sum_{i \neq j} n_{ij} x_i x_j,$$

where $P \succeq 0$ and $n_{ij} \geq 0$ ($P + N$ decomposition). Similarly, $M \in \mathcal{K}_1$ if and only if

$$\left(\sum_{i=1}^{n} x_i \right) (x^T M x) = \sum_{i=1}^{n} x_i \, (x^T Q_i x) + \sum_{i \neq j \neq k} \lambda_{ijk} \, x_i x_j x_k,$$

where $Q_i \succeq 0$ and $\lambda_{ijk} \geq 0$.

Prove the correctness of these statements, and explain why these representations directly prove copositivity of M. What is the relationship between these expressions and Schmüdgen-type certificates of nonnegativity?

Exercise 3.172. Explain how to use the semidefinite relaxations $\mathcal{K}_r$ of the copositive cone $\mathcal{C}_n$ to give outer approximations to the cone $\mathcal{B}_n$ of completely positive matrices. In particular, provide "explicit" SDP characterizations of the first two levels of the hierarchy.

3.6.2 Lyapunov Functions

As we have seen, reformulating conditions for a polynomial to be a sum of squares in terms of semidefinite programming is very useful, since we can use the sos property as a convenient sufficient condition for polynomial nonnegativity. In the context of dynamical systems and control theory, there has been much work applying the sos approach to the problem of finding Lyapunov functions for nonlinear systems [89, 87].

The basic framework of Lyapunov functions was introduced in Section 2.2.1 of the previous chapter for the case of linear systems. The main difference is that now we will allow our system of differential equations to be nonlinear. This approach makes possible searching over affinely parametrized polynomial or rational Lyapunov functions for systems with dynamics of the form

$$\dot{x}_i(t) = f_i(x(t)) \qquad \text{for } i = 1, \ldots, n, \tag{3.45}$$

where the functions f_i are polynomials or rational functions. Recall that, for a system to be globally asymptotically stable, it is sufficient to prove the existence of a Lyapunov function that satisfies

$$V(x) > 0, \qquad \dot{V}(x) = \left(\frac{\partial V}{\partial x}\right)^T f(x) < 0$$

for all $x \in \mathbb{R}^n \setminus \{0\}$, where without loss of generality we have assumed that the dynamical system (3.45) has an equilibrium at the origin (see, e.g., [67]).

As mentioned earlier, we will consider candidate Lyapunov functions that are polynomials (or rational functions). Since polynomial nonnegativity is computationally hard, we will instead impose that the candidate Lyapunov function $V(x)$ and its Lie derivative $\dot{V}(x)$ both satisfy the (possibly stronger) condition:[5]

$$V(x) \text{ is sos}, \qquad -\dot{V}(x) = -\left(\frac{\partial V}{\partial x}\right)^T f(x) \text{ is sos}.$$

Parametrizing a candidate Lyapunov function (e.g., by considering all possible polynomials of degree less than or equal to $2d$), the conditions given above can be expressed as sos constraints in terms of the coefficients of the Lyapunov function. Since both conditions are affine in the coefficients of $V(x)$, using the techniques described earlier in this chapter, these can be easily transformed into a standard semidefinite optimization formulation.

As an example, consider the following nonlinear dynamical system that corresponds to the Moore–Greitzer model of a jet engine with stabilizing feedback operating in the no-stall mode (see, e.g., [71]). The dynamic equations take the form

$$\begin{aligned} \dot{x} &= -y - \frac{3}{2}x^2 - \frac{1}{2}x^3, \\ \dot{y} &= 3x - y. \end{aligned} \tag{3.46}$$

Using SOSTOOLS [101], we easily find a Lyapunov function that is a polynomial of degree 6. The trajectories of the nonlinear system, and the level sets of the found Lyapunov function, are shown in Figure 3.13. Notice that, as expected, $V(x)$ monotonically decreases along trajectories, and thus all trajectories move from larger to smaller level sets of the Lyapunov function for all possible initial conditions.

Similar approaches have been developed for much more complicated problems in systems and control theory. Among others, these include finding Lyapunov functionals for nonpolynomial, time-delayed, stochastic, uncertain, or hybrid systems; see, e.g., [87, 88, 100, 44] and the references therein.

[5] The strict positivity requirement can be easily handled, either by including a strictly positive term, or by relying on the fact that SDP solvers usually compute strictly feasible solutions.

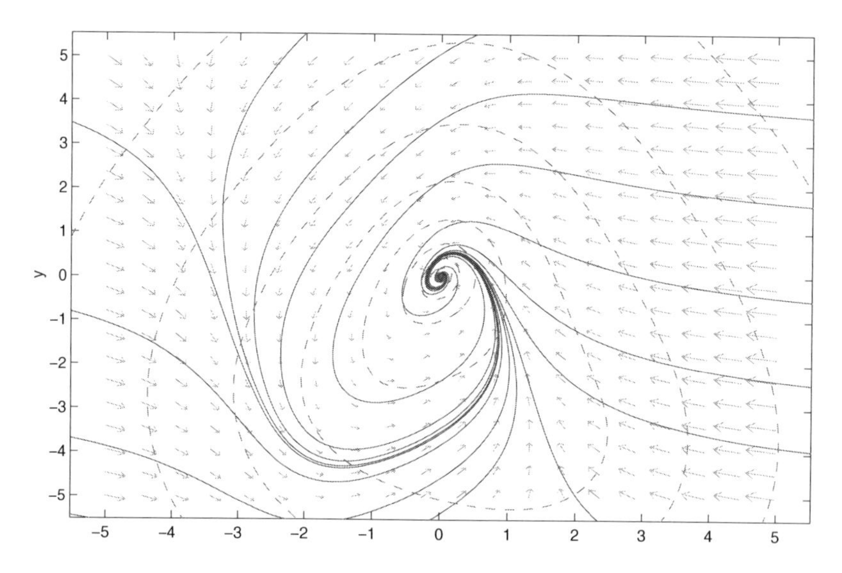

Figure 3.13. *Trajectories of the nonlinear dynamical system* (3.46) *and level sets of a Lyapunov function found using sos techniques.*

Exercise 3.173. Consider the polynomial dynamical system

$$\dot{x} = -x + (1 + x)y,$$
$$\dot{y} = -(1 + x)x.$$

Find a polynomial Lyapunov function of degree 4 that proves global asymptotic stability.

Exercise 3.174. Consider the polynomial dynamical system

$$\dot{x} = -x + xy,$$
$$\dot{y} = -y.$$

1. Show that this system is globally asymptotically stable by considering the (nonpolynomial) Lyapunov function $V(x, y) = \ln(1 + x^2) + y^2$.

2. Using SOSTOOLS (or other software) try to find a polynomial Lyapunov function. Explain your success or failure.

Remark 3.175. *The example in the previous exercise is from* [2]. *Although polynomial Lyapunov functions may fail to exist if global asymptotic stability is desired, it is known that locally exponentially stable polynomial vector fields always have polynomial Lyapunov functions on compact sets; see, e.g.,* [95]. *A suitable modification of the method explained in this section can be used to establish stability for a given compact set of initial conditions.*

3.6.3 Probability Bounds

Two of the most useful results in basic probability theory are the classic *Markov* and *Chebyshev* inequalities. Markov's inequality states that if X is a nonnegative scalar random variable, then, for all $a > 0$,

$$\mathbf{P}(X \geq a) \leq \frac{\mathbf{E}[X]}{a}. \tag{3.47}$$

Similarly, Chebyshev's inequality says that for any random variable X with mean μ and variance σ^2, we have

$$\mathbf{P}(|X - \mu| \geq a) \leq \frac{\sigma^2}{a^2}. \tag{3.48}$$

In fact, Chebyshev's inequality is just Markov's applied to the nonnegative random variable $(X - \mu)^2$.

Both inequalities can be interpreted as producing bounds on the probability of certain events, given partial information about the random variable X expressed in terms of its moments (only the first moment in Markov's, and the first and second moments for Chebyshev).

In this section we describe an important application of polynomial inequalities in probability theory, namely, a technique to generalize Chebyshev-type inequalities to the case of more general events and moment information. For simplicity, we consider only the univariate case; the extensions to the multivariate case are quite straightforward. We refer the reader to [16] for background, extensions, applications, and more details.

The statement of the problem is the following: let X be a scalar random variable with an unknown probability distribution supported on the set $\Omega \subseteq \mathbb{R}$, and for which we know its first $d+1$ moments $(\mu_0, \ldots, \mu_d)$, where $\mu_k = \mathbf{E}[X^k]$. The goal is to find bounds on the probability of an event $S \subseteq \Omega$; i.e., we want to bound $\mathbf{P}(X \in S)$. For simplicity, we assume S and Ω are given intervals.

As we shall see next, we can obtain bounds on this probability via convex optimization. Let $p(x) := \sum_{k=0}^{d} c_k x^k$ be a univariate polynomial, and consider the following optimization problem in the decision variables c_k:

$$\text{minimize } \mathbf{E}[p(X)] \quad \text{subject to} \quad \begin{cases} p(x) \geq 1 & \forall x \in S, \\ p(x) \geq 0 & \forall x \in \Omega, \end{cases} \tag{3.49}$$

or, equivalently,

$$\text{minimize } \sum_{k=0}^{d} c_k \mu_k \qquad \text{subject to} \quad \begin{cases} \sum_{k=0}^{d} c_k x^k \geq 1 & \forall x \in S, \\ \sum_{k=0}^{d} c_k x^k \geq 0 & \forall x \in \Omega. \end{cases} \tag{3.50}$$

Notice that when Ω and S are (unions of) univariate intervals, it follows from the characterizations given in Section 3.3.1 that this is an sos optimization program of the form discussed in Section 3.1.7.

We claim that any feasible solution of (3.49) gives a valid upper bound on $\mathbf{P}(X \in S)$. To see this, notice if $\mathbf{1}_S(x)$ is the indicator function of the set S (i.e., it is equal to 1 if $x \in S$ and 0 otherwise); the constraints in (3.49) imply the inequality $\mathbf{1}_S(x) \leq p(x)$ for all $x \in \Omega$. It then follows that

$$\mathbf{P}(X \in S) = \int_\Omega \mathbf{1}_S(x)\, d\mathbf{P}(x) \leq \int_\Omega p(x)\, d\mathbf{P}(x) = \mathbf{E}[p(X)].$$

In simpler terms, these bounds work by approximating (from above, in the case of upper bounds) the indicator function of the event S by a polynomial. Since we know the moments of X, we can compute in closed form the expectation of this polynomial. By optimizing over the coefficients c_k, we find the best polynomial approximation of the indicator function and thus the best upper bound provable by this method.

Essentially the same techniques apply to much more complicated situations (e.g., the multivariate case, partial moment information, martingale inequalities, etc.). For a detailed treatment, see [16, 17] and references therein.

Exercise 3.176. Show that the Markov and Chebyshev bounds can be interpreted as closed-form solutions of (3.49) for specific sets Ω and S. What are the corresponding optimal polynomials $p(x)$?

Exercise 3.177. Assume that $\Omega = [0, 5]$, $S = [4, 5]$, and the mean and variance of the random variable X are equal to 1 and 1/2, respectively. Give upper and lower bounds on $P(X \in S)$. Are these bounds tight? Can you find the worst-case distributions?

3.6.4 Quantum Separability and Entanglement

The state of a finite-dimensional quantum system can be described in terms of a positive semidefinite Hermitian matrix, called the *density matrix*. A question of interest in quantum information theory is whether a given quantum state can be explained "classically" (i.e., purely in terms of probability theory) or whether the full power of quantum mechanics is needed. In what follows, we explain the core mathematical issues behind this question; see [85] for a detailed treatment of quantum information theory. For simplicity, we consider real symmetric matrices (as opposed to complex Hermitian) and use standard mathematical notation instead of the Dirac formulation used in physics.

Consider a symmetric, positive semidefinite matrix ρ, with trace equal to one. We will refer to ρ as a *density matrix*. An important property of a bipartite quantum state ρ is whether or not it is *separable*, which means that it can be written as a convex combination of tensor products of rank one matrices, i.e.,

$$\rho = \sum_i p_i\, (x_i x_i^T) \otimes (y_i y_i^T), \qquad p_i \geq 0, \quad \sum_i p_i = 1.$$

Here $x_i \in \mathbb{R}^{n_1}$, $y_i \in \mathbb{R}^{n_2}$, and $\rho \in \mathcal{S}_+^{n_1 n_2}$. By construction, the set of separable states is a convex set. If the state is not separable, then it is said to be *entangled*.

The physical interpretation of a separable state corresponds to a probabilistic superposition (with probabilities given by the p_i), where one subsystem is in state x_i and the other subsystem is in state y_i. If no such decomposition is possible, then it is not possible to think of the two subsystems as being "independent" (even though they may be physically separated), and thus actions/measurements on one subsystem may affect the other (i.e., they are "entangled").

The *quantum separability* or *quantum entanglement* question is the following: Given the density matrix ρ of a quantum state, how do we decide whether ρ is entangled or not? If it entangled (or separable), how can we certify this property? It has been shown by Gurvits that in general this is an NP-hard question [58].

As we shall see, quantum entanglement is intimately related to polynomial nonnegativity. A natural mathematical object to study in this context is the set of *positive maps*. These are the linear operators $\Lambda : \mathcal{S}^{n_1} \to \mathcal{S}^{n_2}$ that satisfy $X \succeq 0 \Rightarrow \Lambda(X) \succeq 0$; i.e., they map positive semidefinite matrices into positive semidefinite matrices. Notice that to any such Λ, we can associate a unique "observable" $W_\Lambda \in \mathcal{S}^{n_1 n_2}$ that satisfies $y^T \Lambda(xx^T) y = (x \otimes y)^T W_\Lambda (x \otimes y)$. Furthermore, if Λ is a positive map, then the pairing between the observable W_Λ and any separable state ρ will always give a nonnegative number, since

$$\begin{aligned}\langle W_\Lambda, \rho\rangle &= \operatorname{Tr} W_\Lambda \cdot \left(\sum_i p_i\, (x_i x_i^T) \otimes (y_i y_i^T) \right) = \sum_i p_i \operatorname{Tr} W_\Lambda \cdot (x_i \otimes y_i) \cdot (x_i \otimes y_i)^T \\ &= \sum_i p_i\, (x_i \otimes y_i)^T W_\Lambda (x_i \otimes y_i) = \sum_i p_i\, y_i^T \Lambda(x_i x_i^T) y_i \geq 0.\end{aligned}$$

In other words, every positive map yields a *separating hyperplane* for the convex set of separable states. It can further be shown that every valid inequality corresponds to a positive map, so this yields, in fact, a complete characterization (and thus, the sets of separable states and positive maps are dual to each other). For this reason, the observables W_Λ associated to positive maps Λ are called *entanglement witnesses.*

The set of positive maps (and thus, entanglement witnesses) can be exactly characterized in terms of a multivariate polynomial nonnegativity, since a linear map $\Lambda : \mathcal{S}^{n_1} \to \mathcal{S}^{n_2}$ is positive if and only if the biquadratic form in $n_1 + n_2$ variables $p(x, y) = y^T \Lambda(xx^T) y$ is nonnegative for all x, y (why?). Replacing nonnegativity with sos based conditions, we can obtain a family of efficiently computable criteria that certify entanglement.

Concretely, given a state ρ for which we want to determine whether it is entangled, the first such test corresponds to the optimization problem of finding an entanglement witness W_Λ (or linear map Λ) such that

$$\langle W_\Lambda, \rho\rangle < 0, \qquad y^T \Lambda(xx^T) y \text{ is sos.} \tag{3.51}$$

Interestingly, this corresponds to the well-known "positive partial trace" (PPT) separability criterion. The advantage of sos techniques is that stronger tests can be naturally derived by considering higher-order sos conditions. In particular, we have the parametrized family of tests

$$\langle W_\Lambda, \rho\rangle < 0, \qquad (x^T x)^k \cdot (y^T \Lambda(xx^T) y^T) \text{ is sos} \tag{3.52}$$

for $k \geq 0$ that obviously generalize (3.51) (which corresponds to the case $k = 0$). It should be clear that these sos programs can be numerically solved using semidefinite programming. It can also be shown [40, 41] that this hierarchy is complete in the sense that every entangled state is eventually certified by some value of k.

For more background and details about quantum entanglement and the separability problem, see [40, 41] and the references therein. It has been recently shown [22] that the sos based algorithm described above can be used to provide a quasipolynomial time algorithm for the quantum separability problem.

Exercise 3.178. Consider linear maps between symmetric matrices of the form $\Lambda : \mathcal{S}^{n_1} \to \mathcal{S}^{n_2}$.

1. Show that any linear map of the form $A \mapsto \sum_i P_i^T A P_i$, where $P_i \in \mathbb{R}^{n_1 \times n_2}$, is positive. These maps are known as *decomposable* maps.

2. Consider the polynomial defined by $p(x, y) := y^T \Lambda(xx^T) y$. Show that Λ is a positive map if and only if $p(x, y)$ is nonnegative and that Λ is a decomposable map if and only if $p(x, y)$ is a sum of squares.

3. Show that the linear map $C : \mathcal{S}^3 \to \mathcal{S}^3$ (due to M.-D. Choi) given by

$$C : A \mapsto \begin{bmatrix} 2a_{11} + a_{22} & 0 & 0 \\ 0 & 2a_{22} + a_{33} & 0 \\ 0 & 0 & 2a_{33} + a_{11} \end{bmatrix} - A$$

   is a positive map but is not decomposable.

4. Explain the relationship between this linear map and the Choi matrix discussed earlier in (3.20).

3.6.5 Geometric Theorem Proving

Many geometric statements can be reinterpreted, after a suitable coordinatization, in terms of algebraic inequalities. This opens up the possibility of proving theorems about geometric objects by characterizing the desired properties in terms of algebraic inequalities and then proving these inequalities through sos certificates. We give two concrete examples in what follows. The main value of these simple examples is to illustrate how the process of proving algebraic or geometric inequalities can be made fully algorithmic and how the power of convex optimization can be brought to these questions.

Schur's inequality. This is a classical inequality due to Schur that states that for nonnegative variables x, y, z, we have

$$S(x, y, z) := x^k(x - y)(x - z) + y^k(y - z)(y - x) + z^k(z - x)(z - y) \geq 0, \quad (3.53)$$

where $k \geq 0$ is an integer.

We give next a simple sos proof of this inequality for the case $k = 1$, easily obtainable via semidefinite programming. Define

$$S_1 = \frac{1}{2}\begin{bmatrix} x^2+yz \\ y^2+xz \\ z^2+xy \end{bmatrix}^T \begin{bmatrix} 2 & -1 & -1 \\ -1 & 2 & -1 \\ -1 & -1 & 2 \end{bmatrix} \begin{bmatrix} x^2+yz \\ y^2+xz \\ z^2+xy \end{bmatrix},$$
$$S_2 = yz(y-z)^2 + xz(x-z)^2 + xy(x-y)^2.$$

Since the matrix in the expression above is positive semidefinite, it is clear that both S_1 and S_2 are nonnegative when x, y, z are nonnegative. We have then the easy-to-verify identity

$$(x+y+z)\cdot S(x,y,z) = S_1 + S_2$$

that clearly proves (3.53). Schur's inequality is closely related to the *Robinson form*, one of the first explicit examples of non-sos positive definite forms; see [29] and [106] for background and more details.

Ono's inequality. We present next an sos proof of a geometric inequality due to Ono. This example originally appeared in [117] as a benchmark problem for geometric theorem proving.

Consider a triangle with sides of length a, b, c, and denote its area by K. In 1914, T. Ono [86, 79] conjectured that the inequality

$$(4K)^6 \geq 27\cdot(a^2+b^2-c^2)^2\cdot(b^2+c^2-a^2)^2\cdot(c^2+a^2-b^2)^2 \tag{3.54}$$

holds for all triangles. The statement was subsequently shown to be false in general [11] but proved to hold whenever the triangle in question is *acute* (all angles are less than or equal to $\pi/2$) [12]. Using sos techniques, we will obtain a very concise proof.

For this, we can express the premise that the triangle be acute as the three polynomial inequalities

$$\begin{aligned} t_1 &:= a^2+b^2-c^2 \geq 0, \\ t_2 &:= b^2+c^2-a^2 \geq 0, \\ t_3 &:= c^2+a^2-b^2 \geq 0. \end{aligned} \tag{3.55}$$

It is well known (Heron's formula) that we can rewrite the square of the area K as a polynomial in a, b, c:

$$K^2 = s(s-a)(s-b)(s-c), \qquad s = \frac{a+b+c}{2}.$$

The question, therefore, reduces to verifying that (3.54) holds whenever the inequalities (3.55) are satisfied. A simple proof of Ono's inequality can then be found using the Positivstellensatz and sos methods: define the sos polynomial

$$s(x,y,z) := (x^4+x^2y^2-2y^4-2x^2z^2+y^2z^2+z^4)^2+15\,(x-z)^2(x+z)^2(z^2+x^2-y^2)^2.$$

We have then

$$(4K)^6 - 27 \cdot t_1^2 \cdot t_2^2 \cdot t_3^2 = s(a,b,c) \cdot t_1 \cdot t_2 + s(c,a,b) \cdot t_1 \cdot t_3 + s(b,c,a) \cdot t_2 \cdot t_3, \quad (3.56)$$

therefore proving the inequality.

Another, more complicated, application of these techniques is given in [93]. In that paper, the subadditivity of a geometric quantity for triangles, expressible in terms of its side lengths and an angle, is proved via sos methods. The problem can be reduced to proving the nonnegativity of the polynomial

$$\begin{aligned}
&\alpha^2\beta^2(\alpha-\beta)^2 + \beta^2(1-\alpha\beta)(1+\alpha\beta-2\alpha^2)\gamma^2 \\
&\quad + \alpha^2(1-\alpha\beta)(1+\alpha\beta-2\beta^2)\delta^2 - \alpha\beta(2+\alpha\beta^3-4\alpha\beta+\beta\alpha^3)\gamma\delta \\
&\quad + \beta(1-\alpha\beta)(2\alpha-\beta-\alpha\beta^2)\gamma^3\delta - (\alpha^2+\beta^2+2\alpha^3\beta^3-4\alpha^2\beta^2)\gamma^2\delta^2 \\
&\quad + \alpha(1-\alpha\beta)(2\beta-\alpha-\alpha^2\beta)\gamma\delta^3 + (\alpha-\beta)^2\gamma^3\delta^3 \quad (3.57)
\end{aligned}$$

on the unit box $0 \le \alpha, \beta, \gamma, \delta \le 1$. As shown in [93], it is possible to obtain a concise certificate of its nonnegativity using sos methods.

As generalizations of the well-understood methods of semidefinite programming, sos techniques have proved remarkably powerful in the treatment of geometric problems. A nice example of this is the recent work of Bachoc and Vallentin [10], where the authors have developed improved bounds on kissing numbers. This is the classical question of how many identical n-dimensional nonoverlapping spheres can be simultaneously tangent to a given central sphere of the same radius. It is easy to see that in the plane, this number is six (six disks, surrounding a central disk, in a hexagonal pattern), but determining this number in higher dimensions is a very difficult problem. By combining techniques from harmonic analysis and semidefinite programming (related to the symmetry reduction techniques discussed in Section 3.3.6), the authors of [10] have extended the classical association scheme approach to spherical codes (see, e.g., [37] and the references therein) to obtain the best available upper bounds on kissing numbers.

Sum of squares techniques can also be nicely interfaced with other, more general, methods for automated theorem proving. We refer the reader to the work of Harrison [60] for a discussion of these ideas, some of which have been implemented in the theorem prover HOL Light [61].

An interesting open research question is whether these algebraic proofs or sos certificates can be given "natural" geometric interpretations. For instance, as a concrete question, is there any intrinsic geometric meaning of the polynomial identity proof given in (3.56)?

Exercise 3.179 (Weitzenböck's inequality). Consider a triangle with side lengths equal to a, b, c and area equal to K. Give an sos proof of the inequality

$$a^2 + b^2 + c^2 \ge (4\sqrt{3})K,$$

and show that $4\sqrt{3}$ is the best possible constant.

Exercise 3.180 (Pedoe's inequality). Consider two triangles with side lengths equal to (a_1, b_1, c_1) and (a_2, b_2, c_2) and areas K_1, K_2, respectively. Give an sos proof of the inequality

$$a_1^2(b_2^2 + c_2^2 - a_2^2) + b_1^2(c_2^2 + a_2^2 - b_2^2) + c_1^2(a_2^2 + b_2^2 - c_2^2) \geq 16K_1K_2.$$

Is 16 the best possible constant? What happens if one of the triangles is equilateral?

Exercise 3.181. Prove that the polynomial (3.57) is nonnegative when the variables satisfy $0 \leq \alpha, \beta, \gamma, \delta \leq 1$. Find an sos certificate of this fact.

3.6.6 Polynomial Games

The mathematical theory of games was developed to model and analyze strategic interactions among multiple decision makers with possibly conflicting objectives. Game theory has been successfully used in many domains, including economics, engineering, and biology. Standard modern references include [46, 82]. In this section we present an application of sos methods in game theory, initially described in [92].

We consider two-player zero-sum games, where the payoffs are polynomial functions. This class of polynomial games was originally introduced and studied by Dresher, Karlin, and Shapley in 1950 [42]. In the basic set-up there are two players (which we will denote as Player 1 and Player 2), which simultaneously and independently choose actions parametrized by real numbers x, y, respectively, in the interval $[-1, 1]$. The payoff associated with these choices is given by a polynomial function

$$P(x, y) = \sum_{i=0}^{n} \sum_{j=0}^{m} p_{ij} x^i y^j \qquad (3.58)$$

that assigns payments from Player 2 to Player 1. Thus, Player 1 wants to choose his strategy x to maximize $P(x, y)$, while Player 2 tries to make this expression as small as possible. Players are allowed, and often it is in their interest, to choose their actions randomly according to specific probability distributions; these are called *mixed strategies* (the game of rock-paper-scissors is a simple example of this situation).

The solution concept of interest is called *Nash equilibrium*. This corresponds to a choice of strategies for both players, for which there is no incentive for a player to deviate, assuming the other player keeps their strategy fixed. It is well known that for zero-sum games, this notion reduces to the simpler *minimax* or *saddle-point* equilibrium; see (3.60).

Example 3.182. Consider a polynomial game on $[-1, 1] \times [-1, 1]$, with payoff function given by $P(x, y) = (x - y)^2$. Since Player 2 wants to minimize her payoffs, she should try to "guess" the number chosen by Player 1. Conversely, the first player should try to make his number as difficult to guess as possible (in the sense defined by $P(x, y)$). It is easy to see in this case that the optimal strategy for Player 1 is to randomize between $x = -1$ or $x = 1$ with equal probability, while the optimal strategy for Player 2 is to always choose $y = 0$. Assuming the other player keeps their strategy fixed, no player has incentive to deviate from these strategies, and thus this yields an equilibrium, with the corresponding value of the game being equal to 1. ∎

The question of interest is the following: given a game described by its payoff function $P(x, y)$, how do we efficiently compute its equilibrium solution, i.e., the optimal strategies both players should use?

Recall that players can randomize over their choices, so their strategies will be described by probability measures μ and ν, respectively, supported on $[-1, 1]$. When considering mixed strategies, and similarly to the finite case, we need to consider the expressions

$$\max_{\nu} \min_{\mu} E_{\nu\times\mu}[P(x,y)] \quad \text{and} \quad \min_{\mu} \max_{\nu} E_{\nu\times\mu}[P(x,y)],$$

where $E_{\nu\times\mu}[\cdot]$ denotes the expectation under the product measure. We can rewrite these as bilinear expressions

$$\max_{\nu_i} \min_{\mu_j} \sum_{i=0}^{n}\sum_{j=0}^{m} p_{ij}\nu_i\mu_j, \qquad \min_{\mu_j} \max_{\nu_i} \sum_{i=0}^{n}\sum_{j=0}^{m} p_{ij}\nu_i\mu_j, \tag{3.59}$$

where ν_i, μ_j are the *moments* of the measures ν, μ, i.e.,

$$\nu_i := \int_{-1}^{1} x^i \, d\nu, \qquad \mu_j := \int_{-1}^{1} y^j \, d\mu.$$

Recall from Section 3.5.4 that the moment spaces (i.e., the image of the probability measures under the moment map given above) are compact convex sets in $\mathbb{R}^{n+1}$ and $\mathbb{R}^{m+1}$. Since the objective function in the problems (3.59) is bilinear, and the feasible sets are convex and compact, the minimax theorem (Theorem A.6 in Appendix A) can be used to show that these two quantities are equal. As a consequence, there exist measures $\nu^\star, \mu^\star$ that satisfy the saddle-point condition:

$$\sum_{i=0}^{n}\sum_{j=0}^{m} p_{ij}\nu_i\mu_j^\star \le \sum_{i=0}^{n}\sum_{j=0}^{m} p_{ij}\nu_i^\star\mu_j^\star \le \sum_{i=0}^{n}\sum_{j=0}^{m} p_{ij}\nu_i^\star\mu_j. \tag{3.60}$$

The key fact here is that, due to the separable structure of the payoffs, the optimal strategies can be characterized only in terms of their first m (or n) moments. Higher moments are irrelevant, at least in terms of the payoffs of the players.

From the previous discussion, we have the following result, essentially contained in [42].

Theorem 3.183. *Consider the two-player zero-sum game on* $[-1, 1] \times [-1, 1]$, *with payoffs given by* (3.58). *Then, the value of the game is well defined, and there exist optimal mixed strategies* $\nu^\star, \mu^\star$ *satisfying a saddle-point condition. Furthermore, without loss of generality, the optimal measures can be taken to be discrete, with at most* $\min(n, m) + 1$ *atoms.*

The derivation and computation of the mixed strategies and the value of the game can be done as follows. We first characterize "security strategies" that provide a minimum guaranteed payoff. We can then invoke convex duality to prove that this actually yields the unique value of the game. Proceeding along these lines, by analogy to the finite case, a security strategy of Player 2 can be computed by solving

$$\underset{\gamma,\mu}{\text{minimize}}\ \gamma \quad \text{s.t.} \quad \begin{cases} E_\mu[P(x,y)] & \leq & \gamma \quad \forall x \in [-1,1], \\ \int_{-1}^{1} d\mu(y) & = & 1. \end{cases} \tag{3.61}$$

Indeed, if Player 2 plays the mixed strategy μ obtained from the solution of this problem, the best that Player 1 can do is to choose a value of x that maximizes $E_\mu[P(x,y)]$, thus limiting his gain (and Player 2's loss) to γ.

Since $P(x,y)$ is a polynomial, this expectation can be equivalently written in terms of the first n moments of the measure μ, i.e.,

$$E_\mu[P(x,y)] = \int_{-1}^{1} P(x,y)d\mu(y) = \sum_{i=0}^{n}\sum_{j=0}^{m} p_{ij}\mu_j x^i.$$

Notice that this is a univariate polynomial in the action x of Player 1, with coefficients that depend *affinely* on the moments μ_j of the mixed strategy of Player 2.

Consider now the problem (3.61), but instead of writing it in terms of the decision variable μ (which is a probability measure), let us use instead the moments $\{\mu_j\}_{j=0}^m$. The problem is then reduced to the minimization of the safety level γ, subject to the following conditions:

- The univariate polynomial $\gamma - \sum_{i=0}^n \sum_{j=0}^m p_{ij}\mu_j x^i$ is nonnegative on $[-1,1]$.
- The sequence $\{\mu_j\}_{j=0}^m$ is a valid moment sequence for a probability measure supported in $[-1,1]$.

We can rewrite this in a more compact form, as the optimization problem

$$\text{minimize} \ \ \gamma \quad \text{s.t.} \quad \begin{cases} \gamma - \sum_{i=0}^n \sum_{j=0}^m p_{ij}x^i\mu_j & \in & \mathcal{P}_{1,n}, \\ \mu & \in & \mathcal{M}_m, \end{cases} \tag{3.62}$$

where $\mathcal{P}_{1,n}$ is the set of univariate polynomials of degree n nonnegative in $[-1,1]$, and $\mathcal{M}_m$ is the set of $m+1$ first moments of a probability measure with support on the same interval.

By the characterizations provided in earlier sections, it is clear that both of these conditions can be rewritten in terms of semidefinite programming and thus efficiently solved. Furthermore, using the procedure described in Section 3.5.5, the corresponding optimal mixed strategies can be obtained.

Example 3.184. Consider the guessing game discussed in Example 3.182. In this case, the decision variables (μ_0, μ_1, μ_2) are the moments of the mixed strategy of

Player 1. To compute the optimal strategies, we must then solve (3.62), i.e.,

$$\text{minimize} \quad \gamma \quad \text{s.t.} \quad \begin{cases} \gamma - (x^2\mu_0 - 2x\mu_1 + \mu_2) & = & s_0(x) + s_1 \cdot (1 - x^2), \\ \begin{bmatrix} \mu_0 & \mu_1 \\ \mu_1 & \mu_2 \end{bmatrix} & \succeq & 0, \\ \mu_0 - \mu_2 & \geq & 0, \\ \mu_0 & = & 1, \end{cases}$$

where we have used the sos/semidefinite characterizations of univariate polynomials (Section 3.3.1) and moments constraints (Section 3.5.3) for the interval $[-1, 1]$. The optimal solution of this problem is $\gamma = 1$, $(\mu_0, \mu_1, \mu_2) = (1, 0, 1)$, $s_0(x) = 0$, and $s_1 = 1$. From this, the optimal strategies $\delta(x)$ for Player 1 and $\frac{1}{2}\delta(x-1) + \frac{1}{2}\delta(x+1)$ for Player 2 directly follow. ∎

Exercise 3.185. Consider a two-player game on $[-1, 1] \times [-1, 1]$ with payoff function given by

$$P(x, y) = 5xy - 2x^2 - 2xy^2 - y.$$

Notice this function is neither convex nor concave.

Formulate and solve the corresponding optimization problem to find the optimal solution of this game. Verify that the optimal strategies correspond to Player 1 always choosing $x = 0.2$, and Player 2 choosing $y = 1$ with probability 0.78, and $y = -1$ with probability 0.22.

3.7 Software Implementations

Despite the many advances in theoretical and modeling aspects of SDP and sos methods, much of their impact in applications has undoubtedly been a direct consequence of the efforts of many researchers in producing and making available good quality software implementations. In this section we give pointers to and discuss briefly some of the current computational tools for effectively formulating and solving SDP and sos programs.

Most SDP solvers (e.g., those described in Section 2.3.2) usually take as input either text files containing a problem description or directly the matrices (A_i, b, C) corresponding to the standard primal/dual formulation. This is often inconvenient at the initial modeling and solution stages. A more flexible approach is to formulate the problem using a more natural description, closer to its mathematical formulation, that can later be automatically translated to fit the requirements of each solver. For generic optimization problems, this has indeed been the approach of much of the operations research community, which has developed some well-known standard file formats, such as MPS, or optimization modeling languages like AMPL and GAMS. An important remark to keep in mind, much more critical in the SDP case than for linear optimization, is the extent to which the problem structure can be signaled to the solver.

For sos programs, as we have seen, the conversion process to an SDP formulation is algorithmic, and there are parsers that partially or fully automate this conversion task and can be used from within a problem-solving environment such as MATLAB. The software SOSTOOLS [101] is a free, third-party MATLAB toolbox for formulating and solving general sos programs. The related software Gloptipoly [62] is oriented toward global optimization problems and the associated moment problems. In their current version, both use the SDP solver SeDuMi [118] for numerical computations. Other possibilities include YALMIP [74], a very complete modeling language for convex and nonconvex optimization that includes several sos/moments features, as well as the more specialized toolbox SPOT [78], oriented toward problems in systems and control theory. An interesting new addition to this area is the MATLAB toolbox NCSOStools [25] that specializes in sums of squares in noncommuting variables, a topic that will be discussed extensively in Chapter 8. Any of these parsers can make formulating and solving sos programs a much simpler and more enjoyable task than manual, error-prone methods.

Bibliography

[1] M. Abramowitz and I.A. Stegun, eds. *Handbook of Mathematical Functions*. Dover, New York, 1964.

[2] A.A. Ahmadi, M. Krstić, and P. A. Parrilo. A globally asymptotically stable polynomial vector field with no polynomial Lyapunov function. In *Proceedings of the 50th IEEE Conference on Decision and Control*, IEEE, Washington, DC, 2011.

[3] A.A. Ahmadi, A. Olshevsky, P. A. Parrilo, and J.N. Tsitsiklis. NP-hardness of deciding convexity of quartic polynomials and related problems. *Mathematical Programming*, 1–24, 2011.

[4] A.A. Ahmadi and P. A. Parrilo. A complete characterization of the gap between convexity and sos-convexity. *Mathematical Programming*, to appear. arXiv:1111.4587, 2011.

[5] A.A. Ahmadi and P. A. Parrilo. A convex polynomial that is not sos-convex. *Mathematical Programming*, 135:275–292, 2012.

[6] N. I. Akhiezer. *The Classical Moment Problem*. Hafner Publishing Company, New York, 1965.

[7] C. Andradas. Characterization and description of basic semialgebraic sets. In *Algorithmic and Quantitative Real Algebraic Geometry (Piscataway, NJ, 2001)*, DIMACS Ser. Discrete Math. Theoret. Comput. Sci. 60, Amer. Math. Soc., Providence, RI, 2003, pp. 1–12.

[8] C. Andradas and J.M. Ruiz. Ubiquity of Łojasiewicz's example of a nonbasic semialgebraic set. *The Michigan Mathematical Journal*, 41:465–472, 1994.

[9] E. M. Aylward, S. M. Itani, and P. A. Parrilo. Explicit SOS decomposition of univariate polynomial matrices and the Kalman-Yakubovich-Popov lemma.

In *Proceedings of the 46th IEEE Conference on Decision and Control*, IEEE, Washington, DC, 2007.

[10] C. Bachoc and F. Vallentin. New upper bounds for kissing numbers from semidefinite programming. *J. Amer. Math. Soc*, 21:909–924, 2008.

[11] F. Balitrand. Problem 4417. *Intermed. Math.*, 22:66, 1915.

[12] F. Balitrand. Problem 4417. *Intermed. Math.*, 23:86–87, 1916.

[13] S. Basu, R. Pollack, and M.-F. Roy. *Algorithms in Real Algebraic Geometry*, Algorithms and Computation in Mathematics 10, Springer-Verlag, Berlin, 2003.

[14] A. Ben-Tal, L. El Ghaoui, and A.S. Nemirovski. *Robust Optimization*. Princeton University Press, Princeton, NJ, 2009.

[15] A. Ben-Tal and A. Nemirovski. *Lectures on Modern Convex Optimization*. MPS/SIAM Series on Optimization 2. SIAM, Philadelphia, 2001.

[16] D. Bertsimas and I. Popescu. Optimal inequalities in probability theory: A convex optimization approach. *SIAM J. Optim.*, 15:780–804, 2005.

[17] D. Bertsimas and J. Sethuraman. Moment problems and semidefinite optimization. In *Handbook of Semidefinite Programming*, H. Wolkowicz, R. Saigal, and L. Vandenberghe, eds., Springer, New York, 2000, pp. 469–509.

[18] S.P. Bhattacharyya, H. Chapellat, and L.H. Keel. *Robust Control: The Parametric Approach*. Prentice-Hall, Englewood Cliffs, NJ, 1995.

[19] J. Bochnak, M. Coste, and M-F. Roy. *Real Algebraic Geometry*. Springer, New York, 1998.

[20] J.M. Borwein and H. Wolkowicz. Facial reduction for a cone-convex programming problem. *J. Austral. Math. Soc. Ser. A*, 30:369–380, 1980.

[21] S. Boyd, L. El Ghaoui, E. Feron, and V. Balakrishnan. *Linear Matrix Inequalities in System and Control Theory*, Stud. Appl. Math. 15. SIAM, Philadelphia, 1994.

[22] F.G.S.L. Brandão, M. Christandl, and J. Yard. A quasipolynomial-time algorithm for the quantum separability problem. In *Proceedings of the 43rd Annual ACM Symposium on Theory of Computing*, ACM, New York, 2011, pp. 343–352.

[23] C.W. Brown. *QEPCAD – Quantifier Elimination by Partial Cylindrical Algebraic Decomposition*, 2003. Available from www.cs.usna.edu/~qepcad/B/QEPCAD.html.

[24] S. R. Buss and T. Pitassi. Good degree bounds on Nullstellensatz refutations of the induction principle. *J. Comp. System Sci.*, 57:162–171, 1998.

[25] K. Cafuta, I. Klep, and J. Povh. *NCSOStools: Computer Algebra System for Symbolic and Numerical Computation with Noncommutative Polynomials*, 2011. Available at ncsostools.fis.unm.si.

[26] B. F. Caviness and J. R. Johnson, eds. *Quantifier Elimination and Cylindrical Algebraic Decomposition*, Texts and Monographs in Symbolic Computation, Springer-Verlag, Vienna, 1998.

[27] M. D. Choi. Positive semidefinite biquadratic forms. *Linear Algebra Appl.*, 12:95–100, 1975.

[28] M. D. Choi, T. Y. Lam, and B. Reznick. Real zeros of positive semidefinite forms. I. *Math. Z.*, 171:1–26, 1980.

[29] M. D. Choi, T. Y. Lam, and B. Reznick. Positive sextics and Schur's inequalities. *J. Algebra*, 141:36–77, 1991.

[30] M. D. Choi, T. Y. Lam, and B. Reznick. Sums of squares of real polynomials. *Proceedings of Symposia in Pure Mathematics*, 58:103–126, 1995.

[31] R. W. Cottle, J. S. Pang, and R. E. Stone. *The Linear Complementarity Problem*. Academic Press, New York, 1992.

[32] D. A. Cox, J. B. Little, and D. O'Shea. *Ideals, Varieties, and Algorithms: An Introduction to Computational Algebraic Geometry and Commutative Algebra*. Springer, New York, 1997.

[33] D. A. Cox, J. B. Little, and D. O'Shea. *Using Algebraic Geometry*, Grad. Texts in Math. 185. Springer-Verlag, New York, 1998.

[34] R. E. Curto and L. A. Fialkow. *Solution of the Truncated Complex Moment Problem for Flat Data*. Mem. Amer. Math. Soc. 568. AMS, Providence, RI, 1996.

[35] E. de Klerk and D.V. Pasechnik. Approximating the stability number of a graph via copositive programming. *SIAM J. Optim.*, 12:875–892, 2002.

[36] J. A. De Loera, J. Lee, S. Margulies, and S. Onn. Expressing combinatorial problems by systems of polynomial equations and Hilbert's Nullstellensatz. *Combin., Probab. Comput.*, 18:551, 2009.

[37] P. Delsarte and V. I. Levenshtein. Association schemes and coding theory. *IEEE Trans. Inform. Theory*, 44:2477–2504, 1998.

[38] H. Derksen and G. Kemper. *Computational Invariant Theory*, Encyclopaedia Math. Sci. 130. Springer, Berlin, 2002.

[39] L. Devroye. *Nonuniform Random Variate Generation*. Springer-Verlag, New York, 1986.

[40] A. C. Doherty, P. A. Parrilo, and F. M. Spedalieri. Distinguishing separable and entangled states. *Phys. Rev. Lett.*, 88:187904, 2002.

[41] A. C. Doherty, P. A. Parrilo, and F. M. Spedalieri. Complete family of separability criteria. *Phys. Rev. A*, 69:022308, 2004.

[42] M. Dresher, S. Karlin, and L. S. Shapley. Polynomial games. In *Contributions to the Theory of Games*, Ann. Math. Stud. 24, Princeton University Press, Princeton, NJ, 1950, pp. 161–180.

[43] G. E. Dullerud and F. Paganini. *A Course in Robust Control Theory: A Convex Approach.* Springer-Verlag, New York, 1999.

[44] C. Ebenbauer and F. Allgöwer. Analysis and design of polynomial control systems using dissipation inequalities and sum of squares. *Comput. Chem. Engrg.*, 30:1590–1602, 2006.

[45] A. Fässler and E. Stiefel. *Group Theoretical Methods and Their Applications.* Birkhäuser, Basel, 1992.

[46] D. Fudenberg and J. Tirole. *Game Theory.* MIT Press, Cambridge, MA, 1991.

[47] M. R. Garey and D. S. Johnson. *Computers and Intractability: A Guide to the Theory of NP-Completeness.* W. H. Freeman and Company, New York, 1979.

[48] K. Gatermann and P. A. Parrilo. Symmetry groups, semidefinite programs, and sums of squares. *J. Pure Appl. Algebra*, 192:95–128, 2004.

[49] W. Gautschi. *Orthogonal Polynomials: Computation and Approximation.* Oxford University Press, Oxford, UK, 2004.

[50] Y. Genin, Y. Hachez, Yu. Nesterov, and P. Van Dooren. Optimization problems over positive pseudopolynomial matrices. *SIAM J. Matrix Anal. Appl.*, 25:57–79, 2003.

[51] M. Giesbrecht, G. Labahn, and W. Lee. Symbolic–numeric sparse interpolation of multivariate polynomials. *J. Symbol. Comput.*, 44:943–959, 2009.

[52] G. H. Golub and G. Meurant. *Matrices, Moments and Quadrature with Applications.* Princeton University Press, Princeton, NJ, 2009.

[53] M. Golubitsky, I. Stewart, and D. G. Schaeffer. *Singularities and Groups in Bifurcation Theory II*, Appl. Math. Sci. 69. Springer, New York, 1988.

[54] D. R. Grayson and M. E. Stillman. Macaulay 2: A Software System for Research in Algebraic Geometry. Available at http://www.math.uiuc.edu/Macaulay2.

[55] D. Grigoriev and N. Vorobjov. Complexity of null- and Positivstellensatz proofs. *Ann. Pure Appl. Logic*, 113:153–160, 2002.

[56] M. Grötschel, L. Lovász, and A. Schrijver. *Geometric Algorithms and Combinatorial Optimization*, 2nd ed., *Algorithms and Combinatorics* 2. Springer-Verlag, Berlin, 1993.

[57] J. Guckenheimer, M. Myers, and B. Sturmfels. Computing Hopf bifurcations I. *SIAM J. Numer. Anal.*, 34:1–21, 1997.

[58] L. Gurvits. Classical deterministic complexity of Edmonds' problem and quantum entanglement. In *STOC '03: Proceedings of the* 35*th Annual ACM Symposium on Theory of Computing*, ACM, New York, 2003.

[59] G. H. Hardy, J. E. Littlewood, and G. Pólya. *Inequalities*. Cambridge University Press, Cambridge, UK, 1967.

[60] J. Harrison. Verifying nonlinear real formulas via sums of squares. In *Proceedings of the* 20*th International Conference on Theorem Proving in Higher Order Logics*, Springer-Verlag, New York, 2007, pp. 102–118.

[61] J. Harrison. *The HOL Light Theorem Prover*, 2011. Available at www.cl.cam.ac.uk/~jrh13/hol-light.

[62] D. Henrion and J.-B. Lasserre. *GloptiPoly: Global Optimization over Polynomials with MATLAB and SeDuMi*. Available from http://www.laas.fr/~henrion/software/gloptipoly/.

[63] D. Henrion and J.B. Lasserre. Detecting global optimality and extracting solutions in gloptipoly. In *Positive Polynomials in Control*, D. Henrion and A. Garulli, eds., Lecture Notes in Control and Inform. Sci. 312, Springer, New York, 2005, p. 581.

[64] E. L. Kaltofen, Z. Yang, and L. Zhi. A proof of the monotone column permanent (MCP) conjecture for dimension 4 via sums-of-squares of rational functions. In *Proceedings of the* 2009 *Conference on Symbolic Numeric Computation*, ACM, New York, 2009, pp. 65–70.

[65] E. L. Kaltofen, B. Li, Z. Yang, and L. Zhi. Exact certification in global polynomial optimization via sums-of-squares of rational functions with rational coefficients. *J. Symbol. Comput.*, 2011.

[66] S. Karlin and L. Shapley. *Geometry of Moment Spaces*, Mem. Amer. Math. Soc. 12. AMS, Providence, RI, 1953.

[67] H. Khalil. *Nonlinear Systems*. Macmillan Publishing Company, New York, 1992.

[68] I. Klep and M. Schweighofer. Sums of hermitian squares and the BMV conjecture. *J. Statist. Phys.*, 133:739–760, 2008.

[69] M. Kojima. Sums of Squares Relaxations of Polynomial Semidefinite Programs. Research report B-397, Dept. of Mathematical and Computing Sciences, Tokyo Institute of Technology, Tokyo, 2003.

[70] M. Kojima, S. Kim, and H. Waki. Sparsity in sums of squares of polynomials. *Math. Program.*, 103:45–62, 2005.

[71] M. Krstić, I. Kanellakopoulos, and P. V. Kokotović. *Nonlinear and Adaptive Control Design.* John Wiley & Sons, New York, 1995.

[72] J. B. Lasserre. Global optimization with polynomials and the problem of moments. *SIAM J. Optim.*, 11:796–817, 2001.

[73] M. Laurent. Sums of squares, moment matrices and optimization over polynomials. In *Emerging Applications of Algebraic Geometry*, M. Putinar and S. Sullivant, eds., IMA Vol. Math. Appl. 149, Springer, New York, 2009, pp. 157–270.

[74] J. Löfberg. YALMIP: A toolbox for modeling and optimization in MATLAB. In *Proceedings of the CACSD Conference*, Taipei, Taiwan, 2004. Available at yalmip.org.

[75] J. Löfberg and P. A. Parrilo. From coefficients to samples: A new approach to SOS optimization. In *Proceedings of the 43rd IEEE Conference on Decision and Control*, IEEE, Washington, DC, 2004.

[76] A. Magnani, S. Lall, and S. Boyd. Tractable fitting with convex polynomials via sum-of-squares. In *Proceedings of the 44th IEEE Conference on Decision and Control and the 2005 European Control Conference. CDC-ECC'05*. IEEE, Washington, DC, 2005, pp. 1672–1677.

[77] J. E. Marsden and T. Ratiu. *Introduction to Mechanics and Symmetry*, 2nd ed., *Texts Appl. Math. 17.* Springer-Verlag, New York, 1999.

[78] A. Megretski. *SPOT: Systems Polynomial Optimization Tools*, 2010. MATLAB toolbox, available from web.mit.edu/ameg/www.

[79] D. S. Mitrinović, J. E. Pečarić, and V. Volenec. *Recent Advances in Geometric Inequalities*, Math. Appl. 28. Kluwer Academic Publishers, Dordrecht, The Netherlands, 1989.

[80] T. S. Motzkin and E. G. Straus. Maxima for graphs and a new proof of a theorem of turán. *Canad. J. Math*, 17:533–540, 1965.

[81] K. G. Murty and S. N. Kabadi. Some NP-complete problems in quadratic and nonlinear programming. *Math. Program.*, 39:117–129, 1987.

[82] R. B. Myerson. *Game Theory: Analysis of Conflict.* Harvard University Press, Cambridge, MA, 1991.

[83] Y. Nesterov. Squared functional systems and optimization problems. In *High Performance Optimization*, J. B. G. Frenk, C. Roos, T. Terlaky, and S. Zhang, eds., Kluwer Academic Publishers, Dordrecht, The Netherlands, 2000, pp. 405–440.

[84] Y. E. Nesterov and A. Nemirovski. *Interior Point Polynomial Methods in Convex Programming*, Stud. Appl. Math. 13. SIAM, Philadelphia, 1994.

[85] M. A. Nielsen and I. L. Chuang. *Quantum Computation and Quantum Information*. Cambridge University Press, Cambridge, UK, 2000.

[86] T. Ono. Problem 4417. *Intermed. Math.*, 21:146, 1914.

[87] A. Papachristodoulou and S. Prajna. On the construction of Lyapunov functions using the sum of squares decomposition. In *Proceedings of the 41st IEEE Conference on Decision and Control*, IEEE, Washington, DC, 2002.

[88] A. Papachristodoulou and S. Prajna. Robust stability analysis of nonlinear hybrid systems. *IEEE Trans. Automat. Control*, 54:1035–1041, 2009.

[89] P. A. Parrilo. *Structured Semidefinite Programs and Semialgebraic Geometry Methods in Robustness and Optimization*. Ph.D. thesis, California Institute of Technology, 2000. Available at resolver.caltech.edu/CaltechETD: etd-05062004-055516.

[90] P. A. Parrilo. An Explicit Construction of Distinguished Representations of Polynomials Nonnegative over Finite Sets. Technical Report IfA, Technical Report AUT02-02. Available from www.mit.edu/~parrilo, ETH Zürich, 2002.

[91] P. A. Parrilo. Semidefinite programming relaxations for semialgebraic problems. *Math. Program. Ser. B*, 96:293–320, 2003.

[92] P. A. Parrilo. Polynomial games and sum of squares optimization. In *Proceedings of the 45th IEEE Conference on Decision and Control*, IEEE, Washington, DC, 2006.

[93] P. A. Parrilo and R. Peretz. An inequality for circle packings proved by semidefinite programming. *Discrete Comput. Geom.*, 31:357–367, 2004.

[94] P. A. Parrilo and B. Sturmfels. Minimizing polynomial functions. In *Algorithmic and Quantitative Real Algebraic Geometry*, S. Basu and L. González-Vega, eds., DIMACS Ser. Discrete Math. Theoret. Comput. Sci. 60. AMS, Providence, RI, 2003. Available from arXiv:math.OC/0103170.

[95] M. M. Peet. Exponentially stable nonlinear systems have polynomial Lyapunov functions on bounded regions. *IEEE Trans. Automat. Control*, 54:979–987, 2009.

[96] H. Peyrl and P. A. Parrilo. A Macaulay2 package for computing sum of squares decompositions of polynomials with rational coefficients. In *Proceedings of the 2007 International Workshop on Symbolic-Numeric Computation*, ACM, New York, 2007, pp. 207–208.

[97] H. Peyrl and P. A. Parrilo. `SOS.m2`: A Sum of Squares Package for Macaulay 2. Available from www.control.ee.ethz.ch/~hpeyrl/index.php, 2007.

[98] H. Peyrl and P. A. Parrilo. Computing sum of squares decompositions with rational coefficients. *Theoret. Comput. Sci.*, 409:269–281, 2008.

[99] I. Pólik and T. Terlaky. A survey of the S-lemma. *SIAM Rev.*, 49:371–418, 2007.

[100] S. Prajna, A. Jadbabaie, and G. J. Pappas. Stochastic safety verification using barrier certificates. In *Proceedings of the 43rd IEEE Conference on Decision and Control*, IEEE, Washington, DC, 2004.

[101] S. Prajna, A. Papachristodoulou, and P. A. Parrilo. *SOSTOOLS: Sum of squares optimization toolbox for MATLAB*, 2002–2005. Available from www.cds.caltech.edu/sostools and www.mit.edu/~parrilo/sostools.

[102] M. Putinar. Positive polynomials on compact semi-algebraic sets. *Indiana Univ. Math. J.*, 42:969–984, 1993.

[103] A. J. Quist, E. De Klerk, C. Roos, and T. Terlaky. Copositive relaxation for general quadratic programming. *Optim. Methods Software*, 9:185–208, 1998.

[104] B. Reznick. Extremal PSD forms with few terms. *Duke Math. J.*, 45:363–374, 1978.

[105] B. Reznick. Uniform denominators in Hilbert's seventeenth problem. *Math. Z.*, 220:75–97, 1995.

[106] B. Reznick. Some concrete aspects of Hilbert's 17th problem. In *Real Algebraic Geometry and Ordered Structures*, Contemp. Math. 253, AMS, Providence, RI, 2000, pp. 251–272.

[107] T. Roh and L. Vandenberghe. Discrete transforms, semidefinite programming, and sum-of-squares representations of nonnegative polynomials. *SIAM J. Optim.*, 16:939–964, 2006.

[108] K. Scheiderer. Descending the Ground Field in Sums of Squares Representations, 2012. Preprint, available at arXiv:1209.2976.

[109] C. W. Scherer and C. W. J. Hol. Matrix sum-of-squares relaxations for robust semi-definite programs. *Math. Program. Ser. B*, 107:189–211, 2006.

[110] K. Schmüdgen. The K-moment problem for compact semialgebraic sets. *Math. Ann.*, 289:203–206, 1991.

[111] J.-P. Serre. *Linear Representations of Finite Groups.* Springer-Verlag, New York, 1977.

[112] J. A. Shohat and J. D. Tamarkin. *The Problem of Moments.* Amer. Math. Soc. Math. Surveys 2. AMS, Providence, RI, 1943.

[113] N. Z. Shor. Class of global minimum bounds of polynomial functions. *Cybernet.*, 23:731–734, 1987. (Russian orig.: Kibernetika, No. 6, (1987), 9–11).

[114] G. Stengle. A Nullstellensatz and a Positivstellensatz in semialgebraic geometry. *Math. Ann.*, 207:87–97, 1974.

[115] G. Stengle. Complexity estimates for the Schmüdgen Positivstellensatz. *J. Complexity*, 12:167–174, 1996.

[116] J. Stoer and R. Bulirsch. *Introduction to Numerical Analysis*, Texts Appl. Math. 12. Springer-Verlag, New York, 2002.

[117] A. Strzebonski. Solving algebraic inequalities. *The Mathematica Journal*, 7:525–541, 2000.

[118] J. Sturm, O. Romanko, and I. Pólik. *SeDuMi version* 1.3, 2010. MATLAB toolbox, available from sedumi.ie.lehigh.edu.

[119] B. Sturmfels. *Algorithms in Invariant Theory*, Texts Monogr. Symbol. Comput. 1. Springer, Wien, 1993.

[120] B. Sturmfels. Introduction to resultants. In *Applications of Computational Algebraic Geometry (San Diego, CA,* 1997*)*, Proc. Sympos. Appl. Math. 53, AMS, Providence, RI, 1998, pp. 25–39.

[121] B. Sturmfels. *Solving Systems of Polynomial Equations.* AMS, Providence, RI, 2002.

[122] H. Väliaho. Criteria for copositive matrices. *Linear Algebra Appl.*, 81:19–34, 1986.

[123] F. Vallentin. Symmetry in semidefinite programs. *Linear Algebra Appl.*, 430:360–369, 2009.

[124] H. Waki, S. Kim, M. Kojima, and M. Muramatsu. Sums of squares and semidefinite program relaxations for polynomial optimization problems with structured sparsity. *SIAM J. Optim.*, 17:218–242, 2006.

[125] K. Zhou, K. Glover, and J. C. Doyle. *Robust and Optimal Control.* Prentice–Hall, Englewood Cliffs, NJ, 1995.

Chapter 4

Nonnegative Polynomials and Sums of Squares

Grigoriy Blekherman

A central question, for both practical and theoretical reasons, is how to efficiently test whether a polynomial p is nonnegative. We reformulate this problem in the following way: given a nonnegative polynomial p, how do we efficiently find a representation of p, so that nonnegativity of p is apparent from this representation? In other words, how do we efficiently represent p as an "obviously nonnegative" polynomial? Some polynomials are obviously nonnegative. If we can write p as a sum of squares of polynomials, then it is clear that p is nonnegative just from this presentation. Very importantly, if p is a sum of squares then its sums of squares representation can be efficiently computed via semidefinite programming. This connection was described in detail in Chapter 3. As we will see, the set of sums of squares is a *projected spectrahedron*, while the set of nonnegative polynomials is far more challenging computationally. The main question for this chapter is: what is the relationship between nonnegative polynomials and sums of squares?

4.1 Introduction

Our story begins in 1885, when twenty-three-year-old David Hilbert was one of the examiners in the Ph.D. defense of twenty-one-year-old Hermann Minkowski. During the examination Minkowski claimed that there exist nonnegative polynomials that are not sums of squares. Although he did not provide an example or a proof, his argument must have been convincing, as he defended successfully.

Three years later Hilbert published a paper in which he classified all of the (few) cases, in terms of degree and number of variables, in which nonnegative polynomials are the same as sums of squares. In all other cases Hilbert showed that there exist nonnegative polynomials that are not sums of squares. Interestingly,

Hilbert did not provide an explicit example of such polynomials. The first explicit example was found only seventy years later and is due to Theodore Motzkin. In fact, Motzkin was not aware of what he constructed. Olga Taussky-Todd, who was present during the seminar in which Motzkin described his construction, later notified him that he found the first example of a nonnegative polynomial that is not a sum of squares [22].

We examine the relationship between nonnegativity and sums of squares in two different fundamental ways. We first consider the structures that prevent sums of squares from capturing all nonnegative polynomials, and show that equality occurs precisely when these structures are not present. We then examine in detail the smallest cases where there exist nonnegative polynomials that are not sums of squares and show that the inequalities separating nonnegative polynomials from sums of squares have a simple and elegant structure. Second, we look at the quantitative relationship between nonnegative polynomials and sums of squares. Here we show that when the degree is fixed and the number of variables grows, there are significantly more nonnegative polynomials than sums of squares. We also apply these ideas to studying the relationship between sums of squares and convex polynomials. While the techniques we develop for the two approaches are quite different in nature, the unifying theme is that we examine the sets of nonnegative polynomials and sums of squares geometrically. Algebraic geometry is at the forefront of our examination of fundamental differences between nonnegative polynomials and sums of squares, while convex geometry and analysis are used to examine the quantitative relationship.

The chapter is structured as follows: After discussing Hilbert's theorem and Motzkin's example in Section 4.2, we begin a detailed examination of the underlying causes of differences between nonnegative polynomials and sums of squares in Section 4.3. On the way we will see that nonnegative polynomials and sums of squares form fascinating convex sets. Section 4.4 is devoted to the examination of these objects from the point of view of convex algebraic geometry. We note that many basic questions remain open.

The fundamental reasons for the existence of nonnegative polynomials that are not sums of squares come from Cayley–Bacharach theory in classical algebraic geometry and, in fact, Hilbert's original proof of his theorem already used some of these ideas. We begin developing the necessary techniques in Section 4.5. Duality from convex geometry and its interplay with commutative algebra will play a central role in our investigation. Section 4.6 develops the duality ideas and presents a unified proof of the equality cases of Hilbert's theorem. Sections 4.7 and 4.8 investigate the smallest cases in which there exist nonnegative polynomials that are not sums of squares. We show that this situation fundamentally arises from the existence of Cayley–Bacharach relations and present some consequences.

We proceed by examining the quantitative relationship between nonnegative polynomials and sums of squares in Section 4.9. This is done by establishing bounds on the volume of sets of nonnegative polynomials and sums of squares, and analytic aspects of convex geometry come to the fore in this examination. We will explain that if the degree is fixed and the number of variables is allowed to grow, then there are significantly more nonnegative polynomials than sums of squares [5].

This happens despite the difficulty of constructing explicit examples of nonnegative polynomials that are not sums of squares, and numerical evidence that sums of squares approximate nonnegative polynomials well if the degree and number of variables is small [19]. The question of precisely when nonnegative polynomials begin to significantly overtake sums of squares is currently poorly understood.

Section 4.10 presents an application of the volume ideas to showing that there exist homogeneous polynomials that are convex functions but are not sums of squares. There is no known explicit example of such a polynomial, and this is the only known method of showing their existence.

4.2 A Deeper Look

We first reduce the study of nonnegative polynomials and sums of squares to the case of homogeneous polynomials, which are also called forms. A polynomial $p(x_1, \ldots, x_n)$ of degree d can be made homogeneous by introducing an extra variable x_{n+1} and multiplying every monomial in p by a power of x_{n+1}, so that all monomials have the same degree. More formally, let $\bar{p}$ be the homogenization of p:

$$\bar{p} = x_{n+1}^d p\left(\frac{x_1}{x_{n+1}}, \ldots, \frac{x_n}{x_{n+1}}\right).$$

Exercise 4.1. Let p be a nonnegative polynomial. Show that $\bar{p}$ is a nonnegative form. Also show that if p is a sum of squares, then $\bar{p}$ is a sum of squares as well.

Given a form $\bar{p}$ we can dehomogenize it by setting $x_{n+1} = 1$. Dehomogenization clearly preserves nonnegativity and sums of squares. Therefore the study of nonnegative polynomials and sums of squares in n variables is equivalent to studying forms in $n+1$ variables. From now on we restrict ourselves to the case of forms.

Let $\mathbb{R}[x]_{\mathbf{d}}$ be the vector space of real forms in n variables of degree d. In order to be nonnegative a form must have even degree, and therefore our forms will have even degree $2d$. Inside $\mathbb{R}[x]_{\mathbf{2d}}$ sit two closed convex cones: the cone of nonnegative polynomials,

$$P_{\mathbf{n},\mathbf{2d}} = \{p \in \mathbb{R}[x]_{\mathbf{2d}} \mid p(x) \geq 0 \text{ for all } x \in \mathbb{R}^n\},$$

and the cone of sums of squares,

$$\Sigma_{\mathbf{n},\mathbf{2d}} = \left\{p \in \mathbb{R}[x]_{\mathbf{2d}} \mid p(x) = \sum q_i^2 \text{ for some } q_i \in \mathbb{R}[x]_{\mathbf{d}}\right\}.$$

Exercise 4.2. Show that $P_{\mathbf{n},\mathbf{2d}}$ and $\Sigma_{\mathbf{n},\mathbf{2d}}$ are closed, full-dimensional convex cones in $\mathbb{R}[x]_{\mathbf{2d}}$. (Hint: Consider Exercise 4.17.)

We now come to the first major theorem concerning nonnegative polynomials and sums of squares.

4.2.1 Hilbert's Theorem

The first fundamental result about the relationship between $P_{\mathbf{n,2d}}$ and $\Sigma_{\mathbf{n,2d}}$ was shown by Hilbert in 1888.

Theorem 4.3. *Nonnegative forms are the same as sums of squares, $P_{\mathbf{n,2d}} = \Sigma_{\mathbf{n,2d}}$, in the following three cases: $n = 2$ (univariate nonhomogeneous case), $2d = 2$ (quadratic forms), and $n = 3$, $2d = 4$ (ternary quartics). In all other cases there exist nonnegative forms that are not sums of squares.*

The proof of the three equality cases in Hilbert's theorem usually proceeds by treating each of the three cases separately. For example, it is a simple exercise to show that $P_{\mathbf{n,2}} = \Sigma_{\mathbf{n,2}}$.

Exercise 4.4. Deduce that $P_{\mathbf{n,2}} = \Sigma_{\mathbf{n,2}}$ from diagonalization of symmetric matrices.

We adopt a different approach: We begin by examining the structures that allow the existence of nonnegative forms that are not sums of squares. In Section 4.6.1 we show that the three cases of Hilbert's theorem are the only cases in which these structures do not exist. This provides a unified proof of the three equality cases of Hilbert's theorem, which are usually treated separately.

4.2.2 Motzkin's Example

The first explicit example of a nonnegative form that is not a sum of squares is due to Motzkin:

$$M(x, y, z) = x^4y^2 + x^2y^4 + z^6 - 3x^2y^2z^2.$$

The form M can be seen to be nonnegative by the application of the arithmetic mean-geometric mean inequality. Why is M not a sum of squares?

In the following exercises we develop a general method for showing that a form is not a sum of squares, based on the monomials that occur in the form. This method can also be applied to reduce the size of the semidefinite program that computes the sum of squares decomposition, as explained in Chapter 3. These ideas are originally due to Choi, Lam, and Reznick [22].

Exercise 4.5. For a polynomial p define its *Newton polytope* $\mathcal{N}(p)$ to be the convex hull of the vectors of exponents of monomials that occur in p. For example, if $p = x_1x_2^2 + x_2^2 + x_1x_2x_3$, then $\mathcal{N}(p) = \operatorname{conv}(\{(1, 2, 0), (0, 2, 0), (1, 1, 1)\})$, which is a triangle in $\mathbb{R}^3$.

Show that if $p = \sum q_i^2$, then

$$\mathcal{N}(q_i) \subseteq \frac{1}{2}\mathcal{N}(p).$$

Exercise 4.6. Calculate the Newton polytope of the Motzkin form and use Exercise 4.5 to show that the Motzkin form is not a sum of squares.

For much more on explicit examples of nonnegative polynomials that are not sums of squares see [22].

4.2.3 Quantitative Relationship

While Hilbert's theorem completely settles all cases of equality between $P_{\mathbf{n},\mathbf{2d}}$ and $\Sigma_{\mathbf{n},\mathbf{2d}}$ it does not shed light on whether these cones are close to each other, even if the cone of nonnegative polynomials is strictly larger. Due to the difficulty of constructing explicit examples and numerical evidence for a small number of variables and degrees, it is tempting to assume that $\Sigma_{\mathbf{n},\mathbf{2d}}$ approximates $P_{\mathbf{n},\mathbf{2d}}$ fairly well.

However, it was shown in [5] that if the degree $2d$ is fixed and at least 4, then as the number of variables n grows, there are significantly more nonnegative forms than sums of squares. We will make this statement precise and present a proof in Section 4.9. The main idea is that, although the cones themselves are unbounded, we can slice both cones with the same hyperplane, so that the section of each cone is compact. We then derive separate bounds on the volume of each section.

For now we would like to note that the bounds guarantee that the difference between $P_{\mathbf{n},\mathbf{2d}}$ and $\Sigma_{\mathbf{n},\mathbf{2d}}$ is large only for a very large number of variables n. Whether this is an artifact of the techniques used to derive the bounds is unclear. As we will see, for a small number of variables the distinction between $P_{\mathbf{n},\mathbf{2d}}$ and $\Sigma_{\mathbf{n},\mathbf{2d}}$ is quite delicate, and it is not known at what point $P_{\mathbf{n},\mathbf{2d}}$ becomes much larger than $\Sigma_{\mathbf{n},\mathbf{2d}}$.

We now begin a systematic examination of differences between nonnegative forms and sums of squares. It is actually possible to see that there exist nonnegative forms that are not sums of squares by considering values of forms on finitely many points. The following example will illustrate this idea and explain some of the major themes in our investigation.

4.3 The Hypercube Example

According to Hilbert's theorem the smallest cases where $P_{\mathbf{n},\mathbf{2d}}$ and $\Sigma_{\mathbf{n},\mathbf{2d}}$ differ are forms in 3 variables of degree 6, and forms in 4 variables of degree 4. We take a close look at an explicit example for the case of forms in 4 variables of degree 4. Let $S = \{s_1, \dots, s_8\}$ be the following set of 8 points in $\mathbb{R}^4$:

$$S = \{\pm 1, \pm 1, \pm 1, 1\}.$$

We will see that there is a difference between nonnegative forms and sums of squares by simply looking at the values that nonnegative polynomials and sums of squares take on S. Accordingly, let us define a projection π from $\mathbb{R}[x]_{\mathbf{4},\mathbf{4}}$ to $\mathbb{R}^8$ given by evaluation on S:

$$\pi(f) = (f(s_1), \dots, f(s_8)) \quad \text{for} \quad f \in \mathbb{R}[x]_{\mathbf{4},\mathbf{4}}.$$

We will explicitly describe the images of $P_{\mathbf{4},\mathbf{4}}$ and $\Sigma_{\mathbf{4},\mathbf{4}}$ under this projection. Let

$$P' = \pi(P_{\mathbf{4},\mathbf{4}}) \quad \text{and} \quad \Sigma' = \pi(\Sigma_{\mathbf{4},\mathbf{4}}).$$

As they are images of convex cones under a linear map, it is clear that both P' and Σ' are convex cones in $\mathbb{R}^8$. Although both P' and Σ' will turn out to be closed, projections of closed convex cones do not have to be closed in general.

Exercise 4.7. Construct a closed convex cone C in $\mathbb{R}^3$ and a linear map $\pi : \mathbb{R}^3 \to \mathbb{R}^2$ such that $\pi(C)$ is not closed.

4.3.1 Values of Nonnegative Forms

We first look at values on S that are achievable by nonnegative forms. Let $\mathbb{R}^8_+$ be the nonnegative orthant of $\mathbb{R}^8$:

$$\mathbb{R}^8_+ = \{(x_1, \dots, x_8) \mid x_i \geq 0 \text{ for } i = 1, \dots, 8\}.$$

Since we are evaluating nonnegative polynomials, it is clear that $P' \subseteq \mathbb{R}^8_+$. We claim that, in fact, $P' = \mathbb{R}^8_+$. In other words, any 8-tuple of nonnegative numbers can be attained on S by a globally nonnegative form. By convexity of P' it suffices to show that all the standard basis vectors e_i are in P'. Moreover, substitutions $x_i \mapsto -x_i$ permute the set S, and therefore it is enough to show that $e_i \in P'$ for some i.

Exercise 4.8. Let $p \in \mathbb{R}[x]_{\mathbf{4,4}}$ be the following symmetric form:

$$p = \sum_{i=1}^{4} x_i^4 + 2 \sum_{i \neq j \neq k} x_i^2 x_j x_k + 4x_1x_2x_3x_4.$$

Show that p is nonnegative, and check that p vanishes on exactly 7 points in S. Conclude that $P' = \mathbb{R}^8_+$.

We have seen that all combinations of nonnegative values on S are realizable as values of a nonnegative form. We now look at why some values in $\mathbb{R}^8_+$ are not attainable by sums of squares. In the end we will completely describe the projection Σ'.

4.3.2 Values of Sums of Squares

In order to analyze the values of sums of squares, we need to take a look at the values of the forms that we are squaring. The values of quadratic forms on S are not linearly independent. Here is the unique (up to a constant multiple) linear relation between the values on the points s_i that all quadratic forms in 4 variables satisfy:

$$\sum_{s_i \text{ has even number of 1's}} f(s_i) = \sum_{s_i \text{ has odd number of 1's}} f(s_i). \tag{4.1}$$

Exercise 4.9. Verify that the relation (4.1) holds for all quadratic forms $f \in \mathbb{R}[x]_{\mathbf{4,2}}$ and that it is unique up to a constant multiple.

We are now ready to see how the relation (4.1) prevents sums of squares from attaining all values in $\mathbb{R}^8_+$.

Proposition 4.10 (Hilbert's original insight). *Let e_i be the ith standard basis vector in $\mathbb{R}^8$. Then $e_i \notin \Sigma'$ for all i.*

Proof. Since we did not attach a specific labeling to the points of S it will suffice to show that $e_1 \notin \Sigma' = \pi(\Sigma_{\mathbf{4,4}})$. Suppose that there exists $p \in \Sigma_{\mathbf{4,4}}$ such that $\pi(p) = e_1$. Write $p = \sum_j q_j^2$ for some $q_j \in \mathbb{R}[x]_{\mathbf{4,2}}$. The form p vanishes on $s_2, \dots, s_8$, and it has value 1 on s_1. Since $p = \sum_j q_j^2$ it follows that each q_j vanishes on $s_2, \dots, s_8$. Each q_j is a quadratic form in 4 variables, and therefore each q_j satisfies relation (4.1). From this relation it follows that $q_j(s_1) = 0$ for all j. Therefore $p(s_1) = 0$, which is a contradiction. $\square$

Hilbert's original proof did not use an explicit example to show that the vectors e_i can be realized as values of a nonnegative form, which we did in Exercise 4.8. Instead he provided a recipe for constructing such a form, and proved that the construction works. We largely followed Hilbert's recipe to construct our counterexample. For more information on Hilbert's construction see [23].

4.3.3 Complete Description of Σ'

We can do better than just describing some points that are not in Σ'. Our next goal is to completely describe Σ' and, in particular, we will see how far the points e_i are from being the values of a sum of squares.

We use π to also denote the same evaluation projection on quadratic forms in 4 variables:

$$\pi(f) = (f(s_1), \dots, f(s_8)) \quad \text{for} \quad f \in \mathbb{R}[x]_{\mathbf{4,2}}.$$

Let L be the projection of the entire vector space of quadratic forms:

$$L = \pi(\mathbb{R}[x]_{\mathbf{4,2}}).$$

Using relation (4.1) and Exercise 4.9 we see that L is a hyperplane in $\mathbb{R}^8$. Let C be the set of points that are coordinatewise squares of points in L:

$$C = \{(v_1^2, \dots, v_8^2) \mid v = (v_1, \dots, v_8) \in L\}.$$

We first show the following description of Σ'.

Lemma 4.11. *Σ' is equal to the convex hull of C:*

$$\Sigma' = \operatorname{conv}(C).$$

Proof. Let $v = (v_1, \dots, v_8) \in L$. Then there exists a quadratic form $f \in \mathbb{R}[x]_{\mathbf{4,2}}$ such that $f(s_i) = v_i$ for $i = 1, \dots, 8$. It follows that for the square of f we have $f^2(s_i) = v_i^2$. In other words,

$$\pi(f^2) = (v_1^2, \dots, v_8^2), \quad \text{where} \quad v = (v_1, \dots, v_8) = \pi(f).$$

Therefore we see that $C \subset \Sigma'$ and by convexity of Σ' it follows that $\operatorname{conv}(C) \subseteq \Sigma'$.

To prove the other inclusion, suppose that $p = \sum_i q_i^2 \in \Sigma_{4,4}$. Then $\pi(q_i^2) \in C$ for all i and therefore $\Sigma' \subseteq \operatorname{conv}(C)$. $\square$

Let T_m be the subset of the nonnegative orthant $\mathbb{R}_+^m$ defined by the following m inequalities:

$$T_m = \left\{ (x_1, \dots, x_m) \in \mathbb{R}_+^m \quad \middle| \quad \sum_{i=1}^m \sqrt{x_i} \geq 2\sqrt{x_k} \quad \text{for all} \quad k \right\}.$$

We will show that $\Sigma' = T_8$. We begin with a lemma on the structure on T_m.

Lemma 4.12. *The set T_m is a closed convex cone. Moreover, T_m is the convex hull of the points $x = (x_1, \dots, x_m) \in \mathbb{R}_+^m$, where $\sum_{i=1}^m \sqrt{x_i} = 2\sqrt{x_k}$ for some k.*

Proof. The set T_m is defined as a subset of $\mathbb{R}^m$ by the following $2m$ inequalities: $x_k \geq 0$ and $\sqrt{x_1} + \cdots + \sqrt{x_m} \geq 2\sqrt{x_k}$ for all k. Therefore it is clear that T_m is a closed set.

For $x = (x_1, \dots, x_m) \in \mathbb{R}_+^m$ let $\|x\|_{1/2}$ denote the $L^{1/2}$-norm of x:

$$\|x\|_{1/2} = (\sqrt{x_1} + \cdots + \sqrt{x_m})^2.$$

We can restate the inequalities of T_m as $x_k \geq 0$ and $\|x\|_{1/2} \geq 4x_k$ for all k. Now suppose that $x, y \in T_m$ and let $z = \lambda x + (1-\lambda)y$ for some $0 \leq \lambda \leq 1$. It is clear that $z_k \geq 0$ for all k. It is known by the Minkowski inequality [11, p. 30] that $L^{1/2}$-norm is a concave function: $\|\lambda x + (1-\lambda)y\|_{1/2} \geq \lambda\|x\|_{1/2} + (1-\lambda)\|y\|_{1/2}$. Therefore

$$\|z\|_{1/2} \geq \lambda\|x\|_{1/2} + (1-\lambda)\|y\|_{1/2} \geq 4\lambda x_k + 4(1-\lambda)y_k = 4z_k \quad \text{for all} \quad k.$$

Thus T_m is a convex cone.

To show that T_m is the convex hull of the points where $\|x\|_{1/2} = 4x_k$ for some k we proceed by induction. The base case $m = 2$ is simple since T_2 is just a ray spanned by the point $(1, 1)$. For the induction step we observe that any convex set is the convex hull of its boundary. For any point on the boundary of T_m one of the defining $2m$ inequalities must be sharp. If a point x is on the boundary of T_m and $x_i \neq 0$ for all i, then the inequalities $x_i \geq 0$ are not sharp at x; therefore the inequality $\|x\|_{1/2} \geq 4x_k$ must be sharp for some k, and we are done.

If $x_i = 0$ for some i, then the point x lies in the set T_{m-1} in the subspace spanned by the $m-1$ standard basis vectors excluding e_i, and we are done by induction. $\square$

Exercise 4.13. Show that the cone $T_4 \subset \mathbb{R}^4$ can be transformed via a nonsingular linear transformation into the dual cone of 3×3 positive semidefinite matrices with equal diagonal elements:

$$(x_1, x_2, x_3, x_4) \in \mathbb{R}^4 \quad \text{such that} \quad \begin{bmatrix} x_1 & x_2 & x_3 \\ x_2 & x_1 & x_4 \\ x_3 & x_4 & x_1 \end{bmatrix} \succeq 0.$$

If we restrict x_1 to being 1 then we obtain the *elliptope* $\mathcal{E}_3$, which we have already seen in Chapter 2.

We are now ready to completely describe Σ'.

Theorem 4.14. $\Sigma' = T_8$.

Proof. We rewrite the relation (4.1) in the form

$$\sum_{i=1}^{8} a_i f(s_i) = 0 \quad \text{for} \quad f \in \mathbb{R}[x]_{\mathbf{4,2}}, \tag{4.2}$$

and let $a = (a_1, \ldots, a_8)$ be the vector of coefficients, with $a_i = \pm 1$. It follows that $L = \pi\left(\mathbb{R}[x]_{\mathbf{4,2}}\right)$ is the hyperplane in $\mathbb{R}^8$ perpendicular to a.

Since T_8 is a convex cone, to show the inclusion $\Sigma' \subseteq T_8$ it suffices by Lemma 4.11 to show that $C \subset T_8$. Let $v = (v_1, \ldots, v_8) \in L$ and $t = (v_1^2, \ldots, v_8^2) \in C$. By the relation (4.2) we have $a_1 v_1 + \cdots + a_8 v_8 = 0$ with $a_i = \pm 1$. Without loss of generality, we may assume that v_1 has the maximal absolute value among v_i. Multiplying the relation (4.2) by -1, if necessary, we can make $a_1 = -1$. Then we have $v_1 = a_2 v_2 + \cdots + a_8 v_8$. We can now write $\sqrt{t_1} = \pm\sqrt{t_2} \pm \sqrt{t_3} \pm \cdots \pm \sqrt{t_8}$ with the exact signs depending on a_i and signs of v_i. Therefore we see that $2\sqrt{t_1} \leq \sqrt{t_1} + \cdots + \sqrt{t_8}$. Since v_1 has the largest absolute value among v_i, it follows that $2\sqrt{t_k} \leq \sqrt{t_1} + \cdots + \sqrt{t_8}$ for all $1 \leq k \leq 8$. Hence we see that $\Sigma' \subseteq T_8$.

To show the reverse inclusion $T_8 \subseteq \Sigma'$ we use Lemma 4.12. It suffices to show that all points $x \in T_8$ with $2\sqrt{x_k} = \sqrt{x_1} + \cdots + \sqrt{x_8}$ for some k, are also in Σ'. Without loss of generality we may assume that $k = 1$ and we have $\sqrt{x_1} = \sqrt{x_2} + \cdots + \sqrt{x_8}$. Let $y = (y_1, \ldots, y_8)$ with $y_1 = -\sqrt{x_1}/a_1$ and $y_i = \sqrt{x_i}/a_i$ for $2 \leq i \leq 8$. It follows that $a_1 y_1 + \cdots + a_8 y_8 = 0$. Therefore $y \in \pi(\mathbb{R}[x]_{\mathbf{4,2}})$ and $y = \pi(q)$ for some quadratic form q. Then $\pi(q^2) = x$ and we are done. ◻

We can use Exercise 4.8 and Theorem 4.14 to visualize the discrepancy between P' and Σ'. Let's take a slice of both cones with the hyperplane H given by $x_1 + \cdots + x_8 = 1$. Recall that by Exercise 4.8 we have $P' = \mathbb{R}_8^+$. Therefore the slice of P' with H is the standard simplex. The slice of T_8 with H is the standard simplex with cut off corners. It was Hilbert's observation that the standard basis vectors e_i are not in Σ', and Theorem 4.14 tells us exactly how much is cut off around the corners.

We now take a short break from comparing $P_{\mathbf{n,2d}}$ and $\Sigma_{\mathbf{n,2d}}$ to consider some convexity properties of these cones, such as boundary, facial structure, symmetries, and dual cones.

4.4 Symmetries, Dual Cones, and Facial Structure

4.4.1 Symmetries of $P_{\mathbf{n,2d}}$ and $\Sigma_{\mathbf{n,2d}}$

The cones $P_{\mathbf{n,2d}}$ and $\Sigma_{\mathbf{n,2d}}$ have a lot of built-in symmetries coming from linear changes of coordinates. Suppose that $A \in GL_n(\mathbb{R})$ is a nonsingular linear transformation of $\mathbb{R}^n$.

Exercise 4.15. Show that if $p(x) \in \mathbb{R}[x]_{\mathbf{2d}}$ is a nonnegative form, then $p(Ax)$ is also a nonnegative form in $\mathbb{R}[x]_{\mathbf{2d}}$. Similarly, if $p(x)$ is a sum of squares, then $p(Ax)$ is also a sum of squares.

In more formal terms, a nonsingular linear transformation A of $\mathbb{R}^n$ induces a nonsingular transformation ϕ_A of $\mathbb{R}[x]_{\mathbf{2d}}$, which maps $p(x) \in \mathbb{R}[x]_{\mathbf{2d}}$ to $p(A^{-1}(x))$. We say that the group $GL_n(\mathbb{R})$ acts on $\mathbb{R}[x]_{\mathbf{2d}}$. It follows from Exercise 4.15 that both cones $P_{\mathbf{n,2d}}$ and $\Sigma_{\mathbf{n,2d}}$ are invariant under this action. In other words, $P_{\mathbf{n,2d}}$ and $\Sigma_{\mathbf{n,2d}}$ are invariant under nonsingular linear changes of coordinates.

Exercise 4.16. Show that, up to a constant multiple, $r^{2d} = (x_1^2 + \cdots + x_n^2)^d$ is the only form in $\mathbb{R}[x]_{\mathbf{2d}}$ that is fixed under all orthogonal changes of coordinates; i.e., it is the only form in $\mathbb{R}[x]_{\mathbf{2d}}$ that satisfies

$$p(x) = p(Ax) \quad \text{for all} \quad A \in O_n,$$

where O_n is the group of orthogonal transformations of $\mathbb{R}^n$.

We note that even if a linear transformation A of $\mathbb{R}^n$ is singular, it still induces a linear transformation ϕ_A in the same way. However the linear map ϕ_A will also be singular. The map ϕ_A still sends $P_{\mathbf{n,2d}}$ and $\Sigma_{\mathbf{n,2d}}$ into themselves, but it will no longer preserve the cones. Closed convex cones in $\mathbb{R}[x]_{\mathbf{2d}}$ that are mapped into themselves under any linear change of coordinates are called *blenders* [24].

4.4.2 Dual Cone of $P_{\mathbf{n,2d}}$

Let K be a convex cone in a real vector space V. Let V^* be the dual vector space of linear functionals on V. The dual cone K^* is defined as the set of all linear functionals in V^* that are nonnegative on K:

$$K^* = \{\ell \in V^* \mid \ell(x) \geq 0 \quad \text{for all} \quad x \in K\}.$$

Many general aspects of duality will be discussed in Chapter 5. We examine the specific cases of cones of nonnegative polynomials and sums of squares.

Let's consider the dual space $\mathbb{R}[x]_{\mathbf{2d}}^*$ of linear functionals on $\mathbb{R}[x]_{\mathbf{2d}}$. We first observe that the dual cone of $P_{\mathbf{n,2d}}$ is conceptually simple. For $v \in \mathbb{R}^n$, let ℓ_v be the linear functional in $\mathbb{R}[x]_{\mathbf{2d}}^*$ given by evaluation at v:

$$\ell_v(f) = f(v) \quad \text{for} \quad f \in \mathbb{R}[x]_{\mathbf{2d}}.$$

By homogeneity of forms we know that nonnegativity on the unit sphere is equivalent to global nonnegativity. Therefore it is natural to think that the functionals ℓ_v with $v \in \mathbb{S}^{n-1}$ generate the dual cone $P_{\mathbf{n,2d}}^*$. Before we show that this is in fact the case we need a useful exercise from convexity.

Exercise 4.17. Let $K \subset \mathbb{R}^n$ be a compact convex set with the origin not in K. Show that the conical hull of K, cone(K), is closed. Construct an explicit example that shows that the condition $0 \notin K$ is necessary.

Lemma 4.18. *The dual cone $P^*_{\mathbf{n,2d}}$ of the cone of nonnegative forms is the conical hull of linear functionals ℓ_v with v on the unit sphere:*

$$P^*_{\mathbf{n,2d}} = \operatorname{cone}\left(\ell_v \mid v \in \mathbb{S}^{n-1}\right).$$

Proof. Let $L_{\mathbf{n,2d}} \subset \mathbb{R}[x]^*_{\mathbf{2d}}$ be the conical hull of functionals ℓ_v with $v \in \mathbb{S}^{n-1}$. The dual cone $L^*_{\mathbf{n,2d}}$ is the set of all forms $p \in \mathbb{R}[x]_{\mathbf{2d}}$ such that

$$\ell_v(p) = p(v) \geq 0 \quad \text{for all} \quad v \in \mathbb{S}^{n-1}.$$

Therefore we see that $L^*_{\mathbf{n,2d}} = P_{\mathbf{n,2d}}$. Using *biduality* we see that the dual cone $P^*_{\mathbf{n,2d}}$ is equal to the closure of $L_{\mathbf{n,2d}}$:

$$P^*_{\mathbf{n,2d}} = (L^*_{\mathbf{n,2d}})^* = \overline{L_{\mathbf{n,2d}}}.$$

We now just need to show that the cone $L_{\mathbf{n,2d}}$ is closed and then $\overline{L_{\mathbf{n,2d}}} = L_{\mathbf{n,2d}}$. Consider the set C of all linear functionals ℓ_v with $v \in \mathbb{S}^{n-1}$. The set C is given by a continuous embedding of the unit sphere $\mathbb{S}^{n-1}$ into $\mathbb{R}[x]^*_{\mathbf{2d}}$, and therefore C is compact. If we can show that the convex hull of C does not contain the origin, then we are done by applying Exercise 4.17.

Let $r^{2d} = (x_1^2 + \cdots + x_n^2)^d$ be the form in $\mathbb{R}[x]_{\mathbf{2d}}$ that is constantly 1 on the unit sphere. Suppose that $m = \sum c_v \ell_v \in \operatorname{conv}(C)$. Then it follows that $m(r^{2d}) = \sum c_v = 1$, and therefore m cannot be the zero functional in $\mathbb{R}[x]^*_{\mathbf{2d}}$. It follows that $\operatorname{conv}(C)$ is a compact convex set with $0 \notin C$ and we are done. $\square$

Exercise 4.19. Use the *apolar inner product* from Chapter 3 to identify $\mathbb{R}[x]_{\mathbf{2d}}$ with the dual space $\mathbb{R}[x]^*_{\mathbf{2d}}$. Show that the dual cone $P^*_{\mathbf{n,2d}}$ is identified with the *cone of sums of* $2d$*th powers of linear forms*:

$$\left\{p \in \mathbb{R}[x]_{\mathbf{2d}} \;\middle|\; p = \sum q_i^{2d} \quad \text{with} \quad q_i \in \mathbb{R}[x]_{\mathbf{n,1}}\right\}.$$

Remark 4.20. *The map that sends a point $v \in \mathbb{R}^n$ to the form $(v_1x_1 + \cdots + v_nx_n)^{2d}$ is called the $2d$th Veronese embedding and its image is called the Veronese variety. It follows from Lemma 4.18 that the cone $P^*_{\mathbf{n,2d}}$ is the conical hull of the $2d$th Veronese variety. For more information and for connections to orbitopes we refer to [25].*

By applying spherical symmetries to functionals ℓ_v we obtain the following crucial corollary, which describes the extreme rays of $P^*_{\mathbf{n,2d}}$.

Corollary 4.21. *The functional ℓ_v spans an extreme ray of $P^*_{\mathbf{n,2d}}$ for all $v \in \mathbb{S}^{n-1}$, and the functionals ℓ_v form the complete set of extreme rays of $P^*_{\mathbf{n,2d}}$.*

The extreme rays of the cone $P^*_{\mathbf{n,2d}}$ have a very nice parametrization by points $v \in \mathbb{S}^{n-1}$. However, the cone $P^*_{\mathbf{n,2d}}$ is a very complex object from the computational and convex geometry point of view. For example, given a linear functional $\ell \in \mathbb{R}[x]^*_{\mathbf{2d}}$, determining whether it belongs to the cone $P^*_{\mathbf{n,2d}}$ is known as the *truncated moment problem* in real analysis. Despite a long history, there are very few explicit and computationally feasible criteria for testing membership in $\mathbb{R}[x]^*_{\mathbf{2d}}$. For more on this approach see [15].

Decomposing a given linear functional in $\mathbb{R}[x]^*_{\mathbf{2d}}$ as a linear combination of the functionals ℓ_v, or equivalently by Exercise 4.19, decomposing a given form in $\mathbb{R}[x]_{\mathbf{2d}}$ as a linear combination of forms v^{2d} is known as the *symmetric tensor decomposition problem.* Again, despite a long history, many aspects of symmetric tensor decomposition remain unknown. For more information we refer to [14, 21].

4.4.3 Boundary of the Cone of Nonnegative Polynomials

The boundary and the interior of the cone of nonnegative forms $P_{\mathbf{n,2d}}$ are easy to describe given our knowledge of the dual cone $P^*_{\mathbf{n,2d}}$.

Exercise 4.22. Show that the interior of $P_{\mathbf{n,2d}}$ consists of forms that are strictly positive on $\mathbb{R}^n \setminus \{0\}$ and the boundary of $P_{\mathbf{n,2d}}$ consists of forms with a nontrivial zero.

We note that the situation is slightly different in the nonhomogeneous case. Let $f(x) = x^2 + 1$ be a univariate polynomial, and let P be the cone of nonnegative univariate polynomials of degree at most 4. Clearly $f \in P$ and f is strictly positive on $\mathbb{R}$. However, f lies on the boundary of P. Consider $g_\epsilon = f - \epsilon x^4$. For any $\epsilon > 0$ the polynomial g_ϵ will not be nonnegative. Therefore f is not in the interior of P, and it lies on the boundary of P.

The explanation for this phenomenon is that even though f is strictly positive on $\mathbb{R}$, when viewed as a polynomial of degree 4, f has a *zero at infinity.* The growth of $f(x)$ as x goes to infinity is only of order 2, and therefore we cannot subtract a nonnegative polynomial of degree 4 from f and have the difference remain nonnegative. The easiest way to see the zero at infinity is to homogenize f with an extra variable y: $\bar{f} = x^2y^2 + y^4$.

Note that if we set $y = 1$ in $\bar{f}$ we just recover f. However, $\bar{f}$ is not a strictly positive form on $\mathbb{R}^2 \setminus \{0\}$, since $\bar{f}$ has a nontrivial zero which comes from setting $y = 0$. In general, for a polynomial f in n variables of degree d, let f_d be the degree d component of f consisting of all terms of degree exactly d. *Zeroes at infinity* of f correspond to zeroes of f_d. This can be seen by homogenizing f with an extra variable. When we set this variable equal to 0 we obtain f_d.

4.4.4 Exposed Faces of $P_{\mathbf{n,2d}}$

Exposed faces of $P_{\mathbf{n,2d}}$ are conceptually easy to understand due to our knowledge of the extreme rays of the dual cone $P^*_{\mathbf{n,2d}}$ in Corollary 4.21. Maximal (by inclusion) faces of $P_{\mathbf{n,2d}}$ come from the vanishing of one extreme ray of the dual cone. Therefore it follows that maximal faces $F(v)$ of $P_{\mathbf{n,2d}}$ consist of all nonnegative forms that have a single common zero $v \in \mathbb{S}^{n-1}$:

$$F(v) = \{p \in P_{\mathbf{n,2d}} \mid \ell_v(p) = p(v) = 0\}.$$

We observe that a zero of a nonnegative form p is a local minimum. Therefore, if $p(v) = 0$, this implies that the gradient of p at v is zero as well, $\nabla p(v) = 0$. In other words, p must have a double zero at v.

Exercise 4.23 (Euler's relation). Show that for $p \in \mathbb{R}[x]_{\mathbf{d}}$ and all $v \in \mathbb{R}^n$ the following relation holds:

$$\langle \nabla p(v), v \rangle = d \cdot p(v).$$

From the above exercise it follows that for forms $p \in \mathbb{R}[x]_{\mathbf{2d}}$ the vanishing of the gradient at v, $\nabla p(v) = 0$, forces the form p to vanish at v as well, $p(v) = 0$. Therefore, for a nonnegative form $p \in P_{\mathbf{n,2d}}$ a single zero forces p to satisfy n linear conditions coming from $\nabla p(v) = 0$. It follows that the face $F(v)$ has codimension at least n.

Exercise 4.24. Show that the maximal faces $F(v)$ of $P_{\mathbf{n,2d}}$ have codimension exactly n in $\mathbb{R}[x]_{\mathbf{2d}}$.

All smaller exposed faces $F(v_1, \dots, v_k)$ come from the vanishing of several extreme rays $\ell_{v_1}, \dots, \ell_{v_k}$ of $P^*_{\mathbf{n,2d}}$. The face $F(v_1, \dots, v_k)$ has the form

$$F(v_1, \dots, v_k) = \{p \in P_{\mathbf{n,2d}} \mid p(v_1) = \cdots = p(v_k) = 0, \ v_i \in \mathbb{S}^{n-1}\}.$$

Therefore $F(v_1, \dots, v_k)$ consists of all nonnegative forms with zeroes at prescribed points $v_1, \dots, v_k \in \mathbb{S}^{n-1}$. It is natural to expect that every additional zero increases the codimension of the exposed face by n so that $\operatorname{codim} F(v_1, \dots, v_k) = kn$. However, this intuition fails if the number of zeroes k is sufficiently large. In particular if we prescribe enough zeroes, it is not even clear when the face $F(v_1, \dots, v_k)$ is nonempty. The question of the dimension of $F(v_1, \dots, v_k)$ is quite complicated [6] and it is related to the celebrated Alexander–Hirschowitz theorem [17].

Exposed extreme rays of $P_{\mathbf{n,2d}}$ are also conceptually simple: a nonnegative form $p \in P_{\mathbf{n,2d}}$ is an exposed extreme ray of $P_{\mathbf{n,2d}}$ if and only if the variety defined by p is maximal among all varieties defined by nonnegative polynomials.

Exercise 4.25. Show that $p \in P_{\mathbf{n,2d}}$ is an exposed extreme ray of $P_{\mathbf{n,2d}}$ if and only if for all $q \in P_{\mathbf{n,2d}}$ with $V(p) \subseteq V(q)$ it follows that $q = \lambda p$ for some $\lambda \in \mathbb{R}$.

4.4.5 Nonexposed Faces of $P_{\mathbf{n,2d}}$

The cone $P_{\mathbf{n,2d}}$ has many nonexposed faces. If a form p has a zero at a point $v \in \mathbb{R}$, then it must have a double zero at v. Exposed faces of $P_{\mathbf{n,2d}}$ capture double zeroes on any set of points $v_1, \dots, v_k$, but exposed faces fail to capture zeroes of higher order.

Exercise 4.26. Show that x_1^{2d} is an extreme ray of $P_{\mathbf{n,2d}}$. Use Exercise 4.25 to conclude that x_1^{2d} is not exposed.

More generally, the following construction explains the origins of nonexposed faces of $P_{\mathbf{n,2d}}$. Consider a maximal face $F(v)$ of $P_{\mathbf{n,2d}}$. We can construct an exposed subface of $F(v)$ by considering nonnegative forms with zeroes at v and w for some $w \in \mathbb{S}^{n-1}$. We can also build nonexposed subfaces of $F(v)$ by considering nonnegative forms that are more singular at v.

Let $p \in F(v)$, so that p is a nonnegative form and $p(v) = 0$. Since 0 is the global minimum of p and $\nabla p(v) = 0$, it follows that the Hessian $\nabla^2 p(v)$ must be a positive semidefinite matrix. Let $F_w(v)$ be the set of all nonnegative forms p with zero at v whose Hessian at v is positive semidefinite and w lies in the kernel of $\nabla^2 p(v)$:

$$F_w(v) = \{p \in F(v) \mid \nabla^2 p(v) \cdot w = 0\}.$$

Exercise 4.27. Show that $F_w(v)$ is a face of $P_{\mathbf{n,2d}}$. Use the characterization of exposed faces of $P_{\mathbf{n,2d}}$ to show that $F_w(v)$ is not an exposed face of $P_{\mathbf{n,2d}}$.

4.4.6 Algebraic Boundaries

The boundaries of the cones $P_{\mathbf{n,2d}}$ and $\Sigma_{\mathbf{n,2d}}$ are hypersurfaces in $\mathbb{R}[x]_{\mathbf{2d}}$. Suppose that we would like to describe these hypersurfaces by polynomial equations. This leads to the notion of *algebraic boundary* of the cones $P_{\mathbf{n,2d}}$ and $\Sigma_{\mathbf{n,2d}}$, which is obtained by taking the *Zariski closure* of the boundary hypersurfaces. As explained in Chapter 5, the algebraic boundary of $P_{\mathbf{n,2d}}$ is cut out by a single polynomial, the discriminant. The algebraic boundary of the cone of sums of squares is significantly more complicated.

Exercise 4.28. Show that the hypersurface cut out by the discriminant is a component of the algebraic boundary of $\Sigma_{\mathbf{n,2d}}$.

The above exercise shows that the algebraic boundary of $P_{\mathbf{n,2d}}$ is included in the algebraic boundary of $\Sigma_{\mathbf{n,2d}}$. This seems counterintuitive, but it occurs because we passed to the Zariski closures of the actual boundaries. We will see below that for $\Sigma_{\mathbf{3,6}}$ and $\Sigma_{\mathbf{4,4}}$ the algebraic boundary of the cone of sums of squares has one more component, which is described in Exercise 4.51.

4.5 Generalizing the Hypercube Example

We completely described the values of nonnegative forms and sums of squares on the specific set S of ± 1 vectors in $\mathbb{R}^4$ and we have seen, just from the evaluation on S, that there exist nonnegative forms in $\mathbb{R}[x]_{\mathbf{4,4}}$ that are not sums of squares.

However, these descriptions are limited to the specific set S. We now extend the arguments of Section 4.3 to work in far greater generality. We begin by explaining how the set S was chosen in the first place.

4.5.1 Hypercube Example Revisited

Let q_i be the three quadratic forms

$$q_1 = x_1^2 - x_2^2, \quad q_2 = x_1^2 - x_3^2, \quad q_3 = x_1^2 - x_4^2,$$

and let V be the set of common zeroes of q_i:

$$V = \{x \in \mathbb{R}^4 \mid q_i(x) = 0 \quad \text{for} \quad i = 1, 2, 3\}.$$

Viewed projectively V consists of eight points in the real projective space $\mathbb{RP}^3$. Viewed affinely V consists of eight lines, each line spanned by a point in S. We can extend much of what was proved about the values of nonnegative polynomials to zero-dimensional intersections in $\mathbb{RP}^{n-1}$.

4.5.2 Zero-Dimensional Intersections

Let V be a set of finitely many points in $\mathbb{RP}^{n-1}$:

$$V = \{\bar{s}_1, \ldots, \bar{s}_k\}.$$

Suppose that V is the complete set of real projective zeroes of some forms $q_1, \ldots, q_m$ of degree d:

$$V = \{x \in \mathbb{RP}^{n-1} \mid q_1(x) = \cdots = q_m(x) = 0\}.$$

For each $\bar{s}_i \in V$ let s_i be an affine representative of $\bar{s}_i$ lying on the line spanned by $\bar{s}_i$. Now let $S = \{s_1, \ldots, s_k\}$, be the set of affine representatives corresponding to the common zeroes of q_i.

Let's consider the values of nonnegative forms of degree $2d$ on S. Let π_S: $\mathbb{R}[x]_{\mathbf{2d}} \to \mathbb{R}^k$ be the evaluation projection:

$$\pi_S(f) = (f(s_1), \ldots, f(s_k)) \quad \text{for} \quad f \in \mathbb{R}[x]_{\mathbf{2d}}.$$

Let H be the image of $\mathbb{R}[x]_{\mathbf{2d}}$ and let P' be the image of $P_{\mathbf{n,2d}}$ under π_S:

$$H = \pi_S(\mathbb{R}[x]_{\mathbf{2d}}), \qquad P' = \pi_S(P_{\mathbf{n,2d}}).$$

We have an additional complication that H does not have to equal $\mathbb{R}^k$. We know, however, that P' must lie in H, and since we are evaluating nonnegative forms it follows that P' lies inside the nonnegative orthant of $\mathbb{R}^k$: $P' \subseteq \mathbb{R}^k_+$. Therefore it follows that P' lies inside the intersection of H and $\mathbb{R}^k_+$:

$$P' \subseteq H \cap \mathbb{R}^k_+.$$

The following theorem shows that this inclusion is almost an equality.

Theorem 4.29. *Let $\mathbb{R}^k_{++}$ be the positive orthant of $\mathbb{R}^k$. The intersection of H with the positive orthant is contained in P':*

$$H \cap \mathbb{R}^k_{++} \subset P'.$$

Before proving Theorem 4.29 we make some remarks. As we know from Exercise 4.7 we cannot simply conclude that $P' = H \cap \mathbb{R}^k_+$ using a closure argument, since a projection of a closed cone does not have to be closed. We now show that this occurs for evaluation projections as well.

Exercise 4.30. Let $S \subset \mathbb{R}^5$ be the set of 16 points $S = \{\pm 1, \pm 1, \pm 1, \pm 1, 1\}$. Show that S can be defined as a common zero set of four quadratic forms in $\mathbb{R}[x]_{\mathbf{5},\mathbf{2}}$, and use Theorem 4.29 to show that $\mathbb{R}^{16}_{++} \subseteq \pi_S(P_{\mathbf{5},\mathbf{4}})$. Show that the standard basis vectors $e_i \in \mathbb{R}^{16}$ are not in the image $\pi_S(P_{\mathbf{5},\mathbf{4}})$. In other words, the vectors e_i are not realized as values on S of a nonnegative form of degree 4 in 5 variables, but all strictly positive points in $\mathbb{R}^{16}_{++}$ are realized.

***Proof of Theorem* 4.29.** Let $v = (v_1, \dots, v_k) \in H \cap \mathbb{R}^k_{++}$. Since $v \in H$ there exists a form $f \in \mathbb{R}[x]_{\mathbf{2d}}$ such that $f(s_i) = v_i$. Let $g = q_1^2 + \cdots + q_m^2$, where q_i are the forms defining V. We claim that for large enough $\lambda \in \mathbb{R}$ the form $\bar{f} = f + \lambda g$ will be nonnegative, and since each q_i is zero on S we will also have $\pi_S\left(\bar{f}\right) = v$.

By homogeneity of $\bar{f}$ it suffices to show that it is nonnegative on the unit sphere $\mathbb{S}^{n-1}$. Furthermore, we may assume that the evaluation points s_i lie on the unit sphere. Since we are dealing with forms, evaluation on the points outside of the unit sphere amounts to rescaling of the values on $\mathbb{S}^{n-1}$.

Let $B_\epsilon(S)$ be the open epsilon neighborhood of S in the unit sphere $\mathbb{S}^{n-1}$. Since $f(s_i) > 0$ for all i, it follows that for sufficiently small ϵ the form f is strictly positive on $B_\epsilon(S)$:

$$f(x) > 0 \quad \text{for all} \quad x \in B_\epsilon(S).$$

The complement of $B_\epsilon(S)$ in $\mathbb{S}^{n-1}$ is compact, and therefore we can let m_1 be the minimum of g and m_2 be the minimum of f on $\mathbb{S}^{n-1} \setminus B_\epsilon(S)$. If $m_2 \geq 0$, then f itself is nonnegative and we are done. Therefore, we may assume $m_2 < 0$. We also note that since g vanishes on S only, it follows that m_1 is strictly positive.

Now let $\lambda \geq -\frac{m_2}{m_1}$. The form $\bar{f} = f + \lambda g$ is positive on $B_\epsilon(S)$. By construction of $B_\epsilon(S)$ we also see that the minimum of $\bar{f}$ on the complement of $B_\epsilon(S)$ is at least 0. Therefore $\bar{f}$ is nonnegative on the unit sphere $\mathbb{S}^{n-1}$, and we are done. □

We proved in Theorem 4.29 that any set of strictly positive values on the finite set S, coming from real zeroes of forms of degree d, can be achieved by a globally nonnegative form of degree $2d$. We now look at the values that sums of squares can take on such sets S.

4.5.3 Values of Sums of Squares

We recall from Section 4.3 that the reason that sums of squares could not achieve all the possible nonnegative values on the hypercube was that the values of quadratic forms on the hypercube satisfied a linear relation. The points of the hypercube come from common zeroes of the quadratic forms, as we have seen in Section 4.5.1.

There is a general theory in algebraic geometry on the number of relations that values of forms of certain degree have to satisfy on finite sets of points. These relations are known as Cayley–Bacharach relations. For more details we refer the reader to [10].

At first glance it is surprising that there should be any linear relation at all. If the points were chosen generically then the values of forms of degree d on these points would be linearly independent, at least until we have as many points as the dimension of the vector space of forms of degree d. However, our choice of points is not generic; point sets that come from common zeroes are special.

For the cases $\mathbb{R}[x]_{\mathbf{4,4}}$ and $\mathbb{R}[x]_{\mathbf{3,6}}$ it is easy to establish the existence of the linear relation by simple dimension counting. We explain the case of $\mathbb{R}[x]_{\mathbf{4,4}}$.

Since common zeroes of real forms do not have to be real, for this section we will work with complex forms. Suppose that $q_1, q_2, q_3 \in \mathbb{C}[x]_{\mathbf{4,2}}$ are complex quadratic forms in 4 variables. As before let V be the complete set of projective zeroes of some forms q_1, q_2, q_3:

$$V = \{\bar{x} \in \mathbb{CP}^3 \quad | \quad q_1(\bar{x}) = q_2(\bar{x}) = q_3(\bar{x}) = 0\}.$$

Three quadratic forms in $\mathbb{C}[x]_{\mathbf{4,2}}$ are expected to generically have $2^3 = 8$ common zeroes. Suppose that this is the case and let $V = \{\bar{s}_1, \ldots, \bar{s}_8\}$.

For each $\bar{s}_i \in V$ let s_i be an affine representative of $\bar{s}_i$ lying on the line corresponding to $\bar{s}_i$. Let $S = \{s_1, \ldots, s_8\}$, be the set of affine representatives corresponding to the common zeroes of q_i. Define $\pi_S : \mathbb{C}[x]_{\mathbf{4,2}} \to \mathbb{C}^8$ to be the evaluation projection.

Lemma 4.31. *The values of quadratic forms in $\mathbb{C}[x]_{\mathbf{4,2}}$ satisfy a linear relation on the points of S. In other words there exist $\mu_1, \ldots \mu_8 \in \mathbb{C}$ such that*

$$\mu_1 f(s_1) + \cdots + \mu_8 f(s_8) = 0 \quad \textit{for all} \quad f \in \mathbb{C}[x]_{\mathbf{4,2}}. \tag{4.3}$$

Proof. The dimension of $\mathbb{C}[x]_{\mathbf{4,2}}$ is 10. Note that the kernel of π_S contains the three forms q_i, since each q_i evaluates to 0 on S. Therefore the dimension of the kernel of π_S is at least 3. It follows that the image of π_S has dimension at most $10 - 3 = 7$. Since the image of π_S lies inside $\mathbb{C}^8$, it follows that there exists a linear functional that vanishes on the image of π_S. This linear functional gives us the desired linear relation. □

Remark 4.32. *It is possible to show in the above proof that the dimension of the kernel of π_S is exactly* 3 *and therefore the linear relation* (4.3) *is unique. Furthermore, it can be shown that each $\mu_i \neq 0$, or, in other words, the unique linear relation has to involve all of the points of S.*

Exercise 4.33. Suppose that $q_1, q_2 \in \mathbb{C}[x]_{3,3}$ are two cubic forms intersecting in $3^2 = 9$ points in $\mathbb{CP}^2$. Let S be the set of affine representatives of the common zeroes of q_1 and q_2. Use the argument of Lemma 4.31 to show that the values of cubic forms on S satisfy a linear relation.

Exercise 4.34. The Robinson form

$$R(x, y, z) = x^6 + y^6 + z^6 - (x^4y^2 + x^2y^4 + x^4z^2 + x^2z^4 + y^4z^2 + y^2z^4) + 3x^2y^2z^2$$

is an explicit example of a nonnegative polynomial that is not a sums of squares. Let $q_1 = x(x+z)(x-z)$ and $q_2 = y(y+z)(y-z)$. Calculate the 9 common zeroes of q_1 and q_2. Show that $R(x, y, z)$ vanishes on 8 of the 9 zeroes. Use Exercise 4.33 to show that $R(x, y, z)$ is not a sum of squares.

We have examined in detail what happens to values of nonnegative forms and sums of squares on finite sets of points coming from common zeroes of forms. However, this still seems to be a very special construction. We now move to show that the difference in values on such sets is in fact the fundamental reason that there exists nonnegative polynomials that are not sums of squares.

4.6 Dual Cone of $\Sigma_{\mathbf{n,2d}}$

We gave a simple description of the extreme rays of the dual cone $P^*_{\mathbf{n,2d}}$ in Corollary 4.21. The description of the extreme rays of the dual cone $\Sigma^*_{\mathbf{n,2d}}$ is significantly more complicated. We will see that evaluation on the special finite point sets we described in Section 4.5 will naturally lead to extreme rays of $\Sigma^*_{\mathbf{n,2d}}$.

We first describe the connection between $\Sigma^*_{\mathbf{n,2d}}$ and the cone of positive semidefinite matrices that lies at the heart of semidefinite programming approaches to polynomial optimization. To every linear functional $\ell \in \mathbb{R}[x]^*_{\mathbf{2d}}$ we can associate a quadratic form Q_ℓ defined on $\mathbb{R}[x]_{\mathbf{d}}$ by setting

$$Q_\ell(f) = \ell(f^2) \quad \text{for all} \quad f \in \mathbb{R}[x]_{\mathbf{d}}.$$

The cone $\Sigma^*_{\mathbf{n,2d}}$ can be thought of as a section of the cone of positive semidefinite quadratic forms. We now show how this description arises.

Lemma 4.35. *Let ℓ be a linear functional in $\mathbb{R}[x]^*_{\mathbf{2d}}$. Then $\ell \in \Sigma^*_{\mathbf{n,2d}}$ if and only if the quadratic form Q_ℓ is positive semidefinite.*

Proof. Suppose that $\ell \in \Sigma^*_{\mathbf{n,2d}}$. Then $\ell(f^2) \geq 0$ for all $f \in \mathbb{R}[x]_{\mathbf{d}}$. Therefore $Q_\ell(f) \geq 0$ for all $f \in \mathbb{R}[x]_{\mathbf{d}}$ and Q_ℓ is positive semidefinite.

Now suppose that Q_ℓ is positive semidefinite. Then $\ell(f^2) \geq 0$ for all $f \in \mathbb{R}[x]_{\mathbf{d}}$. Let $g = \sum f_i^2 \in \Sigma_{\mathbf{n,2d}}$. Then $\ell(g) = \sum \ell(f_i^2) \geq 0$ and $\ell \in \Sigma^*_{\mathbf{n,2d}}$. $\square$

An Aside: The Monomial Basis and Moment Matrices

Suppose that we fix the monomial basis for $\mathbb{R}[x]_{\mathbf{d}}$. Given a linear functional $\ell \in \mathbb{R}[x]^*_{\mathbf{2d}}$ we can write an explicit matrix $M(\ell)$ for the quadratic form Q_ℓ using the monomial basis of $\mathbb{R}[x]_{\mathbf{d}}$. The matrix $M(\ell)$ is known as the *moment matrix* or *generalized Hankel matrix*. The entries of $M(\ell)$ are indexed by monomials $x^\alpha, x^\beta \in \mathbb{R}[x]_{\mathbf{d}}$. The entry $M(\ell)_{\alpha,\beta}$ is given by evaluating ℓ on $x^\alpha x^\beta = x^{\alpha+\beta}$:

$$M(\ell)_{\alpha,\beta} = \ell(x^{\alpha+\beta}).$$

For example, consider the linear functional $\ell_v : \mathbb{R}[x]_{\mathbf{2,4}} \to \mathbb{R}$ given by evaluation on $v = (1, 2)$. The monomial basis of $\mathbb{R}[x]_{\mathbf{2,2}}$ is given by x^2, xy, y^2 and the moment matrix of $M(\ell_v)$ reads as

$$M(\ell_v) = \begin{bmatrix} 1 & 2 & 4 \\ 2 & 4 & 8 \\ 4 & 8 & 16 \end{bmatrix}.$$

The rank of the quadratic form Q_ℓ is the same as the rank of its moment matrix $M(\ell)$, and Q_ℓ being nonnegative is equivalent to having a positive semidefinite moment matrix $M(\ell)$. However, the moment approach is tied to the specific choice of the monomial basis. Below we prefer to keep a basis independent approach with emphasis on the underlying geometry, but we note that the results are readily translatable into the terminology of moments. ■

Let $S^{n,d}$ be the vector space of real quadratic forms on $\mathbb{R}[x]_{\mathbf{d}}$. We can view the dual space $\mathbb{R}[x]^*_{\mathbf{2d}}$ as a subspace of $S^{n,d}$ by identifying the linear functional $\ell \in \mathbb{R}[x]^*_{\mathbf{2d}}$ with its quadratic form Q_ℓ. Let $S^{n,d}_+$ be the cone of positive semidefinite forms in $S^{n,d}$:

$$S^{n,d}_+ = \left\{ Q \in S^{n,d} \;\middle|\; Q(f) \geq 0 \quad \text{for all} \quad f \in \mathbb{R}[x]_{\mathbf{d}} \right\}.$$

We can restate Lemma 4.35 as follows.

Corollary 4.36. *The cone $\Sigma^*_{\mathbf{n,2d}}$ is the section of the cone of positive semidefinite matrices $S^{n,d}_+$ with the subspace $\mathbb{R}[x]^*_{\mathbf{2d}}$:*

$$\Sigma^*_{\mathbf{n,2d}} = S^{n,d}_+ \cap \mathbb{R}[x]^*_{\mathbf{2d}}.$$

Note that this shows that the cone $\Sigma^*_{\mathbf{n,2d}}$ is a *spectrahedron.*

The following exercise establishes the connection between the cone of sums of squares $\Sigma_{\mathbf{n,2d}}$ and the cone of positive semidefinite matrices $S^{n,d}_+$. This allows us to formulate sums of squares questions in terms of semidefinite programming.

Exercise 4.37. Use the result of Corollary 4.36 to show that the cone $\Sigma_{\mathbf{n,2d}}$ is a projection of the cone $S^{n,d}_+$ of positive semidefinite matrices on $\mathbb{R}[x]_{\mathbf{d}}$. Use the monomial basis of $\mathbb{R}[x]_{\mathbf{d}}$ to describe this projection explicitly. Conclude that the cone $\Sigma_{\mathbf{n,2d}}$ is a *projected spectrahedron.* (Hint: See Chapter 5 for a general discussion of the relationship between duality and projections.)

Remark 4.38. *In order to apply the result of Exercise 4.37 to actual computation we need to work with an explicit basis of $\mathbb{R}[x]_{\mathbf{d}}$. See Chapter 3 for the discussion of possible basis choices and their impact on computational performance. We note that the size of the positive semidefinite matrices we work with is the dimension of $\mathbb{R}[x]_{\mathbf{d}}$, which is equal to $\binom{n+d-1}{d}$. Therefore the size of the underlying positive semidefinite matrices increases rather rapidly as a function of n and d. This is one of the main computational limitations of semidefinite programming approaches to polynomial optimization.*

We would like to see what separates sums of squares from nonnegative forms. The extreme rays of $\Sigma^*_{\mathbf{n,2d}}$ cut out the cone of sums of squares. Therefore we would like to find extreme rays of $\Sigma^*_{\mathbf{n,2d}}$ that are not in the dual cone $P^*_{\mathbf{n,2d}}$, since these are the functionals that distinguish the cone of sums of squares from the cone of nonnegative forms.

Formally the dual cone $\Sigma^*_{\mathbf{n,2d}}$ is defined as the cone of linear functionals nonnegative on $\Sigma_{\mathbf{n,2d}}$, which is equivalent to being nonnegative on squares. One way of constructing linear functionals nonnegative on squares is to consider point evaluation functionals ℓ_v with $v \in \mathbb{R}^n$ that send $p \in \mathbb{R}[x]_{\mathbf{2d}}$ to $p(v)$. However, as we have seen in Corollary 4.21, point evaluation functionals are precisely the extreme rays of $P^*_{\mathbf{n,2d}}$. Therefore, these linear functionals are not helpful in distinguishing between $\Sigma^*_{\mathbf{n,2d}}$ and $P^*_{\mathbf{n,2d}}$. Our goal now is to find a new way of constructing functionals nonnegative on squares and also to understand why such functionals do not exist when $\Sigma_{\mathbf{n,2d}} = P_{\mathbf{n,2d}}$.

We showed in Corollary 4.36 that the cone $\Sigma^*_{\mathbf{n,2d}}$ is a spectrahedron. We now prove a general lemma about spectrahedra that states that extreme rays of a spectrahedron are quadratic forms with maximal kernel [20]. The examination of the kernels of extreme rays of $\Sigma^*_{\mathbf{n,2d}}$ will provide a crucial tool for our understanding of $\Sigma^*_{\mathbf{n,2d}}$.

Let $\mathcal{S}$ be the vector space of quadratic forms on a real vector space V. Let $\mathcal{S}_+$ be the cone of psd forms in $\mathcal{S}$.

Lemma 4.39. *Let L be a linear subspace of $\mathcal{S}$ and let K be the section of $\mathcal{S}_+$ with L:*

$$K = \mathcal{S}_+ \cap L.$$

Suppose that a quadratic form Q spans an extreme ray of K. Then the kernel of Q is maximal for all quadratic forms in L: if $P \in L$ and $\ker Q \subseteq \ker P$ then $P = \lambda Q$ for some $\lambda \in \mathbb{R}$.

Proof. Suppose not, so that there exists an extreme ray Q of K and a quadratic form $P \in L$ such that $\ker Q \subseteq \ker P$ and $P \neq \lambda Q$. Since $\ker Q \subseteq \ker P$ it follows that all eigenvectors of both Q and P corresponding to nonzero eigenvalues lie in the orthogonal complement $(\ker Q)^\perp$ of $\ker Q$. Furthermore, Q is positive definite on $(\ker Q)^\perp$.

It follows that Q and P can be simultaneously diagonalized to matrices Q' and P' with the additional property that whenever the diagonal entry Q'_{ii} is 0 the corresponding entry P'_{ii} is also 0. Therefore, for sufficiently small $\epsilon \in \mathbb{R}$ we have that $Q+\epsilon P$ and $Q-\epsilon P$ are positive semidefinite and therefore $Q+\epsilon P, Q-\epsilon P \in K$. Then Q is not an extreme ray of K, which is a contradiction. □

We now apply Lemma 4.39 to the case $\Sigma^*_{\mathbf{n,2d}}$. This gives us a crucial tool for studying extreme rays of $\Sigma^*_{\mathbf{n,2d}}$.

Corollary 4.40. *Suppose that Q spans an extreme ray of $\Sigma^*_{\mathbf{n,2d}}$. Then either* $\operatorname{rank} Q = 1$ *or the forms in the kernel of Q have no common zeroes, real or complex.*

Proof. Let $W \subset \mathbb{R}[x]_{\mathbf{d}}$ be the kernel of Q and suppose that the forms in W have a common real zero $v \neq 0$. Let $\ell \in \mathbb{R}[x]_{\mathbf{2d}}^*$ be the linear functional given by evaluation at v: $\ell(f) = f(v)$ for all $f \in \mathbb{R}[x]_{\mathbf{2d}}$. Then Q_ℓ is a rank 1 positive semidefinite quadratic form and $\ker Q \subseteq \ker Q_\ell$. By Lemma 4.39 it follows that $Q = \lambda Q_\ell$ and thus Q has rank 1.

Now suppose that the forms in W have a common complex zero $z \neq 0$. Let $\ell \in \mathbb{R}[x]_{\mathbf{2d}}^*$ be the linear functional given by taking the real part of the value at z: $\ell(f) = \operatorname{Re} f(z)$ for all $f \in \mathbb{R}[x]_{\mathbf{2d}}$. It is easy to check that the kernel of Q_ℓ includes all forms that vanish at z and therefore $W \subseteq \ker Q_\ell$. Therefore by applying Lemma 4.39 we again see that $Q = \lambda Q_\ell$. However, we claim that Q_ℓ is not a positive semidefinite form.

The quadratic form Q_ℓ is given by $Q_\ell(f) = \operatorname{Re} f^2(z)$ for $f \in \mathbb{R}[x]_{\mathbf{d}}$. However, there exist $f \in \mathbb{R}[x]_{\mathbf{d}}$ such that $f(z)$ is purely imaginary and therefore $Q_\ell(f) < 0$. The corollary now follows. $\square$

Corollary 4.40 shows that extreme rays of $\Sigma_{\mathbf{n},\mathbf{2d}}^*$ are of two types: either they are rank 1 quadratic forms or they have a kernel with no common zeroes. We now deal with the rank 1 extreme rays of $\Sigma_{\mathbf{n},\mathbf{2d}}^*$. For $v \in \mathbb{R}^n$ let ℓ_v be the linear functional in $\mathbb{R}[x]_{\mathbf{2d}}^*$ given by evaluation at v,

$$\ell_v(f) = f(v) \text{ for } f \in \mathbb{R}[x]_{\mathbf{2d}},$$

and let Q_v be the quadratic form associated to ℓ_v: $Q_v(f) = f^2(v)$. In this case we say that Q_v (or ℓ_v) corresponds to point evaluation. Recall that the inequalities $\ell_v \geq 0$ are the defining inequalities of the cone of nonnegative forms $P_{\mathbf{n},\mathbf{2d}}$. The following lemma shows that all rank 1 forms in $\mathbb{R}[x]_{\mathbf{2d}}^*$ correspond to point evaluations. Since we are interested in the inequalities that are valid on $\Sigma_{\mathbf{n},\mathbf{2d}}$ but not valid on $P_{\mathbf{n},\mathbf{2d}}$ it allows us to disregard rank 1 extreme rays of $\Sigma_{\mathbf{n},\mathbf{2d}}^*$ and focus on the case of a kernel with no common zeros.

Lemma 4.41. *Suppose that Q is a rank 1 quadratic form in $\mathbb{R}[x]_{\mathbf{2d}}^*$. Then $Q = \lambda Q_v$ for some $v \in \mathbb{R}^n$ and $\lambda \in \mathbb{R}$.*

Proof. Let Q be a rank 1 form in $\mathbb{R}[x]_{\mathbf{2d}}^*$. Then $Q(f) = \lambda s^2(f)$ for some linear functional $s \in \mathbb{R}[x]_{\mathbf{d}}^*$. Therefore it suffices to show that if $Q = s^2(f)$ for some $s \in \mathbb{R}[x]_{\mathbf{d}}^*$, then $Q = Q_v$ for some $v \in \mathbb{R}^n$.

Since $Q \in \mathbb{R}[x]_{\mathbf{2d}}^*$ we know that Q is defined by $Q(f) = \ell(f^2)$ for a linear functional $\ell \in \mathbb{R}[x]_{\mathbf{2d}}^*$ and therefore $\ell(f^2) = s^2(f)$ for all $f \in \mathbb{R}[x]_{\mathbf{d}}$. We have $Q(f+g) = \ell((f+g)^2) = \ell(f^2)+2\ell(fg)+\ell(g^2) = (s(f)+s(g))^2 = s^2(f)+2s(f)s(g)+s^2(g)$ and it follows that $\ell(fg) = s(f)s(g)$ for all $f, g \in \mathbb{R}[x]_{\mathbf{d}}$.

Let x^α denote the monomial $x_1^{\alpha_1} \cdots x_n^{\alpha_n}$. If we take monomials x^α, x^β, x^γ, x^δ in $\mathbb{R}[x]_{\mathbf{d}}$ such that $x^\alpha x^\beta = x^\gamma x^\delta$, then we must have $s(x^\alpha)s(x^\beta) = s(x^\gamma)s(x^\delta)$.

Suppose that $s(x_i^d) = 0$ for all i. Then we see that

$$s(x_i^{d-1}x_j)^2 = s(x_i^d)s(x_i^{d-2}x_j^2) = 0,$$

and continuing in similar fashion we have $s(x^\alpha) = 0$ for all monomials in $\mathbb{R}[x]_{\mathbf{d}}$. Then ℓ is the zero functional and Q does not have rank one which is a contradiction.

We may assume without loss of generality that $s(x_1^d) \neq 0$. Since we are interested in $\ell(f^2) = s^2(f)$ we can work with $-s$ if necessary, and thus we may assume that $s(x_1^d) > 0$. Let $s_i = s(x_1^{d-1}x_i)$ for $1 \leq i \leq n$. We will express $s(x^\alpha)$ in terms of s_i for all $x^\alpha \in \mathbb{R}[x]_{\mathbf{d}}$. Since $(x_1^d)(x_1^{d-2}x_ix_j) = (x_1^{d-1}x_i)(x_1^{d-1}x_j)$ we have $s(x_1^{d-2}x_ix_j) = s_is_j/s_1$. Continuing in this fashion we find that

$$s(x_1^{\alpha_1}\cdots x_n^{\alpha_n}) = \frac{s_2^{\alpha_2}\cdots s_n^{\alpha_n}}{s_1^{d-1-\alpha_1}}.$$

Now let $v \in \mathbb{R}^n$ be the following vector:

$$v = (s_1^{1/d}, s_1^{-(d-1)/d}s_2, \ldots, s_1^{-(d-1)/d}s_n).$$

Let s_v be the linear functional on $\mathbb{R}[x]_{\mathbf{d}}$ defined by evaluating a form at v: $s_v(f) = f(v)$. Then we have $s_v(x_1^{d-1}x_i) = s_i$ and

$$s_v(x_1^{\alpha_1}\cdots x_n^{\alpha_n}) = s_2^{\alpha_2}\cdots s_n^{\alpha_n}s_1^{\alpha_1/d-(d-1)(d-\alpha_1)/d} = \frac{s_2^{\alpha_2}\cdots s_n^{\alpha_n}}{s_1^{d-1-\alpha_1}}.$$

Since s agrees with s_v on monomials it follows that $s = s_v$ and thus $\ell(f^2) = s^2(f) = f(v)^2 = f^2(v)$. Therefore ℓ indeed corresponds to point evaluation and we are done. □

Suppose that Q_ℓ spans an extreme ray of $\Sigma_{\mathbf{n,2d}}^*$ that does not correspond to point evaluation. Let W_ℓ be the kernel of Q_ℓ. Then by Corollary 4.40 and Lemma 4.41 we know that the forms in W_ℓ have no common zeroes real or complex. This condition gives us a lot of dimensional information about W_ℓ and places strong restrictions on the linear functionals ℓ. As we will see, for the three equality cases of Hilbert's theorem the dimensional restrictions on W_ℓ will allow us to derive nonexistence of the extreme rays of $\Sigma_{\mathbf{n,2d}}^*$ with kernel W_ℓ, thus proving the equality between nonnegative forms and sums of squares.

Let W be a linear subspace of $\mathbb{R}[x]_{\mathbf{d}}$ and define $W^{\langle 2\rangle}$ to be the degree $2d$ part of the ideal generated by W:

$$W^{\langle 2\rangle} = \langle W\rangle_{2d}.$$

We use $V_{\mathbb{C}}(W)$ to denote the set of common zeroes (real and complex) of forms in W.

We next show that there is a strong relation between the linear functional ℓ and the kernel W_ℓ of the quadratic form Q_ℓ. Namely, we show that ℓ vanishes on all of $W_\ell^{\langle 2\rangle}$:

$$\ell(p) = 0 \quad \text{for all} \quad p \in W_\ell^{\langle 2\rangle}. \tag{4.4}$$

We will write the condition (4.4) as $\ell(W_\ell^{\langle 2\rangle}) = 0$ for short. We also now show that W_ℓ is the maximal subspace among all W such that $\ell(W^{\langle 2\rangle}) = 0$.

Lemma 4.42. *Let Q_ℓ be a quadratic form in $\Sigma_{\mathbf{n,2d}}^*$ and let $W_\ell \subset \mathbb{R}[x]_{\mathbf{d}}$ be the kernel of Q_ℓ. Then $p \in W_\ell$ if and only if $\ell(pq) = 0$ for all $q \in \mathbb{R}[x]_{\mathbf{d}}$.*

Proof. In order to investigate W_ℓ, we need to define the associated bilinear form B_ℓ:

$$B_\ell(p,q) = \frac{Q_\ell(p+q) - Q_\ell(p) - Q_\ell(q)}{2} \quad \text{for} \quad p,q \in \mathbb{R}[x]_{\mathbf{d}}.$$

By definition of Q_ℓ we have $Q_\ell(p) = \ell(p^2)$. Therefore it follows that

$$B_\ell(p,q) = \ell(pq).$$

A form $p \in \mathbb{R}[x]_{\mathbf{d}}$ is in the kernel of Q_ℓ if and only if $B_\ell(p,q) = 0$ for all $q \in \mathbb{R}[x]_{\mathbf{d}}$. Since $B_\ell(p,q) = \ell(pq)$, the lemma follows. □

We note that $V_{\mathbb{C}}(W) = \emptyset$ implies that the dimension of W_ℓ is at least n and we can find forms $p_1, \ldots, p_n \in W_\ell$ that have no common zeroes. We need a dimensional lemma from algebraic geometry which we will use without proof.

Lemma 4.43. *Suppose that $p_1, \ldots, p_n \in \mathbb{R}[x]_{\mathbf{d}}$ are forms such that $V_{\mathbb{C}}(p_1, \ldots, p_n) = \emptyset$ and let $I = \langle p_1, \ldots, p_n \rangle$ be the ideal generated by the forms p_i. Then*

$$\dim I_{2d} = n \cdot \dim \mathbb{R}[x]_{\mathbf{d}} - \binom{n}{2}.$$

Remark 4.44. *The forms $p_1, \ldots, p_n \in \mathbb{R}[x]_{\mathbf{d}}$ such that $V_{\mathbb{C}}(p_1, \ldots, p_n) = \emptyset$ form a complete intersection. The dimensional information of the ideal $\langle p_1, \ldots, p_n \rangle$ is well understood via the Koszul complex. The statement of Lemma 4.43 is an easy consequence of the powerful techniques developed for complete intersections* [12].

4.6.1 Equality Cases of Hilbert's Theorem

We have obtained enough information on the dual cone $\Sigma^*_{\mathbf{n,2d}}$ to give a unified proof of the equality cases of Hilbert's theorem.

Proof of equality cases in Hilbert's theorem. Suppose that $\Sigma_{\mathbf{n,2d}} \neq P_{\mathbf{n,2d}}$. Then there exists an extreme ray of $\Sigma^*_{\mathbf{n,2d}}$ that does not come from point evaluation. Let ℓ be such an extreme ray and let W_ℓ be the kernel of Q_ℓ. By Lemma 4.41 it follows that $\operatorname{rank} Q_\ell > 1$, and therefore by Corollary 4.40 we see that $V_{\mathbb{C}}(W_\ell) = \emptyset$.

Therefore $\dim W_\ell \geq n$ and we can find forms $p_1, \ldots, p_n \in W_\ell$ such that $V_{\mathbb{C}}(p_1, \ldots, p_n) = \emptyset$. Let $I = \langle p_1, \ldots, p_n \rangle$ be the ideal generated by p_i. It follows that $W_\ell^{\langle 2 \rangle}$ includes I_{2d} and $\dim I_{2d} = n \cdot \dim \mathbb{R}[x]_{\mathbf{d}} - \binom{n}{2}$ by Lemma 4.43. Therefore we see that

$$\dim W_\ell^{\langle 2 \rangle} \geq n \cdot \dim \mathbb{R}[x]_{\mathbf{d}} - \binom{n}{2}.$$

However, by (4.4) we must also have

$$\dim W_\ell^{\langle 2 \rangle} \leq \dim \mathbb{R}[x]_{\mathbf{2d}} - 1,$$

since a nontrivial linear functional $\ell \in \mathbb{R}[x]^*_{\mathbf{2d}}$ vanishes on $W_\ell^{\langle 2\rangle}$. We now go case by case and derive a contradiction from these dimensional facts in each of the equality cases.

Suppose that $n = 2$. Then $\dim \mathbb{R}[x]_{\mathbf{2,d}} = d+1$ and thus $\dim W_\ell^{\langle 2\rangle} \geq 2(d+1)-1 = 2d+1 = \dim \mathbb{R}[x]_{\mathbf{2,2d}}$, which is a contradiction.

Suppose that $2d = 2$. Then $\dim \mathbb{R}[x]_{\mathbf{n,1}} = n$ and $\dim W_\ell^{\langle 2\rangle} \geq n^2 - \binom{n}{2} = \binom{n+1}{2} = \dim \mathbb{R}[x]_{\mathbf{n,2}}$, leading to the same contradiction.

Finally suppose that $n = 3$ and $2d = 4$. Then $\dim \mathbb{R}[x]_{\mathbf{3,2}} = 6$ and $\dim W_\ell^{\langle 2\rangle} \geq 6 \cdot 3 - \binom{3}{2} = 15 = \dim \mathbb{R}[x]_{\mathbf{3,4}}$, which again leads to the same dimensional contradiction. $\square$

We now turn our attention to the structure of extreme rays of $\Sigma^*_{\mathbf{n,2d}}$ in the smallest cases where there exist nonnegative polynomials that are not sums of squares: 3 variables, degree 6, and 4 variables, degree 4.

4.7 Ranks of Extreme Rays of $\Sigma^*_{3,6}$ and $\Sigma^*_{4,4}$

We first examine, in the cases $(3,6)$ and $(4,4)$, the structure of linear functionals $\ell \in \mathbb{R}[x]^*_{\mathbf{2d}}$ with a given kernel W such that $V_{\mathbb{C}}(W) = \emptyset$.

Proposition 4.45. *Let W be a three-dimensional subspace of $\mathbb{R}[x]_{\mathbf{3,3}}$ such that $V_{\mathbb{C}}(W) = \emptyset$. Then* $\dim W^{\langle 2\rangle} = 27$ *and there exists a unique quadratic form $Q_\ell \in \mathbb{R}[x]_{\mathbf{3,6}}$ containing W in its kernel. Furthermore* $\ker Q_\ell = W$.

Before we prove Proposition 4.45 we note that the unique form Q_ℓ with kernel W need not be positive semidefinite. The investigation of positive definiteness of Q_ℓ will lead us to evaluation on finite point sets in the next section.

***Proof of Proposition* 4.45.** By applying Lemma 4.43 we see that

$$\dim W^{\langle 2\rangle} = 3 \cdot \dim \mathbb{R}[x]_{\mathbf{3,3}} - 3 = 27.$$

Since $\dim \mathbb{R}[x]_{\mathbf{3,6}} = 28$ it follows that $W^{\langle 2\rangle}$ is a hyperplane in $\mathbb{R}[x]_{\mathbf{3,6}}$ and therefore there is a unique linear functional ℓ vanishing on W. By Lemma 4.42 it follows that Q_ℓ is the unique (up to a constant multiple) quadratic form with W in its kernel.

We leave the part that the dimension of the kernel of Q_ℓ cannot be more than 3 as an exercise. $\square$

There is also the corresponding proposition for the case $(4,4)$ with the same proof.

Proposition 4.46. *Let W be a four-dimensional subspace of $\mathbb{R}[x]_{\mathbf{4,2}}$ such that $V_{\mathbb{C}}(W) = \emptyset$. Then* $\dim W^{\langle 2\rangle} = 34$ *and there exists a unique quadratic form $Q_\ell \in \mathbb{R}[x]_{\mathbf{4,4}}$ containing W in its kernel. Furthermore* $\ker Q_\ell = W$.

We obtain the following interesting corollaries.

Corollary 4.47. *Suppose that ℓ spans an extreme ray of $\Sigma^*_{\mathbf{3,6}}$ and ℓ does not correspond to point evaluation. Then* $\operatorname{rank} Q_\ell = 7$. *Conversely, suppose that Q_ℓ is a psd form of rank 7 in $S^{3,3}_+$ and let W_ℓ be the kernel of Q_ℓ. If $V_{\mathbb{C}}(W_\ell) = \emptyset$, then Q_ℓ spans an extreme ray of $\Sigma^*_{\mathbf{3,6}}$.*

Proof. Suppose that ℓ spans an extreme ray of $\Sigma^*_{\mathbf{3,6}}$ and ℓ does not correspond to point evaluation. Let W_ℓ be the kernel of Q_ℓ. We know that $V(W_\ell) = \emptyset$ and $\dim W_\ell \geq 3$. We can then find a three-dimensional subspace W of W_ℓ such that $V(W) = \emptyset$. Applying Proposition 4.45 we see that there exists a unique quadratic form Q containing W in its kernel. Then it must happen that Q_ℓ is a scalar multiple of Q, and since $\ker Q = W$ we see that the kernel of Q_ℓ has dimension 3 and thus Q_ℓ has rank 7.

Conversely suppose that Q_ℓ is a positive semidefinite form of rank 7 and $V_{\mathbb{C}}(W_\ell) = \emptyset$. Then by Proposition 4.45 Q_ℓ is the unique quadratic form in $\mathbb{R}[x]^*_{\mathbf{3,6}}$ with kernel W_ℓ. Suppose that $Q_\ell = Q_1 + Q_2$ with $Q_1, Q_2 \in \Sigma^*_{\mathbf{3,6}}$. Then Q_1 and Q_2 are positive semidefinite forms by Lemma 4.35 and therefore $\ker Q_\ell \subseteq \ker Q_i$. Then Q_1 and Q_2 are scalar multiples of Q_ℓ and therefore Q_ℓ spans an extreme ray of $\Sigma^*_{\mathbf{3,6}}$. ☐

The above corollary has a couple of interesting consequences. If the quadratic form Q_ℓ is in $\Sigma^*_{\mathbf{3,6}}$ and its rank is at most 6, then it must be a convex combination of rank 1 forms in $\Sigma^*_{\mathbf{3,6}}$, which we know are point evaluations. Restated in measure and moment language, this says that if a positive semidefinite moment matrix in $\mathbb{R}[x]^*_{\mathbf{3,6}}$ has rank at most 6, then the linear functional can be written as a combination of point evaluations, and therefore the linear functional has a representing measure. However, there are rank 7 positive semidefinite moment matrices that do not admit a representing measure.

Another consequence can be stated in optimization terms. Suppose that we would like to optimize a linear functional over a compact base of the $\Sigma^*_{\mathbf{3,6}}$. Then the point where the optimum is achieved will have rank 1 or rank 7.

Corollary 4.48. *Suppose that $p \in \Sigma_{\mathbf{3,6}}$ lies on the boundary of the cone of sums of squares and p is a strictly positive form. Then p is a sum of exactly 3 squares.*

Proof. Let p be as above. Since p lies in the boundary of $\Sigma_{\mathbf{3,6}}$ there exists an extreme ray ℓ of the dual cone $\Sigma^*_{\mathbf{3,6}}$ such that $\ell(p) = 0$. Now suppose that $p = \sum f_i^2$ for some $f_i \in \mathbb{R}[x]_{\mathbf{3,3}}$. It follows that $Q_\ell(f_i) = 0$ for all i, and since Q_ℓ is a positive semidefinite quadratic form, we see that all f_i lie in the kernel W_ℓ of Q_ℓ. By Corollary 4.47 we know that $\dim W_\ell = 3$ and therefore p is a sum of squares of forms coming from a three-dimensional subspace of $\mathbb{R}[x]_{\mathbf{3,3}}$. It follows that p is a sum of at most 3 squares. Since any two ternary cubics have a common real zero and p is strictly positive, it follows that p cannot be a sum of two or fewer squares. ☐

The equivalent corollaries hold for the case $(4,4)$, although the proof of Corollary 4.50 requires slightly more work, while the proof of Corollary 4.49 is exactly the same. For complete details see [7].

Corollary 4.49. *Suppose that ℓ spans an extreme ray of $\Sigma^*_{4,4}$ and ℓ does not correspond to point evaluation. Then* $\operatorname{rank} Q_\ell = 6$. *Conversely, suppose that Q_ℓ is a positive semidefinite form of rank 6 in $S^{4,2}_+$ and let W_ℓ be the kernel of Q_ℓ. If $V_{\mathbb{C}}(W_\ell) = \emptyset$, then Q_ℓ spans an extreme ray of $\Sigma^*_{4,4}$.*

Corollary 4.50. *Suppose that $p \in \Sigma_{4,4}$ lies on the boundary of the cone of sums of squares and p is a strictly positive form. Then p is a sum of exactly 4 squares.*

Corollaries 4.48 and 4.50 were used to study the algebraic boundary of the cones $\Sigma_{3,6}$ and $\Sigma_{4,4}$ in [8].

Exercise 4.51. Show that all forms in $\mathbb{R}[x]_{3,6}$ that can be written as linear combinations of squares of 3 cubics form an irreducible hypersurface in $\mathbb{R}[x]_{3,6}$. Similarly, show that all forms in $\mathbb{R}[x]_{4,4}$ that are linear combinations of squares of 4 quadratics also form an irreducible hypersurface in $\mathbb{R}[x]_{4,4}$. (Hint: Use Terracini's lemma.) Use Corollaries 4.48 and 4.50 to show that the algebraic boundary of $\Sigma_{3,6}$ and $\Sigma_{4,4}$ has a single component in addition to the discriminant hypersurface.

It was shown in [8] that despite their simple definition the hypersurfaces of Exercise 4.51 have very high degree: 83200 in the case $(3,6)$ and 38475 in the case $(4,4)$. This shows that the boundary of the cone of sums of squares is quite complicated from the algebraic point of view.

4.8 Extracting Finite Point Sets

We have established in the previous section that the "interesting" extreme rays of $\Sigma^*_{3,6}$ have rank 7 and those of $\Sigma^*_{4,4}$ have rank 6. Let's consider the case of 4 variables of degree 4. We have shown that a four-dimensional subspace W leads to a unique form Q_ℓ of rank 6 such that the kernel of Q_ℓ contains W. However, the form Q_ℓ does not have to lie in $\Sigma^*_{4,4}$, since the form Q_ℓ is not necessarily positive semidefinite.

In order to examine positive semidefiniteness of Q_ℓ we reduce the problem to looking at an evaluation on finite point sets.

Exercise 4.52. Let W be a subspace of $\mathbb{R}[x]_{\mathbf{d}}$ such that $V_{\mathbb{C}}(W) = \emptyset$. Show that there exist forms $q_1, \dots, q_{n-1} \in W$ that intersect in d^{n-1} projective points in $\mathbb{CP}^{n-1}$:

$$V_{\mathbb{C}}(q_1, \dots, q_{n-1}) = \{\bar{s}_1, \dots, \bar{s}_{d^{n-1}} \mid \bar{s}_i \in \mathbb{CP}^{n-1}\}.$$

We apply this result to our case of $W \subset \mathbb{R}[x]_{4,4}$ and obtain forms $q_1, q_2, q_3 \in W$ intersecting in $2^3 = 8$ projective points $\bar{s}_i \in \mathbb{CP}^3$. We can take their affine representatives $s_1, \dots, s_8 \in \mathbb{C}^n$. Unfortunately, even though the forms $q_i \in W$ are real, their points of intersection may be complex.

However, as was shown in [7], the fact that the form Q_ℓ is positive semidefinite restricts the number of complex zeroes. Since complex zeroes of real forms come in conjugate pairs, the fewest number of complex zeroes that the forms q_i may have is 2.

Theorem 4.53. *Suppose that $\ell \in \mathbb{R}[x]^*_{\mathbf{4,4}}$ is an extreme ray of $\Sigma^*_{\mathbf{4,4}}$ that does not correspond to point evaluation and let W_ℓ be the kernel of Q_ℓ. Let $q_1, q_2, q_3 \in W_\ell$ be any three forms intersecting in $2^3 = 8$ projective points in $\mathbb{CP}^3$. Then the forms q_i have at most 2 common complex zeroes. Conversely, given $q_1, q_2, q_3 \in \mathbb{R}[x]_{\mathbf{4,2}}$ intersecting in 8 points with at most 2 of them complex, there exists an extreme ray of $\Sigma^*_{\mathbf{4,4}}$ whose kernel contains q_1, q_2, q_3.*

There is an equivalent theorem for the case $(3, 6)$.

Theorem 4.54. *Suppose that $\ell \in \mathbb{R}[x]^*_{\mathbf{3,6}}$ is an extreme ray of $\Sigma^*_{\mathbf{3,6}}$ that does not correspond to point evaluation and let W_ℓ be the kernel of Q_ℓ. Let $q_1, q_2 \in W_\ell$ be any two forms intersecting in $3^2 = 9$ projective points in $\mathbb{CP}^2$. Then the forms q_i have at most 2 common complex zeroes. Conversely, given $q_1, q_2, q_3 \in \mathbb{R}[x]_{\mathbf{3,3}}$ intersecting in 9 points with at most 2 of them complex, there exists an extreme ray of $\Sigma^*_{\mathbf{3,6}}$ whose kernel contains q_1, q_2.*

It is possible to apply the Cayley–Bacharach machinery explained in Section 4.5 to completely describe the structure of the extreme rays of $\Sigma^*_{\mathbf{n,2d}}$ for the cases $(4, 4)$ and $(3, 6)$ using the coefficients of the unique Cayley–Bacharach relation that exists on the points of intersection of the forms q_i.

We have now come full circle, from using a finite point set to establish that there exist nonnegative forms that are not sums of squares in Section 4.3 to showing that these sets underlie all linear inequalities that separate $\Sigma_{\mathbf{n,2d}}$ from $P_{\mathbf{n,2d}}$.

4.9 Volumes

We now switch gears completely and turn to the question of the quantitative relationship between $P_{\mathbf{n,2d}}$ and $\Sigma_{\mathbf{n,2d}}$. Our goal is to compare the relative sizes of the cones $P_{\mathbf{n,2d}}$ and $\Sigma_{\mathbf{n,2d}}$. While the cones themselves are unbounded objects, we can take a section of each cone with the same hyperplane so that both sections are compact.

Let $L_{\mathbf{n,2d}}$ be an affine hyperplane in $\mathbb{R}[x]_{\mathbf{2d}}$ consisting of all forms with integral (average) 1 on the unit sphere $\mathbb{S}^{n-1}$ in $\mathbb{R}^n$:

$$L_{\mathbf{n,2d}} = \left\{ p \in \mathbb{R}[x]_{\mathbf{2d}} \;\middle|\; \int_{\mathbb{S}^{n-1}} p\, d\sigma = 1 \right\},$$

where σ is the rotation invariant probability measure on $\mathbb{S}^{n-1}$. Let $\bar{P}_{\mathbf{n,2d}}$ and $\bar{\Sigma}_{\mathbf{n,2d}}$ be the sections of $P_{\mathbf{n,2d}}$ and $\Sigma_{\mathbf{n,2d}}$ with $L_{\mathbf{n,2d}}$:

$$\bar{P}_{\mathbf{n,2d}} = P_{\mathbf{n,2d}} \cap L_{\mathbf{n,2d}} \quad \text{and} \quad \bar{\Sigma}_{\mathbf{n,2d}} = \Sigma_{\mathbf{n,2d}} \cap L_{\mathbf{n,2d}}.$$

Let $r^{2d} = (x_1^2 + \cdots + x_n^2)^d$ be the form in $\mathbb{R}[x]_{\mathbf{2d}}$ that is constantly 1 on the unit sphere. Convex bodies $\bar{P}_{\mathbf{n,2d}}$ and $\bar{\Sigma}_{\mathbf{n,2d}}$ lie in the affine hyperplane $L_{\mathbf{n,2d}}$ of forms of integral 1 on the unit sphere. We now translate them to lie in the linear hyperplane $M_{\mathbf{n,2d}}$ of forms of integral 0 on the unit sphere by subtracting r^{2d}:

$$\tilde{P}_{\mathbf{n,2d}} = \bar{P}_{\mathbf{n,2d}} - r^{2d} = \{p \in \mathbb{R}[x]_{\mathbf{2d}} \mid p + r^{2d} \in \bar{P}_{\mathbf{n,2d}}\}$$

and

$$\tilde{\Sigma}_{\mathbf{n,2d}} = \bar{\Sigma}_{\mathbf{n,2d}} - r^{2d} = \{p \in \mathbb{R}[x]_{\mathbf{2d}} \mid p + r^{2d} \in \bar{\Sigma}_{\mathbf{n,2d}}\}.$$

The estimation of the volumes of $\tilde{P}_{\mathbf{n,2d}}$ and $\tilde{\Sigma}_{\mathbf{n,2d}}$ will be done separately. Before proceeding we make a short note on the proper way to measure the size of a convex set. Let $K \subset \mathbb{R}^n$ be a convex body. Suppose that we expand K by a constant factor α. Then the volume changes as follows:

$$\operatorname{Vol}(\alpha K) = \alpha^n \operatorname{Vol} K.$$

We would like to think of K and αK as similar in size, but if the ambient dimension n grows, then αK is significantly larger in volume. Therefore the proper measure of volume that takes care of the dimensional effects is

$$(\operatorname{Vol} K)^{\frac{1}{n}}.$$

4.9.1 Volume of Nonnegative Forms

Let $M_{\mathbf{n,2d}}$ be the linear hyperplane of forms of integral 0 on the unit sphere:

$$M_{\mathbf{n,2d}} = \left\{ p \in \mathbb{R}[x]_{\mathbf{2d}} \;\middle|\; \int_{\mathbb{S}^{n-1}} p \, d\sigma = 0 \right\}.$$

Both convex bodies $\tilde{P}_{\mathbf{n,2d}}$ and $\tilde{\Sigma}_{\mathbf{n,2d}}$ live inside $M_{\mathbf{n,2d}}$, so our calculations will involve the unit sphere and the unit ball in $M_{\mathbf{n,2d}}$.

We equip $\mathbb{R}[x]_{\mathbf{2d}}$ with the L^2 inner product:

$$\langle p, q \rangle = \int_{\mathbb{S}^{n-1}} pq \, d\sigma.$$

We note that with this metric we have

$$\|p\|^2 = \langle p, p \rangle = \int_{\mathbb{S}^{n-1}} p^2 \, d\sigma = \|p\|_2^2.$$

We also let $||p||_\infty$ denote the L^∞-norm of p:

$$||p||_\infty = \max_{x \in \mathbb{S}^{n-1}} |p(x)|.$$

Let N be the dimension of $M_{\mathbf{n,2d}}$. Since $M_{\mathbf{n,2d}}$ is a hyperplane in $\mathbb{R}[x]_{\mathbf{2d}}$ we know that $N = \dim \mathbb{R}[x]_{\mathbf{2d}} - 1 = \binom{n+2d-1}{2d} - 1$. Let $\mathbb{S}^{N-1}$ and B^N denote the unit sphere and the unit ball in $M_{\mathbf{n,2d}}$ with respect to the L^2 inner product.

Our goal is to show the following estimate on the volume of $\tilde{P}_{\mathbf{n,2d}}$.

Theorem 4.55.

$$\left(\frac{\operatorname{Vol}\tilde{P}_{\mathbf{n,2d}}}{\operatorname{Vol}B^N}\right)^{1/N} \geq \frac{1}{2\sqrt{4d+2}}n^{-1/2}.$$

We first develop a general way of estimating the volume of a convex set, starting from simply writing out the integral for the volume in polar coordinates. We refer to [11] for the relevant analytic inequalities.

Exercise 4.56. Let $K \subset \mathbb{R}^n$ be a convex body with the origin in its interior and let χ_K be the characteristic function of K: $\chi_K(x) = 1$ if $x \in K$ and $\chi_K(x) = 0$ otherwise. The volume of K is given by the following integral:

$$\operatorname{Vol}K = \int_{\mathbb{R}^n} \chi_K \, d\mu,$$

where μ is the Lebesgue measure.

Let G_K be the gauge of K. Rewrite the above integral in polar coordinates to show that

$$\frac{\operatorname{Vol}K}{\operatorname{Vol}B^n} = \int_{\mathbb{S}^{n-1}} G_K^{-n} \, d\sigma,$$

where B^n and $\mathbb{S}^{n-1}$ are the unit ball and the unit sphere in $\mathbb{R}^n$ and σ is the rotation invariant probability measure on $\mathbb{S}^{n-1}$.

Exercise 4.57. Use Exercise 4.56 and Hölder's inequality to show that

$$\left(\frac{\operatorname{Vol}K}{\operatorname{Vol}B^n}\right)^{1/n} \geq \int_{\mathbb{S}^{n-1}} G_K^{-1} \, d\sigma.$$

Exercise 4.58. Use Exercise 4.57 and Jensen's inequality to show that

$$\left(\frac{\operatorname{Vol}K}{\operatorname{Vol}B^n}\right)^{1/n} \geq \left(\int_{\mathbb{S}^{n-1}} G_K \, d\sigma\right)^{-1}.$$

Now we apply the results of Exercises 4.56–4.58 to the case of $\tilde{P}_{\mathbf{n,2d}}$.

Lemma 4.59.

$$\left(\frac{\operatorname{Vol}\tilde{P}_{\mathbf{n,2d}}}{\operatorname{Vol}B^N}\right)^{1/N} \geq \left(\int_{\mathbb{S}^{N-1}} \|p\|_\infty \, d\sigma_p\right)^{-1}.$$

Proof. We observe that $\bar{P}_{\mathbf{n,2d}}$ consists of all forms of integral 1 on $\mathbb{S}^{n-1}$ whose minimum on $\mathbb{S}^{n-1}$ is at least 0. Therefore $\tilde{P}_{\mathbf{n,2d}}$ consists of all forms of integral 0 on $\mathbb{S}^{n-1}$ whose minimum on the unit sphere is at least -1:

$$\tilde{P}_{\mathbf{n,2d}} = \left\{ p \in \mathbb{R}[x]_{\mathbf{2d}} \;\middle|\; \int_{\mathbb{S}^{n-1}} p\, d\sigma = 0 \text{ and } \min_{x \in \mathbb{S}^{n-1}} p(x) \geq -1 \right\}.$$

It follows that the gauge of $\tilde{P}_{\mathbf{n,2d}}$ is given by $-\min_{\mathbb{S}^{n-1}}$:

$$G_{\tilde{P}_{\mathbf{n,2d}}}(p) = -\min_{x \in \mathbb{S}^{n-1}} p(x). \tag{4.5}$$

Using Exercise 4.58 we can bound the volume of $\tilde{P}_{\mathbf{n,2d}}$ from below:

$$\left(\frac{\operatorname{Vol} \tilde{P}_{\mathbf{n,2d}}}{\operatorname{Vol} B^N} \right)^{1/N} \geq \left(\int_{\mathbb{S}^{N-1}} -\min(p)\, d\sigma_p \right)^{-1}.$$

Since $-\min_{x \in \mathbb{S}^{n-1}} p(x)$ is bounded above by $||p||_\infty$ we obtain

$$\left(\frac{\operatorname{Vol} \tilde{P}_{\mathbf{n,2d}}}{\operatorname{Vol} B^N} \right)^{1/N} \geq \left(\int_{\mathbb{S}^{N-1}} ||p||_\infty \; d\sigma_p \right)^{-1}$$

as desired. $\square$

From Lemma 4.59 we see that in order to obtain a lower bound on the volume of $\tilde{P}_{\mathbf{n,2d}}$ we need to find an upper bound on the average L^∞-norm of forms in $\mathbb{S}^{N-1}$:

$$\int_{\mathbb{S}^{N-1}} ||p||_\infty \; d\sigma_p.$$

It is easy to see that the L^∞-norm of any polynomial is bounded from below by any of its L^{2k}-norms:

$$||p||_\infty \geq ||p||_{2k}$$

for all k. Finding upper bounds on the L^∞-norm of forms in $\mathbb{R}[x]_{\mathbf{2d}}$ in terms of their L^{2k}-norms is significantly more challenging.

Exercise 4.60. It was shown by Barvinok in [3] that the following inequality holds for all $p \in \mathbb{R}[x]_{\mathbf{2d}}$ and all k:

$$||p||_\infty \leq \binom{2kd + n - 1}{2kd}^{\frac{1}{2k}} ||p||_{2k}.$$

Show that for $k = n$ we have

$$||p||_\infty \leq 2\sqrt{2d+1}||p||_{2n}$$

for all $p \in \mathbb{R}[x]_{\mathbf{2d}}$.

Remark 4.61. *It is possible to obtain slightly better bounds for our purposes by using $k = n\log(2d+1)$ in the above inequality. See [4] for details.*

We use Barvinok's inequality to convert the problem of bounding the average L^∞-norm on $\mathbb{S}^{N-1}$ into bounding the average L^{2n}-norm. In order for this to be useful we need lower bounds on the average L^{2k}-norms. We will show the following bound.

Lemma 4.62.

$$\int_{\mathbb{S}^{N-1}} \|p\|_{2k}\, d\sigma_p \leq \sqrt{2k}.$$

Before we proceed with the proof we need some preliminary results.

Exercise 4.63. Let Γ denote the gamma function. Show that for $k \in \mathbb{N}$

$$\int_{\mathbb{S}^{n-1}} x_1^{2k}\, d\sigma = \frac{\Gamma\left(\frac{n}{2}\right)\Gamma\left(k+\frac{1}{2}\right)}{\Gamma\left(\frac{1}{2}\right)\Gamma\left(k+\frac{n}{2}\right)}. \tag{4.6}$$

Now let $\ell : \mathbb{R}^n \to \mathbb{R}$ be a linear form given by $\ell(x) = \langle x, \xi\rangle$ for some vector $\xi \in \mathbb{R}^n$. Use (4.6) to show that

$$\int_{\mathbb{S}^{n-1}} \ell^{2k}(x)\, d\sigma_x = \|\xi\|^{2k} \frac{\Gamma\left(\frac{n}{2}\right)\Gamma\left(k+\frac{1}{2}\right)}{\Gamma\left(\frac{1}{2}\right)\Gamma\left(k+\frac{n}{2}\right)}. \tag{4.7}$$

In order to apply the result of Exercise 4.63 we will need to know the L^2-norm of a special form in $M_{\mathbf{n},\mathbf{2d}}$.

Lemma 4.64. *Let $v \in \mathbb{S}^{n-1}$ be a unit vector and let $\xi_v \in M_{\mathbf{n},\mathbf{2d}}$ be the form such that*

$$\langle p, \xi_v\rangle = p(v) \quad \textit{for all} \quad p \in M_{\mathbf{n},\mathbf{2d}}.$$

Then

$$\|\xi_v\| = \sqrt{\dim M_{\mathbf{n},\mathbf{2d}}} = \sqrt{\binom{n+2d-1}{2d} - 1}.$$

Proof. Consider the following average:

$$\int_{\mathbb{S}^{N-1}} p^2(v)\, d\sigma_p = \int_{\mathbb{S}^{N-1}} \langle p, \xi_v\rangle^2\, d\sigma_p.$$

On one hand it is the average of a quadratic form on the unit sphere and by Exercise 4.63 we have

$$\int_{\mathbb{S}^{N-1}} p^2(v)\, d\sigma_p = \frac{\|\xi_v\|^2}{\dim M_{\mathbf{n},\mathbf{2d}}}.$$

On the other hand, by symmetry, this average is independent of the choice of $v \in \mathbb{S}^{n-1}$. Therefore we may introduce an extra average over the unit sphere:

$$\int_{\mathbb{S}^{N-1}} p^2(v)\, d\sigma_p = \int_{\mathbb{S}^{n-1}} \int_{\mathbb{S}^{N-1}} p^2(v)\, d\sigma_p\, d\sigma_v.$$

Now we switch the order of integration:

$$\int_{\mathbb{S}^{N-1}} p^2(v)\, d\sigma_p = \int_{\mathbb{S}^{N-1}} \int_{\mathbb{S}^{n-1}} p^2(v)\, d\sigma_v\, d\sigma_p.$$

We observe that $\int_{\mathbb{S}^{n-1}} p^2(v)\, d\sigma_v = 1$ for all $p \in \mathbb{S}^{N-1}$ and therefore

$$\int_{\mathbb{S}^{N-1}} p^2(v)\, d\sigma_p = 1.$$

The lemma now follows. $\square$

We are now ready to estimate the average L^{2k}-norm on $\mathbb{S}^{N-1}$.

***Proof of Lemma* 4.62.**

$$\int_{\mathbb{S}^{N-1}} \|p\|_{2k}\, d\sigma_p = \int_{\mathbb{S}^{N-1}} \left(\int_{\mathbb{S}^{n-1}} p^{2k}(x)\, d\sigma_x \right)^{\frac{1}{2k}} d\sigma_p.$$

By applying the Hölder inequality we can move the exponent $\frac{1}{2k}$ outside and obtain

$$\int_{\mathbb{S}^{N-1}} \|p\|_{2k}\, d\sigma_p \leq \left(\int_{\mathbb{S}^{N-1}} \int_{\mathbb{S}^{n-1}} p^{2k}(x)\, d\sigma_x\, d\sigma_p \right)^{\frac{1}{2k}}.$$

Now we exchange the order of integration:

$$\int_{\mathbb{S}^{N-1}} \|p\|_{2k}\, d\sigma_p \leq \left(\int_{\mathbb{S}^{n-1}} \int_{\mathbb{S}^{N-1}} p^{2k}(x)\, d\sigma_p\, d\sigma_x \right)^{\frac{1}{2k}}.$$

Consider the inner integral

$$\int_{\mathbb{S}^{N-1}} p^{2k}(x)\, d\sigma_p. \tag{4.8}$$

By rotational invariance it does not depend on the choice of the point $x \in \mathbb{S}^{n-1}$. Therefore the outer integral over $\mathbb{S}^{n-1}$ is redundant and we obtain

$$\int_{\mathbb{S}^{N-1}} \|p\|_{2k}\, d\sigma_p \leq \left(\int_{\mathbb{S}^{N-1}} p^{2k}(v)\, d\sigma_p \right)^{\frac{1}{2k}} \quad \text{for any} \quad v \in \mathbb{S}^{n-1}. \tag{4.9}$$

We can rewrite this as

$$\int_{\mathbb{S}^{N-1}} \|p\|_{2k}\, d\sigma_p \leq \left(\int_{\mathbb{S}^{N-1}} \langle p, \xi_v \rangle^{2k}\, d\sigma_p\right)^{\frac{1}{2k}}.$$

Now we see that the integral in (4.8) is actually just the average of the $2k$th power of a linear form and we can apply Exercise 4.63 to see that

$$\int_{\mathbb{S}^{N-1}} \langle p, \xi_v \rangle^{2k}\, d\sigma_p = \|\xi_v\|^{2k} \frac{\Gamma\left(\frac{N}{2}\right)\Gamma\left(k+\frac{1}{2}\right)}{\Gamma\left(\frac{1}{2}\right)\Gamma\left(k+\frac{N}{2}\right)}.$$

By Lemma 4.64 we know that $\|\xi_v\|^2 = \dim M_{\mathbf{n},\mathbf{2d}} = N$.

Putting it all together with (4.9) we see that

$$\int_{\mathbb{S}^{N-1}} \|p\|_{2k}\, d\sigma_p \leq \sqrt{N}\left(\frac{\Gamma\left(\frac{N}{2}\right)\Gamma\left(k+\frac{1}{2}\right)}{\Gamma\left(\frac{1}{2}\right)\Gamma\left(k+\frac{N}{2}\right)}\right)^{\frac{1}{2k}}.$$

We now use the following two estimates to finish the proof:

$$\left(\frac{\Gamma\left(\frac{N}{2}\right)}{\Gamma\left(k+\frac{N}{2}\right)}\right)^{\frac{1}{2k}} \leq \sqrt{\frac{2}{N}} \qquad \text{and} \qquad \left(\frac{\Gamma\left(k+\frac{1}{2}\right)}{\Gamma\left(\frac{1}{2}\right)}\right)^{\frac{1}{2k}} \leq \sqrt{k}.$$

We remark that asymptotically the second estimate is an overestimate by a factor of $\sqrt{e}$. □

***Proof of Theorem* 4.55.** We first use Lemma 4.59 to see that

$$\left(\frac{\operatorname{Vol} \tilde{P}_{\mathbf{n},\mathbf{2d}}}{\operatorname{Vol} B^N}\right)^{1/N} \geq \left(\int_{\mathbb{S}^{N-1}} ||p||_\infty\, d\sigma_p\right)^{-1}.$$

By Exercise 4.60 we know that for all $p \in \mathbb{R}[x]_{\mathbf{2d}}$

$$||p||_\infty \leq 2\sqrt{2d+1}||p||_{2n}.$$

Therefore we see that

$$\left(\frac{\operatorname{Vol} \tilde{P}_{\mathbf{n},\mathbf{2d}}}{\operatorname{Vol} B^N}\right)^{1/N} \geq \frac{1}{2\sqrt{2d+1}}\left(\int_{\mathbb{S}^{N-1}} ||p||_{2n}\, d\sigma_p\right)^{-1}.$$

Now we can apply Lemma 4.62 with $k = n$ and obtain

$$\left(\frac{\operatorname{Vol} \tilde{P}_{\mathbf{n},\mathbf{2d}}}{\operatorname{Vol} B^N}\right)^{1/N} \geq \frac{1}{2\sqrt{4d+2}} n^{-1/2}$$

as desired. □

4.9.2 Volume of Sums of Squares

We now turn our attention to the cone of sums of squares $\Sigma_{\mathbf{n},\mathbf{2d}}$. Although it will be somewhat obscured by our presentation, the main reason for our ability to derive bounds on the volume of $\bar{\Sigma}_{\mathbf{n},\mathbf{2d}}$ comes from the fact that the dual cone $\Sigma^*_{\mathbf{n},\mathbf{2d}}$ is a section of the cone of positive semidefinite matrices.

We have just seen how to derive lower bounds on the volume of the cone of nonnegative forms. These bounds, of course, apply to quadratic forms, and they can be extended to work for sections of the cone. This gives us a lower bound on the volume of the dual cone, which can be turned around into an upper bound on the volume of $\tilde{\Sigma}_{\mathbf{n},\mathbf{2d}}$. The approach to bounding the volume of $\tilde{\Sigma}_{\mathbf{n},\mathbf{2d}}$ is therefore very similar to what we did for nonnegative forms. In fact, the technique in the proofs of the main bounds in Lemma 4.70 and Lemma 4.62 is nearly identical.

Let D be the dimension of $\mathbb{R}[x]_{\mathbf{d}}$. Our main result on the volume of $\tilde{\Sigma}_{\mathbf{n},\mathbf{2d}}$ is as follows.

Theorem 4.65.

$$\left(\frac{\operatorname{Vol}\tilde{\Sigma}_{\mathbf{n},\mathbf{2d}}}{\operatorname{Vol}B^N}\right)^{1/N} \leq 2^{4d+1}\sqrt{\frac{6D}{N}}.$$

Remark 4.66. *Recall that*

$$N = \binom{n+2d-1}{2d} - 1 \quad \textit{and} \quad D = \binom{n+d-1}{d}.$$

Therefore, for fixed degree d our upper bound on the volume of $\tilde{\Sigma}_{\mathbf{n},\mathbf{2d}}$ is of the order $n^{-d/2}$. In Theorem 4.55 we proved a lower bound on the volume of $\tilde{P}_{\mathbf{n},\mathbf{2d}}$ that is of the order $n^{-1/2}$. Therefore, when the total degree $2d$ is at least 4, the lower bound on the volume of $\tilde{P}_{\mathbf{n},\mathbf{2d}}$ is asymptotically much larger than the upper bound on the volume of $\tilde{\Sigma}_{\mathbf{n},\mathbf{2d}}$. Thus we see that if the degree $2d$ is fixed and at least 4, there are significantly more nonnegative forms than sums of squares.

It is possible to show that the bounds of Theorems 4.55 and 4.65 are asymptotically tight for the case of fixed degree $2d$. See [5] for more details.

In Exercises 4.56–4.58 we showed how to bound the volume of a convex body K from below using the average of its gauge over the unit sphere $\mathbb{S}^{n-1}$. As we explained above, we are now dealing with the dual situation, and we need a related dual inequality that bounds the volume of K from above by the average gauge of its dual body K°.

Exercise 4.67. Let $K \subset \mathbb{R}^n$ be a convex body with 0 in its interior and let K° be the dual convex body defined as

$$K^\circ = \{x \in \mathbb{R}^n \mid \langle x, y\rangle \leq 1 \quad \text{for all} \quad y \in K\}.$$

Show that the gauge of K° is given by the following formula:

$$G_{K^\circ}(x) = \max_{y\in K} \langle x, y\rangle.$$

The following is known as Urysohn's inequality [26].

Lemma 4.68.

$$\left(\frac{\operatorname{Vol} K}{\operatorname{Vol} B^n}\right)^{1/n} \leq \int_{\mathbb{S}^{n-1}} G_{K^\circ}(x)\, d\sigma_x.$$

In order to apply Lemma 4.68 we need a description of the gauge of $\tilde{\Sigma}^*_{\mathbf{n,2d}}$. Let $\mathbb{S}^{D-1}$ be the unit sphere in $\mathbb{R}[x]_{\mathbf{d}}$ with respect to the L^2 inner product.

Lemma 4.69. *We have the following description of the gauge of $\tilde{\Sigma}^\circ_{\mathbf{n,2d}}$:*

$$G_{\tilde{\Sigma}^\circ_{\mathbf{n,2d}}}(p) = \max_{q\in\mathbb{S}^{D-1}} \langle p, q^2\rangle.$$

Proof. By Exercise 4.67 the gauge of $\tilde{\Sigma}^\circ_{\mathbf{n,2d}}$ is given by

$$G_{\tilde{\Sigma}^\circ_{\mathbf{n,2d}}}(p) = \max_{q\in\tilde{\Sigma}_{\mathbf{n,2d}}} \langle p, q\rangle.$$

We observe that the maximal inner product $\max_{q\in\tilde{\Sigma}_{\mathbf{n,2d}}}\langle p, q\rangle$ always occurs at an extreme point of $\tilde{\Sigma}_{\mathbf{n,2d}}$. Extreme points of $\bar{\Sigma}_{\mathbf{n,2d}}$ are all squares, and therefore extreme point of $\tilde{\Sigma}_{\mathbf{n,2d}}$ are translates of squares and have the form

$$q^2 - r^{2d} \quad \text{with} \quad q \in \mathbb{R}[x]_{\mathbf{d}} \quad \text{and} \quad \int_{\mathbb{S}^{n-1}} q^2\, d\sigma = 1.$$

The condition $\int_{\mathbb{S}^{n-1}} q^2\, d\sigma = 1$ corresponds exactly to q lying in the unit sphere of $\mathbb{R}[x]_{\mathbf{d}}$. Since forms $p \in M_{\mathbf{n,2d}}$ have integral zero on the unit sphere $\mathbb{S}^{n-1}$, it follows that

$$\langle p, r^{2d}\rangle = 0 \quad \text{for all} \quad p \in M_{\mathbf{n,2d}}.$$

Combining with the description of the extreme points of $\tilde{\Sigma}_{\mathbf{n,2d}}$ we see that

$$G_{\tilde{\Sigma}^\circ_{\mathbf{n,2d}}}(p) = \max_{q\in\mathbb{S}^{D-1}} \langle p, q^2\rangle. \qquad \square$$

Given a form $p \in \mathbb{R}[x]_{\mathbf{2d}}$ we define the associated quadratic form Q_p on $\mathbb{R}[x]_{\mathbf{d}}$:

$$Q_p(q) = \langle p, q^2\rangle \quad \text{for} \quad q \in \mathbb{R}[x]_{\mathbf{d}}.$$

By Lemma 4.69 we see that the gauge of $\tilde{\Sigma}_{\mathbf{n,2d}}$ is given by the maximum of Q_p on the unit sphere $\mathbb{S}^{D-1}$ in $\mathbb{R}[x]_{\mathbf{d}}$:

$$G_{\tilde{\Sigma}_{\mathbf{n,2d}}}(p) = \max_{q\in\mathbb{S}^{D-1}} Q_p(q).$$

Since Q_p is a quadratic form on $\mathbb{R}[x]_{\mathbf{d}}$, its L^∞-norm is the maximal value it takes on the unit sphere $\mathbb{S}^{D-1}$:

$$||Q_p||_\infty = \max_{q\in\mathbb{S}^{D-1}} |Q_p(q)|.$$

Applying Lemma 4.68 we see that

$$\left(\frac{\operatorname{Vol}\tilde{\Sigma}_{\mathbf{n,2d}}}{\operatorname{Vol}B^N}\right)^{1/N} \leq \int_{\mathbb{S}^{N-1}} ||Q_p||_\infty \, d\sigma_p.$$

Now we can apply Barvinok's inequality to bound $||Q_p||_\infty$ by high L^{2k}-norms. Using Exercise 4.60 with $k = D$ we see that

$$||Q_p||_\infty \leq 2\sqrt{3}||Q_p||_{2D}.$$

Therefore we obtain

$$\left(\frac{\operatorname{Vol}\tilde{\Sigma}_{\mathbf{n,2d}}}{\operatorname{Vol}B^N}\right)^{1/N} \leq 2\sqrt{3}\int_{\mathbb{S}^{N-1}} ||Q_p||_{2D} \, d\sigma_p.$$

The proof is now finished with the following estimate, which proceeds in nearly the same way as the proof of Lemma 4.62.

Lemma 4.70.

$$\int_{\mathbb{S}^{N-1}} ||Q_p||_{2D} \, d\sigma_p \leq 2^{4d}\sqrt{\frac{2D}{N}}.$$

Proof. We first write out the integral we would like to estimate:

$$\int_{\mathbb{S}^{N-1}} ||Q_p||_{2D} \, d\sigma_p = \int_{\mathbb{S}^{N-1}} \left(\int_{\mathbb{S}^{D-1}} \langle p, q^2\rangle^{2D} \, d\sigma_q\right)^{1/2D} d\sigma_p.$$

Using the Hölder inequality we move the exponent $1/2D$ outside:

$$\int_{\mathbb{S}^{N-1}} ||Q_p||_{2D} \, d\sigma_p \leq \left(\int_{\mathbb{S}^{N-1}} \int_{\mathbb{S}^{D-1}} \langle p, q^2\rangle^{2D} \, d\sigma_q \, d\sigma_p\right)^{1/2D}.$$

Next we interchange the order of integration:

$$\int_{\mathbb{S}^{N-1}} ||Q_p||_{2D} \, d\sigma_p \leq \left(\int_{\mathbb{S}^{D-1}} \int_{\mathbb{S}^{N-1}} \langle p, q^2\rangle^{2D} \, d\sigma_p \, d\sigma_q\right)^{1/2D}. \tag{4.10}$$

Consider the inner integral

$$\int_{\mathbb{S}^{N-1}} \langle p, q^2\rangle^{2D} \, d\sigma_p. \tag{4.11}$$

We apply Exercise 4.63 to see that

$$\int_{\mathbb{S}^{N-1}} \langle p, q^2\rangle^{2D} \, d\sigma_p \leq ||q^2||^{2D}\frac{\Gamma\left(\frac{N}{2}\right)\Gamma\left(D+\frac{1}{2}\right)}{\Gamma\left(\frac{1}{2}\right)\Gamma\left(D+\frac{N}{2}\right)}.$$

The reason that we have an inequality, instead of equality, above is that q^2 does not lie in the hyperplane $M_{\mathbf{n},\mathbf{2d}}$, and for equality we should use the norm of the projection of q^2 into $M_{\mathbf{n},\mathbf{2d}}$. We now observe that

$$\|q^2\| = \|q\|_4^2.$$

Since q lies in the unit sphere of $\mathbb{S}^{D-1}$ it follows that $\|q\| = 1$. By a result of Duoandikoetxea in [9] we know that

$$\|q\|_4 \leq 4^{2d}\|q\|.$$

Putting it all together we get

$$\int_{\mathbb{S}^{N-1}} \langle p, q^2\rangle^{2D}\, d\sigma_p \leq 4^{4dD}\frac{\Gamma\left(\frac{N}{2}\right)\Gamma\left(D+\frac{1}{2}\right)}{\Gamma\left(\frac{1}{2}\right)\Gamma\left(D+\frac{N}{2}\right)}.$$

We note that this estimate is independent of q and therefore the outer integral in (4.10) is redundant and we obtain

$$\int_{\mathbb{S}^{N-1}} \|Q_p\|_{2D}\, d\sigma_p \leq 4^{2d}\left(\frac{\Gamma\left(\frac{N}{2}\right)\Gamma\left(D+\frac{1}{2}\right)}{\Gamma\left(\frac{1}{2}\right)\Gamma\left(D+\frac{N}{2}\right)}\right)^{1/2D}.$$

As in the proof of Lemma 4.62 we use the estimates

$$\left(\frac{\Gamma\left(\frac{N}{2}\right)}{\Gamma\left(D+\frac{N}{2}\right)}\right)^{\frac{1}{2D}} \leq \sqrt{\frac{2}{N}} \qquad \text{and} \qquad \left(\frac{\Gamma\left(D+\frac{1}{2}\right)}{\Gamma\left(\frac{1}{2}\right)}\right)^{\frac{1}{2D}} \leq \sqrt{D}.$$

Therefore we have

$$\int_{\mathbb{S}^{N-1}} \|Q_p\|_{2D} \leq 2^{4d}\sqrt{\frac{2D}{N}}. \qquad \square$$

4.10 Convex Forms

There is another very interesting convex cone inside $\mathbb{R}[x]_{\mathbf{2d}}$, the cone of convex forms $C_{\mathbf{n},\mathbf{2d}}$. A form $p \in \mathbb{R}[x]_{\mathbf{2d}}$ is called convex if p is a convex function on $\mathbb{R}^n$:

$$p\left(\frac{x+y}{2}\right) \leq \frac{p(x)+p(y)}{2} \quad \text{for all} \quad x, y \in \mathbb{R}^n.$$

It is an easy exercise to show that $C_{\mathbf{n},\mathbf{2d}}$ is contained in the cone of nonnegative forms.

Exercise 4.71. Show that if a form $p \in \mathbb{R}[x]_{\mathbf{2d}}$ is convex, then p is nonnegative. Show that $x_1^2x_2^2 \in P_{\mathbf{2},\mathbf{4}}$ is not convex.

The relationship between convex forms and sums of squares is significantly harder to understand. An equivalent definition of convexity is that a form $p \in \mathbb{R}[x]_{\mathbf{2d}}$ is convex if and only if its Hessian $\nabla^2 p$ is a positive semidefinite matrix on all of $\mathbb{R}^n$. We can associate with p its Hessian form H_p, which is a form in $2n$ variables, with old variables $\mathbf{x} = (x_1, \dots, x_n)$ and new variables $\mathbf{y} = (y_1, \dots, y_n)$. The Hessian form $H_p(x, y)$ is given by

$$H_p(x, y) = y^T \left(\nabla^2 p(x)\right) y.$$

We note that H_p is a bihomogeneous form; it is quadratic in y and of degree $2d - 2$ in x. A form p is convex if and only if its Hessian form H_p is nonnegative on $\mathbb{R}^{2n}$.

A form $p \in \mathbb{R}[x]_{\mathbf{2d}}$ is called *sos-convex* if H_p is a sum of squares. Sos-convexity is a more restrictive condition than being a sum of squares.

Exercise 4.72. Let $p \in \mathbb{R}[x]_{\mathbf{2d}}$ be an sos-convex form. Show that p is a sum of squares.

An explicit example of a convex form that is not sos-convex was constructed in [1]. We will explain below that there exist convex forms that are not sums of squares. In fact, we will show using volume arguments that asymptotically there are significantly more convex forms than sums of squares. However, it is still an open question to find an explicit example of a convex form that is not a sum of squares.

4.10.1 Volumes of Convex Forms

As before we can take a compact section of $C_{\mathbf{n,2d}}$ with the hyperplane $L_{\mathbf{n,2d}}$ of forms of integral 1 on $\mathbb{S}^{n-1}$:

$$\bar{C}_{\mathbf{n,2d}} = C_{\mathbf{n,2d}} \cap L_{\mathbf{n,2d}}.$$

We also let $\tilde{C}_{\mathbf{n,2d}}$ be $\bar{C}_{\mathbf{n,2d}}$ translated by subtracting r^{2d}:

$$\tilde{C}_{\mathbf{n,2d}} = \bar{C}_{\mathbf{n,2d}} - r^{2d}.$$

The convex body $\tilde{C}_{\mathbf{n,2d}}$ lies in the hyperplane $M_{\mathbf{n,2d}}$ of forms of average 0 on the unit sphere $\mathbb{S}^{n-1}$. We will show the following estimate on the volume of $\tilde{C}_{\mathbf{n,2d}}$ that, together with Theorems 4.55 and 4.65, implies that if the degree $2d$ is fixed and the number of variables grows then there are significantly more convex forms than sums of squares. This is the only currently known method of establishing existence of convex forms that are not sums of squares.

Theorem 4.73.

$$\left(\frac{\operatorname{Vol} \tilde{C}_{\mathbf{n,2d}}}{\operatorname{Vol} \tilde{P}_{\mathbf{n,2d}}}\right)^{1/N} \geq \frac{1}{2(2d-1)}.$$

Remark 4.74. *From Exercise* 4.71 *it follows that* $\bar{C}_{\mathbf{n,2d}} \subseteq \bar{P}_{\mathbf{n,2d}}$. *Therefore the estimate of Theorem* 4.73 *is asymptotically tight for the case of fixed degree* $2d$.

Our first goal is to show that if a form $p \in \mathbb{R}[x]_{\mathbf{2d}}$ is sufficiently close to being constant on the unit sphere, then p must be convex.

Theorem 4.75. *Let p be a form in $\mathbb{R}[x]_{\mathbf{2d}}$. If for all $v \in \mathbb{S}^{n-1}$*

$$1 - \frac{1}{2d-1} \leq p(v) \leq 1 + \frac{1}{2d-1},$$

then p is convex.

For a point $\xi \in \mathbb{S}^{n-1}$ we can think of ξ as a direction. We will use

$$\frac{\partial p}{\partial \xi} = \langle \nabla p, \xi \rangle$$

to denote the derivative of p in the direction ξ. A function $f : \mathbb{R}^n \to \mathbb{R}$ is convex if and only if for all $v \in \mathbb{R}^n$ and all $\xi \in \mathbb{S}^{n-1}$ we have

$$\frac{\partial^2 f}{\partial \xi^2}(v) \geq 0.$$

Since we are working with forms it suffices to restrict our attention to $v \in \mathbb{S}^{n-1}$. We use $|\nabla p|$ to denote the length of the gradient of p. We will need the following theorem of Kellogg [13].

Theorem 4.76. *Let p be a form in $\mathbb{R}[x]_{\mathbf{d}}$. For all $v \in \mathbb{S}^{n-1}$*

$$|\nabla p(v)| \leq d \, ||p||_\infty .$$

Theorem 4.76 implies that for any $v \in \mathbb{S}^{n-1}$

$$\left| \frac{\partial p}{\partial \xi}(v) \right| \leq d \, ||p||_\infty .$$

This follows since

$$\frac{\partial p}{\partial \xi} = \langle \nabla p, \xi \rangle \leq |\nabla p| \cdot |\xi| = |\nabla p|$$

by applying the Cauchy–Schwarz inequality.

We extend Theorem 4.76 to cover the case of higher derivatives, which is necessary since convexity is a condition on second derivatives:

Lemma 4.77. *Let p be a form in $\mathbb{R}[x]_{\mathbf{d}}$. For any v and $\xi_1, \ldots \xi_k \in \mathbb{S}^{n-1}$*

$$\left| \frac{\partial^k p}{\partial \xi_1 \cdots \partial \xi_k}(v) \right| \leq \frac{d!}{(d-k)!} \, ||p||_\infty .$$

Proof. We proceed by induction on the order of partial derivatives k. The base case $k = 1$ is covered by Theorem 4.76. Now we need to show the induction step. We assume that the statement holds for all derivatives of order at most k and consider

$$\left|\frac{\partial^{k+1} p}{\partial \xi_1 \cdots \partial \xi_{k+1}}(v)\right|$$

for some $\xi_1, \ldots \xi_{k+1} \in \mathbb{S}^{n-1}$.

Let

$$q = \frac{\partial p}{\partial \xi_1}.$$

Using the base case we see that

$$||q||_\infty \leq d\, ||p||_\infty. \tag{4.12}$$

Also, we know that q is a form in n variables of degree $d-1$. Therefore by the induction assumption

$$\left|\frac{\partial^k q}{\partial \xi_2 \cdots \partial \xi_{k+1}}(v)\right| \leq \frac{(d-1)!}{(d-k-1)!}\, ||q||_\infty. \tag{4.13}$$

Putting together (4.12) and (4.13), the lemma follows. □

We are now ready to prove Theorem 4.75, which provides a sufficient condition for a form to be convex.

Proof of Theorem 4.75. Let p be as in the statement of the theorem, and let $q = p - r^{2d}$. By the assumptions of the theorem it follows that, for all $v \in \mathbb{S}^{n-1}$,

$$-\frac{1}{2d-1} \leq q(v) \leq \frac{1}{2d-1}.$$

In other words

$$||q||_\infty \leq \frac{1}{2d-1}.$$

Then by Lemma 4.77 we know that for any v and $\xi \in \mathbb{S}^{n-1}$

$$\left|\frac{\partial^2 q}{\partial \xi^2}(v)\right| \leq 2d.$$

In particular, it follows that

$$\frac{\partial^2 q}{\partial \xi^2}(v) \geq -2d$$

for all v and $\xi \in \mathbb{S}^{n-1}$.

It is easy to check that

$$\frac{\partial^2 r^{2d}}{\partial \xi^2}(v) = 2d + 4d(d-1)\langle v, \xi \rangle^2 \geq 2d.$$

Since we know that $p = q + r^{2d}$ it follows that for all v and $\xi \in \mathbb{S}^{n-1}$

$$\frac{\partial^2 p}{\partial \xi^2}(v) \geq 0,$$

and therefore p is convex. $\square$

We need one more result from convexity to help us with the volume bounds (see [16]).

Exercise 4.78. Let K be a convex body in $\mathbb{R}^n$. The *barycenter* of K is defined to be a vector $b = (b_1, \ldots, b_n) \in K$ given by

$$b_i = \int_{\mathbb{R}^n} x_i \chi_K \; d\mu,$$

where χ_K is the characteristic function of K and μ is the Lebesgue measure. Let K' be the reflection of K through the barycenter b: $K' = b - (K - b)$. Show that

$$\left(\frac{\operatorname{Vol} K \cap K'}{\operatorname{Vol} K}\right)^{\frac{1}{n}} \geq \frac{1}{2}.$$

Exercise 4.79. The set $\tilde{P}_{\mathbf{n,2d}}$ is a convex body in the hyperplane $M_{\mathbf{n,2d}}$ of all forms of integral 0 on the unit sphere. Use invariance of $\tilde{P}_{\mathbf{n,2d}}$ under orthogonal changes of coordinates to show that 0 is the barycenter of $\tilde{P}_{\mathbf{n,2d}}$. Let $-\tilde{P}_{\mathbf{n,2d}}$ be the reflection of $\tilde{P}_{\mathbf{n,2d}}$ through the origin. Show that $\tilde{P}_{\mathbf{n,2d}} \cap -\tilde{P}_{\mathbf{n,2d}}$ consists of all forms in $M_{\mathbf{n,2d}}$ whose values on the unit are between -1 and 1, i.e., the forms with L^∞-norm at most 1:

$$\tilde{P}_{\mathbf{n,2d}} \cap -\tilde{P}_{\mathbf{n,2d}} = \{p \in M_{\mathbf{n,2d}} \;\mid\; ||p||_\infty \leq 1\}.$$

***Proof of Theorem* 4.73.** Let $K_{\mathbf{n,2d}}$ be the set of forms that take values only between $1 - \frac{1}{2d-1}$ and $1 + \frac{1}{2d-1}$ on the unit sphere:

$$K_{\mathbf{n,2d}} = \left\{ p \in \mathbb{R}[x]_{\mathbf{2d}} \;\middle|\; 1 - \frac{1}{2d-1} \leq p(v) \leq 1 + \frac{1}{2d-1} \quad \text{for all} \quad v \in \mathbb{S}^{n-1} \right\}.$$

We note that $K_{\mathbf{n,2d}}$ is a compact convex set. We let $\bar{K}_{\mathbf{n,2d}}$ be the section of $K_{\mathbf{n,2d}}$ with $L_{\mathbf{n,2d}}$,

$$\bar{K}_{\mathbf{n,2d}} = K_{\mathbf{n,2d}} \cap L_{\mathbf{n,2d}},$$

and let $\tilde{K}_{\mathbf{n,2d}}$ be the translated section:

$$\tilde{K}_{\mathbf{n,2d}} = \bar{K}_{\mathbf{n,2d}} - r^{2d}.$$

It follows that $\tilde{K}_{\mathbf{n,2d}}$ consists of all the forms in $M_{\mathbf{n,2d}}$ that take values between $-\frac{1}{2d-1}$ and $\frac{1}{2d-1}$ on the unit sphere, so forms with L^∞-norm at most $\frac{1}{2d-1}$:

$$\tilde{K}_{\mathbf{n,2d}} = \left\{ p \in M_{\mathbf{n,2d}} \;\middle|\; ||p||_\infty \leq \frac{1}{2d-1} \right\}.$$

By Exercise 4.79 it follows that

$$\frac{1}{2d-1}\left(\tilde{P}_{\mathbf{n,2d}} \cap -\tilde{P}_{\mathbf{n,2d}}\right) \subseteq \tilde{K}_{\mathbf{n,2d}}.$$

Using Exercise 4.78 we see that

$$\left(\frac{\operatorname{Vol} \tilde{P}_{\mathbf{n,2d}} \cap -\tilde{P}_{\mathbf{n,2d}}}{\operatorname{Vol} \tilde{P}_{\mathbf{n,2d}}}\right)^{1/N} \geq \frac{1}{2}.$$

Therefore it follows that

$$\left(\frac{\operatorname{Vol} \tilde{K}_{\mathbf{n,2d}}}{\operatorname{Vol} \tilde{P}_{\mathbf{n,2d}}}\right)^{1/N} \geq \frac{1}{2(2d-1)}.$$

On the other hand, by Theorem 4.75 we know that $\tilde{K}_{\mathbf{n,2d}}$ is contained in $\tilde{C}_{\mathbf{n,2d}}$, and the theorem follows. □

Bibliography

[1] A. A. Ahmadi and P. A. Parrilo. A convex polynomial that is not sos-convex. *Math. Program. Ser. A*, 135:275–292, 2012.

[2] A. Barvinok. *A Course in Convexity.* American Mathematical Society, Providence, RI, 2002.

[3] A. Barvinok. Estimating L^∞ norms by L^{2k} norms for functions on orbits. *Found. Comput. Math.*, 2:393–412, 2002.

[4] A. Barvinok and G. Blekherman. Convex geometry of orbits. In *Combinatorial and Computational Geometry*, Math. Sci. Res. Inst. Publ. 52, Cambridge University Press, Cambridge, UK, 2005, pp. 51–77.

[5] G. Blekherman. There are significantly more nonnegative polynomials than sums of squares. *Israel J. Math.*, 183:355–380, 2006.

[6] G. Blekherman. *Dimensional differences between nonnegative polynomials and sums of squares.* Submitted for publication, arXiv:0907.1339.

[7] G. Blekherman. Nonnegative polynomials and sums of squares. *J. Amer. Math. Soc.*, 25:617–635, 2012.

[8] G. Blekherman, J. Hauenstein, J. C. Ottem, K. Ranestad, and B. Sturmfels. Algebraic boundaries of Hilbert's SOS cones. To appear in *Compositio Mathematica.* arXiv:1107.1846.

[9] J. Duoandikoetxea. Reverse Hölder inequalities for spherical harmonics. Proc. Amer. Math. Soc., 101:487–491, 1987.

[10] D. Eisenbud, M. Green, and J. Harris. Cayley-Bacharach theorems and conjectures. *Bull. Amer. Math. Soc.*, 33:295–324, 1996.

[11] G. Hardy, E. Littlewood, and G. Polya. *Inequalities.* Cambridge University Press, Cambridge, UK, 1988.

[12] J. Harris. *Algebraic Geometry. A First Course,* Grad. Texts in Math. 133. Springer-Verlag, New York, 1995.

[13] O. D. Kellogg. On bounded polynomials in several variables. *Math. Z.*, 27:55–64, 1928.

[14] J. M. Landsberg and Z. Teitler. On the ranks and border ranks of symmetric tensors. *Found. Comput. Math.*, 10:339–366, 2010.

[15] J. B. Lasserre, *Moments, Positive Polynomials and Their Applications.* Imperial College Press, London, 2009.

[16] V. D. Milman and A. Pajor. Entropy and asymptotic geometry of non-symmetric convex bodies. *Adv. Math.*, 152:314–335, 2000.

[17] R. Miranda. Linear systems of plane curves. *Notices Amer. Math. Soc.*, 46:192–202, 1999.

[18] P. A. Parrilo. Semidefinite programming relaxations for semialgebraic problems. *Math. Program. Ser. B*, 96:293–320, 2000/01.

[19] P. A. Parrilo and B. Sturmfels. Minimizing polynomial functions. In *Algorithmic and Quantitative Aspects of Real Algebraic Geometry*, DIMACS Ser. Discrete Math. Theoret. Comput. Sci. 60, American Mathmatical Society, Providence, RI, 2003, pp. 83–100.

[20] M. Ramana and A. J. Goldman. Some geometric results in semidefinite programming. *J. Global Optim.*, 7:33–50, 1995.

[21] B. Reznick, *Sums of Even Powers of Real Linear Forms*, Mem. Amer. Math. Soc. 96, American Mathematical Society, Providence, RI, 1992.

[22] B. Reznick. Some concrete aspects of Hilbert's 17th problem. In *Real Algebraic Geometry and Ordered Structures*, Contemp. Math. 253. American Mathematical Society, Providence, RI, 2000, pp. 251–272.

[23] B. Reznick. *On Hilbert's construction of positive polynomials.* Submitted for publication, arXiv:0707.2156.

[24] B. Reznick. Blenders. In *Notions of Positivity and the Geometry of Polynomials*, P. Branden, M. Passare, and M. Putinar, eds., Trends in Math. Birkhäuser, Basel, 2011, pp. 345–373.

[25] R. Sanyal, F. Sottile, and B. Sturmfels. Orbitopes. *Mathematika*, 57:275–314, 2011.

[26] R. Schneider. *Convex Bodies: The Brunn-Minkowski Theory.* Cambridge University Press, Cambridge, UK, 1993.

Chapter 5

Dualities

Philipp Rostalski[†] and Bernd Sturmfels[‡]

Dualities are ubiquitous in mathematics and its applications. This chapter compares several notions of duality that are central to the connections between convexity, optimization, and algebraic geometry developed in this book. It is meant as a first introduction and is intended for a diverse audience ranging from graduate students in mathematics to practitioners of optimization who are based in engineering.

5.1 Introduction

Convex algebraic geometry concerns the interplay between optimization theory and real algebraic geometry. Its objects of study include convex semialgebraic sets that arise in semidefinite programming and from sums of squares. This chapter compares three notions of duality that are relevant in these contexts: duality of convex bodies, duality of projective varieties, and the Karush–Kuhn–Tucker conditions derived from Lagrange duality. We show that the optimal value of a polynomial program is an algebraic function whose minimal polynomial is expressed by the hypersurface projectively dual to the constraint set. We give an introduction to the algebraic geometry in the boundary of the convex hull of a compact variety. Our focus lies on making the polynomials that vanish on that boundary explicit, in contrast to the representation of convex bodies as projected spectrahedra. We also explore the geometric underpinnings of semidefinite programming duality.

Duality for vector spaces lies at the heart of linear algebra and functional analysis. Duality in convex geometry is essentially an involution on the set of

[†]Philipp Rostalski was supported by the Alexander-von-Humboldt Foundation through a Feodor Lynen postdoctoral fellowship.

[‡]Bernd Sturmfels was supported by NSF grants DMS-0757207 and DMS-0968882.

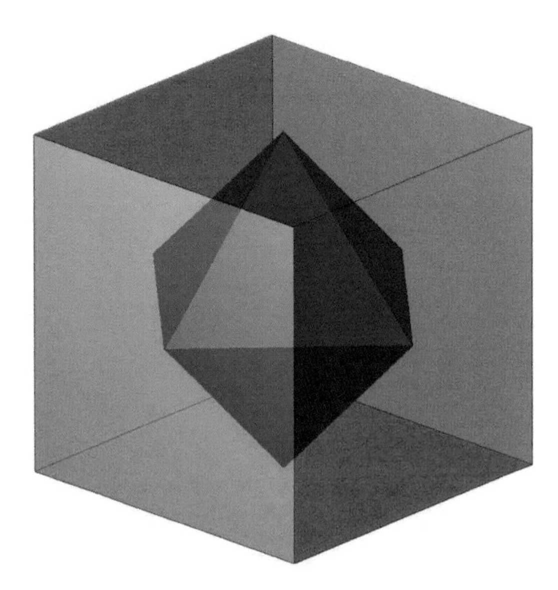

Figure 5.1. *The cube is dual to the octahedron.*

convex bodies: for instance, it maps the cube to the octahedron and vice versa (Figure 5.1). Duality in optimization, known as *Lagrange duality*, plays a key role in designing efficient algorithms for the solution of various optimization problems. In projective geometry, points are dual to hyperplanes, and this leads to a natural notion of *projective duality* for algebraic varieties. Our aim here is to explore these dualities and their interconnections in the context of polynomial optimization and semidefinite programming. Toward the end of the introduction, we shall discuss the context and organization of this chapter. At this point, however, we jump right in and present a concrete three-dimensional example that illustrates our perspective on these topics.

5.1.1 How to Dualize a Pillow

We consider the following symmetric matrix with three indeterminate entries:

$$Q(x,y,z) \quad = \quad \begin{pmatrix} 1 & x & 0 & x \\ x & 1 & y & 0 \\ 0 & y & 1 & z \\ x & 0 & z & 1 \end{pmatrix}. \tag{5.1}$$

This symmetric 4×4 matrix specifies a three-dimensional compact convex body

$$P \;=\; \big\{ (x,y,z) \in \mathbb{R}^3 \mid Q(x,y,z) \succeq 0 \big\}. \tag{5.2}$$

The notation "$\succeq 0$" means that the matrix is *positive semidefinite*, i.e., all four eigenvalues are nonnegative real numbers. Such a *linear matrix inequality* always

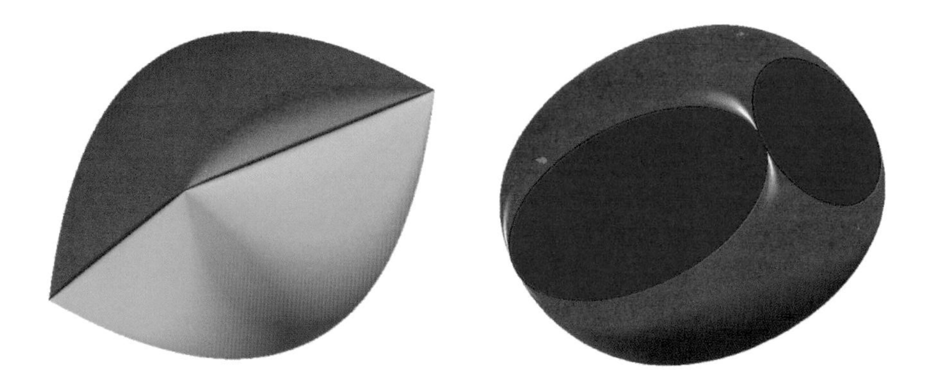

Figure 5.2. *A three-dimensional spectrahedron P and its dual convex body P°.*

defines a closed convex set (as in (5.2)) which is referred to as a *spectrahedron*. Positive semidefinite matrices and spectrahedra appear in all chapters of this book.

Our spectrahedron P looks like a pillow. It is shown on the left in Figure 5.2. The *algebraic boundary* of P is the surface specified by the determinant

$$\det(Q(x,y,z)) \quad = \quad x^2(y-z)^2 - 2x^2 - y^2 - z^2 + 1 \quad = \quad 0.$$

At this point we pause to emphasize that Subsection 5.1.1 is intended to be a first welcome to our readers. The objects of study are introduced here informally, by way of one concrete example in three dimensions, which may guide the reader through the following sections. Precise definitions of the general concepts, such as "algebraic boundary," "algebraic degree," etc., will be furnished in the later sections.

The interior of the spectrahedron P in (5.1) represents all matrices $Q(x,y,z)$ whose four eigenvalues are positive. At all smooth points on the boundary of P, precisely one eigenvalue vanishes, and the rank of the matrix $Q(x,y,z)$ drops from 4 to 3. However, the rank drops further to 2 at the four singular points

$$(x,y,z) \quad = \quad \frac{1}{\sqrt{2}}(1,1,-1),\ \frac{1}{\sqrt{2}}(-1,-1,1),\ \frac{1}{\sqrt{2}}(1,-1,1),\ \frac{1}{\sqrt{2}}(-1,1,-1). \qquad (5.3)$$

We find these from a *Gröbner basis* of the ideal of 3×3 minors of $Q(x,y,z)$:

$$\{\, 2x^2 - 1,\ 2z^2 - 1,\ y + z \,\}.$$

The linear polynomial $y+z$ in this Gröbner basis defines the symmetry plane of the pillow P. The four singular points form a square in that plane. Its edges are also edges of P. All other faces of P are exposed points. These come in two families, sometimes called *protrusions*, one above the plane $y+z=0$ and one below it.

The protrusions are drawn in two different colors on the left in Figure 5.2. Note that the surface ∂P is smooth along the four edges that separate the two protrusions. To be more precise, the four points (5.3) are the only singular points in ∂P. All points in the relative interiors of the four edges are nonsingular in ∂P.

Like all convex bodies, our pillow P has an associated *dual convex body*

$$P^\circ \;=\; \left\{ (a,b,c) \in \mathbb{R}^3 \mid ax+by+cz \leq 1 \text{ for all } (x,y,z) \in P \right\}, \tag{5.4}$$

consisting of all linear forms that evaluate to at most one on the convex body P.

The dual pillow P° is shown on the right in Figure 5.2. Note the association of faces under duality. The pillow P has four one-dimensional faces, four singular zero-dimensional faces, and two smooth families of zero-dimensional faces. The corresponding dual faces of P° have dimensions 0, 2, and 0, respectively.

Semidefinite programming was introduced in Chapter 2 as the computational problem of optimizing a linear function over a spectrahedron. For our pillow P, this optimization problem takes the form

$$\begin{aligned} p^\star(a,b,c) \;=\; & \max_{(x,y,z)\in\mathbb{R}^3} \quad ax+by+cz \\ & \text{subject to} \quad Q(x,y,z) \succeq 0. \end{aligned} \tag{5.5}$$

We regard this as a parametric optimization problem: we are interested in the optimal value and optimal solution of (5.5) as a function of $(a,b,c) \in \mathbb{R}^3$. This function can be expressed in terms of the dual body P° as follows:

$$d^\star(a,b,c) \;=\; \min_{\gamma\in\mathbb{R}} \gamma \quad \text{subject to} \quad \frac{1}{\gamma}\cdot(a,b,c) \;\in\; P^\circ. \tag{5.6}$$

We distinguish this formulation from the duality in semidefinite programming. The *dual* to (5.5) is the following program with 7 decision variables:

$$\begin{aligned} d^\star(a,b,c) \;=\; & \min_{u\in\mathbb{R}^7} \quad u_1+u_4+u_6+u_7 \\ & \text{subject to} \quad \begin{pmatrix} 2u_1 & 2u_2 & 2u_3 & -2u_2-a \\ 2u_2 & 2u_4 & -b & 2u_5 \\ 2u_3 & -b & 2u_6 & -c \\ -2u_2-a & 2u_5 & -c & 2u_7 \end{pmatrix} \succeq 0. \end{aligned} \tag{5.7}$$

The derivation of such a dual formulation will be explained in Section 5.5. Since (5.5) and (5.7) are both strictly feasible, strong duality holds [5, Subsection 5.2.3]; i.e., the two programs attain the same optimal value: $p^\star(a,b,c) = d^\star(a,b,c)$. Hence, problem (5.7) can be derived from (5.6), as we shall see in Section 5.5.

We write $M(u;a,b,c)$ for the 4×4 matrix in (5.7). The following equations and inequalities, known as the *Karush–Kuhn–Tucker conditions* (KKT), are necessary and sufficient for any pair of optimal solutions:

$$\begin{aligned} Q(x,y,z)\cdot M(u;a,b,c) \;&=\; 0, \qquad \text{(complementary slackness)} \\ Q(x,y,z) \;&\succeq\; 0, \\ M(u;a,b,c) \;&\succeq\; 0. \end{aligned}$$

We relax the inequality constraints and consider the system of equations

$$\gamma = ax+by+cz \quad \text{and} \quad Q(x,y,z)\cdot M(u;a,b,c) \;=\; 0.$$

This is a system of 17 equations. Using computer algebra, we eliminate the 10 unknowns $x, y, z, u_1, \dots, u_7$. The result is a polynomial in a, b, c, and γ. Its factors, shown in (5.8)–(5.9), express the optimal value $\gamma^\star$ in terms of a, b, c.

At the optimal solution, the product of the two 4×4 matrices $Q(x, y, z)$ and $M(u; a, b, c)$ is zero, and hence the pair $(\text{rank}(Q), \text{rank}(M))$ equals either $(3, 1)$ or $(2, 2)$. In the former case the optimal value $\gamma^\star$ is one of the two solutions of

$$(b^2 + 2bc + c^2) \cdot \gamma^2 - a^2b^2 - a^2c^2 - b^4 - 2b^2c^2 - 2bc^3 - c^4 - 2b^3c = 0. \tag{5.8}$$

In the latter case it comes from the four corners of the pillow, and it satisfies

$$\begin{array}{rl} & (2\gamma^2 - a^2 + 2ab - b^2 + 2bc - c^2 - 2ac) \\ \cdot & (2\gamma^2 - a^2 - 2ab - b^2 + 2bc - c^2 + 2ac) \quad = \quad 0. \end{array} \tag{5.9}$$

These two equations describe the algebraic boundary of the dual body P°. Namely, after setting $\gamma = 1$, the irreducible polynomial in (5.8) describes the quartic surface that makes up the curved part of the boundary of P°, as seen in Figure 5.2. In addition, there are four planes spanned by flat two-dimensional faces of P°. The product of the four corresponding affine linear forms is the expression (5.9). Indeed, each of the two quadrics in (5.9) factors into two linear factors. These two characterize the planes spanned by opposite 2-faces of P°.

The two equations (5.8) and (5.9) also offer a first glimpse of the concept of *projective duality* in algebraic geometry, defined precisely in Subsection 5.2.4. Namely, consider the surface in projective space $\mathbb{P}^3$ defined by $\det(Q(x, y, z)) = 0$ after replacing the ones along the diagonals by a homogenization variable. Then (5.8) is its dual surface in the dual projective space $(\mathbb{P}^3)^*$. The surface (5.9) in $(\mathbb{P}^3)^*$ is dual to the zero-dimensional variety in $\mathbb{P}^3$ cut out by the 3×3 minors of $Q(x, y, z)$.

The optimal value function of the optimization problem (5.5) is represented, in the sense of Section 5.3, by the algebraic surfaces dual to the boundary of P and its singular locus. We have seen two different ways of dualizing (5.5): the dual optimization problem (5.7) and the optimization problem (5.6) on P°. These two formulations are related as follows. If we regard (5.7) as specifying a 10-dimensional spectrahedron, then the dual pillow P° is a projection of that spectrahedron:

$$P^\circ \;=\; \big\{(a, b, c) \in \mathbb{R}^3 \,|\, \exists u \in \mathbb{R}^7 \,:\, M(u; a, b, c) \succeq 0 \text{ and } u_1 + u_4 + u_6 + u_7 \,=\, 1\big\}.$$

Linear projections of spectrahedra, so-called *projected spectrahedra*, were introduced in Chapter 2. They are at the heart of several parts of this book, most notably, Chapters 6 and 7. The dual of a spectrahedron is generally not a spectrahedron, but it is always a projected spectrahedron. We shall see this in Theorem 5.57.

5.1.2 Context and Outline

Duality is a central concept in convexity and convex optimization, and numerous authors have written about their connections and their interplay with other notions of duality and polarity. Relevant references include Barvinok's textbook [1, Section 4] and the survey by Luenberger [24]. The latter focuses on dualities used in engineering, such as duality of vector spaces, polytopes, graphs, and control systems. The objective of this chapter is to revisit the theme of duality in the context of convex algebraic geometry and semidefinite optimization. In algebraic geometry, there is a natural notion of projective duality, which associates to every algebraic variety a dual variety. One of our main goals is to explore the meaning of projective duality for optimization theory. It is precisely this deeper connection with algebra which distinguishes this chapter from other treatments of duality in convex optimization.

Our presentation is organized as follows. In Section 5.2 we cover preliminaries needed for the rest of the chapter. Here the various dualities are carefully defined and their basic properties are illustrated by means of examples. In Section 5.3 we derive the result that the optimal value function of a polynomial program is represented by the defining equation of the hypersurface projectively dual to the manifold describing the boundary of all feasible solutions. This highlights the important fact that the duality best known to algebraic geometers arises very naturally in convex optimization. Section 5.4 concerns the convex hull of a compact algebraic variety in $\mathbb{R}^n$. We discuss work of Ranestad and Sturmfels [31, 32] on the hypersurfaces in the boundary of such a convex body, and we present several examples and applications.

In Section 5.5 we focus on semidefinite programming (SDP), and we offer a concise geometric introduction to SDP duality. This leads us to the concept of algebraic degree of SDP [12, 27] or, more geometrically, to projective duality for varieties defined by rank constraints on symmetric matrices of linear forms.

A *projected spectrahedron* is the image of a spectrahedron under a linear projection. Its dual body is a linear section of the dual body to the spectrahedron. In Section 5.6 we examine this situation in the context of sums-of-squares programming, and we discuss linear families of nonnegative polynomials. The figures in this chapter were made with the software package Bermeja [34], which specializes in computations in convex algebraic geometry.

We now come to the first round of exercises in this chapter. They are meant for our readers to "get their hands dirty" right away. The problems can be approached from first principles. No knowledge of any general algorithms or theorems is needed. The use of both numerical software and computer algebra tools is encouraged.

Exercises

Exercise 5.1. Maximize the function $2x + 3y + 7y$ over the spectrahedron P given in (5.2). Express the optimal solution in exact arithmetic. Locate the cost function on the right in Figure 5.2 and locate the optimal solution on the left.

Exercise 5.2. Compute the projections of the spectrahedron P into the (x, y)-plane and into the (y, z)-plane. Determine polynomials $f(x, y)$ and $g(y, z)$ that vanish on the boundaries of these two planar convex bodies.

Exercise 5.3. Project P into a random plane and compute the irreducible polynomial of degree eight in two variables that vanishes on the boundary of image.

Exercise 5.4. Does there exist a projected spectrahedron that is not a spectrahedron?

Exercise 5.5. A *correlation matrix* is a positive semidefinite real symmetric $n \times n$ matrix whose n diagonal entries are all equal to 1.

(a) Maximize the sum of the off-diagonal entries over correlation matrices with $n = 3$. Solve this optimization problem also for $n = 4$.

(b) Minimize the sum of the off-diagonal entries over correlation matrices with $n = 3$. Solve this optimization problem also for $n = 4$.

(c) Does there exist a correlation matrix, of any size n, whose determinant is larger than 1? Find a proof or counterexample.

5.2 Ingredients

In this section we review the mathematical preliminaries needed for the rest of the chapter, we give precise definitions, and we fix more of the notation. We begin with the notion of duality for vector spaces and cones therein; then we move on to convex bodies, polytopes, Lagrange duality in optimization, the KKT conditions, and projective duality in algebraic geometry, and we conclude with discriminants.

5.2.1 Vector Spaces and Cones

We fix an ordered field K. The primary example is the field of real numbers, $K = \mathbb{R}$, but we also allow other fields, such as the rational numbers $K = \mathbb{Q}$ or the real Puiseux series $K = \mathbb{R}\{\{\epsilon\}\}$. The examples in this chapter have their algebraic representation over the rationals $\mathbb{Q}$, but we consider the corresponding geometric objects over the reals $\mathbb{R}$. However, special geometric features naturally lead to intermediate fields, e.g., the singular points in (5.3) live over the field $\mathbb{Q}(\sqrt{2})$. Puiseux series come in handy when one needs a deformation parameter ϵ to deal with degeneracies. This is standard for algorithms in real algebraic geometry [2].

Fix a finite-dimensional vector space V over an ordered field K. The dual vector space is the set $V^* = \mathrm{Hom}(V, K)$ of all linear forms on V. Let V and W be vector spaces and $\varphi : V \to W$ a linear map. The *adjoint* $\varphi^* : W^* \to V^*$ is the linear map defined by $\varphi^*(w^*) = w^* \circ \varphi \in V^*$ for every $w^* \in W^*$. If we fix bases of both V and W, then φ is represented by a matrix A. The adjoint φ^* is represented, relative to the dual bases for W^* and V^*, by the *transpose* A^T of the matrix A.

A subset $C \subset V$ is a *cone* if it is closed under multiplication with positive scalars. A cone C need not be convex, but its *dual cone*

$$C^* \;=\; \{\, l \in V^* \mid \text{for all } x \in C : l(x) \geq 0 \,\} \tag{5.10}$$

is always closed and convex in V^*. If C is a *convex cone*, then the second dual $(C^*)^*$ is the closure of C. Thus, if C is a closed convex cone in V, then

$$(C^*)^* \;=\; C. \tag{5.11}$$

This important relationship is referred to as *biduality*.

Every linear subspace $L \subset V$ is also a closed convex cone. The dual of L, when viewed as a cone, is the orthogonal complement of L, viewed as a subspace:

$$L^* \;=\; L^\perp \;=\; \{\, l \in V^* \mid \text{for all } x \in L : l(x) = 0 \,\}.$$

The adjoint to the inclusion $L \subset V$ is the projection $\pi_L : V^* \to V^*/L^\perp$. Given any (convex) cone $C \subset V$, the intersection $C \cap L$ is a (convex) cone in L. Its dual cone $(C \cap L)^*$ is the projection of the cone C^* into $V^*/L^\perp$. More precisely,

$$(C \cap L)^* \;=\; \overline{C^* + L^\perp} \quad \text{in } V^*. \tag{5.12}$$

Now, it makes sense to consider this convex set modulo $L^\perp$. We can thus identify

$$(C \cap L)^* \;=\; \overline{\pi_L(C^*)} \quad \text{in } V^*/L^\perp. \tag{5.13}$$

This formula expresses the fact that projection and intersection are dual operations.

Example 5.6. It is necessary to take the closure of $\pi_L(C^*)$ in (5.12) and (5.13) because projections of closed convex cones need not be closed. The following simple example is derived from [18, Example 3.5, p. 196]. Consider the closed convex cone

$$C^* \;=\; \big\{(u,x,y,z) \in \mathbb{R}^4 \,:\, u \geq 0,\; u+x \geq 0,\; y \geq 0,\; z \geq 0, \;\text{ and } (u+x)y \geq z^2\big\},$$

and fix the hyperplane $L = \{(0,x,y,z) : x,y,z \in \mathbb{R}\} \simeq \mathbb{R}^3$. Then π_L is the projection from $\mathbb{R}^4$ to $\mathbb{R}^3$ given by dropping the u-coordinate. We claim that the image $\pi_L(C^*)$ is not closed. To see this, we note that for every $\epsilon > 0$ the vector $(1/\epsilon, 0, \epsilon, 1)$ lies in C^*, and hence $(0, \epsilon, 1)$ lies in $\pi_L(C^*)$. On the other hand, $(0,0,1)$ does not lie in $\pi_L(C^*)$ because $z = 1$ implies $(u+x)y \geq 1$ and hence $y > 0$. ∎

The results summarized above are fundamental in convex analysis. For proofs and details we refer to the textbook by Rockafellar [33, Section 16]. The space $V^*/L^\perp$ is the space $\mathrm{Hom}(L, K)$ of linear functionals on L. In applications one often identifies this space with L itself, by means of an inner product on the ambient space V. The linear map π_L then becomes the orthogonal projection from V onto L, and (5.13) is the closure of the image of C under that orthogonal projection.

A subset $F \subseteq C$ of a convex set C is a *face* if F is itself convex and contains any line segment $L \subset C$ whose relative interior intersects F. We say that F is an *exposed face* if there exists a linear functional l that attains its minimum over C precisely at F. Clearly, every exposed face of C is a face, but the converse does not hold. For instance, the edges of the triangle on the top in Figure 5.6 are nonexposed faces of the three-dimensional convex body shown there.

An exposed face F of a cone C determines a face of the dual cone C^* via

$$F^\diamond \;=\; \{\, l \in C^* \mid l \text{ attains its minimum over } C \text{ at } F \,\}.$$

The dimensions of the faces F of C and $F^\diamond$ of C^* satisfy the inequality

$$\dim(F) \,+\, \dim(F^\diamond) \;\leq\; \dim(V). \tag{5.14}$$

If C is a polyhedral cone, then C^* is also polyhedral. In that case, the number of faces F and $F^\diamond$ is finite and equality holds in (5.14). On the other hand, most convex cones considered in this chapter are not polyhedral; they have infinitely many faces, and the inequality in (5.14) is usually strict. For instance, the *second-order cone* $C = \{(x,y,z) \in \mathbb{R}^3 : \sqrt{x^2+y^2} \leq z\}$ is self-dual, each proper face F of C is one-dimensional, and the formula (5.14) says that $1+1 \leq 3$.

5.2.2 Convex Bodies and Their Algebraic Boundary

A *convex body* in V is a full-dimensional convex set that is closed and bounded. If C is a cone and $\varphi \in \text{int}(C^*)$, then the hyperplane $\{\varphi(x) = z\}$ intersects C for all $z \geq 0$ and yields a convex body. In this manner, every pointed r-dimensional cone gives rise to an $(r-1)$-dimensional convex body by fixing $z = 1$. The convex body forms the base of the cone. The cone can be recovered from its base up to a linear isomorphism. These transformations, known as *homogenization* and *dehomogenization* with homogenization variable z, respect faces and algebraic boundaries. They allow us to go back and forth between convex bodies and cones in the next higher dimension. For instance, the three-dimensional body P in (5.2) is the base of the cone in $\mathbb{R}^4$ we get by multiplying the constants 1 on the diagonal in (5.1) with a new variable.

Let P be a full-dimensional convex body in V and assume that $0 \in \text{int}(P)$. Dehomogenizing the definition for cones, we obtain the *dual convex body*

$$P^\circ \;=\; \{\,\ell \in V^* \mid \text{for all } x \in P : \ell(x) \leq 1\,\}. \tag{5.15}$$

This is derived from (5.10) using the identification $l(x) = z - \ell(x)$ for $z = 1$. We note that the dual of a convex body (as opposed to the dual of a cone) is not an intrinsic construction, but it depends on the position of P relative to the origin.

Just as in the case of convex cones, if P is closed, then biduality holds:

$$(P^\circ)^\circ \;=\; P.$$

The definition (5.15) makes sense for arbitrary subsets P of V. That is, P need not be convex or closed. A standard fact from convex analysis [33, Corollary 12.1.1 and Section 14] says that the double dual is the closure of the convex hull with the origin:

$$(P^\circ)^\circ \;=\; \overline{\text{conv}(P \cup 0)}.$$

All convex bodies discussed in this chapter are *semialgebraic*, that is, they can be described by Boolean combinations of polynomial inequalities. We note that if P is semialgebraic then its dual body P° is also semialgebraic. This is a consequence of Tarski's theorem on quantifier elimination in real algebraic geometry [2, 4].

The *algebraic boundary* of a semialgebraic convex body P, denoted $\partial_a P$, is the smallest complex algebraic variety that contains the boundary ∂P. In geometric language, $\partial_a P$ is the *Zariski closure* of ∂P. It is identified with the squarefree polynomial f_P that vanishes on ∂P. Namely, $\partial_a P = V_{\mathbb{C}}(f_P)$ is the zero set of the polynomial f_P. Note that f_P is unique up to a multiplicative constant. Thus $\partial_a P$ is the smallest complex algebraic hypersurface which contains the boundary ∂P.

A *polytope* is the convex hull of a finite subset of V. If P is a polytope, then so is its dual P° [37]. The boundary of P consists of finitely many *facets* F. These are the faces $F = v^\diamond$ dual to the *vertices* v of P°. The algebraic boundary $\partial_a P$ is the arrangement of hyperplanes spanned by the facets of P. Its defining polynomial f_P is the product of the linear polynomials $\langle v, x\rangle - 1$.

Example 5.7. A polytope known to everyone is the three-dimensional cube

$$P = \operatorname{conv}\big(\{(\pm1,\pm1,\pm1)\}\big) = \{-1 \le x,y,z \le 1\}.$$

Figure 5.1 illustrates the familiar fact that its dual polytope is the octahedron

$$P^\circ = \{-1 \le a \pm b \pm c \le 1\} = \operatorname{conv}\big(\{\pm e_1, \pm e_2, \pm e_3\}\big).$$

Here e_i denotes the ith unit vector. The eight vertices of P correspond to the facets of P°, and the six facets of P correspond to the vertices of P°. The algebraic boundary of the cube is described by a degree 6 polynomial

$$\partial_a P = V_{\mathbb{C}}\left((x^2-1)(y^2-1)(z^2-1)\right).$$

The algebraic boundary of the octahedron is given by a degree 8 polynomial

$$\partial_a P^\circ = V_{\mathbb{C}}\left(\prod(1-a\pm b\pm c)\prod(a\pm b\pm c+1)\right).$$

Note that P and P° are the unit balls for the norms L_∞ and L_1 on $\mathbb{R}^3$. ∎

Recall that the L_p*-norm* on $\mathbb{R}^n$ is defined by $\|x\|_p = (\sum_{i=1}^n |x_i|^p)^{1/p}$ for $x \in \mathbb{R}^n$. The *dual norm* to the L_p-norm is the L_q-norm for $\frac{1}{p}+\frac{1}{q}=1$, that is,

$$\|y\|_q = \sup\{\langle y,x\rangle \mid x \in \mathbb{R}^n,\ \|x\|_p \le 1\}.$$

Geometrically, the unit balls for these norms are dual as convex bodies.

Example 5.8. Consider the case $n=2$ and $p=4$. Here the unit ball equals

$$P = \{\, (x,y) \in \mathbb{R}^2 : x^4+y^4 \le 1\,\}.$$

This planar convex set is shown in Figure 5.3. The ordinary boundary ∂P of this convex set is the real curve defined by the quartic polynomial $x^4+y^4=1$. In this example, the ordinary boundary coincides with the algebraic boundary $\partial_a P$.

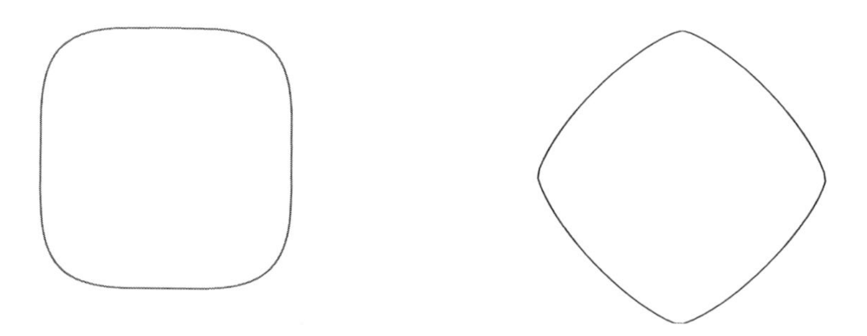

Figure 5.3. *The unit balls for the L_4-norm and the $L_{4/3}$-norm are dual. The curve on the left has degree 4, while its dual curve on the right has degree 12.*

The dual body is the unit ball for the $L_{4/3}$-norm on $\mathbb{R}^2$:

$$P^\circ \;=\; \{(a,b)\in\mathbb{R}^2 \,:\, |a|^{4/3}+|b|^{4/3}\leq 1\}\,.$$

The algebraic boundary of P° is an irreducible algebraic curve of degree 12,

$$\partial_a P^\circ \;=\; V\big(a^{12}+3a^8b^4+3a^4b^8+b^{12}-3a^8+21a^4b^4-3b^8+3a^4+3b^4-1\big), \qquad (5.16)$$

which again coincides precisely with the (geometric) boundary ∂P°. This dual polynomial is easily produced by the following one-line program in the computer algebra system `Macaulay2` due to Grayson and Stillman [13]:

```
R = QQ[x,y,a,b]; eliminate({x,y},ideal(x^4+y^4-1,x^3-a,y^3-b))
```

In Subsection 5.2.4 we shall introduce the general algebraic framework for performing such duality computations, not just for curves, but for arbitrary varieties. ∎

5.2.3 Lagrange Duality in Optimization

We now come to a standard concept of duality in optimization theory. The treatment here is more general than duality in convex optimization, which was presented in Chapter 2. Let us consider the following general nonlinear polynomial optimization problem:

$$\begin{aligned} \underset{x\in\mathbb{R}^n}{\text{minimize}} \quad & f(x) \\ \text{subject to} \quad & g_i(x)\leq 0, \quad i=1,\ldots,m, \\ & h_j(x)=0, \quad j=1,\ldots,p. \end{aligned} \qquad (5.17)$$

Here the $g_1,\ldots,g_m,h_1,\ldots,h_p$ and f are polynomials in $\mathbb{R}[x_1,\ldots,x_n]$. The *Lagrangian* associated with the optimization problem (5.17) is the function

$$\begin{aligned} L\,:\,\mathbb{R}^n\times\mathbb{R}^m_+\times\mathbb{R}^p &\to \quad \mathbb{R}^n, \\ (x,\lambda,\mu) &\mapsto \quad f(x)+\textstyle\sum_{i=1}^m \lambda_i g_i(x)+\sum_{j=1}^p \mu_j h_j(x). \end{aligned}$$

The scalars $\lambda_i\in\mathbb{R}_+$ and $\mu_j\in\mathbb{R}$ are the *Lagrange multipliers* for the constraints $g_i(x)\leq 0$ and $h_j(x)=0$. The Lagrangian $L(x,\lambda,\mu)$ can be interpreted as an augmented cost function with penalty terms for the constraints. For more information on the above formulation see [5, Section 5.1].

One can show that the problem (5.17) is equivalent to finding

$$u^\star \;=\; \min_{x\in\mathbb{R}^n}\;\max_{\mu\in\mathbb{R}^p \text{ and } \lambda\geq 0}\; L(x,\lambda,\mu).$$

The key observation here is that any positive evaluation of one of the polynomials $g_i(x)$, or any nonzero evaluation of one of the polynomials $h_j(x)$, would render the inner optimization problem unbounded.

The *dual optimization problem* to (5.17) is obtained by exchanging the order of the two nested optimization subproblems in the above formulation:

$$v^\star \;=\; \max_{\mu\in\mathbb{R}^p \text{ and } \lambda\geq 0}\; \underbrace{\min_{x\in\mathbb{R}^n}\; L(x,\lambda,\mu)}_{\phi(\lambda,\mu)}\,.$$

The function $\phi(\lambda, \mu)$ is known as the *Lagrange dual function* to our problem. This function is always concave, so the dual is always a convex optimization problem. It follows from the definition of the dual function that $\phi(\lambda, \mu) \leq u^\star$ for all λ, μ. Hence the optimal values satisfy the inequality

$$v^\star \;\leq\; u^\star.$$

If equality occurs, $v^\star = u^\star$, then we say that *strong duality* holds. A necessary condition for strong duality is $\lambda_i^\star g_i(x^\star) = 0$ for all $i = 1, \ldots, m$, where $(x^\star, \lambda^\star, \mu^\star)$ denote a primal and dual optimizer. We see this by evaluating the Lagrangian at an optimizer and taking into account the fact that $h_j(x) = 0$ for all feasible x.

Collecting all inequality and equality constraints in the primal and dual optimization problems yields the following optimality conditions.

Theorem 5.9 (KKT conditions). *Let $(x^\star, \lambda^\star, \mu^\star)$ be primal and dual optimal solutions with $u^\star = v^\star$ (strong duality). Then*

$$\begin{aligned} \nabla_x f\Big|_{x^\star} + \sum_{i=1}^{m} \lambda_i^\star \cdot \nabla_x g_i\Big|_{x^\star} + \sum_{j=1}^{p} \mu_j^\star \cdot \nabla_x h_j\Big|_{x^\star} &= 0, \\ g_i(x^\star) &\leq 0 \quad \textit{for } i = 1, \ldots, m, \\ \lambda_i^\star &\geq 0 \quad \textit{for } i = 1, \ldots, m, \\ h_j(x^\star) &= 0 \quad \textit{for } j = 1, \ldots, p, \\ \textit{Complementary slackness:} \quad \lambda_i^\star \cdot g_i(x^\star) &= 0 \quad \textit{for } i = 1, \ldots, m. \end{aligned} \tag{5.18}$$

For a derivation of this theorem see [5, Subsection 5.5.2]. Several comments on the KKT conditions are in order. First, we note that complementary slackness amounts to a case distinction between active ($g_i = 0$) and inactive inequalities ($g_i < 0$). For any index i with $g_i(x^\star) \neq 0$ we need $\lambda_i = 0$, so the corresponding inequality does not play a role in the gradient condition. On the other hand, if $g_i(x^\star) = 0$, then this can be treated as an equality constraint.

From an algebraic point of view, it is natural to relax the inequalities and to focus on the *KKT equations.* These are the polynomial equations in (5.18):

$$\begin{aligned} \nabla_x f\Big|_{x^\star} + \sum_{i=1}^{m} \lambda_i^\star \cdot \nabla_x g_i\Big|_{x^\star} + \sum_{j=1}^{p} \mu_j^\star \cdot \nabla_x h_j\Big|_{x^\star} &= 0, \\ h_1(x) = \cdots = h_p(x) = \lambda_1 g_1(x) = \cdots = \lambda_m g_m(x) &= 0. \end{aligned} \tag{5.19}$$

If we wish to solve our optimization problem exactly, then we must compute the algebraic variety in $\mathbb{R}^n \times \mathbb{R}^m \times \mathbb{R}^p$ that is defined by these equations.

In what follows we explore Lagrange duality and the KKT conditions in two special cases, namely in optimizing a linear function over an algebraic variety (Section 5.3) and in semidefinite programming (Section 5.5).

5.2.4 Projective Varieties and Their Duality

In algebraic geometry, it is customary to work over an algebraically closed field, such as the complex numbers $\mathbb{C}$. All our varieties will be defined over a subfield K

of the real numbers $\mathbb{R}$, and their points have coordinates in $\mathbb{C}$. It is also customary to work in projective space $\mathbb{P}^n$ rather than affine space $\mathbb{C}^n$, i.e., we work with equivalence classes $x \sim \alpha x$ for all $\alpha \in \mathbb{C}\backslash\{0\}$, $x \in \mathbb{C}^{n+1}\backslash\{0\}$. Points $(x_0 : x_1 : \cdots : x_n)$ in projective space $\mathbb{P}^n$ are lines through the origin in $\mathbb{C}^{n+1}$, and the usual affine coordinates are obtained by *dehomogenization* with respect to x_0 (i.e., setting $x_0 = 1$). All points with $x_0 = 0$ are then considered as points at infinity. We refer to [8, Chapter 8] for an elementary introduction to projective algebraic geometry.

Let $I = \langle h_1, \ldots, h_p \rangle$ be a homogeneous ideal in the ring $K[x_0, x_1, \ldots, x_n]$ of polynomials in $n+1$ unknowns with coefficients in K. We write $X = V_{\mathbb{C}}(I)$ for its variety in the projective space $\mathbb{P}^n$ over $\mathbb{C}$. The *singular locus* $\mathrm{Sing}(X)$ is a proper subvariety of X. It is defined inside X by the vanishing of the $c \times c$ minors of the $p{\times}(n{+}1)$ Jacobian matrix $\mathrm{Jac}(X) = \big(\partial h_i / \partial x_j\big)$, where $c = \mathrm{codim}(X)$. See [8, Section 9.6] for background on singularities and dimension. While the matrix $\mathrm{Jac}(X)$ depends on our choice of ideal generators h_i, the singular locus of X is independent of that choice. Points in $\mathrm{Sing}(X)$ are called *singular points* of X. We write $X_{\mathrm{reg}} = X\backslash\mathrm{Sing}(X)$ for the set of *regular points* in X. We say that the projective variety X is smooth if $\mathrm{Sing}(X) = \emptyset$ or, equivalently, if $X = X_{\mathrm{reg}}$.

The dual projective space $(\mathbb{P}^n)^*$ parametrizes hyperplanes in $\mathbb{P}^n$. A point $(u_0 : u_1 : \cdots : u_n) \in (\mathbb{P}^n)^*$ represents the hyperplane $\big\{x \in \mathbb{P}^n \mid \sum_{i=0}^n u_i x_i = 0\big\}$. We say that u is *tangent* to X at a regular point $x \in X_{\mathrm{reg}}$ if x lies in that hyperplane and its representing vector $(u_0, u_1, \ldots, u_n)$ lies in the row space of the Jacobian matrix $\mathrm{Jac}(X)$ at the point x.

We define the *conormal variety* $\mathrm{CN}(X)$ of X to be the closure of the set

$$\big\{(x,u) \in \mathbb{P}^n \times (\mathbb{P}^n)^* \mid x \in X_{\mathrm{reg}} \text{ and } u \text{ is tangent to } X \text{ at } x\big\}.$$

The projection of $\mathrm{CN}(X)$ onto the second factor is denoted X^* and is called the dual variety. More precisely, the *dual variety* X^* is the closure of the set

$$\big\{\, u \in (\mathbb{P}^n)^* \mid \text{the hyperplane } u \text{ is tangent to } X \text{ at some regular point}\,\big\}.$$

In our definitions of conormal variety and dual variety, the word "closure" can mean either Zariski closure or the classical strong closure over the complex numbers. Both will lead to the same complex projective variety in the situations considered here.

Proposition 5.10. *The conormal variety* $\mathrm{CN}(X)$ *has dimension* $n-1$.

Proof sketch. We may assume that X is irreducible. Let $c = \mathrm{codim}(X)$. There are $n{-}c$ degrees of freedom in picking a point x in X_{reg}. Once the regular point x is fixed, the possible tangent vectors u to X at x form a linear space of dimension $c{-}1$. Hence the dimension of $\mathrm{CN}(X)$ is $(n{-}c) + (c{-}1) = n{-}1$. □

Since the dual variety X^* is a linear projection of the conormal variety $\mathrm{CN}(X)$, Proposition 5.10 implies that the dimension of X^* is at most $n-1$. We expect X^* to have dimension $n-1$. In other words, regardless of the dimension of X, the dual variety X^* is typically a hypersurface in the dual projective space $(\mathbb{P}^n)^*$. We shall see many examples of such dual hypersurfaces throughout this chapter.

To compute the dual X^* of a given projective variety X, we set up a system of polynomial equations, and we eliminate some of the variables. This can be done using Gröbner bases [8, 13]. We first illustrate this for a familiar example.

Example 5.11 (Example 5.8 continued). Fix coordinates $(x : y : z)$ on $\mathbb{P}^2$ and consider the ideal $I = \langle x^4 + y^4 - z^4 \rangle$. Then $X = V(I)$ is the projective version of the quartic curve in Example 5.8. The dual curve X^* is the projective version of the curve $\partial_a P^\circ$ in (5.16). Hence, X^* is a curve of degree 12 in $(\mathbb{P}^2)^*$.

The equations used to compute X^* algebraically consist of the given quartic $x^4 + y^4 - z^4$ together with the 2×2 minors of the augmented Jacobian matrix

$$\overline{\mathrm{Jac}} \;=\; \begin{pmatrix} a & b & c \\ 4x^3 & 4y^3 & -4z^3 \end{pmatrix}.$$

We write J' for the ideal generated by these four polynomials in $\mathbb{Q}[x, y, z, a, b, c]$. We then replace J' by its saturation

$$J \;=\; J' : \langle x, y, z \rangle^\infty. \tag{5.20}$$

This has the effect of removing an extraneous component of $V_{\mathbb{C}}(J')$ that corresponds to the origin $(0, 0, 0)$ in (x, y, z)-space. We now eliminate the three unknowns x, y, z from J, that is, we compute $J \cap \mathbb{Q}[a, b, c]$. This elimination ideal is the principal ideal generated by the homogenization of the degree 12 polynomial in (5.16). ∎

The steps we described in Example 5.11 to compute the degree 12 curve dual to the given quartic can be extended to arbitrary instances. The role of the ideal $\langle x, y, z \rangle$ in (5.20) is then played by the equations defining the singular locus of X. This results in the following general algorithm for dualizing projective varieties.

Algorithm 5.1. Computing the dual variety X^*.

Require: The input is the homogeneous ideal I of a projective variety $X = V(I)$.
Ensure: The output is the ideal I_{dual} representing the dual variety $X^* = V(I_{\mathrm{dual}})$.

1: Determine the codimension c of the variety X in $\mathbb{P}^n$.
2: Generate the augmented Jacobian matrix

$$\overline{\mathrm{Jac}}(X) \;=\; \begin{pmatrix} u_0 \; u_1 \; \cdots \; u_n \\ \mathrm{Jac(X)} \end{pmatrix}$$

3: Compute $J' = I + \langle\, (c+1) \times (c+1)$ minors of $\overline{\mathrm{Jac}}(X) \,\rangle \subset K[x, u]$.
4: Remove the singular locus by computing the *saturation ideal*

$$J \;:=\; \big(J' \;:\; \langle\, c \times c \text{ minors of } \mathrm{Jac}(X) \,\rangle^\infty\big).$$

5: Compute the desired ideal $I_{\mathrm{dual}} = J \cap K[u]$ by elimination.
6: **return** Dual variety $X^* = V(I_{\mathrm{dual}})$.

The steps in this algorithm can be executed either using exact arithmetic in a computer algebra system, such as `Macaulay2`, or using floating point arithmetic in the framework of numerical algebraic geometry. Such a numerical implementation in the software `Bertini` [3] is currently being developed by Jonathan Hauenstein.

Remark 5.12. *The ideal J in step 3 above is bihomogeneous in x and u, respectively. Its zero set in $\mathbb{P}^n \times (\mathbb{P}^n)^*$ is the conormal variety* $\mathrm{CN}(X)$.

Theorem 5.13 (Biduality, [11, Theorem 1.1]). *Every irreducible projective variety $X \subset \mathbb{P}^n$ satisfies*

$$(X^*)^* \;=\; X.$$

Proof sketch. The main step in proving this important theorem is that the conormal variety is self-dual, in the sense that $\mathrm{CN}(X) = \mathrm{CN}(X^*)$. In this identity, the roles of $x \in \mathbb{P}^n$ and $u \in (\mathbb{P}^n)^*$ are swapped. It implies $(X^*)^* = X$. A proof for the self-duality of the conormal variety is found in [11, Subsection I.1.3]. □

Example 5.14. Suppose that $X \subset \mathbb{P}^n$ is a general smooth hypersurface of degree d. Then X^* is a hypersurface of degree $d(d-1)^{n-1}$ in $(\mathbb{P}^n)^*$. A concrete instance for $d = 4$ and $n = 2$ was seen in Examples 5.8 and 5.11. When X is a hypersurface that is not smooth, then the dual variety X^* is either a hypersurface of degree less than $d(d-1)^{n-1}$, or X^* is a variety of codimension at least 2. ∎

Example 5.15. Let X be the variety of symmetric $m \times m$ matrices of rank at most r. Then X^* is the variety of symmetric $m \times m$ matrices of rank at most $m - r$ [11, Subsection I.1.4]. Here the conormal variety $\mathrm{CN}(X)$ consists of pairs of symmetric matrices A and B such that $A \cdot B = 0$. This conormal variety will be important for our discussion of duality in semidefinite programming in Section 5.5. ∎

An important class of examples, arising from toric geometry, is featured in the book by Gel'fand, Kapranov, and Zelevinsky [11]. A *projective toric variety* X_A in $\mathbb{P}^n$ is specified by an integer matrix A of format $r \times (n+1)$ and rank r with columns $a_0, a_1, \ldots, a_n$ and whose row space contains the vector $(1, 1, \ldots, 1)$. We define X_A as the closure in $\mathbb{P}^n$ of the set $\{(t^{a_0} : t^{a_1} : \cdots : t^{a_n}) \,|\, t \in (\mathbb{C}\backslash\{0\})^r\}$.

The dual variety X_A^* is called the *A-discriminant.* It is usually a hypersurface, in which case we identify the *A-discriminant* with the irreducible polynomial Δ_A that vanishes on X_A^*. The A-discriminant is indeed a discriminant in the sense that its vanishing characterizes Laurent polynomials

$$p(t) \;=\; \sum_{j=0}^{n} c_j \cdot t_1^{a_{1j}} t_2^{a_{2j}} \cdots t_r^{a_{rj}}$$

with the property that the hypersurface $\{p(t) = 0\}$ has a singular point in $(\mathbb{C}\backslash\{0\})^r$. In other words, we can define (and compute) the A-discriminant as the unique irreducible polynomial Δ_A that vanishes on the hypersurface

$$X_A^* = \overline{\left\{c \in (\mathbb{P}^n)^* \mid \exists t \in (\mathbb{C}\backslash\{0\})^r \text{ with } p(t) = \frac{\partial p}{\partial t_1} = \cdots = \frac{\partial p}{\partial t_r} = 0\right\}}.$$

Example 5.16. Let $r = 2$, $n = 4$, and fix the matrix

$$A = \begin{pmatrix} 4 & 3 & 2 & 1 & 0 \\ 0 & 1 & 2 & 3 & 4 \end{pmatrix}.$$

The associated toric variety is the rational normal curve

$$\begin{aligned} X_A &= \{(t_1^4 : t_1^3t_2 : t_1^2t_2^2 : t_1t_2^3 : t_2^4) \in \mathbb{P}^4 \mid (t_1 : t_2) \in \mathbb{P}^1\} \\ &= V(x_0x_2 - x_1^2, x_0x_3 - x_1x_2, x_0x_4 - x_2^2, x_1x_3 - x_2^2, x_1x_4 - x_2x_3, x_2x_4 - x_3^2). \end{aligned}$$

A hyperplane $\{\sum_{j=0}^4 c_jx_j = 0\}$ is tangent to X_A if and only if the binary form

$$p(t_1, t_2) = c_0t_2^4 + c_1t_1t_2^3 + c_2t_1^2t_2^2 + c_3t_1^3t_2 + c_4t_1^4$$

has a linear factor of multiplicity ≥ 2. This is controlled by the A-discriminant

$$\Delta_A = \frac{1}{c_4} \cdot \det \begin{pmatrix} c_0 & c_1 & c_2 & c_3 & c_4 & 0 & 0 \\ 0 & c_0 & c_1 & c_2 & c_3 & c_4 & 0 \\ 0 & 0 & c_0 & c_1 & c_2 & c_3 & c_4 \\ c_1 & 2c_2 & 3c_3 & 4c_4 & 0 & 0 & 0 \\ 0 & c_1 & 2c_2 & 3c_3 & 4c_4 & 0 & 0 \\ 0 & 0 & c_1 & 2c_2 & 3c_3 & 4c_4 & 0 \\ 0 & 0 & 0 & c_1 & 2c_2 & 3c_3 & 4c_4 \end{pmatrix}, \tag{5.21}$$

given here in the form of the determinant of a Sylvester matrix, see [9, Section 3]. The sextic hypersurface $X_A^* = V(\Delta_A)$ is the dual variety of the curve X_A. ∎

Exercises

Exercise 5.17. Let P be a convex body in $\mathbb{R}^3$ obtained by intersecting a ball and a cube, where neither of these bodies contains the other. Describe the dual convex body P°. Can you draw pictures of P and P°?

Exercise 5.18. Determine the irreducible polynomial that vanishes on the $L_{4/3}$-unit sphere in $\mathbb{R}^3$. In other words, extend Example 5.8 from $n = 2$ to $n = 3$.

Exercise 5.19. Let X be the variety of symmetric $m \times m$ matrices of rank at most r. Determine the dimension of X and describe the singular locus $\mathrm{Sing}(X)$.

Exercise 5.20. Find an example of a surface X in $\mathbb{P}^3$ whose dual variety X^* is a curve.

Exercise 5.21. Study the equation $X \cdot Y = 0$ when X and Y are unknown symmetric 4×4 matrices. This constraint translates into 16 bilinear equations in the

20 unknown matrix entries. Decompose the algebraic variety defined by these 16 equations into its irreducible components. What is the dimension of each component? How do you know that it is irreducible?

5.3 The Optimal Value Function

A fundamental question concerning any optimization problem is how the output depends on the input. The optimal solution and the optimal value of the problem are functions of the parameters, and it is important to understand the nature of these functions. For instance, for a linear programming problem,

$$\text{maximize } \langle w, x \rangle \text{ subject to } A \cdot x = b \text{ and } x \geq 0, \tag{5.22}$$

the optimal solution depends in a convex and piecewise linear manner on the cost vector w and the right hand side b, and it is a piecewise rational function of the entries of the matrix A. The area of mathematics which studies these functions is geometric combinatorics, specifically the theory of matroids for the dependence on A, and the theory of regular polyhedral subdivisions for the dependence on w and b. Exercise 5.30 at the end of this section asks for a further exploration. If we replace (5.22) with the corresponding integer programming problem, where the coordinates of x are required to be integers, then the dependence on w and b becomes more subtle and finite Abelian groups enter the picture. The optimal value function of an integer program has a certain arithmetic behavior, in addition to the polyhedral structures which govern the parametric versions of the linear programming problem.

For a second example, consider the following basic question in game theory:

$$\text{Given a game, compute its Nash equilibria.} \tag{5.23}$$

If there are only two players and one is interested in fully mixed Nash equilibria, then this is a linear problem and in fact closely related to linear programming. On the other hand, if the number of players is more than two, then the problem (5.23) is universal in the sense of real algebraic geometry: Datta [10] showed that every real algebraic variety is isomorphic to the set of Nash equilibria of some three-person game. A corollary of her construction is that, if the Nash equilibria are discrete, then their coordinates can be arbitrary algebraic functions of the given input data.

Our third motivating example concerns maximum likelihood estimation in statistical models for discrete data. Here the optimization problem is as follows:

$$\text{maximize } p_1(\theta)^{u_1} p_2(\theta)^{u_2} \cdots p_n(\theta)^{u_n} \text{ subject } \textit{to } \theta \in \Theta, \tag{5.24}$$

where Θ is an open subset of $\mathbb{R}^m$, the $p_i(\theta)$ are polynomial functions that sum to one, and the u_i are positive integers (these are the data). The optimal solution $\hat{\theta}$, which is the maximum likelihood estimator, depends algebraically on the data:

$$(u_1, \ldots, u_n) \mapsto \hat{\theta}(u_1, \ldots, u_n). \tag{5.25}$$

Catanese et al. [7] give a formula for the degree of this algebraic function under certain hypotheses on the polynomials $p_i(\theta)$ which specify the statistical model.

In this section we study this issue for the polynomial optimization problem (5.17). We shall assume throughout that the cost function $f(x)$ is linear and that there are no inequality constraints $g_i(x)$. The purpose of these restrictions is to simplify the presentation and focus on the key ideas. Also, this is compatible with Chapter 7, which offers an algebraic method for the important problem of computing lower bounds on the optimal value function. Our analysis can be extended to the general problem (5.17), and we discuss this briefly at the end of this section.

To be precise, we consider the problem of optimizing a linear cost function over a compact real algebraic variety X in $\mathbb{R}^n$. This is written formally as follows:

$$\begin{aligned} c_0^\star &= \min_x \quad \langle c, x\rangle \\ &\text{subject to} \quad x \in X = \{v \in \mathbb{R}^n \mid h_1(v) = \cdots = h_p(v) = 0\}. \end{aligned} \tag{5.26}$$

Here $h_1, h_2, \ldots, h_p$ are fixed polynomials in n unknowns $x_1, \ldots, x_n$. The expression $\langle c, x\rangle = c_1x_1 + \cdots + c_nx_n$ is a linear form whose coefficients $c_1, \ldots, c_n$ are unspecified parameters. Our aim is to compute the *optimal value function* $c_0^\star$. Thus, we regard the optimal value $c_0^\star$ as a function $\mathbb{R}^n \to \mathbb{R}$ of the parameters $c_1, \ldots, c_n$. We seek to derive an exact symbolic representation of this algebraic function.

The hypothesis that X be compact has been included to ensure that the optimal value function $c_0^\star$ is well-defined on all of $\mathbb{R}^n$. Again, also this hypothesis can be relaxed. We assume compactness here just for convenience.

Our problem is equivalent to that of describing the dual convex body P° of the convex hull $P = \operatorname{conv}(X)$, assuming that the latter contains the origin in its interior. Indeed, P° is precisely the set of points $(c_1, \ldots, c_n)$ at which the value of the function c_0^* is less than or equal to 1. Hence the optimal value function of P computes the *gauge* of the dual body P°. A small instance of this was seen in (5.6).

Since our convex body P is a semialgebraic set, Tarski's theorem on quantifier elimination in real algebraic geometry [2, 4] ensures that the dual body P° is also semialgebraic. This implies that the optimal value function $c_0^\star$ is an algebraic function, i.e., there exists a polynomial $\Phi(c_0, c_1, \ldots, c_n)$ in $n+1$ variables such that

$$\Phi(c_0^\star, c_1, \ldots, c_n) \quad = \quad 0. \tag{5.27}$$

Our aim is to compute such a polynomial Φ of least possible degree. The input consists of the polynomials $h_1, \ldots, h_p$ that cut out the variety X. The degree of Φ in the unknown c_0 is called the *algebraic degree* of the optimization problem (5.17). This number is an intrinsic algebraic complexity measure for the problem of optimizing a linear function over X. For instance, if $c_1, \ldots, c_n$ are rational numbers, then the algebraic degree indicates the degree of the field extension K over $\mathbb{Q}$ that contains the coordinates of the optimal solution.

We illustrate our discussion by computing the optimal value function and its algebraic degree for the trigonometric space curve featured in [31, Section 1].

Example 5.22. Let X be the curve in $\mathbb{R}^3$ with parametric representation

$$(x_1, x_2, x_3) \quad = \quad (\cos(\theta), \sin(2\theta), \cos(3\theta)).$$

In terms of equations, our curve can be written as $X = V(h_1, h_2)$, where

$$h_1 \;=\; x_1^2 - x_2^2 - x_1x_3 \quad \text{and} \quad h_2 \;=\; x_3 - 4x_1^3 + 3x_1.$$

The optimal value function for maximizing $c_1x_1+c_2x_2+c_3x_3$ over X is given by

$$
\begin{aligned}
\Phi \;=\; & (11664c_3^4)\cdot c_0^6 \;+\; (864c_1^3c_3^3 + 1512c_1^2c_2^2c_3^2 - 19440c_1^2c_3^4 \\
& +576c_1c_2^4c_3 - 1296c_1c_2^2c_3^3 + 64c_2^6 - 25272c_2^2c_3^4 - 34992c_3^6)\cdot c_0^4 \\
& + (16c_1^6c_3^2 + 8c_1^5c_2^2c_3 - 1152c_1^5c_3^3 - 1920c_1^4c_2^2c_3^2 + 8208c_1^4c_3^4 - 724c_1^3c_2^4c_3 + 144c_1^3c_2^2c_3^3 \\
& +c_1^4c_2^4 - 17280c_1^3c_3^5 - 80c_1^2c_2^6 - 2802c_1^2c_2^4c_3^2 - 3456c_1^2c_2^2c_3^4 + 3888c_1^2c_3^6 - 1120c_1c_2^6c_3 \\
& +540c_1c_2^4c_3^3 + 55080c_1c_2^2c_3^5 - 128c_2^8 - 208c_2^6c_3^2 + 15417c_2^4c_3^4 + 15552c_2^2c_3^6 + 34992c_3^8)\cdot c_0^2 \\
& + (-16c_1^8c_3^2 - 8c_1^7c_2^2c_3 + 256c_1^7c_3^3 - c_1^6c_2^4 + 328c_1^6c_2^2c_3^2 - 1600c_1^6c_3^4 + 114c_1^5c_2^4c_3 \\
& -2856c_1^5c_2^2c_3^3 + 4608c_1^5c_3^5 + 12c_1^4c_2^6 - 1959c_1^4c_2^4c_3^2 + 9192c_1^4c_2^2c_3^4 - 4320c_1^4c_3^6 \\
& -528c_1^3c_2^6c_3 + 7644c_1^3c_2^4c_3^3 - 7704c_1^3c_2^2c_3^5 - 6912c_1^3c_3^7 - 48c_1^2c_2^8 + 3592c_1^2c_2^6c_3^2 \\
& -4863c_1^2c_2^4c_3^4 - 13608c_1^2c_2^2c_3^6 + 15552c_1^2c_3^8 + 800c_1c_2^8c_3 - 400c_1c_2^6c_3^3 - 10350c_1c_2^4c_3^5 \\
& +16200c_1c_2^2c_3^7 + 64c_2^{10} + 80c_2^8c_3^2 - 1460c_2^6c_3^4 + 135c_2^4c_3^6 + 9720c_2^2c_3^8 - 11664c_3^{10}).
\end{aligned}
$$

The optimal value function $c_0^\star$ is the algebraic function of c_1, c_2, c_3 obtained by solving $\Phi = 0$ for the unknown c_0. Since c_0 has degree 6 in Φ, we see that the algebraic degree of this optimization problem is 6. Note that there are no odd powers of c_0 in Φ. Thus, Φ is a cubic polynomial in c_0^2, and this implies that we can write the optimal value function $c_0^\star$ as an expression in radicals in c_1, c_2, c_3. ∎

We now come to the main result in this section. It will explain what the polynomial Φ means and how it was computed in the previous example. For the sake of simplicity, we shall first assume that the given variety X is smooth, i.e. $X = X_{\text{reg}}$, where the set X_{reg} denotes all regular points on X.

Theorem 5.23. *Let $X^* \subset (\mathbb{P}^n)^*$ be the dual variety to the projective closure of a real affine variety X in $\mathbb{R}^n$. If X is irreducible, smooth, and compact in $\mathbb{R}^n$, then X^* is an irreducible hypersurface, and its defining polynomial equals $\Phi(-c_0, c_1, \dots, c_n)$ where Φ represents the optimal value function as in (5.27) of the optimization problem (5.26). In particular, the algebraic degree of (5.26) is the degree in c_0 of the irreducible polynomial that vanishes on the dual hypersurface X^*.*

Here the change of sign in the coordinate c_0 is needed because the equation $c_0 = c_1x_1 + \cdots + c_nx_n$ for the objective function value in $\mathbb{R}^n$ becomes the homogenized equation $(-c_0)x_0 + c_1x_1 + \cdots + c_nx_n = 0$ when we pass to $\mathbb{P}^n$.

Proof. Since X is compact, for every cost vector c there exists an optimal solution $x^\star$. Our assumption that X is smooth ensures that $x^\star$ is a regular point of X, and c lies in the span of the gradient vectors $\nabla_x h_i\big|_{x^\star}$ for $i = 1, \dots, p$. In other words, the KKT conditions are necessary at the point $x^\star$:

$$
\begin{aligned}
c &= \sum_{i=1}^{p} \mu_i^\star \cdot \nabla_x h_i\big|_{x^\star}, \\
h_i(x^\star) &= 0 \qquad \text{for } i = 1, 2, \dots, p.
\end{aligned}
$$

The scalars $\mu_1^\star, \dots, \mu_p^\star$ express c as a vector in the orthogonal complement of the tangent space of X at $x^\star$. In other words, the hyperplane $\{x \in \mathbb{R}^n : \langle c, x\rangle = c_0^\star\}$ contains the tangent space of X at $x^\star$. This means that the pair

$$\big(x^\star\,,\,(-c_0^\star : c_1 : \cdots : c_n)\big)$$

lies in the conormal variety $\mathrm{CN}(X) \subset \mathbb{P}^n \times (\mathbb{P}^n)^*$ of the projective closure of X. By projection onto the second factor, we see that $(-c_0^\star : c_1 : \cdots : c_n)$ lies in the dual variety X^*.

Our argument shows that the boundary of the dual body P° is a subset of X^*. Since that boundary is a semialgebraic set of dimension $n-1$, we conclude that X^* is a hypersurface. If we write its defining equation as $\Phi(-c_0, c_1, \ldots, c_n) = 0$, then the polynomial Φ satisfies (5.27), and the statement about the algebraic degree follows as well. $\square$

Theorem 5.23 tells us that the minimal polynomial Φ which represents the desired optimal value function $c_0^\star$ can be computed using Algorithm 5.1.

The KKT condition for the optimization problem (5.26) involves three sets of variables, two of which are dual variables, to be carefully distinguished:

1. Primal variables $x_1, \ldots, x_n$ to describe the set X of feasible solutions.
2. (Lagrange) dual variables $\mu_1, \ldots, \mu_p$ to parametrize the linear space of all hyperplanes that are tangent to X at a fixed point $x^\star$.
3. (Projective) dual variables $c_0, c_1, \ldots, c_n$ for the space of all hyperplanes. These are coordinates for the dual variety X^* and the dual body P°.

We can compute the equation Φ that defines the dual hypersurface X^* by eliminating the first two groups of variables $x = (x_1, \ldots, x_n)$ and $\mu = (\mu_1, \ldots, \mu_p)$ from the following system of polynomial equations:

$$c_0 = \langle c, x \rangle \text{ and } h_1(x) = \cdots = h_p(x) = 0 \text{ and } c = \mu_1 \nabla_x h_1 + \cdots + \mu_p \nabla_x h_p.$$

Example 5.24 (Example 5.8 continued). We consider (5.26) with $n = 2, p = 1$, and $h_1 = x_1^4 + x_2^4 - 1$. The KKT equations for maximizing the function

$$c_0 = c_1 x_1 + c_2 x_2 \tag{5.28}$$

over the "TV screen" curve $X = V(h_1)$ are

$$c_1 = \mu_1 \cdot 4x_1^3, \quad c_2 = \mu_1 \cdot 4x_2^3, \quad x_1^4 + x_2^4 = 1. \tag{5.29}$$

We eliminate the three unknowns x_1, x_2, μ_1 from the system of four polynomial equations in (5.28) and (5.29). The result is the polynomial $\Phi(-c_0, c_1, c_2)$ of degree 12 which expresses the optimal value $c_0^\star$ as an algebraic function of c_1 and c_2. We note that $\Phi(1, c_1, c_2)$ is precisely the polynomial in (5.16). ∎

It is natural to ask what happens with Theorem 5.23 when X fails to be smooth or compact or if there are additional inequality constraints. Let us first consider the case when X is no longer smooth, but still compact. Now, X_{reg} is a proper (open, dense) subset of X. The optimal value function $c_0^\star$ for the problem (5.26) is still perfectly well-defined on all of $\mathbb{R}^n$, and it is still an algebraic function of $c_1, \ldots, c_n$. However, the polynomial Φ that represents $c_0^\star$ may now have more factors than just the equation of the dual variety X^*.

Example 5.25. Let $n = 2$ and $p = 1$ as in Example 5.24, but now we consider a singular quartic. The *bicuspid curve*, shown in Figure 5.4, is defined by

$$h_1 = (x_1^2 - 1)(x_1 - 1)^2 + (x_2^2 - 1)^2 = 0.$$

The algebraic degree of optimizing a linear function $c_1x_1 + c_2x_2$ over $X = V(h_1)$ equals 8. The optimal value function $c_0^\star = c_0^\star(c_1, c_2)$ is represented by

$$\begin{aligned}\Phi \;=\; & (c_0 - c_1 + c_2) \cdot (c_0 - c_1 - c_2) \cdot \big(16c_0^6 - 48(c_1^2 + c_2^2)c_0^4 \\ & +(24c_1^2c_2^2 + 21c_2^4 + 64c_1^4)c_0^2 + (54c_1c_2^4 + 32c_1^5)c_0 + 8c_1^4c_2^2 - 3c_1^2c_2^4 + 11c_2^6\big).\end{aligned}$$

The first two linear factors correspond to the singular points of the bicuspid curve X, and the larger factor of degree six represents the dual curve X^*. ∎

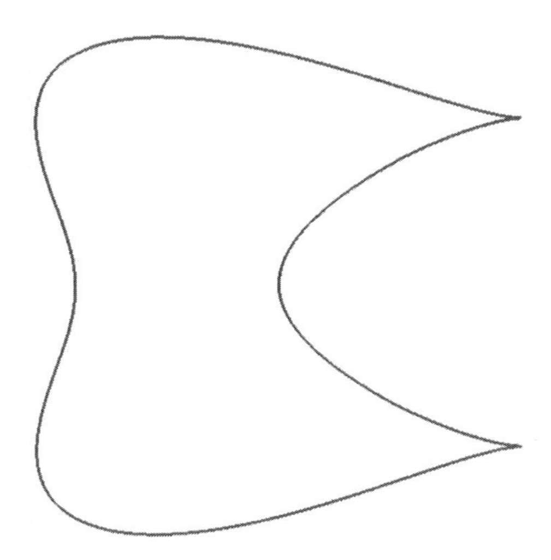

Figure 5.4. *The bicuspid curve in Example* 5.25.

This example shows that, when X has singularities, it does not suffice to just dualize the variety X but we must also dualize the singular locus of X. This process is recursive, and we must also consider the singular locus of the singular locus etc. We believe that, in order to characterize the value function Φ, it always suffices to dualize all irreducible varieties occurring in a *Whitney stratification* of X but this has not been worked out yet. In our view, this topic requires more research, both on the theoretical side and on the computational side.

The following result is valid for any variety X in $\mathbb{R}^n$.

Corollary 5.26. *If the dual variety of X is a hypersurface then its defining polynomial contributes a factor to the value function of the problem* (5.26).

This result can be extended to an arbitrary optimization problem of the form (5.17). We obtain a similar characterization of the optimal value $c_0^\star$ as a semialgebraic function of $c_1, c_2, \ldots, c_n$ by eliminating all primal variables $x_1, \ldots, x_n$ and all dual (optimization) variables $\lambda_1, \ldots, \lambda_m, \mu_1, \ldots, \mu_p$ from the KKT equations.

Again, the optimal value function is represented by a unique square-free polynomial $\Phi(c_0, c_1, \ldots, c_n)$, and each factor of this polynomial is the dual hypersurface Y^* of some variety Y that is obtained from X by setting $g_i(x) = 0$ for some of the inequality constraints, by recursively passing to singular loci. In Section 5.5 we shall explore this for semidefinite programming.

We close this section with a simple example involving A-discriminants.

Example 5.27. Consider the calculus exercise of minimizing a polynomial

$$q(t) \quad = \quad c_1 t + c_2 t^2 + c_3 t^3 + c_4 t^4$$

of degree four over the real line $\mathbb{R}$. Equivalently, we wish to minimize

$$c_0 \;=\; c_1 x_1 + c_2 x_2 + c_3 x_3 + c_4 x_4$$

over the *rational normal curve* $X_A \cap \{x_0 = 1\} = V(x_1^2 - x_2, x_1^3 - x_3, x_1^4 - x_4)$, seen in Example 5.16. The optimal value function $c_0^\star$ is given by the equation $\Delta_A(-c_0, c_1, c_2, c_3, c_4) = 0$, where Δ_A is the discriminant in (5.21). Hence the algebraic degree of this optimization problem is equal to three. ∎

Exercises

Exercise 5.28. Consider the plane curve $Y = \{(\sin(2\theta), \cos(3\theta) : \theta \in \mathbb{R}\}$ obtained from Example 5.22 by projection onto the last two coordinates. Determine the optimal value function for maximizing a linear function over Y.

Exercise 5.29. Maximize $2x + 3y + 7z$ subject to $x^4 + y^4 + z^4 = 1$. Can you express the optimal solution and the optimal value in terms of radicals?

Exercise 5.30. What is the algebraic degree of finding the global minimum of a polynomial function of degree 4 in two variables?

Exercise 5.31. Characterize the optimal value functions arising in linear programming.

Exercise 5.32. Let X denote the *Veronese surface* in five-dimensional projective space $\mathbb{P}^5$ that has the parametric representation $(1 : x : y : x^2 : xy : y^2)$. Compute the conormal variety $\mathrm{CN}(X)$ and the dual variety X^*. Verify the biduality theorem $(X^*)^* = X$ for this example.

5.4 An Algebraic View of Convex Hulls

The problem of optimizing arbitrary linear functions over a given subset of $\mathbb{R}^n$, discussed in the previous section, leads naturally to the geometric question of how to represent the convex hull of that subset. In this section we explore this question from the perspective of algebraic geometry. To be precise, we shall study the algebraic boundary $\partial_a P$ of the convex hull $P = \mathrm{conv}(X)$ of a compact real algebraic variety X in $\mathbb{R}^n$. Biduality of projective varieties (Theorem 5.13) will play an important role in understanding the structure of $\partial_a P$. The results to be presented are drawn from [31, 32]. In Section 5.6 we shall briefly discuss the alternative representation of P as a projected spectrahedron, a topic much further elaborated in Chapter 7.

We begin with the seemingly easy example of a plane quartic curve.

Example 5.33. We consider the following smooth compact plane curve:

$$X = \big\{(x,y) \in \mathbb{R}^2 \mid 144x^4 + 144y^4 - 225(x^2+y^2) + 350x^2y^2 + 81 = 0\big\}. \quad (5.30)$$

This curve is known as the *Trott curve.* It was first constructed by Michael Trott in [36], and is illustrated above in Figure 5.5. A classical result of algebraic geometry states that a general quartic curve in the complex projective plane $\mathbb{P}^2$ has 28 bitangent lines, and the Trott curve X is an instance where all 28 lines are real and have a coordinatization in terms of radicals over $\mathbb{Q}$. Four of the 28 bitangents form edges of $\mathrm{conv}(X)$. These special bitangents are

$$\{(x,y) \in \mathbb{R}^2 \mid \pm\, x \pm\, y = \gamma\}, \quad \text{where } \gamma = \frac{\sqrt{48050 + 434\sqrt{9889}}}{248} = 1.2177\ldots.$$

The boundary of $\mathrm{conv}(X)$ alternates between these four edges and pieces of the curve X. The eight transition points have the floating point coordinates

$$(\pm 0.37655\ldots, \pm 0.84122\ldots)\,, \quad (\pm 0.84122\ldots, \pm 0.37655\ldots).$$

These coordinates lie in the field $\mathbb{Q}(\gamma)$ and we invite the reader to write them in the form $q_1 + q_2\gamma$, where $q_i \in \mathbb{Q}$. The $\mathbb{Q}$-Zariski closure of the 4 edge lines of $\mathrm{conv}(X)$ is a curve Y of degree 8. Its equation has two irreducible factors:

$$(992x^4 - 3968x^3y + 5952x^2y^2 - 3968xy^3 + 992y^4 - 1550x^2 + 3100xy - 1550y^2 + 117),$$
$$(992x^4 + 3968x^3y + 5952x^2y^2 + 3968xy^3 + 992y^4 - 1550x^2 - 3100xy - 1550y^2 + 117).$$

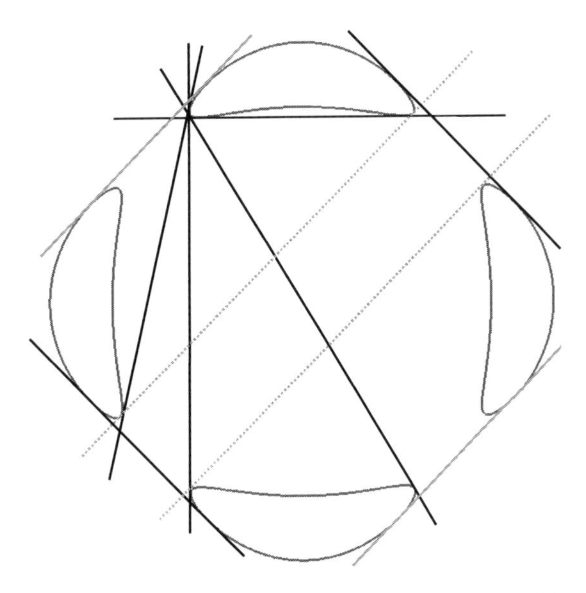

Figure 5.5. *A quartic curve in the plane can have up to 28 real bitangents.*

Each reduces over $\mathbb{R}$ to four parallel lines (cf. Figure 5.5), two of which contribute to the boundary. The point of this example is to stress the role of the (arithmetic of) bitangents in any exact description of the convex hull of a plane curve. ∎

We now present a general formula for the algebraic boundary of the convex hull of a compact variety X in $\mathbb{R}^n$. The key observation is that the algebraic boundary of $P = \mathrm{conv}(X)$ will consist of different types of components, resulting from planes that are simultaneously tangent at k different points of X, for various values of the integer k. For the Trott curve X in Example 5.33, the relevant integers were $k = 1$ and $k = 2$, and we demonstrated that the algebraic boundary of its convex hull P is a reducible curve of degree 12:

$$\partial_a(P) \quad = \quad X \cup Y. \tag{5.31}$$

In the following definitions we regard X as a complex projective variety in $\mathbb{P}^n$. Let $X^{[k]}$ be the variety in the dual projective space $(\mathbb{P}^n)^*$ which is the closure of the set of all hyperplanes that are tangent to X at k regular points which span a $(k-1)$-plane in $\mathbb{P}^n$. This definition makes sense for $k = 1, 2, \ldots, n$. Note that $X^{[1]}$ coincides with the dual variety X^*, and $X^{[2]}$ parametrizes all hyperplanes that are tangent to X at two distinct points. Typically, $X^{[2]}$ is an irreducible component of the singular locus of $X^* = X^{[1]}$. We have the following nested chain of projective varieties in the dual space:

$$X^{[n]} \subseteq X^{[n-1]} \subseteq \cdots \subseteq X^{[2]} \subseteq X^{[1]} \subseteq (\mathbb{P}^n)^*.$$

We now dualize each of the varieties in this chain. The resulting varieties $(X^{[k]})^*$ live in the primal projective space $\mathbb{P}^n$. For $k = 1$ we return to our original variety, i.e., we have $(X^{[1]})^* = X$ by biduality (Theorem 5.13). In the following result we assume that X is smooth as a complex variety in $\mathbb{P}^n$, and we require one technical hypothesis concerning tangency of hyperplanes.

Theorem 5.34 ([32, Theorem 1.1]). *Let X be a smooth and compact real algebraic variety that affinely spans $\mathbb{R}^n$, and such that only finitely many hyperplanes are tangent to X at infinitely many points. The algebraic boundary $\partial_a P$ of its convex hull, P = conv(X), can be computed by biduality as follows:*

$$\partial_a P \subseteq \bigcup_{k=1}^{n} (X^{[k]})^*. \tag{5.32}$$

Since $\partial_a P$ is pure of codimension one, in the union we need only indices k having the property that $(X^{[k]})^*$ is a hypersurface in $\mathbb{P}^n$. As argued in [32], this leads to the following lower bound on the relevant values to be considered:

$$k \geq \left\lceil \frac{n}{\dim(X)+1} \right\rceil. \tag{5.33}$$

The formula (5.32) computes the algebraic boundary $\partial_a P$ in the following sense. For each relevant k we check whether $(X^{[k]})^*$ is a hypersurface, and if so, we determine its irreducible components (over the field K of interest). For each component we then check, usually by means of numerical computations, whether it meets the boundary ∂P in a regular point. The irreducible hypersurfaces which survive this test are precisely the components of $\partial_a X$.

Example 5.35. When X is a plane curve in $\mathbb{R}^2$, (5.32) says that

$$\partial_a P \subseteq X \cup (X^{[2]})^*. \tag{5.34}$$

Here $X^{[2]}$ is the set of points in $(\mathbb{P}^2)^*$ that are dual to the bitangent lines of X, and $(X^{[2]})^*$ is the union of those lines in $\mathbb{P}^2$. If we work over $K = \mathbb{Q}$ and the curve X is general enough then we expect equality to hold in (5.34). For special curves the inclusion can be strict. This happens for the Trott curve (5.30) since Y is a proper subset of $(X^{[2]})^*$. Namely, Y consists of two of the six $\mathbb{Q}$-components of $(X^{[2]})^*$. However, a small perturbation of the coefficients in (5.30) leads to a curve X with equality in (5.34), as the relevant Galois group acts transitively on the 28 points in $X^{[2]}$ for general quartics X. See [28] for more details. We conclude that the algebraic boundary of X over $\mathbb{Q}$ is a reducible curve of degree $32 = 28 + 4$. ∎

If we are given the variety X in terms of equations or in parametric form, then we can compute equations for $X^{[k]}$ by an elimination process similar to the computation of the dual variety X^* in Algorithm 5.1. However, expressing the tangency condition at k different points requires a larger number of additional variables (which need to be eliminated afterwards) and thus the computations are quite involved. The subsequent step of dualizing $X^{[k]}$ to get the right-hand side of (5.32) is even more forbidding. The resulting hypersurfaces $(X^{[k]})^*$ tend to have high degree and their defining polynomials are very large when $n \geq 3$.

The article [31] offers a detailed study of the case when X is a space curve in $\mathbb{R}^3$. Here the lower bound (5.33) tells us that $\partial_a X \subseteq (X^{[2]})^* \cup (X^{[3]})^*$. The surface $(X^{[2]})^*$ is the *edge surface* of the curve X, and $(X^{[3]})^*$ is the union of all *tritangent planes* of X. The following example illustrates these objects.

Example 5.36. We consider the trigonometric curve X in $\mathbb{R}^3$ parametrized by $x = \cos(\theta)$, $y = \cos(2\theta)$, $z = \sin(3\theta)$. This is an algebraic curve of degree six. Its implicit representation equals $X = V(h_1, h_2)$, where

$$h_1 \;=\; 2x^2 - y - 1 \text{ and } h_2 \;=\; 4y^3 + 2z^2 - 3y - 1.$$

The edge surface $(X^{[2]})^*$ has three irreducible components. Two of the components are the quadric $V(h_1)$ and the cubic $V(h_2)$. The third and most interesting component of $(X^{[2]})^*$ is the surface of degree 16 with equation $h_3 =$

$$\begin{aligned}
&-419904x^{14}y^2 + 664848x^{12}y^4 - 419904x^{10}y^6 + 132192x^8y^8 - 20736x^6y^{10} + 1296x^4y^{12}\\
&-46656x^{14}z^2 + 373248x^{12}y^2z^2 - 69984x^{10}y^4z^2 - 22464x^8y^6z^2 + 4320x^6y^8z^2 + 31104x^{12}z^4\\
&+5184x^{10}y^2z^4 + 4752x^8y^4z^4 + 1728x^{10}z^6 + 699840x^{14}y - 46656x^{12}y^3 - 902016x^{10}y^5\\
&+694656x^8y^7 - 209088x^6y^9 - 1150848x^{10}y^3z^2 + 279936x^8y^5z^2 + 17280x^6y^7z^2 - 4032x^4y^9z^2\\
&-98496x^{10}yz^4 + 27072x^4y^{11} - 1152x^2y^{13} - 419904x^{12}yz^2 - 25920x^8y^3z^4 - 4608x^6y^5z^4
\end{aligned}$$

$$- 1728x^8yz^6 - 291600x^{14} - 169128x^{12}y^2 - 256608x^{10}y^4 + 956880x^8y^6 - 618192x^6y^8$$
$$+ 148824x^4y^{10} - 13120x^2y^{12} + 256y^{14} + 392688x^{12}z^2 + 671976x^{10}y^2z^2 + 1454976x^8y^4z^2$$
$$- 292608x^6y^6z^2 - 4272x^4y^8z^2 + 1016x^2y^{10}z^2 - 116208x^{10}z^4 + 135432x^8y^2z^4 + 18144x^6y^4z^4$$
$$+ 1264x^4y^6z^4 - 5616x^8z^6 + 504x^6y^2z^6 - 1108080x^{12}y + 925344x^{10}y^3 + 215136x^8y^5$$
$$- 672192x^6y^7 + 331920x^4y^9 - 54240x^2y^{11} + 2304y^{13} + 273456x^{10}yz^2 + 282528x^8y^3z^2$$
$$- 1185408x^6y^5z^2 + 149376x^4y^7z^2 - 368x^2y^9z^2 - 32y^{11}z^2 + 273456x^8yz^4 - 67104x^6y^3z^4$$
$$- 4704x^4y^5z^4 - 64x^2y^7z^4 + 4752x^6yz^6 - 32x^4y^3z^6 + 747225x^{12} + 636660x^{10}y^2$$
$$- 908010x^8y^4 - 65340x^6y^6 + 291465x^4y^8 - 101712x^2y^{10} + 8256y^{12} - 818100x^{10}z^2$$
$$- 1405836x^8y^2z^2 - 905634x^6y^4z^2 + 583824x^4y^6z^2 - 39318x^2y^8z^2 + 368y^{10}z^2 + 193806x^8z^4$$
$$- 282996x^6y^2z^4 + 15450x^4y^4z^4 + 716x^2y^6z^4 + y^8z^4 + 6876x^6z^6 - 1140x^4y^2z^6 + 2x^2y^4z^6$$
$$+ x^4z^8 + 507384x^{10}y - 809568x^8y^3 + 569592x^6y^5 - 27216x^4y^7 - 71648x^2y^9 + 13952y^{11}$$
$$+ 555768x^8yz^2 + 869040x^6y^3z^2 + 688512x^4y^5z^2 - 154128x^2y^7z^2 + 4416y^9z^2 - 343224x^6yz^4$$
$$+ 127360x^4y^3z^4 - 1656x^2y^5z^4 - 64y^7z^4 - 4536x^4yz^6 + 48x^2y^3z^6 - 775170x^{10} - 191808x^8y^2$$
$$+ 599022x^6y^4 - 245700x^4y^6 + 31608x^2y^8 + 7872y^{10} + 765072x^8z^2 + 589788x^6y^2z^2$$
$$- 66066x^4y^4z^2 - 234252x^2y^6z^2 + 16632y^8z^2 - 173196x^6z^4 + 248928x^4y^2z^4 - 26158x^2y^4z^4$$
$$- 32y^6z^4 - 3904x^4z^6 + 804x^2y^2z^6 + 2y^4z^6 - 2x^2z^8 + 5832x^8y + 98280x^6y^3 - 219456x^4y^5$$
$$+ 72072x^2y^7 - 8064y^9 - 724032x^6yz^2 - 515760x^4y^3z^2 - 99672x^2y^5z^2 + 29976y^7z^2$$
$$+ 225048x^4yz^4 - 76216x^2y^3z^4 + 1912y^5z^4 + 1696x^2yz^6 - 32y^3z^6 + 411345x^8 - 66096x^6y^2$$
$$- 62532x^4y^4 + 29388x^2y^6 - 11856y^8 - 365346x^6z^2 + 19812x^4y^2z^2 + 104922x^2y^4z^2 + 24636y^6z^2$$
$$+ 85090x^4z^4 - 104580x^2y^2z^4 + 8282y^4z^4 + 1014x^2z^6 - 144y^2z^6 + z^8 - 39744x^6y + 61992x^4y^3$$
$$+ 2304x^2y^5 + 576y^7 + 305328x^4yz^2 + 86640x^2y^3z^2 + 960y^5z^2 - 73480x^2yz^4 + 16024y^3z^4$$
$$- 200yz^6 - 114966x^6 + 24120x^4y^2 - 5958x^2y^4 + 6192y^6 + 85494x^4z^2 - 39696x^2y^2z^2$$
$$- 11970y^4z^2 - 21610x^2z^4 + 16780y^2z^4 - 94z^6 - 3672x^4y - 11024x^2y^3 + 272y^5$$
$$- 46904x^2yz^2 - 4632y^3z^2 + 9368yz^4 + 15246x^4 - 84x^2y^2 - 1908y^4 - 6892x^2z^2$$
$$+ 2204y^2z^2 + 2215z^4 + 3216x^2y + 168y^3 + 904yz^2 - 664x^2 + 292y^2 - 282z^2 - 96y + 9.$$

The boundary of $P = \text{conv}(X)$ contains patches from all three surfaces $V(h_1)$, $V(h_2)$, and $V(h_3)$. There are also two triangles, with vertices at $(\sqrt{3}/2, 1/2, \pm 1)$, $(\sqrt{3}/2, 1/2, \pm 1)$, and $(0, -1, \pm 1)$. They span two of the tritangent planes of X, namely, $z = 1$ and $z = -1$. The union of all tritangent planes equals $(X^{[3]})^*$. Only one triangle is visible in Figure 5.6. It is colored yellow. The curved blue patch adjacent to one of the edges of the triangle is given by the cubic h_2, while the other two edges of the triangle lie in the degree 16 surface $V(h_3)$. The curve X has two singular points at $(x, y, z) = (\pm 1/2, -1/2, 0)$. Around these two singular points, the boundary is given by four alternating patches from the quadric $V(h_1)$ highlighted in red and the degree 16 surface $V(h_3)$ in green. We conclude that the edge surface $(X^{[2]})^* = V(h_1 h_2 h_3)$ is reducible of degree $21 = 2 + 3 + 16$, and the algebraic boundary $\partial_a(P)$ is a reducible surface of degree $23 = 2 + 21$. ∎

In our next example we examine the convex hull of space curves of degree four that are obtained as the intersection of two quadratic surfaces in $\mathbb{R}^3$.

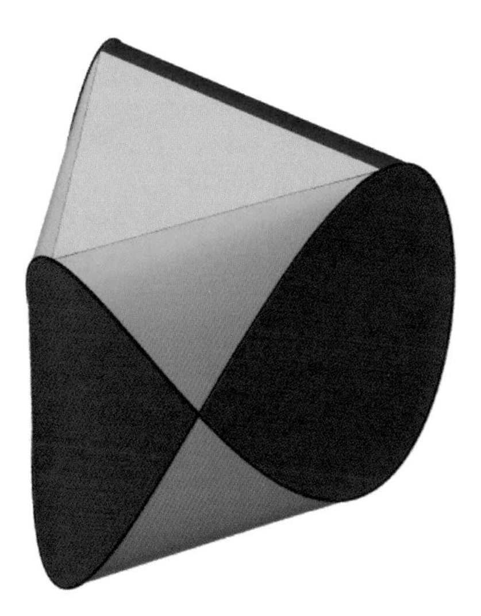

Figure 5.6. *The convex hull of the curve* $(\cos(\theta), \cos(2\theta), \sin(3\theta))$ *in* $\mathbb{R}^3$.

Example 5.37. Let $X = V(h_1, h_2)$ be the intersection of two quadratic surfaces in 3-space. We assume that X has no singularities in $\mathbb{P}^3$. Then X is a curve of genus one. According to recent work of Scheiderer [35], the convex body $P = \mathrm{conv}(X)$ can be represented exactly using Lasserre relaxations, a topic we shall return to when discussing projected spectrahedron in Section 5.6. If we are willing to work over $\mathbb{R}$, then P is in fact a spectrahedron, as shown in [31, Example 2.3]. We here derive that representation for a concrete example.

Lazard et al. [23, Section 8.2] examine the curve X cut out by the two quadrics

$$h_1 = x^2 + y^2 + z^2 - 1 \quad \text{and} \quad h_2 = 19x^2 + 22y^2 + 21z^2 - 20.$$

Figure 5.7 shows the two components of X on the unit sphere $V(h_1)$.

The dual variety X^* is a surface of degree 8 in $(\mathbb{P}^3)^*$. The singular locus of X^* contains the curve $X^{[2]}$ which is the union of four quadratic curves. The duals of these four plane curves are the singular quadratic surfaces defined by

$$h_3 = x^2 - 2y^2 - z^2, \; h_4 = 2x^2 - y^2 - 1, \; h_5 = 3y^2 + 2z^2 - 1, \; h_6 = 3x^2 + z^2 - 2.$$

The edge surface of X is the union of these four quadrics:

$$(X^{[2]})^* = V(h_3) \cup V(h_4) \cup V(h_5) \cup V(h_6).$$

The algebraic boundary of P consists of the last two among these quadrics:

$$\partial_a P = V(h_5) \cup V(h_6).$$

Figure 5.7. *The curve on the unit sphere discussed in Examples* 5.37 *and* 5.61.

These two quadrics are convex. From this we derive a representation of P as a spectrahedron by applying Schur complements to the quadrics h_5 and h_6:

$$P \;=\; \left\{ (x,y,z) \in \mathbb{R}^3 \;\middle|\; \begin{pmatrix} 1+\sqrt{3}y & \sqrt{2}z & 0 & 0 \\ \sqrt{2}z & 1-\sqrt{3}y & 0 & 0 \\ 0 & 0 & \sqrt{2}+z & \sqrt{3}x \\ 0 & 0 & \sqrt{3}x & \sqrt{2}-z \end{pmatrix} \succeq 0 \right\}.$$

An extension of this example is suggested in Exercise 5.42 below. ∎

Exercises

Exercise 5.38. Give an example of a compact algebraic curve of degree six in the plane $\mathbb{R}^2$ whose convex hull has more than 8 straight edges in its boundary. It is an interesting problem to determine the maximal number $\mu(d)$ of edges in the convex hull of any curve of degree d in $\mathbb{R}^2$. For instance, $\mu(6) \geq 9$.

Exercise 5.39. If X is a surface in 3-space, then its algebraic boundary consists of three surfaces $(X^{[1]})^*$, $(X^{[2]})^*$, and $(X^{[3]})^*$. Describe the geometric meaning of these surfaces. Show that all three of them are needed for some X.

Exercise 5.40. Let P be the convex hull of the union of two circles in three-dimensional space, where the first circle is defined by $x^2+y^2 = 5/4$ and $z = 0$, and the second circle is defined by $x^2 + z^2 = 1$ and $y = 0$. Compute the irreducible polynomial in x, y, z that vanishes on the boundary of P.

Exercise 5.41. Describe an algorithm for computing the variety $X^{[k]}$ from the equations of X. Apply your algorithm to the curve $X = V(h_1, h_2)$ in Example 5.22.

Exercise 5.42. Intersect the unit sphere in 3-space with a general quadratic surface. Show that the convex hull of the resulting curve is a spectrahedron.

5.5 Spectrahedra and Semidefinite Programming

Spectrahedra and semidefinite programming (SDP) have already surfaced numerous times throughout this book. In this section we take a systematic look at these topics from the point of view of duality. We write $\mathcal{S}^n$ for the space of real symmetric $n\times n$-matrices and $\mathcal{S}^n_+$ for the cone of positive semidefinite matrices in $\mathcal{S}^n \simeq \mathbb{R}^{\binom{n+1}{2}}$. This cone is self-dual with respect to the inner product $\langle U, V\rangle = \mathrm{Tr}(U \cdot V)$.

A *spectrahedron* is the intersection of the cone $\mathcal{S}^n_+$ with an affine subspace

$$\mathcal{K} \;=\; C + \underbrace{\mathrm{Span}(A_1, A_2, \ldots, A_m)}_{\mathcal{W}}.$$

Here $C, A_1, \ldots, A_m$ are symmetric $n \times n$ matrices, and we assume that $\mathcal{W}$ is a linear subspace of dimension m in $\mathcal{S}^n$. Recall from Chapter 2 that we may also think of a spectrahedron as a set in the Euclidean space as follows:

$$P \;=\; \left\{ x \in \mathbb{R}^m \;\middle|\; C - \sum_{i=1}^{m} x_i A_i \succeq 0 \right\} \;\simeq\; \mathcal{K} \cap \mathcal{S}^n_+. \tag{5.35}$$

We shall assume that C is positive definite or, equivalently, that $0 \in \mathrm{int}(P)$. The dual body to our spectrahedron is written in the coordinates on $\mathbb{R}^m$ as

$$P^\circ \;=\; \{ y \in \mathbb{R}^m \mid \langle y, x\rangle \leq 1 \text{ for all } x \in P \}.$$

We can express P° as a projection of the $\binom{n+1}{2}$-dimensional spectrahedron

$$Q \;=\; \{ U \in \mathcal{S}^n_+ \mid \langle U, C\rangle \leq 1 \}. \tag{5.36}$$

While Q is not literally a spectrahedron when regarded as a set of $n\times n$ matrices, we will identify it with the spectrahedron consisting of all symmetric positive semidefinite $(n+1) \times (n+1)$ matrices $\tilde{U} = \left(\begin{smallmatrix} U & 0 \\ 0 & x \end{smallmatrix}\right)$ that satisfy the equation $\langle U, C\rangle + x = 1$.

To write P° as a projection of the spectrahedron Q, we consider the linear map dual to the inclusion of the linear subspace $\mathcal{W} = \mathrm{Span}(A_1, A_2, \ldots, A_m)$ in the $\binom{n+1}{2}$-dimensional real vector space $\mathcal{S}^n$:

$$\begin{aligned} \pi_{\mathcal{W}} : \mathcal{S}^n \;&\to\; \mathcal{S}^n/\mathcal{W}^\perp \;\simeq\; \mathbb{R}^m \\ U \;&\mapsto\; \big(\langle U, A_1\rangle, \langle U, A_2\rangle, \ldots, \langle U, A_m\rangle\big). \end{aligned}$$

Remark 5.43. *The convex body P° dual to the spectrahedron P is affinely isomorphic to the closure of the image of the spectrahedron Q in* (5.36) *under the linear map $\pi_{\mathcal{W}}$, i.e., $P^\circ \simeq \overline{\pi_{\mathcal{W}}(Q)}$.*

This result in Remark 5.43 is due to Ramana and Goldman [30]. In summary, while the dual to a spectrahedron is generally not a spectrahedron, it is always a *projected spectrahedron.* We shall return to this issue in Theorem 5.57.

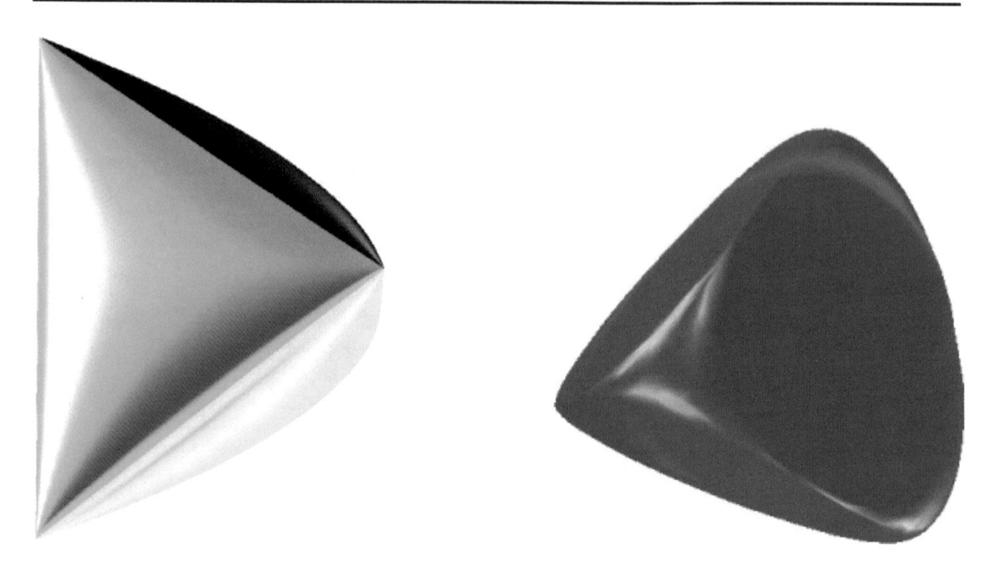

Figure 5.8. *The elliptope $P = \mathcal{E}_3$ and its dual convex body P°.*

Example 5.44. The *elliptope* $\mathcal{E}_n$ is the spectrahedron consisting of all *correlation matrices* of size n; see [20]. These are the positive semidefinite symmetric $n\times n$ matrices whose diagonal entries are 1. We consider the case $n = 3$:

$$\mathcal{E}_3 \;=\; \left\{ (x,y,z) \in \mathbb{R}^3 \;\middle|\; \begin{pmatrix} 1 & x & y \\ x & 1 & z \\ y & z & 1 \end{pmatrix} \succeq 0 \right\}. \tag{5.37}$$

This spectrahedron of dimension $m = 3$ is shown on the left in Figure 5.8. The algebraic boundary of $\mathcal{E}_3$ is the cubic surface X defined by the vanishing of the 3×3 determinant in (5.37). That surface has four isolated singular points

$$X_{\mathrm{sing}} = \{(1,1,1), (1,-1,-1), (-1,1,-1), (-1,-1,1)\}.$$

The six edges of the tetrahedron $\mathrm{conv}(X_{\mathrm{sing}})$ are edges of the elliptope $\mathcal{E}_3$. The dual body, shown on the right of Figure 5.8, is the projected spectrahedron

$$\mathcal{E}_3^\circ \;=\; \left\{ (a,b,c) \in \mathbb{R}^3 \;\middle|\; \exists u,v \in \mathbb{R} : \begin{pmatrix} u & -a & -b \\ -a & v & -c \\ -b & -c & 2{-}u{-}v \end{pmatrix} \succeq 0 \right\}. \tag{5.38}$$

The algebraic boundary of $\mathcal{E}_3^\circ$ can be computed by the following method. We form the ideal generated by the determinant in (5.38) and its derivatives with respect to u and v, and we eliminate u, v. This results in the polynomial

$$(a^2b^2 + b^2c^2 + a^2c^2 + 2abc)(a+b+c-1)(a-b-c-1)(a-b+c+1)(a+b-c+1).$$

The first factor is the equation of *Steiner's quartic surface* X^*, which is dual to *Cayley's cubic surface* $X = \partial_a \mathcal{E}_3$. The four linear factors represent the arrangement $(X_{\mathrm{sing}})^*$ of the four planes dual to the four singular points.

Thus the algebraic boundary of the dual body $\mathcal{E}_3^\circ$ is the reducible surface

$$\partial_a \mathcal{E}_3^\circ \;=\; X^* \cup (X_{\mathrm{sing}})^* \quad \subset \; (\mathbb{P}^3)^*. \tag{5.39}$$

We note that $\mathcal{E}_3^\circ$ is not a spectrahedron as it fails to be a *basic semialgebraic set*; i.e., it cannot be described by a conjunction of polynomial inequalities $g_i \geq 0$. Since the algebraic boundary of $\mathcal{E}_3^\circ$ is uniquely defined by (the irreducible polynomial) equation $\phi(a,b,c) = 0$, such a description would contain the inequality $\phi(a,b,c) \geq 0$. This is not possible since the Steiner surface has a regular point in the interior of the dual convex body $\mathcal{E}_3^\circ$. ■

Semidefinite programming (SDP) is the branch of convex optimization that is concerned with maximizing a linear function b over a spectrahedron:

$$p^\star \;:=\; \max_x \, \langle b, x\rangle \text{ subject to } x \in P. \tag{5.40}$$

Here P is as in (5.35). As the semidefiniteness of a matrix is equivalent to the simultaneous nonnegativity of its principal minors, SDP is an instance of the polynomial optimization problem (5.17). Lagrange duality theory applies here by [5, Section 5]. We shall derive the optimization problem dual to (5.40) from

$$d^\star \;:=\; \underset{\gamma}{\text{minimize}} \; \gamma \text{ subject to } \frac{1}{\gamma} b \in P^\circ. \tag{5.41}$$

Since we assumed $0 \in \mathrm{int}(P)$, strong duality holds and we have $p^\star = d^\star$.

The fact that P° is a projected spectrahedron implies that the dual optimization problem is again a semidefinite optimization problem. In light of Remark 5.43, the condition $\frac{1}{\gamma}b \in P^\circ$ can be expressed as follows:

$$\exists\, U : U \succeq 0,\; \langle C, U\rangle \leq 1 \text{ and } b_i = \gamma\langle A_i, U\rangle \text{ for } i = 1, 2, \ldots, m.$$

Since the optimal value of (5.41) is attained at the boundary of P°, we can here replace the condition $\langle C, U\rangle \leq 1$ with $\langle C, U\rangle = 1$. Indeed, assume that $\langle C, U^\star\rangle = \alpha < 1$ at the optimum, then we could scale $U^\star$ by $\frac{1}{\alpha}$ and the optimal cost $\gamma^\star$ by the factor α and obtain a feasible solution with a smaller cost function value, a contradiction.

This is in fact what was done to obtain (5.38). If we now set $Y = \gamma U$, then (5.41) translates into

$$\begin{aligned} d^\star \;:=\; \underset{Y}{\text{minimize}} \quad & \langle C, Y\rangle \\ \text{subject to} \quad & \langle A_i, Y\rangle \;=\; b_i \text{ for } i = 1, \ldots, m \\ & \text{and } Y \in \mathcal{S}_+^n. \end{aligned} \tag{5.42}$$

We recall that $\mathcal{W} = \mathrm{Span}(A_1, A_2, \ldots, A_m)$ and we fix any matrix $B \in \mathcal{S}^n$ with $\langle A_i, B\rangle = b_i$ for $i = 1, \ldots, m$. Then (5.42) can be written as follows:

$$d^\star \;:=\; \underset{Y}{\text{minimize}}\, \langle C, Y\rangle \text{ subject to } Y \in (B + \mathcal{W}^\perp) \cap \mathcal{S}_+^n. \tag{5.43}$$

The following reformulation of (5.40) highlights the symmetry between the primal and dual formulations of our SDP problem:

$$p^\star := \max_X \langle B, C - X\rangle \text{ subject to } X \in (C + \mathcal{W}) \cap \mathcal{S}^n_+ \tag{5.44}$$

Then the following variant of the KKT conditions holds.

Theorem 5.45 ([5, Section 5.9.2]). *If both the primal problem* (5.44) *and its dual* (5.43) *are strictly feasible, then the KKT conditions take the following form:*

$$\begin{aligned} X &\in (C + \mathcal{W}) \cap \mathcal{S}^n_+, \\ Y &\in (B + \mathcal{W}^\perp) \cap \mathcal{S}^n_+, \\ X \cdot Y &= 0 \quad \textit{(complementary slackness).} \end{aligned}$$

These conditions characterize all the pairs (X, Y) *of optimal solutions.*

This theorem can be related to the general optimality conditions (5.18) by regarding the entries of $Y \in \mathcal{S}^n$ as the (Lagrangian) dual variables to the positive semidefinite constraint $X = C - \sum_{i=1}^m x_i A_i \succeq 0$. The three KKT conditions in Theorem 5.45 are both necessary and sufficient for optimality. This holds because SDP is a convex problem and every local optimum is also a global optimal solution.

In order to study algebraic and geometric properties of SDP, we will relax the conic inequalities $X, Y \in \mathcal{S}^n_+$ and focus only on the *KKT equations*

$$X \in C + \mathcal{W}, \; Y \in B + \mathcal{W}^\perp, \text{ and } X \cdot Y = 0. \tag{5.45}$$

Given the data B, C, and $\mathcal{W}$, our problem is to solve the polynomial equations (5.45). The theorem ensures that, among its solutions (X, Y), there is precisely one pair of positive semidefinite matrices. That pair is the one desired in SDP.

Example 5.46. Consider the problem of minimizing a linear function $Y \mapsto \langle C, Y\rangle$ over the set of all correlation matrices Y, that is, over the elliptope $\mathcal{E}_n$ of Example 5.44. Here $m = n$, B is the identity matrix, C is any symmetric matrix, $\mathcal{W}$ is the space of all diagonal matrices, and $\mathcal{W}^\perp$ consists of matrices with zero diagonal. This problem is dual to maximizing the trace of $C - X$ over all matrices $X \in \mathcal{S}^n_+$ such that $C - X$ is diagonal. Equivalently, we seek to find the minimum trace $t^\star$ of any positive semidefinite matrix that agrees with C in its off-diagonal entries.

For $n = 4$, the KKT equations (5.45) can be written in the form

$$X \cdot Y = \begin{pmatrix} x_1 & c_{12} & c_{13} & c_{14} \\ c_{12} & x_2 & c_{23} & c_{24} \\ c_{13} & c_{23} & x_3 & c_{34} \\ c_{14} & c_{24} & c_{34} & x_4 \end{pmatrix} \cdot \begin{pmatrix} 1 & y_{12} & y_{13} & y_{14} \\ y_{12} & 1 & y_{23} & y_{24} \\ y_{13} & y_{23} & 1 & y_{34} \\ y_{14} & y_{24} & y_{34} & 1 \end{pmatrix} = 0. \tag{5.46}$$

This is a system of 16 quadratic equations in 10 unknowns. For general values of the 6 parameters c_{ij}, these equations have 14 solutions. Eight of these solutions have $\text{rank}(X) = 3$ and $\text{rank}(Y) = 1$ and they are defined over $\mathbb{Q}(c_{ij})$. The other six solutions form an irreducible variety over $\mathbb{Q}(c_{ij})$ and they satisfy $\text{rank}(X) = \text{rank}(Y) = 2$. This case distinction reflects the boundary structure of the dual body to the six-dimensional elliptope $\mathcal{E}_4$:

$$\partial_a \mathcal{E}_4^\circ \quad = \quad \{\text{rank}(Y) \leq 2\}^* \cup \{\text{rank}(Y) = 1\}^*. \tag{5.47}$$

Indeed, the boundary of $\mathcal{E}_4$ is the quartic hypersurface $\{\text{rank}(Y) \leq 3\}$, its singular locus is the degree 10 threefold $\{\text{rank}(Y) \leq 2\}$, and, finally, the singular locus of that threefold consists of eight matrices of rank 1:

$$\{\text{rank}(Y) = 1\} \quad = \quad \big\{ \, (u_1, u_2, u_3, u_4)^T \cdot (u_1, u_2, u_3, u_4) \; : \; u_i \in \{-1, +1\} \, \big\}.$$

The last two strata are dual to the hypersurfaces in (5.47). The second component in (5.47) consists of eight hyperplanes, while the first component is irreducible of degree 18. The corresponding projective hypersurface is defined by an irreducible homogeneous polynomial of degree 18 in seven unknowns $c_{12}, c_{13}, c_{14}, c_{23}, c_{24}, c_{34}, t^\star$. That polynomial has degree 6 in the special unknown $t^\star$. Hence, the algebraic degree of our SDP, i.e., the degree of the optimal value function, is 6 when $\text{rank}(Y) = 2$.

We note that $\{\text{rank}(Y) \leq 3\}^*$ does not appear as a component in the union (5.47) since it is not a hypersurface. Nevertheless, it is still a subset of $\partial_a \mathcal{E}_4^\circ$. ∎

In algebraic geometry, it is natural to regard the matrix pairs (X, Y) as points in the product of projective spaces $\mathbb{P}(\mathcal{S}^n) \times \mathbb{P}(\mathcal{S}^n)^*$. This has the advantage that solutions of (5.45) are invariant under scaling, i.e., whenever (X, Y) is a solution, then so is $(\alpha X, \beta Y)$ for any nonzero $\alpha, \beta \in \mathbb{R}$. In that setting, there are no worries about complications due to solutions at infinity.

For the algebraic formulation we assume that, without loss of generality,

$$b_1 = 1, \quad b_2 = 0, \quad b_3 = 0, \; \ldots \; , b_m = 0.$$

This means that $\langle A_1, X \rangle = 1$ plays the role of the homogenizing variable. Our SDP instance is specified by two linear subspaces of symmetric matrices:

$$\mathcal{L} = \text{Span}(A_2, A_3, \ldots, A_m) \; \subset \; \mathcal{U} = \text{Span}(C, A_1, A_2, \ldots, A_m) \; \subset \; \mathcal{S}^n.$$

Note that we have the following identifications:

$$\mathbb{R}C + \mathcal{W} = \mathcal{U} \quad \text{and} \quad \mathbb{R}B + \mathcal{W}^\perp \; = \; \mathbb{R}B + (\mathcal{L}^\perp \cap A_1^\perp) \; = \; \mathcal{L}^\perp.$$

With the linear spaces $\mathcal{L} \subset \mathcal{U}$, we write the *homogeneous KKT equations* as

$$X \in \mathcal{U}, \;\; Y \in \mathcal{L}^\perp \text{ and } X \cdot Y = 0. \tag{5.48}$$

Here is an *abstract definition of SDP* that might appeal to some of our algebraically inclined readers: Given two nested linear subspaces $\mathcal{L} \subset \mathcal{U} \subset \mathcal{S}^n$ with $\dim(\mathcal{U}/\mathcal{L}) = 2$, locate the unique semidefinite point in the variety (5.48).

For instance, in Example 5.46 the space $\mathcal{L}$ consists of traceless diagonal matrices and $\mathcal{U}/\mathcal{L}$ is spanned by the unit matrix B and one off-diagonal matrix C. We seek to solve the matrix equation $X \cdot Y = 0$ where the diagonal entries of X are constant and the off-diagonal entries of Y are proportional to C.

The formulation (5.48) suggests that we study the variety $\{XY = 0\}$ for pairs of symmetric matrices X and Y. In [27, Equation (3.9)] it was shown that this variety has the following decomposition into irreducible components:

$$\{XY = 0\} = \bigcup_{r=1}^{n-1} \{XY = 0\}^r \subset \mathbb{P}(\mathcal{S}^n) \times \mathbb{P}(\mathcal{S}^n)^*.$$

Here $\{XY = 0\}^r$ denotes the subvariety consisting of pairs (X, Y) where $\operatorname{rank}(X) \leq r$ and $\operatorname{rank}(Y) \leq n-r$. This is irreducible because, by Example 5.15, it is the conormal variety of the variety of symmetric matrices of rank $\leq r$. See also Exercise 5.19 at the end of Section 5.2.

The KKT equations describe sections of these conormal varieties:

$$\{XY = 0\}^r \cap \big(\mathbb{P}(\mathcal{U}) \times \mathbb{P}(\mathcal{L}^\perp)\big). \tag{5.49}$$

All solutions of a semidefinite optimization problem (and thus also the boundary of a spectrahedron and its dual) can be characterized by rank conditions. The main result in [27] describes the case when the section in (5.49) is generic:

Theorem 5.47 ([27, Theorem 7]). *For generic subspaces $\mathcal{L} \subset \mathcal{U} \subset \mathcal{S}^n$ with $\dim(\mathcal{L}) = m - 1$ and $\dim(\mathcal{U}) = m + 1$, the variety (5.49) is empty unless*

$$\binom{n-r+1}{2} \leq m \quad \text{and} \quad \binom{r+1}{2} \leq \binom{n+1}{2} - m. \tag{5.50}$$

In that case, the variety (5.49) is reduced, nonempty, and zero-dimensional and at each point the rank of X and Y is r and $n-r$, respectively (strict complementarity). The cardinality of this variety depends only on m, n, and r.

The generic choice of nested subspaces $\mathcal{L} \subset \mathcal{U}$ corresponds to the assumption that our matrices $A_1, A_2, \ldots, A_m, B, C$ lie in a certain dense open subset in the space of all SDP instances. The inequalities (5.50) are known as *Pataki's inequalities.* If m and n are fixed, then they give a lower bound and an upper bound for the possible ranks r of the optimal matrix of a generic SDP instance. The variety (5.49) represents all complex solutions of the KKT equations for such a generic SDP instance. Its cardinality, denoted $\delta(m, n, r)$, is known as the *algebraic degree of SDP.*

Corollary 5.48. *Consider the variety of symmetric $n \times n$ matrices of rank $\leq r$ that lie in the generic m-dimensional linear subspace $\mathbb{P}(\mathcal{U})$ of $\mathbb{P}(\mathcal{S}^n)$. Its dual variety is a hypersurface if and only if Pataki's inequalities (5.50) hold, and the degree of that hypersurface is $\delta(m, n, r)$, the algebraic degree of SDP.*

Proof. The genericity of $\mathcal{U}$ ensures that $\{XY = 0\}^r \cap (\mathbb{P}(\mathcal{U}) \times \mathbb{P}(\mathcal{U})^*)$ is the conormal variety of the given variety. We obtain its dual by projection onto the second factor $\mathbb{P}(\mathcal{U})^* = \mathbb{P}(\mathcal{S}^n/\mathcal{U}^\perp)$. The degree of the dual hypersurface is found by intersecting with a generic line. The line we take is $\mathbb{P}(\mathcal{L}^\perp/\mathcal{U}^\perp)$. That intersection corresponds to the second factor $\mathbb{P}(\mathcal{L}^\perp)$ in (5.49). □

We note that the symmetry in the equations (5.48) implies the duality

$$\delta\big(m, n, r\big) \quad = \quad \delta\left(\binom{n+1}{2} - m, n, n-r\right),$$

first shown in [27, Proposition 9]. See also [27, Table 2]. Bothmer and Ranestad [12] derived an explicit combinatorial formula for the algebraic degree of SDP. Their result implies that $\delta(m, n, r)$ is a polynomial of degree m in n when $n - r$ is fixed. For example, in addition to [27, Theorem 11], we have

$$\delta(6, n, n-2) \quad = \quad \frac{1}{72}\big(11n^6 - 81n^5 + 185n^4 - 75n^3 - 196n^2 + 156n\big).$$

The algebraic degree of SDP is important because it represents a universal upper bound on the intrinsic algebraic complexity of optimizing a linear function over any m-dimensional spectrahedron of $n\times n$ matrices. The algebraic degree can be much smaller for families of instances involving special matrices A_i, B, or C.

Example 5.49. Fix $n = 4$ and $m = 6 = \dim(\mathcal{E}_4)$. Pataki's inequalities (5.50) state that the rank of the optimal matrix is $r = 1$ or $r = 2$, and this was indeed observed in Example 5.46. For $r = 2$ we had found the algebraic degree six when solving (5.46). However, here B is the identity matrix and A_1, A_2, A_3, A_4 are diagonal. When these are replaced by generic symmetric matrices, then the algebraic degree jumps from six to $\delta(6, 4, 2) = 30$. ■

We now state a result that elucidates the decompositions in (5.39) and (5.47).

Theorem 5.50. *If the matrices $A_1, \ldots, A_m$ and C in the definition* (5.35) *of the spectrahedron P are sufficiently generic, then the algebraic boundary of the dual body P° is the following union of dual hypersurfaces:*

$$\partial_a P^\circ \quad \subseteq \quad \bigcup_{r \text{ as in } (5.50)} \{X \in \mathcal{L} \mid \operatorname{rank}(X) \leq r\}^*. \tag{5.51}$$

Proof. Let $\mathcal{Y}$ be any irreducible component of $\partial_a P^\circ \subset (\mathbb{P}^m)^*$. Then $\mathcal{Y} \cap \partial P^\circ$ is a semialgebraic subset of codimension 1 in P°. We consider a general point in that set. The corresponding hyperplane H in the primal $\mathbb{R}^m$ supports the spectrahedron P at a unique point Z. Then $r = \operatorname{rank}(Z)$ satisfies Pataki's inequalities, by Theorem 5.47. Moreover, the genericity in our choices of $A_1, \ldots, A_m, C, H$ ensure that Z is a regular point in $\{X \in \mathcal{L} \,|\, \operatorname{rank}(X) \leq r\}$. Bertini's theorem ensures that this determinantal variety is irreducible and that its singular locus consists only of matrices of rank $< r$. This implies that $\{X \in \mathcal{L} \,|\, \operatorname{rank}(X) \leq r\}$ is the Zariski closure of $\{X \in P \mid \mathrm{rank}(X) = r\}$ and hence also of a neighborhood of Z in that rank stratum. Likewise, $\mathcal{Y}$ is the Zariski closure in $(\mathbb{P}^m)^*$ of $\mathcal{Y} \cap \partial P^\circ$. An open dense subset of points in $\mathcal{Y} \cap \partial P^\circ$ corresponds to hyperplanes that support P at a rank r matrix. We conclude $\mathcal{Y}^* = \{X \in \mathcal{L} \mid \mathrm{rank}(X) \le r\}$. Biduality completes the proof. $\square$

Theorem 5.50 is similar to Theorem 5.34 in that it characterizes the algebraic boundary in terms of dual hypersurfaces. Just as in Section 5.4, we can apply this result to compute $\partial_a P^\circ$. For each rank r in the Pataki range (5.50), we need to check whether the corresponding dual hypersurface meets the boundary of P°. The indices r which survive this test determine $\partial_a P^\circ$.

When the data that specify the spectrahedron P are not generic but special then the computation of $\partial_a P^\circ$ is more subtle and we know of no formula as simple as (5.51). This issue certainly deserves further research.

We close this section with an interesting three-dimensional example.

Example 5.51. The *cyclohexatope* is a spectrahedron with $m = 3$ and $n = 5$ that arises in the study of chemical conformations [14]. Consider the following *Schönberg matrix* for the pairwise distances $\sqrt{D_{ij}}$ among six carbon atoms:

$$\begin{pmatrix} 2D_{12} & D_{12}+D_{13}-D_{23} & D_{12}+D_{14}-D_{24} & D_{12}+D_{15}-D_{25} & D_{12}+D_{16}-D_{26} \\ D_{12}+D_{13}-D_{23} & 2D_{13} & D_{13}+D_{14}-D_{34} & D_{13}+D_{15}-D_{35} & D_{13}+D_{16}-D_{36} \\ D_{12}+D_{14}-D_{24} & D_{13}+D_{14}-D_{34} & 2D_{14} & D_{14}+D_{15}-D_{45} & D_{14}+D_{16}-D_{46} \\ D_{12}+D_{15}-D_{25} & D_{13}+D_{15}-D_{35} & D_{14}+D_{15}-D_{45} & 2D_{15} & D_{15}+D_{56}-D_{56} \\ D_{12}+D_{16}-D_{26} & D_{13}+D_{16}-D_{36} & D_{14}+D_{16}-D_{46} & D_{15}+D_{56}-D_{56} & 2D_{16} \end{pmatrix}.$$

The D_{ij} are the squared distances among six points in $\mathbb{R}^3$ if and only if this matrix is positive semidefinite of rank ≤ 3. The points represent the carbon atoms in *cyclohexane* C_6H_{12} if and only if $D_{i,i+1} = 1$ and $D_{i,i+2} = 8/3$ for all indices i, understood cyclically. The three "diagonal" distances $x = D_{14}, y = D_{25}$, and $z = D_{36}$ are unknowns, so, for cyclohexane conformations, the above Schönberg matrix equals

$$\mathbf{C}_6(x,y,z) \quad = \quad \begin{pmatrix} 2 & 8/3 & x-5/3 & 11/3-y & -2/3 \\ 8/3 & 2 & 5/3+x & 8/3 & 11/3-z \\ x-5/3 & 5/3+x & 16/3 & x+5/3 & x-5/3 \\ 11/3-y & 8/3 & x+5/3 & 2y & 8/3 \\ -2/3 & 11/3-z & x-5/3 & 8/3 & 16/3 \end{pmatrix}.$$

The *cyclohexatope* Cyc_6 is the spectrahedron in $\mathbb{R}^3$ defined by $\mathbf{C}_6(x,y,z) \succeq 0$. Its algebraic boundary decomposes as $\partial_a \mathrm{Cyc}_6 = V(f) \cup V(g)$, where

$$\begin{aligned} f &= 27xyz - 75x - 75y - 75z - 250 \qquad \text{and} \\ g &= 3xy + 3xz + 3yz - 22x - 22y - 22z + 121. \end{aligned}$$

The conformation space of cyclohexane is the real algebraic variety

$$\big\{(x,y,z) \in \mathrm{Cyc}_6 \mid \mathrm{rank}(\mathbf{C}_6(x,y,z)) \le 3\big\} \quad = \quad V(f,g) \ \cup \ V(g)_{\mathrm{sing}}.$$

The first component is the closed curve of all *chair conformations*. The second component is the *boat conformation* point $(x, y, z) = \left(\frac{11}{3}, \frac{11}{3}, \frac{11}{3}\right)$. These are well-known to chemists [14]. Remarkably, the cyclohexatope coincides with the convex hull of these two components. This spectrahedron is another example of a convex hull of a space curve, now with an isolated point. SDP over the cyclohexatope means computing the conformation which minimizes a linear function in the squared distances D_{ij}. ∎

Exercises

Exercise 5.52. Maximize the sum of the off-diagonal entries over all positive semidefinite 4×4 matrices with trace 1. Formulate this as a pair of primal and dual problems and solve the KKT equations. For both primal and dual, determine the set of all optimal solutions, and verify Theorem 5.45.

Exercise 5.53. A result in classical algebraic geometry states that every smooth cubic surface contains 27 lines and is obtained by blowing up $\mathbb{P}^2$ at six points. Are these statements still true for Cayley's cubic surface $X = \partial_a \mathcal{E}_3$ as in Example 5.44?

Exercise 5.54. Determine the positive integer $\delta(5, 7, 5)$. Explain in your own words what this number means for SDP on 7×7 matrices.

Exercise 5.55. Compute the right-hand side of (5.51) for the spectrahedron P in (5.2).

Exercise 5.56. The *analytic center* of a spectrahedron is the symmetric matrix in its interior that maximizes the determinant function. Compute the analytic center of the three-dimensional spectrahedron

$$\begin{pmatrix} x & z+1 & x+y+z \\ z+1 & y & x-y \\ x+y+z & x-y & 1-x-y \end{pmatrix} \succeq 0.$$

Determine the values $x^\star, y^\star$, and $z^\star$ for the optimal matrix as floating point numbers. Make sure that you have at least twenty accurate digits. If this is possible, write $x^\star, y^\star$, and $z^\star$ in terms of radicals over $\mathbb{Q}$.

5.6 Projected Spectrahedra

A *projected spectrahedron* is the image of a spectrahedron under a linear map. The class of projected spectrahedra is much larger than the class of spectrahedra. In fact, it has even been conjectured that every convex basic semialgebraic set in $\mathbb{R}^n$ is a projected spectrahedron [17]. See Chapter 6 for a detailed discussion.

Our point of departure is the result that the convex body dual to a projected spectrahedron is again a projected spectrahedron [15, Proposition 3.3].

Theorem 5.57. *The class of projected spectrahedra is closed under duality.*

Proof (Construction). A projected spectrahedron can be written in the form

$$P = \left\{ x \in \mathbb{R}^m \,\middle|\, \exists y \in \mathbb{R}^p \text{ with } C + \sum_{i=1}^{m} x_i A_i + \sum_{j=1}^{p} y_j B_j \succeq 0 \right\}.$$

An expression for the dual body P° is obtained by the following variant of the construction in Remark 5.43. We consider the same linear map as before:

$$\pi : \mathcal{S}_+^n \to \mathbb{R}^m, \ U \mapsto (\langle A_1, U\rangle, \ldots, \langle A_m, U\rangle).$$

We apply this linear map π to the spectrahedron

$$Q = \{ U \in \mathcal{S}_+^n \mid \langle C, U\rangle \le 1 \text{ and } \langle B_1, U\rangle = \cdots = \langle B_p, U\rangle = 0 \}.$$

The closure of the projected spectrahedron $\pi(Q)$ equals the dual convex body P°. This closure is itself a projected spectrahedron, e.g., by using the "extended Lagrange–Slater dual" formulation proposed by Ramana [29]. □

We now consider the following problem: Given a real variety $X \subset \mathbb{R}^n$, find a representation of its convex hull $\mathrm{conv}(X)$ as a projected spectrahedron. A systematic approach to computing such representations was introduced by Lasserre [21] and further developed by Gouveia et al. [16]. It is based on the relaxation of nonnegative polynomial functions on X as sums of squares in the coordinate ring $\mathbb{R}[X]$. This approach is known as *moment relaxation* (also Lasserre relaxation; see Chapter 7) in light of the duality between positive polynomials and moments of measures.

We shall begin by exploring these ideas for homogeneous polynomials of even degree $2d$ that are nonnegative on $\mathbb{R}^n$. These form a cone in a real vector space of dimension $\binom{2d+n-1}{2d}$. Inside that cone lies the smaller *sos cone* of polynomials p that are sums of squares of polynomials of degree d:

$$p = q_1^2 + q_2^2 + \cdots + q_r^2. \tag{5.52}$$

By Hilbert's theorem [25, Theorem 1.2.6], this inclusion of convex cones is strict unless $(n, 2d)$ equals $(1, 2d)$ or $(n, 2)$ or $(2, 4)$. The sos cone is easily seen to be a projected spectrahedron. Indeed, consider an unknown symmetric matrix $Q \in \mathcal{S}^N$ and write $p = v^T Q v$, where v is the vector of all N monomials of degree d. The matrix Q is positive semidefinite if it has a Cholesky factorization $Q = C^T C$. The resulting identity $p = (Cv)^T(Cv)$ can be rewritten as (5.52). Hence the sos cone is the image of $\mathcal{S}_+^N$ under the linear map $Q \mapsto v^T Q v$.

The boundaries of our two cones and their duals have been described in detail already in Chapter 4, and here we want only to briefly make some connections to our previous discussion about dualities. In the work of Nie [26] the structure of these boundaries was approached by computations with discriminants, encountered at the end of Section 5.2.4.

Proposition 5.58 (Theorem 4.1 in [26]). *The algebraic boundary of the cone of homogeneous polynomials p of degree $2d$ that are nonnegative on $\mathbb{R}^n$ is given* *by the discriminant of a polynomial p whose coefficients are indeterminates. This discriminant is the irreducible hypersurface dual to the Veronese embedding*

$$\mathbb{P}^{n-1} \hookrightarrow \mathbb{P}^{N-1},\ (x_1 : \cdots : x_n) \mapsto (x_1^{2d} : x_1^{2d-1}x_2 : \cdots : x_n^{2d}).$$

The degree of this discriminant is $n(2d-1)^{n-1}$.

Proof. The discriminant of p vanishes if and only if there exists $x \in \mathbb{P}^{n-1}$ with $p(x) = 0$ and $\nabla p\big|_x = 0$. If p is in the boundary of the cone of positive polynomials then such a real point x exists. For the degree formula, see [11]. □

Results similar to Proposition 5.58 hold when we restrict ourselves to polynomials p that lie in linear subspaces. This is why the *A-discriminants* Δ_A from Section 5.2.4 are relevant. We show this for a two-dimensional family of polynomials.

Example 5.59. Consider the two-dimensional family of ternary quartics

$$f_{a,b}(x,y,z) \quad = \quad x^4 + y^4 + ax^3z + ay^2z^2 + by^3z + bx^2z^2 + (a+b)z^4.$$

Here a and b are parameters. Such a polynomial is nonnegative on $\mathbb{R}^3$ if and only if it is a sum of squares, by Hilbert's theorem. This condition defines a closed convex region $\mathcal{C}$ in the (a,b)-plane $\mathbb{R}^2$. It is nonempty because $(0,0) \in \mathcal{C}$. Its boundary $\partial_a \mathcal{C}$ is derived from the *A-discriminant* Δ_A, where

$$A \quad = \quad \begin{pmatrix} 4 & 0 & 3 & 0 & 0 & 2 & 0 \\ 0 & 4 & 0 & 2 & 3 & 0 & 0 \\ 0 & 0 & 1 & 2 & 1 & 2 & 4 \end{pmatrix}. \tag{5.53}$$

This A-discriminant is an irreducible homogeneous polynomial of degree 24 in the seven coefficients. What we are interested in here is the *specialized discriminant* which is obtained from Δ_A by substituting the vector of coefficients $(1,1,a,a,b,b,a+b)$ corresponding to our polynomial $f_{a,b}$. The specialized discriminant is an inhomogeneous polynomial of degree 24 in the two unknowns a and b, and it is no longer irreducible. A computation reveals that it is the product of four irreducible factors whose degrees are 1, 5, 5, and 13.

The linear factor equals $a+b$. The two factors of degree 5 are

$$256a^2-27a^5+512ab+144a^3b-27a^4b+256b^2-128ab^2+144a^2b^2-128b^3-4a^2b^3+16b^4,$$
$$256a^2-128a^3+16a^4+512ab-128a^2b+256b^2+144a^2b^2-4a^3b^2+144ab^3-27ab^4-27b^5.$$

Finally, the factor of degree 13 in the specialized discriminant equals

$$\begin{gathered} 2916a^{11}b^2 + 19683a^9b^4 + 19683a^8b^5 + 2916a^7b^6 + 2916a^6b^7 + 19683a^5b^8 \\ +19683a^4b^9 + 2916a^2b^{11} - 11664a^{12} - 104976a^{10}b^2 - 136080a^9b^3 - 27216a^8b^4 \\ -225504a^7b^5 - 419904a^6b^6 - 225504a^5b^7 - 27216a^4b^8 - 136080a^3b^9 \\ -104976a^2b^{10} - 11664b^{12} + 93312a^{11} + 217728a^{10}b + 76032a^9b^2 \\ +1133568a^8b^3 + 1976832a^7b^4 + 891648a^6b^5 + 891648a^5b^6 + 1976832a^4b^7 \\ +1133568a^3b^8 + 76032a^2b^9 + 217728ab^{10} + 93312b^{11} - 241920a^{10} \\ -1368576a^9b - 2674944a^8b^2 - 1511424a^7b^3 - 4729600a^6b^4 - 9369088a^5b^5 \\ -4729600a^4b^6 - 1511424a^3b^7 - 2674944a^2b^8 - 1368576ab^9 - 241920b^{10} \end{gathered}$$

$$+663552a^9 + 2949120a^8b + 10539008a^7b^2 + 17727488a^6b^3 + 9981952a^5b^4$$
$$+9981952a^4b^5 + 17727488a^3b^6 + 10539008a^2b^7 + 2949120ab^8 + 663552b^9$$
$$-2719744a^8 - 8847360a^7b - 14974976a^6b^2 - 36503552a^5b^3 - 56360960a^4b^4$$
$$-36503552a^3b^5 - 14974976a^2b^6 - 8847360ab^7 - 2719744b^8 + 4587520a^7$$
$$+25821184a^6b + 52035584a^5b^2 + 50724864a^4b^3 + 50724864a^3b^4 + 52035584a^2b^5$$
$$+25821184ab^6 + 4587520b^7 - 6291456a^6 - 31457280a^5b - 94371840a^4b^2$$
$$-138412032a^3b^3 - 94371840a^2b^4 - 31457280ab^5 - 6291456b^6 + 16777216a^5$$
$$+50331648a^4b + 67108864a^3b^2 + 67108864a^2b^3 + 50331648ab^4 + 16777216b^5$$
$$-16777216a^4 - 67108864a^3b - 100663296a^2b^2 - 67108864ab^3 - 16777216b^4.$$

The relevant pieces of these four curves in the (a, b)-plane are depicted in Figure 5.9. The line $a + b = 0$ is seen in the lower left, the degree 13 curve is the swallowtail in the upper right, and the two quintic curves form the upper-left and lower-right boundaries of the enclosed convex region $\mathcal{C}$.

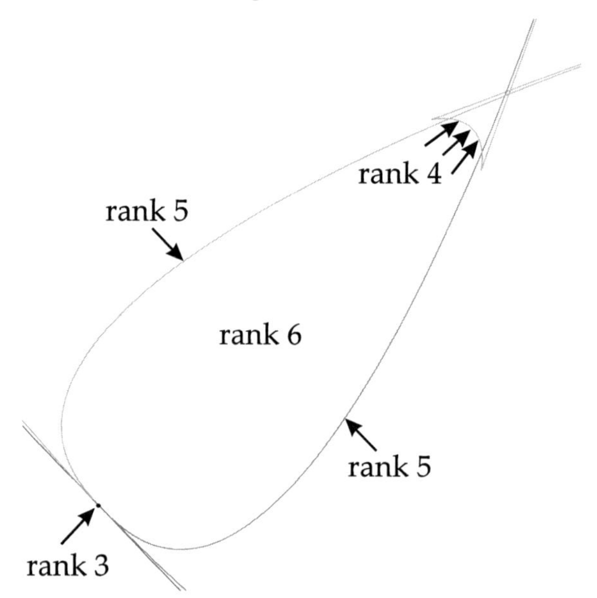

Figure 5.9. *The discriminant in Example* 5.59 *defines a curve in the* (a, b)*-plane. The* projected spectrahedron $\mathcal{C}$ *is the set of points where the ternary quartic* $f_{a,b}$ *is sos. The ranks of the corresponding sos matrices* Q *are indicated.*

For each $(a, b) \in \mathcal{C}$, the ternary quartic $f_{a,b}$ has an sos representation

$$f_{a,b}(x, y, z) \;=\; (x^2, xy, y^2, xz, yz, z^2) \cdot Q \cdot (x^2, xy, y^2, xz, yz, z^2)^T, \qquad (5.54)$$

where Q is a positive semidefinite $6{\times}6$ matrix. This identity gives 15 independent linear constraints which, together with $Q \succeq 0$, define an eight-dimensional spectrahedron in the $(21 + 2)$-dimensional space of parameters (Q, a, b). The projection of this spectrahedron onto the (a, b)-plane is our convex region $\mathcal{C}$. This proves that $\mathcal{C}$ is a projected spectrahedron. If (a, b) lies in the interior of $\mathcal{C}$, then the fiber of the projection is a six-dimensional spectrahedron. If (a, b) lies in the boundary $\partial\mathcal{C}$, then the fiber consists of a single point. The ranks of these unique matrices are indicated in Figure 5.9. Notice that $\partial\mathcal{C}$ has three singular points, at which the rank drops from 5 to 4 and 3, respectively. ∎

We now turn our attention to the question of approximating the convex hull of a variety by a nested family of projected spectrahedra. Let I be an ideal in $\mathbb{R}[x_1,\ldots,x_n]$ and $V_{\mathbb{R}}(I)$ the variety it defines in $\mathbb{R}^n$. Consider the set of affine-linear polynomials that are nonnegative on $V_{\mathbb{R}}(I)$:

$$P_1(I) \quad = \quad \{\, f \in \mathbb{R}[x_1,\ldots,x_n]_1 \mid f(x) \geq 0 \text{ for all } x \in V_{\mathbb{R}}(I)\}.$$

In light of the biduality theorem for convex sets (cf. Section 5.2.2), we can characterize the (closure of) the convex hull of our variety as follows:

$$\overline{\operatorname{conv}(V_{\mathbb{R}}(I))} \quad = \quad \{x \in \mathbb{R}^n \mid f(x) \geq 0 \text{ for all } f \in P_1(I)\}.$$

The geometry behind this formula is shown in Figure 5.10.

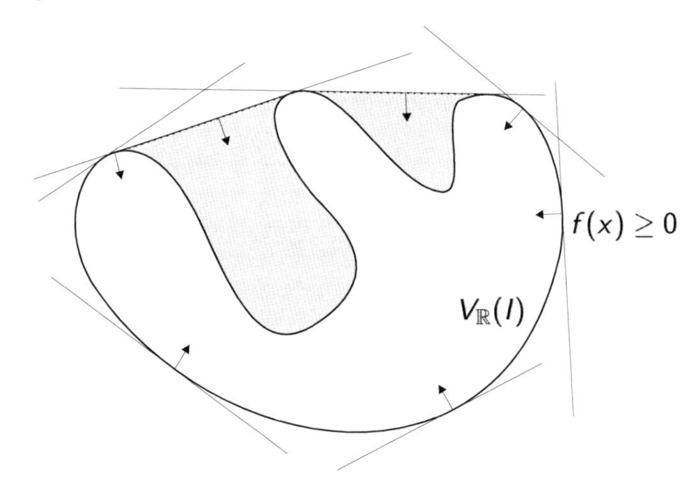

Figure 5.10. *Convex hull as intersection of half spaces.*

The hard constraint that $f(x)$ be nonnegative on $V_{\mathbb{R}}(I)$ can now be relaxed to the (hopefully easier) constraint that $f(x)$ be a sum of squares in the coordinate ring $\mathbb{R}[x_1,\ldots,x_n]/I$; see [16]. Introducing a parameter d that indicates the degree of the polynomials allowed in that sos representation, we consider the following set of affine linear polynomials:

$$\Sigma_1^d(I) \;=\; \bigl\{\, f \mid f - q_1^2 - \cdots - q_r^2 \,\in\, I \;\text{ for some } q_i \in \mathbb{R}[x_1,\ldots,x_n]_d \,\bigr\}.$$

The following chain of inclusions holds:

$$\Sigma_1^1(I) \subseteq \Sigma_1^2(I) \subseteq \Sigma_1^3(I) \subseteq \;\cdots\; \subseteq P_1(I). \tag{5.55}$$

We now dualize the situation by considering the subsets of $\mathbb{R}^n$ where the various f are nonnegative. The *dth theta body* of the ideal I is the set

$$\mathrm{TH}_d(I) \;=\; \bigl\{\, x \in \mathbb{R}^n \mid f(x) \geq 0 \text{ for all } f \in \Sigma_1^d(I)\bigr\}.$$

The following reverse chain of inclusions holds among subsets in $\mathbb{R}^n$:

$$\mathrm{TH}_1(I) \supseteq \mathrm{TH}_2(I) \supseteq \mathrm{TH}_3(I) \supseteq \cdots \supseteq \overline{\mathrm{conv}(V_{\mathbb{R}}(I))}. \qquad (5.56)$$

This chain of outer approximations can fail to converge in general, but there are various convergence results when the geometry is nice. For instance, if the real variety $V_{\mathbb{R}}(I)$ is compact then *Schmüdgen's Positivstellensatz* [35, Section 3] ensures asymptotic convergence. When $V_{\mathbb{R}}(I)$ is a finite set, so that $\mathrm{conv}(V_{\mathbb{R}}(I))$ is a polytope, then finite convergence follows from [19], that is, $\exists d : \mathrm{TH}_d(I) = \mathrm{conv}(V_{\mathbb{R}}(I))$. More information on theta bodies and related constructions is given in Chapter 7. The main point we wish to record here is the following:

Theorem 5.60 ([16, 22]). *Each theta body* $\mathrm{TH}_d(I)$ *is a projected spectrahedron.*

Proof. We may assume, without loss of generality, that the origin 0 lies in the interior of $\mathrm{conv}(V_{\mathbb{R}}(I))$. Then $\Sigma_1^d(I)$ is the cone over the convex set dual to $\mathrm{TH}_d(I)$. Since the class of projected spectrahedra is closed under duality, and under intersection with affine hyperplanes, it suffices to show that $\Sigma_1^d(I)$ is a projected spectrahedron. But this follows from the formula $f - q_1^2 - \cdots - q_r^2 \in I$ by an argument similar to that given after (5.52). □

In this chapter we have seen two rather different representations of the convex hull of a real variety, namely, the characterization of the algebraic boundary in Section 5.4, and the representation as a theta body suggested above. The relationship between these two is not yet well understood. A specific question is how to efficiently compute the algebraic boundary of a projected spectrahedron. This leads to problems in elimination theory that seem to be particularly challenging for current computer algebra systems.

We conclude by revisiting one of the examples we had seen in Section 5.4.

Example 5.61 (Example 5.37 continued). We revisit the curve $X = V(h_1, h_2)$ with

$$h_1 = x^2 + y^2 + z^2 - 1,$$
$$h_2 = 19x^2 + 21y^2 + 22z^2 - 20.$$

Scheiderer [35] proved that finite convergence holds in (5.56) whenever I defines a curve of genus 1, such as X. We will show that $d = 1$ suffices in our example; i.e., we will show that $\mathrm{TH}_1(I) = \mathrm{conv}(X)$ for the ideal $I = \langle h_1, h_2 \rangle$.

We are interested in affine linear forms f that admit a representation

$$f \;=\; 1 + ux + vy + wz \;=\; \mu_1 h_1 + \mu_2 h_2 + \sum_i q_i^2. \qquad (5.57)$$

Here μ_1 and μ_2 are real parameters. Moreover, we want f to lie in $\Sigma_1^1(I)$, so we require $\deg q_i = 1$ for all i. The sum of squares can be written as

$$\sum_i q_i^2 \;=\; (1, x, y, z) \cdot Q \cdot (1, x, y, z)^T, \qquad \text{where } Q \in \mathcal{S}_+^4.$$

After matching coefficients in (5.57), we obtain the projected spectrahedron

$$\Sigma_1^1(I) = \left\{ (u,v,w) \in \mathbb{R}^3 \mid \exists \mu_1, \mu_2 : \right.$$

$$\left. \begin{pmatrix} 1+\mu_1+20\mu_2 & u & v & w \\ u & -\mu_1 - 19\mu_2 & 0 & 0 \\ v & 0 & -\mu_1 - 21\mu_2 & 0 \\ w & 0 & 0 & -\mu_1 - 22\mu_2 \end{pmatrix} \succeq 0 \right\}.$$

Dual to this is the theta body $\mathrm{TH}_1(I) = \Sigma_1^1(I)^\circ$. It has the representation

$$\mathrm{TH}_1(I) = \left\{ (x,y,z) \in \mathbb{R}^3 \mid \exists u_1,u_2,u_3,u_4 : \begin{pmatrix} 1 & x & y & z \\ x & \frac{2}{3} - \frac{1}{3}u_4 & u_1 & u_2 \\ y & u_1 & \frac{1}{3} - \frac{2}{3}u_4 & u_3 \\ z & u_2 & u_3 & u_4 \end{pmatrix} \succeq 0 \right\}.$$

To show that $\mathrm{TH}_1(I) = \mathrm{conv}(X)$, we use the general approach outlined in Remark 5.62 below. We consider the ideal generated by this 4×4 determinant and its derivatives with respect to u_1, u_2, u_3, u_4, we saturate by the ideal of 3×3 minors, and then we eliminate u_1, u_2, u_3, u_4. The result is the principal ideal $\langle h_4 h_5 h_6 \rangle$, with h_i as in Example 5.37. This computation reveals that the algebraic boundary of $\mathrm{conv}(X)$ consists of quadrics, and we can conclude that $\mathrm{TH}_1(I) = \mathrm{conv}(X)$.

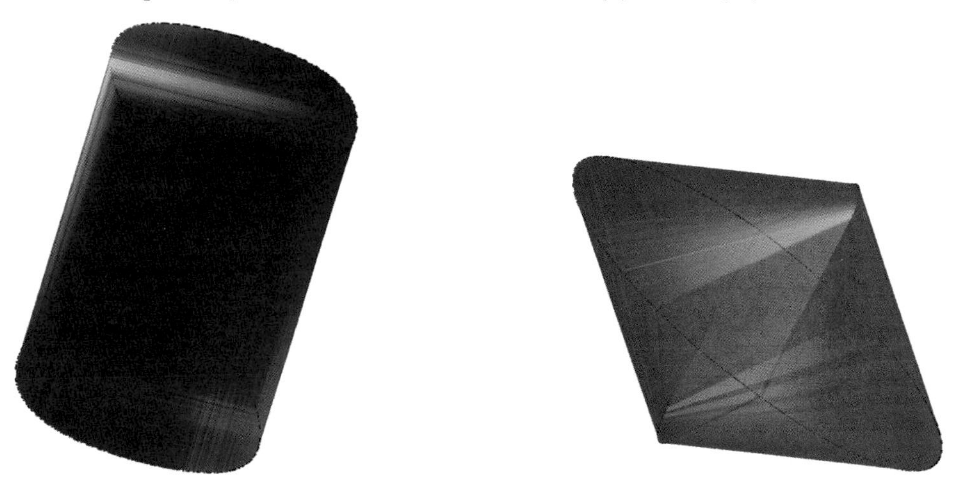

Figure 5.11. *Convex hull of the curve in Figure 5.7 and its dual convex body.*

Pictures of our convex body and its dual are shown in Figure 5.11. Diagrams such as these can be drawn fairly easily for any projected spectrahedron in $\mathbb{R}^3$. To be precise, the matrix representation of $\mathrm{TH}_1(I)$ and $\Sigma_1^1(I)$ given above can be used to rapidly sample the boundaries of these convex bodies, by maximizing many linear functions via SDP. ∎

Remark 5.62. *It would be desirable to develop a practical algorithm for computing the algebraic boundary of a projected spectrahedron. After a linear change of* *coordinates, we may assume that the given spectrahedron is represented by a symmetric matrix whose entries are linear forms in some unknowns, and our task is to eliminate a subset of these unknowns. To do this, we consider the ideal generated by the determinant and its partial derivative with respect to the unknowns to be eliminated. The variety of this ideal contains the ramification locus of the projection, but it also contains the singular locus of the determinantal hypersurface. The main difficulty in the computation is that we need to remove that singular locus before we eliminate the unknowns. Frequently, like in the previous example, the singular locus is given by the vanishing of the comaximal minors. However, this need not always be the case. A concrete example is discussed below in Example* 5.63. *Thus, one issue is how to best represent the singular locus of the algebraic boundary of a spectrahedron, in order to perform the saturation step. Once we have the correct ideal for the ramification locus, then we can compute the branch locus by elimination, and the result will be the desired hypersurface.*

Example 5.63. Consider the surface in 3-space defined by

$$\det \begin{pmatrix} x & y+z & x & y \\ y+z & 1 & y & 1 \\ x & y & z & x \\ y & 1 & x & 1 \end{pmatrix} = 0.$$

Its singular locus is the line $x - y = z = 0$. This does not coincide, in this example, with the variety defined by the vanishing of the (comaximal) 3×3 minors which consist only of the two points $(0, 0, 0)$ and $(1, 1, 0)$. ■

Exercises

Exercise 5.64. Find an explicit symmetric 6×6 matrix Q, with entries that are linear in a and b, that satisfies the identify (5.54). Is your matrix Q unique?

Exercise 5.65. The polynomial $p(x) = 1+x+x^2+x^3+x^4+x^5+x^6$ is nonnegative on the real line. What is its minimum value? Write $p(x)$ as a sum of squares. The set of all sums of squares representations of $p(x)$ is a three-dimensional spectrahedron. Draw a picture of this spectrahedron. Determine all possible representations of $p(x)$ as a sum of *two* squares.

Exercise 5.66. Let C denote the convex set of all points $(u, v) \in \mathbb{R}^2$ such that $f_{u,v}(x) = x^4 + ux^2 + vx + 1$ is a sum of squares. Draw a picture of C, express C as a projected spectrahedron, and compute a polynomial $g(u, v)$ that vanishes on the boundary of C.

Exercise 5.67. Let $I = \langle h_1 \rangle$, where $h_1 = (x_1^2-1)(x_1-1)^2+(x_2^2-1)^2$ is the bicuspid curve in Example 5.25. Compute and draw the second theta body $\mathrm{TH}_2(I)$.

Exercise 5.68. The A-discriminant Δ_A of the 3×7 matrix in (5.53) is a homogeneous polynomial of degree 24 in seven indeterminates. Can you compute Δ_A explicitly? How many monomials appear in the expansion of Δ_A?

Notes. This chapter grew out of the notes for three lectures given by Bernd Sturmfels on March 22–24, 2010, at the spring school on Linear Matrix Inequalities and Polynomial Optimization (LMIPO) at UC San Diego. Later that spring, Bernd Sturmfels lectured on convex algebraic geometry at the Università de Roma 3. This led to the publication of a first version of the material in this chapter under the title "Dualities in Convex Algebraic Geometry" in *Rendiconti di Matematica, Serie VII*, 30:285–327, 2010.

Bibliography

[1] A. I. Barvinok. *A Course in Convexity*, Grad. Stud. in Math. 54. American Mathematical Society, Providence, RI, 2002.

[2] S. Basu, R. Pollack, and M.-F. Roy. *Algorithms in Real Algebraic Geometry*. Springer, Berlin, 2006.

[3] D. Bates, J. Hauenstein, A. Sommese, and C. Wampler. `Bertini`: Software for Numerical Algebraic Geometry. Available at http://www.nd.edu/~sommese/bertini.

[4] J. Bochnak, M. Coste, and M.-F. Roy. *Géométrie Algébraique Réelle*, Ergebn. Math. Grenzgeb. 12. Springer, Berlin, 1987.

[5] S. Boyd and L. Vandenberghe. *Convex Optimization*. Cambridge University Press, Cambridge, UK, 2004.

[6] S. Boyd and L. Vandenberghe. Semidefinite programming. *SIAM Rev.*, 38:49–95, 1996.

[7] F. Catanese, S. Hoşten, A. Khetan, and B. Sturmfels. The maximum likelihood degree. *Amer. J. Math.*, 128:671–697, 2006.

[8] D. Cox, J. Little, and D. O'Shea. *Ideals, Varieties and Algorithms*, 3rd edition, Undergrad. Texts Math. Springer, New York, 2007.

[9] D. Cox, J. Little, and D. O'Shea. *Using Algebraic Geometry*, 2nd edition, Grad. Texts in Math. Springer, New York, 2005.

[10] R. Datta. Universality of Nash equilibria. *Math. Oper. Res.*, 28:424–432, 2003.

[11] I. Gelfand, M. Kapranov, and A. Zelevinsky: *Discriminants, Resultants and Multidimensional Determinants*. Birkhäuser, Boston, 1994.

[12] H.-C. Graf von Bothmer and K. Ranestad. A general formula for the algebraic degree in semidefinite programming. *Bull. Lond. Math. Soc.*, 41:193–197, 2009.

[13] D. Grayson and M. Stillman: Macaulay2, a software system for research in algebraic geometry. Available at http://www.math.uiuc.edu/Macaulay2/.

[14] N. Go and H. A. Scheraga. Ring closure in chain molecules with C_n,I, and S_{2n} symmetry. *Macromolecules*, 6:273–281, 1973.

[15] J. Gouveia and T. Netzer. Positive polynomials and projections of spectrahedra. *SIAM J. Optim.*, 21:960–976, 2011.

[16] J. Gouveia, P. A. Parrilo, and R. R. Thomas. Theta bodies for polynomial ideals. *SIAM J. Optim.*, 20:2097–2118, 2010.

[17] J. W. Helton and J. Nie. Semidefinite representation of convex sets. *Math. Program. Ser. A*, 122:21–64, 2010.

[18] M. Hestenes. *Optimization Theory: The Finite Dimensional Case.* Wiley & Sons, New York, 1975.

[19] M. Laurent, J. B. Lasserre, and P. Rostalski. Semidefinite characterization and computation of zero-dimensional real radical ideals. *Found. Comp. Math.*, 8:607–647, 2008.

[20] M. Laurent and S. Poljak. On the facial structure of the set of correlation matrices. *SIAM J. Matrix Anal. Appl.*, 17:530–547, 1996.

[21] J. B. Lasserre. Global optimization with polynomials and the problem of moments. *SIAM J. Optim.*, 11:796–817, 2001.

[22] J. B. Lasserre. *Moments, Positive Polynomials and Their Applications.* Imperial College Press, London, 2010.

[23] S. Lazard, L. M. Peñaranda, and S. Petitjean. Intersecting quadrics: an efficient and exact implementation. *Comput. Geom.*, 35:74–99, 2006.

[24] D. G. Luenberger. A double look at duality. *IEEE Trans. Automat. Control*, 73:1474–1482, 1992.

[25] M. Marshall. *Positive Polynomials and Sums of Squares.* American Mathematical Society, Providence, RI, 2008.

[26] J. Nie. Discriminants and nonnegative polynomials. *J. Symbolic Comput.*, 47:167–191, 2012.

[27] J. Nie, K. Ranestad, and B. Sturmfels. The algebraic degree of semidefinite programming. *Math. Program.*, 122:379–405, 2010.

[28] D. Plaumann, B. Sturmfels, and C. Vinzant. Quartic curves and their bitangents. *J. Symbolic Comput.*, 46:712–733, 2011.

[29] M. Ramana. An exact duality theory for semidefinite programming and its complexity implications. *Math. Program.*, 77:129–162, 1997.

[30] M. Ramana and A. J. Goldman. Some geometric results in semidefinite programming. *J. Global Optim.*, 7:33–50, 1995.

[31] K. Ranestad and B. Sturmfels. On the convex hull of a space curve. *Adv. Geom.*, 12:157–178, 2012.

[32] K. Ranestad and B. Sturmfels. The convex hull of a variety. In P. Bränden, M. Passare, and M. Putinar, editors, *Notions of Positivity and the Geometry of Polynomials.* Trends Math. Springer-Verlag, Basel, 2011, pp. 331–344.

[33] R. T. Rockafeller. *Convex Analysis.* Princeton University Press, Princeton, NJ, 1970.

[34] P. Rostalski. Bermeja, Software for Convex Algebraic Geometry. Available at http://math.berkeley.edu/~philipp/cagwiki.

[35] C. Scheiderer. Convex hulls of curves of genus one. *Adv. Math.*, 228:2606–2622, 2011.

[36] M. Trott. Applying GroebnerBasis to three problems in geometry. *Mathematica in Education and Research*, 6:15–28, 1997.

[37] G. Ziegler. *Lectures on Polytopes.* Grad. Texts in Math. Springer, New York, 1995.

Chapter 6

Semidefinite Representability

Jiawang Nie[†]

It is natural to ask which convex optimization problems can be formulated as semidefinite programs. If such a formulation exists, how can we find it? The answer to these questions is equivalent to finding an exact representation of a convex set as a spectrahedron or projected spectrahedron. Whenever this can be done, we say that the convex set has a *semidefinite representation* or it is *semidefinite representable.*

6.1 Introduction

We begin by examining the question of when a convex set S is a spectrahedron. Since a spectrahedron is defined by a linear matrix inequality, the points on the boundary of S must satisfy a polynomial equation given by the determinant of its linear pencil. Therefore, only convex sets whose boundaries have a polynomial description can be spectrahedra. In fact, being a spectrahedron is even more restrictive and we will examine some of these restrictions in this chapter. In particular, we will present a complete characterization of two-dimensional spectrahedra due to Helton and Vinnikov. In higher dimensions a full characterization of which convex sets are spectrahedra is unknown.

The class of projected spectrahedra is considered next. We will provide some natural necessary conditions for a set to be a projected spectrahedron. Deriving sufficient conditions brings us to explicit construction methods for semidefinite representations. A general technique for constructing such representations and approximations of a convex set S given by polynomial equations and inequalities is to use *moments.* The basic idea is that we introduce an independent variable for

[†]Jiawang Nie was supported by NSF grants DMS-0757212 and DMS-0844775.

every monomial, so that the defining inequalities of S become linear inequalities in the new variables. We then consider a set consisting of points satisfying the defining inequalities of S in the new variables, and some extra positive semidefiniteness conditions coming from moment matrices. The moment approach is equivalent, via duality, to showing that every linear polynomial nonnegative on S has a weighted sum-of-squares representation with uniform degree bounds. Therefore, the sum-of-squares theory will naturally appear in studying semidefinite representability. We will examine in detail the power of the moment approach to provide exact representations of convex sets. In particular, under some local boundary conditions we obtain exact semidefinite representations.

Another approach for constructing semidefinite representations is called *localization*. If we can divide a convex set S into several parts and find a semidefinite representation for each piece, then these representations can be glued together to provide a semidefinite representation for S. The main tool for this approach is building a single semidefinite representation for the convex hull of the union of several projected spectrahedra.

Sufficient conditions will follow from combining localization with the moment approach. While at this time we do not have a full understanding of semidefinite representability of convex semialgebraic sets, the necessary and sufficient conditions derived in this chapter are reasonably close to each other.

6.2 Spectrahedra

Recall from Chapter 2 that a set $S \subseteq \mathbb{R}^n$ is called a *spectrahedron* if it can be described by a *linear matrix inequality* as

$$S = \{x \in \mathbb{R}^n : A_0 + x_1 A_1 + \cdots + x_n A_n \succeq 0\}. \tag{6.1}$$

Here, each A_i is a constant symmetric matrix, and if the origin is in the interior of S, then A_0 can be chosen to be positive definite. Furthermore, if $A_0 \succ 0$, we can apply a congruence transformation to the matrices $A_1, \ldots, A_n$ and make $A_0 = I$. For instance, if $A_0 = BB^T$ with B nonsingular, then S can be described by

$$I + x_1 B^{-1} A_1 B^{-T} + \cdots + x_n B^{-1} A_n B^{-T} \succeq 0.$$

When $A_0 = I$, the linear matrix inequality in (6.1) is said to be *monic* and the origin is in the interior of S. Conversely, if S defined by (6.1) has nonempty interior, we may assume A_0 is positive definite by translating an interior point to the origin. The expression $A_0 + x_1 A_1 + \cdots + x_n A_n$ is called a *symmetric linear matrix polynomial* or a *linear pencil.*

6.2.1 Examples of Spectrahedra

We begin by giving examples of spectrahedra that naturally arise in optimization.

- **Ellipsoids.** An *ellipsoid* $\mathcal{E}$ is a set in $\mathbb{R}^n$ that can be described as

$$\mathcal{E} = \{x \in \mathbb{R}^n : (x - c)^T E^{-1} (x - c) \leq 1\}$$

for a symmetric positive definite matrix $E \in \mathcal{S}^n_{++}$ and a vector $c \in \mathbb{R}^n$. The vector c is called the center of $\mathcal{E}$, and E is called the shape matrix of $\mathcal{E}$. An ellipsoid $\mathcal{E}$ is a spectrahedron because a point x is in $\mathcal{E}$ if and only if it satisfies the linear matrix inequality

$$\begin{bmatrix} E & x-c \\ (x-c)^T & 1 \end{bmatrix} = \begin{bmatrix} E & -c \\ -c^T & 1 \end{bmatrix} + \sum_{i=1}^n x_i \begin{bmatrix} 0 & e_i \\ e_i^T & 0 \end{bmatrix} \succeq 0.$$

We can use *Schur complement* to verify that the above linear matrix inequality describes $\mathcal{E}$. Ellipsoids have wide applications in optimization [3, 7, 8, 32].

- **Second order cones.** The set $\{(x,t) \in \mathbb{R}^n \times \mathbb{R}_+ : \|x\|_2 \leq t\}$ is called the *second order cone* (also *Lorentz cone* or *ice cream cone*). We have already seen this cone in Chapter 2. It is a spectrahedron, because it is defined by the linear matrix inequality

$$\begin{bmatrix} tI_n & x \\ x^T & t \end{bmatrix} = \sum_{i=1}^n x_i \begin{bmatrix} 0 & e_i \\ e_i^T & 0 \end{bmatrix} + t \begin{bmatrix} I_n & 0 \\ 0 & 1 \end{bmatrix} \succeq 0.$$

  Second order cones also have wide applications in optimization (cf. [2]).

- **Convex quadratic sets.** More general convex sets than ellipsoids and second order cones are defined by quadratic inequalities. Let $Q := \{x \in \mathbb{R}^n : q(x) \leq 0\}$ be a nonempty set, with

$$q(x) := x^T B x + b^T x + c$$

  being a quadratic function. Here B is a symmetric matrix. It is interesting to note that the set Q is convex if and only if it is a spectrahedron. We leave this as an exercise to the readers.

- **Matrices with bounded eigenvalues or singular values.** Denote by $\lambda_{min}(\cdot)$ and $\lambda_{max}(\cdot)$, respectively, the minimum and maximum eigenvalues of a symmetric matrix. Let $X \in \mathbb{R}^{n\times n}$. If X is symmetric, then $\lambda_{max}(X) \leq t$ if and only if

$$tI - X \succeq 0$$

  and $\lambda_{min}(X) \geq t$ if and only if

$$X - tI \succeq 0.$$

  If X is not symmetric, then its maximum singular value $\sigma_{max}(X) \leq t$ if and only if

$$\begin{bmatrix} tI & X \\ X^T & tI \end{bmatrix} \succeq 0.$$

  These linear matrix inequalities all define spectrahedra in the space of (X,t).

- **Fractional linear-quadratic inequalities [3].** Fractional linear-quadratic inequalities can be used to define interesting convex sets. Let

$$F = \left\{ x \in \mathbb{R}^n \;\middle|\; \frac{\|Bx+f\|_2^2}{a^Tx+b} \le c^Tx+d,\; a^Tx+b>0 \right\}$$

be a nonempty set defined by a, b, c, d, f, B. Note that the denominator a^Tx+b is positive on F. The closure of F is a spectrahedron since by Schur complement it can be described by the linear matrix inequality

$$\begin{bmatrix} (a^Tx+b)I & Bx+f \\ (Bx+f)^T & c^Tx+d \end{bmatrix} \succeq 0.$$

- **Quadratic matrix inequalities [3].** Let V be a symmetric positive definite matrix and $\mathcal{L}(X) : \mathbb{R}^{m\times n} \to \mathcal{S}^k$ be a linear operator. Consider the following quadratic matrix inequality on matrix pairs (X, Y) with Y symmetric:

$$XV^{-1}X^T + \mathcal{L}(X) \preceq Y.$$

By Schur complement, it is equivalent to the linear matrix inequality

$$\begin{bmatrix} V & X^T \\ X & Y - \mathcal{L}(X) \end{bmatrix} \succeq 0$$

which defines a spectrahedron in the space of (X, Y).

- **Matrix cubes parameterized by eigenvalues [30].** Consider the linear matrix polynomials $B_1(x), \ldots, B_m(x)$. Let

$$\mathcal{C} \;=\; \left\{ (x,d) \in \mathbb{R}^n \times \mathbb{R} \;\middle|\; \begin{array}{r} d\cdot A_0 + \sum_{k=1}^m t_k A_k \succeq 0 \text{ whenever} \\ \lambda_{min}(B_k(x)) \le t_k \le \lambda_{max}(B_k(x)) \\ \text{for } k = 1,2,\ldots,m \end{array} \right\},$$

where every A_k is a constant symmetric matrix. The set $\mathcal{C}$ is a spectrahedron (cf. [30]), because there exists a symmetric linear matrix polynomial $\mathcal{L}(x,d)$ in (x,d) such that

$$\mathcal{C} \;=\; \{\, (x,d) \in \mathbb{R}^n \times \mathbb{R} : \mathcal{L}(x,d) \succeq 0 \,\}.$$

The construction of $\mathcal{L}(x,d)$ is given in [30].

A special case of matrix cubes is the *k-ellipse*, which consists of all points in the plane that have a constant sum of distances to a set of given foci (cf. [31]). We have already encountered the k-ellipse in Section 2.1.3. For instance, the 3-ellipse with foci $(0,0), (1,0), (0,1)$ and radius $d = 5$ is defined by the equation

$$\sqrt{x_1^2+x_2^2} + \sqrt{(x_1-1)^2+x_2^2} + \sqrt{x_1^2+(x_2-1)^2} = 5.$$

The region surrounded by this 3-ellipse is convex and can be described by the linear matrix inequality:

$$\begin{bmatrix} 6-3x_1 & x_2 & x_2-1 & 0 & x_2 & 0 & 0 & 0 \\ x_2 & 6-x_1 & 0 & x_2-1 & 0 & x_2 & 0 & 0 \\ x_2-1 & 0 & 6-x_1 & x_2 & 0 & 0 & x_2 & 0 \\ 0 & x_2-1 & x_2 & 6+x_1 & 0 & 0 & 0 & x_2 \\ x_2 & 0 & 0 & 0 & 4-x_1 & x_2 & x_2-1 & 0 \\ 0 & x_2 & 0 & 0 & x_2 & 4+x_1 & 0 & x_2-1 \\ 0 & 0 & x_2 & 0 & x_2-1 & 0 & 4+x_1 & x_2 \\ 0 & 0 & 0 & x_2 & 0 & x_2-1 & x_2 & 4+3x_1 \end{bmatrix} \succeq 0.$$

Therefore, this convex region is a spectrahedron. A defining polynomial for this 3-ellipse is given by the determinant of the above matrix.

6.2.2 Spectrahedra and Algebraic Interiors

Let S be a spectrahedron defined as in (6.1), and $p_I(x)$ denote the principal minor of the linear pencil

$$A(x) := A_0 + A_1 x_1 + \cdots + A_n x_n,$$

whose rows and columns are indexed by a nonempty set $I \subseteq \{1, 2, \ldots, m\}$, where m is the size of the matrices A_i. Then, a point $x \in S$ if and only if all the principal minors are nonnegative at x:

$$p_I(x) \geq 0 \quad \text{for all } I \subseteq \{1, 2, \ldots, n\}.$$

Therefore, S is a *basic closed semialgebraic* set (defined by finitely many weak polynomial inequalities). The boundary of S lies on the determinantal hypersurface

$$\det A(x) \;=\; 0.$$

If $A_0 \succ 0$ (the origin is in the interior of S), then S is the closure of the connected component of the set

$$\{x : \det A(x) > 0\}$$

containing the origin.

The above observation leads to the definition of *algebraic interior*, which was introduced by Helton and Vinnikov [17]. A subset T of $\mathbb{R}^n$ is an algebraic interior if it equals the closure of a connected component of the set $\{x : p(x) > 0\}$ for some polynomial p. The polynomial p is called a *defining polynomial* of T. The defining polynomial of an algebraic interior is not unique. However, the one of the smallest degree is unique up to a positive constant factor, and divides all the defining polynomials of T. Its degree is called the degree of T.

Example 6.1. Consider the spectrahedron defined by

$$\begin{bmatrix} 1 & x_1 & x_2 \\ x_1 & 1 & x_3 \\ x_2 & x_3 & 1 \end{bmatrix} \succeq 0.$$

It is the elliptope $\mathcal{E}_3$, which we have previously seen in Chapter 2 and Chapter 5, and an algebraic interior defined by the cubic polynomial inequality

$$p_{\{1,2,3\}}(x) := 2x_1x_2x_3 - x_1^2 - x_2^2 - x_3^2 + 1 > 0.$$

This spectrahedron is a basic closed semialgebraic set defined by the four polynomial inequalities:

$$p_{\{1,2,3\}}(x) \geq 0, \quad p_{\{1,2\}} = 1 - x_1^2 \geq 0, \quad p_{\{1,3\}} = 1 - x_2^2 \geq 0, \quad p_{\{2,3\}} = 1 - x_3^2 \geq 0.$$

A picture of this elliptope is shown in Chapter 5, Figure 5.8. ∎

A spectrahedron defined by a monic linear pencil is convex and an algebraic interior. We will now consider the converse of this statement. Suppose a set $S \subset \mathbb{R}^n$ is convex and equals the closure of a connected component of the set

$$\{x : p(x) > 0\}$$

for some polynomial p. Does it follow that S is a spectrahedron? As we will see, a spectrahedron satisfies a stronger condition called *rigid convexity*.

6.2.3 Rigid Convexity

Suppose S is a spectrahedron defined by a monic linear pencil $A(x)$. Then S is an algebraic interior with defining polynomial $p(x) = \det A(x)$. Given an arbitrary real direction $0 \neq w \in \mathbb{R}^n$, consider the line $x(t) := tw$ passing through 0. Note that

$$p(x(t)) = \det(I + tW), \quad W = \sum_i w_i A_i.$$

Since W is symmetric, the equation $p(x(t)) = 0$ has only real roots. This is an important property satisfied by spectrahedra.

A polynomial $p \in \mathbb{R}[x]$ is called *real zero* with respect to a point u with $p(u) > 0$ if for every $0 \neq w \in \mathbb{R}^n$ the univariate polynomial $p(u + tw) \in \mathbb{R}[t]$ has only real zeros. If $u = 0$, we simply say that p is real zero. Real zero polynomials are nonhomogeneous versions of hyperbolic polynomials. A homogeneous polynomial $h(x)$ is *hyperbolic* with respect to a direction $u \in \mathbb{R}^n$ with $h(u) > 0$ if for every $0 \neq w \in \mathbb{R}^n$ the univariate polynomial $h(u + tw) \in \mathbb{R}[t]$ has only real zeros. If a form $h(x)$ is hyperbolic with respect to $u = (1, u_2, \ldots, u_n)$, then the dehomogenized polynomial $h(1, x_2, \ldots, x_n)$ is real zero with respect to $(u_2, \ldots, u_n)$.

Example 6.2. (i) The cubic polynomial from Example 6.1,

$$2x_1x_2x_3 - x_1^2 - x_2^2 - x_3^2 + 1,$$

is real zero, because it is the determinant of a monic linear pencil.

(ii) The polynomial $p(x) = 1 - (x_1^4 + x_2^4)$ is not real zero [17]. For every $0 \neq (w_1, w_2) \in \mathbb{R}^2$, the univariate polynomial in t

$$p(tw) = \left(1 - t^2(w_1^4 + w_2^4)^{1/2}\right)\left(1 + t^2(w_1^4 + w_2^4)^{1/2}\right)$$

has two nonreal zeros. The origin lies in the interior of $\{x : p(x) > 0\}$. ∎

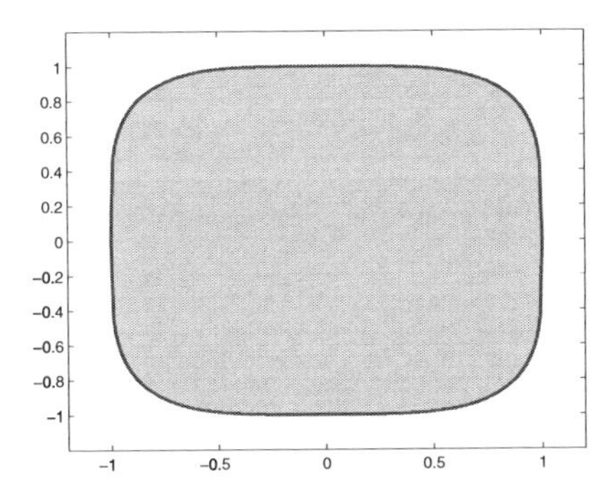

Figure 6.1. *The TV screen* $\{(x_1, x_2) : x_1^4 + x_2^4 \leq 1\}$.

Suppose an algebraic interior $S \subset \mathbb{R}^n$ is defined by a polynomial p. Then S is called *rigidly convex* if p is real zero with respect to an interior point u of S. If so, we say S passes the *line test* with respect to u; i.e., every real line ℓ passing through u intersects the hypersurface $p(x) = 0$ at *only* real points. The properties of rigidly convex sets are summarized in the following theorem due to Helton and Vinnikov.

Theorem 6.3 ([17]). *Suppose S is an algebraic interior.*

(i) *If S passes the line test with respect to a point $u \in \text{int}(S)$, then it must be convex.*

(ii) *If S is rigidly convex with respect to a point $u \in \text{int}(S)$, then S is rigidly convex with respect to every point $v \in \text{int}(S)$.*

Not all convex algebraic interiors are rigidly convex. As we saw above, the *TV screen* (see Figure 6.1)

$$\{(x_1, x_2) : 1 - x_1^4 - x_2^4 \geq 0\}$$

does not pass the line test and hence is not rigidly convex (cf. Example 6.2). Here is another such example.

Example 6.4 ([17]). Consider the polynomial

$$p(x) = x_1^3 - 3x_2^2 x_1 - (x_1^2 + x_2^2)^2.$$

The inequality $p(x) > 0$ defines three bounded convex components shown in Figure 6.2. Let S be the closure of the component lying in the half space $x_1 \geq 0$. It is an algebraic interior of degree 4 and is shaded in Figure 6.2. The point $u = (0.5, 0)$ lies in the interior of S. Figure 6.2 shows a line passing through u and intersecting the curve $p(x) = 0$ in only two real points. Thus, S is not rigidly convex. ∎

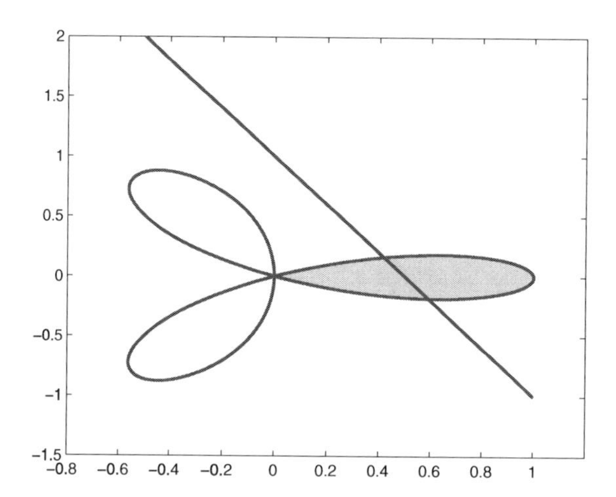

Figure 6.2. *A line passing through* $(0.5, 0)$ *intersects the curve* $x_1^3 - 3x_2^2x_1 - (x_1^2 + x_2^2)^2 = 0$ *in only* 2 *real points.*

The relationship between spectrahedra and rigid convexity is described by the following fundamental result of Helton and Vinnikov. It completely characterizes two-dimensional spectrahedra.

Theorem 6.5 ([17]). *If an algebraic interior* $S \subset \mathbb{R}^n$ *is a spectrahedron, then* S *is rigidly convex. When* $n = 2$, *the converse is also true, and* S *can be represented by a monic linear matrix inequality whose size equals the degree of its boundary* ∂S.

The first statement in Theorem 6.5 has been shown at the beginning of this subsection: the determinant of a monic linear pencil must be real zero. In the two-dimensional case ($n = 2$), the converse statement is established by showing that every real zero bivariate polynomial of degree d is the determinant of a monic linear pencil of size $d \times d$. Finding a spectrahedral representation of a rigidly convex algebraic interior S is equivalent to finding a representation of a defining polynomial of S as the determinant of a monic linear pencil. This naturally leads to studying determinantal representations of polynomials.

6.2.4 Symmetric Determinantal Representations

Given a polynomial $p \in \mathbb{R}[x]$, we say it has a *symmetric determinantal representation* if there exists a *linear pencil*

$$L(x) := L_0 + x_1 L_1 + \cdots + x_n L_n$$

such that $p = \det L(x)$ and every L_i is symmetric. If $L_0 \succ 0$, we say that p admits a *monic symmetric determinantal representation* . An important result due to Helton, McCullough, and Vinnikov is that every polynomial p can be expressed as the determinant of a linear pencil (not necessarily monic).

Theorem 6.6 ([16]). *Every polynomial p (with $p(0) \neq 0$) admits a symmetric determinantal representation of the form*

$$p(x) = c \cdot \det(L_0 + L_1 x_1 + \cdots + L_n x_n), \tag{6.2}$$

where L_0 is a "signature matrix" (L_0 is diagonal and $L_0^2 = I$) and c is a nonzero constant.

Clearly, if $\deg(p) = d$, the size of matrices L_j should be at least d. When $n > 2$, typically the size of L_j has to be larger than d. This can be shown by a dimension comparison. Suppose L_j has dimension $N \times N$ and L_0 is diagonal. The dimension of the space of degree d polynomials is $\binom{n+d}{d}$, while the dimension of the space of pencils $L(x)$ is $N + nN(N+1)/2$. For any fixed $n > 2$, the former dimension grows significantly faster than the latter if $N = O(d)$. So we should expect $N > O(d)$ when $n > 2$.

Example 6.7. (i) The polynomial $1 + x_1^2 + x_2^2$ has the symmetric determinantal representation

$$1 + x_1^2 + x_2^2 = -\det \begin{bmatrix} 1 & 0 & x_1 \\ 0 & 1 & x_2 \\ x_1 & x_2 & -1 \end{bmatrix}.$$

This linear pencil is clearly not monic.
(ii) Consider the following bivariate quartic polynomial:

$$1 + x_1^2 + x_2^2 + 4x_1^2 x_2 - 4x_1 x_2^2 + x_1^4 - 2x_1^3 x_2 - 2x_1 x_2^3 - x_1^2 x_2^2 + x_2^4.$$

It is the determinant of the following linear pencil which is also not monic:

$$\begin{bmatrix} 1 & x_1 & x_2 & x_2 \\ x_1 & -1 & x_1 & x_2 \\ x_2 & x_1 & -1 & x_1 \\ x_2 & x_2 & x_1 & 1 \end{bmatrix}. \qquad \blacksquare$$

In the context of semidefinite representations it is natural to ask whether a real zero polynomial admits a monic symmetric determinantal representation. For the general case $n > 2$, a counterexample was found by Brändén [9]. He further showed that there are real zero polynomials p for which there is no power $k > 0$ such that p^k admits a monic symmetric determinantal representation. Simpler counterexamples were found by Netzer and Thom [26]. For instance, for every $n \geq 4$, the simple quadratic polynomial $(1 + x_1)^2 - x_2^2 - \cdots - x_n^2$ does not admit a monic symmetric determinantal representation (cf. [26, Example 3.5]).

It follows from Theorem 6.3 that for $n = 2$, a degree d real zero polynomial always has a monic symmetric determinantal representation of size $d \times d$. The proof uses complexification of projective algebraic curves and the constructions are mostly theoretical. Computational aspects of these constructions are discussed in [35].

When $S \subset \mathbb{R}^n$ $(n > 2)$ is an algebraic interior that is rigidly convex, its minimum degree defining polynomial p might not admit a monic symmetric determinantal representation. However, this does not exclude the possibility of a multiple of p having a monic symmetric determinantal representation. If this is true, then S would be a spectrahedron. Indeed, Helton and Vinnikov [17] conjectured that *every rigidly convex algebraic interior of* $\mathbb{R}^n$ *is a spectrahedron.*

6.2.5 Exercises

Exercise 6.8. Let $C = \{x \in \mathbb{R}^n : f(x) \le 0\}$ be a nonempty convex set defined by a smooth function $f : \mathbb{R}^n \to \mathbb{R}$. Suppose u lies on the boundary of C and $\nabla f(u) \neq 0$. Show that the following is true:

(i) The Hessian $\nabla^2 f(u)$ is positive semidefinite in the tangent space of C at u, i.e., $v^T \nabla^2 f(u) v \ge 0$ for all $v \in \nabla f(u)^\perp := \{w : \nabla f(u)^T w = 0\}$.

(ii) The set C belongs to the half space $\nabla f(u)^T (x - u) \le 0$.

Exercise 6.9. Let $Q = \{x \in \mathbb{R}^n : q_1(x) \ge 0, \dots, q_m(x) \ge 0\}$ be a nonempty set with each q_i being a quadratic polynomial. Show that Q is convex if and only if it is a spectrahedron.

Exercise 6.10. Decide whether the following polynomials are real zero or not with respect to the vector $(1, \dots, 1)$ of all ones:

(a) $x_1 \cdots x_n - 1/2$;

(b) $x_1 \cdots x_n$;

(c) $x_1 \cdots x_n (1/x_1 + \cdots + 1/x_n)$;

(d) $(n+1)x_{n+1}^p - x_1^p - \cdots - x_n^p$, ($p > 1$ is an integer).

Exercise 6.11. Find a *smallest* size symmetric determinantal representation for the following polynomials:

(a) $1 - x_1^2 - x_2^2 - x_3^2$;

(b) $1 + x_1^3 + x_2^3$;

(c) $1 - x_1^4 - x_2^4$;

(d) $1 + x_1^6 + x_2^6$.

Exercise 6.12. Consider the 3-ellipse with foci $(0,0), (-1,0), (0,-1)$ and radius 3:

$$\sqrt{x_1^2 + x_2^2} + \sqrt{(x_1+1)^2 + x_2^2} + \sqrt{x_1^2 + (x_2+1)^2} = 3.$$

Represent the convex region surrounded by this 3-ellipse by a linear matrix inequality in variables x_1 and x_2 only. What is the polynomial of smallest degree (up to a constant factor) vanishing on this 3-ellipse?

Exercise 6.13. Suppose S is a spectrahedron. Show that every face of S is exposed. (A face F of S is called *exposed* if either $F = S$ or there exists a supporting hyperplane H of S such that $H \cap S = F$.)

6.3 Projected Spectrahedra

A set $S \subseteq \mathbb{R}^n$ is called a *projected spectrahedron* if there exists a spectrahedron $P \subseteq \mathbb{R}^{n+k}$ such that

$$S = \left\{ x \in \mathbb{R}^n \,\middle|\, (x, y) \in P \text{ for some } y \in \mathbb{R}^k \right\}. \tag{6.3}$$

In the above, y is called a *lifting vector* and P a *lifting spectrahedron* of S. Using the linear matrix inequality defining P, we can write S as

$$S = \left\{ x \in \mathbb{R}^n \,\middle|\, A_0 + \sum_{i=1}^{n} x_i A_i + \sum_{j=1}^{k} y_j B_j \succeq 0 \text{ for some } y \in \mathbb{R}^k \right\}. \tag{6.4}$$

Projected spectrahedra are a much larger class of convex sets than spectrahedra, with significantly greater modeling power. Unlike in the case of spectrahedra where rigid convexity is a natural requirement, no nontrivial obstructions to being a projected spectrahedron are known. In the remainder of this chapter we discuss representability of convex sets as projected spectrahedra.

6.3.1 Examples of Projected Spectrahedra

We now give several examples of projected spectrahedra, many of which are important in applications.

- The TV screen $\{(x_1, x_2) : 1 - x_1^4 - x_2^4 \geq 0\}$ of Example 6.2 is a projected spectrahedron since it admits the semidefinite representation

$$\text{BlockDiag}\left(\begin{bmatrix} 1 + y_1 & y_2 \\ y_2 & 1 - y_1 \end{bmatrix}, \begin{bmatrix} 1 & x_1 \\ x_1 & y_1 \end{bmatrix}, \begin{bmatrix} 1 & x_2 \\ x_2 & y_2 \end{bmatrix} \right) \succeq 0.$$

  It has two lifting variables, and we have seen that the TV screen is not a spectrahedron.

- The three-dimensional *hyperboloid* $H = \{x \in \mathbb{R}^3_+ : x_1 x_2 x_3 \geq 1\}$ is a projected spectrahedron, since it admits the semidefinite representation

$$\text{BlockDiag}\left(\begin{bmatrix} x_1 & y_1 \\ y_1 & x_2 \end{bmatrix}, \begin{bmatrix} x_3 & y_2 \\ y_2 & 1 \end{bmatrix}, \begin{bmatrix} y_1 & 1 \\ 1 & y_2 \end{bmatrix} \right) \succeq 0.$$

There are two lifting variables. The hyperboloid H is not a spectrahedron, because its defining polynomial $x_1x_2x_3 - 1$ is not real zero with respect to $(1, 1, 2)$, an interior point of H.

For any rational $r \in [0, 1/m]$ the set

$$H(m, r) := \{(x, t) \in \mathbb{R}^m_+ \times \mathbb{R} : t \leq (x_1 \cdots x_m)^r\} \tag{6.5}$$

is a projected spectrahedron (cf. [3, Section 3.3]). As we will see below, the sets $H(m, r)$ are useful in constructing semidefinite representations for convex sets.

- **Sums of largest eigenvalues [3].** In optimization one often needs to minimize the sum of k largest eigenvalues over an affine subspace of symmetric matrices. This optimization problem is convex and can be formulated as a semidefinite program. For $X \in \mathcal{S}^n$, let $\lambda_i(X)$ be the ith largest eigenvalue of X. Define $s_k(X) := \lambda_1(X) + \cdots + \lambda_k(X)$ to be the sum of k largest eigenvalues of X. Denote the set

$$S^n_k := \{(X, t) \in \mathcal{S}^n \times \mathbb{R} : s_k(X) \leq t\}.$$

Note that $s_k(X) \leq t$ if and only if there exists $(Z, \tau) \in \mathcal{S}^n \times \mathbb{R}$ such that [3, Section 4.2]

$$\begin{aligned} t - k\tau - \mathrm{Tr}(Z) &\geq 0, \\ Z &\succeq 0, \\ Z - X + \tau I_n &\succeq 0. \end{aligned} \tag{6.6}$$

It can be checked that (6.6) implies $s_k(X) \leq t$. Conversely, if $s_k(X) \leq t$, then we can find a pair $(Z, \tau) \in \mathcal{S}^n \times \mathbb{R}$ satisfying (6.6). To see this, we may assume that X is diagonal (up to an orthogonal transformation) and choose

$$\tau = \lambda_k(X), \quad Z = \mathrm{Diag}(\lambda_1(X) - \tau, \ldots, \lambda_{k-1}(X) - \tau, 0, \ldots, 0).$$

Hence, (6.6) is a semidefinite representation of S^n_k, and S^n_k is a projected spectrahedron.

A semidefinite representation similar to (6.6) can be constructed for the set of all pairs $(X, t) \in \mathcal{S}^n \times \mathbb{R}$ satisfying

$$\lambda_{n-k+1}(X) + \cdots + \lambda_n(X) \geq t$$

by using the relation $\lambda_{n-i}(X) = -\lambda_{i+1}(-X)$. This means that maximizing the sum of k smallest eigenvalues over an affine subspace of symmetric matrices can also be formulated as a semidefinite program.

- **Sums of largest singular values [3].** Another frequently encountered optimization problem is to minimize the sum of k largest singular values of matrices in an affine subspace. This problem can also be formulated as a semidefinite program in a similar way. For $X \in \mathbb{R}^{m \times n}$, denote by $\sigma_i(X)$ the

ith largest singular value of X. Note that

$$\sigma_i(X) = \lambda_i\left(\begin{bmatrix} 0 & X \\ X^T & 0 \end{bmatrix}\right).$$

A semidefinite representation as in (6.6) can be similarly constructed for the set of all pairs (X, t) satisfying

$$\sigma_1(X) + \cdots + \sigma_k(X) \leq t.$$

- **Powers of determinants [3].** In many applications, such as matrix completion problems, one often needs to maximize the determinant of a positive semidefinite matrix over an affine subspace. This problem can also be formulated as a semidefinite program. For a rational number $r \in [0, 1/n]$, the set

$$\mathcal{D}_r^n := \left\{(X, t) \in \mathcal{S}_+^n \times \mathbb{R} : (\det X)^r \geq t\right\}$$

is a projected spectrahedron, because $(X, t) \in \mathcal{D}_r^n$ if and only if there exists a *lower triangular* matrix $L \in \mathbb{R}^{n \times n}$ satisfying

$$\begin{bmatrix} X & L \\ L^T & \operatorname{Diag}(L) \end{bmatrix} \succeq 0, \quad (\operatorname{diag}(L), t) \in H(n, r).$$

Here $H(n, r)$ is as in (6.5). This was shown in [3, Section 4.2].

- **Sums of squares polynomials.** In Chapter 3, we have seen that sos polynomials are very useful in global optimization of polynomial functions. Recall that $\Sigma_{n,2d}$ is the set of sos polynomials of degree $2d$ in n variables. We already know that a polynomial $f \in \Sigma_{n,2d}$ if and only if there exists a Gram matrix $X \succeq 0$ such that

$$f(x) = [x]_d^T X [x]_d, \quad X \succeq 0,$$

where $[x]_d$ denotes the column vector of monomials of degree at most d:

$$[x]_d = \begin{bmatrix} 1 & x_1 & \cdots & x_1^2 & x_1x_2 & \cdots & x_n^d \end{bmatrix}^T.$$

Note that the Gram matrix X is usually not unique for a given f. The above implies $\Sigma_{n,2d}$ is a projected spectrahedron, which we have also seen in Chapter 4.

For instance, the set $\Sigma_{1,4}$ of univariate quartic sos polynomials is the set

$$\left\{(f_0, f_1, f_2, f_3, f_4) \in \mathbb{R}^5 \,\middle|\, \sum_{i=0}^{4} f_i x^i \geq 0 \; \forall x \in \mathbb{R}\right\}.$$

It admits the following semidefinite representation with one lifting variable μ:

$$\begin{bmatrix} f_0 & \frac{1}{2}f_1 & \frac{1}{3}f_2 - \mu \\ \frac{1}{2}f_1 & \frac{1}{3}f_2 + 2\mu & \frac{1}{2}f_3 \\ \frac{1}{3}f_2 - \mu & \frac{1}{2}f_3 & f_4 \end{bmatrix} \succeq 0.$$

- **Truncated quadratic modules and preordering.** In constrained polynomial optimization, weighted sos polynomials are very useful in representing polynomials that are nonnegative on a set. For a tuple of polynomials $g := (g_1, \ldots, g_m)$, its kth order *truncated quadratic module* is defined as

$$\mathbf{qmodule}_k(g) = \left\{ \sum_{i=0}^{m} \sigma_i g_i \;\middle|\; \begin{array}{c} \deg(\sigma_i g_i) \leq 2k \text{ for all } i \\ \sigma_0, \ldots, \sigma_m \text{ are sos} \end{array} \right\}, \tag{6.7}$$

and its kth order *truncated preorder* is defined as

$$\mathbf{preorder}_k(g) = \left\{ \sum_{\nu \in \{0,1\}^m}^{m} \sigma_\nu g_\nu \;\middle|\; \begin{array}{c} \deg(\sigma_\nu g_\nu) \leq 2k, \\ \sigma_\nu \text{ is sos for every } \nu \end{array} \right\}. \tag{6.8}$$

In the above, we denote $g_\nu := g_1^{\nu_1} \cdots g_m^{\nu_m}$ and $g_0 = 1$. The set of all sos polynomials with a fixed degree is a projected spectrahedron, as shown in the preceding example. Therefore, both $\mathbf{qmodule}_k(g)$ and $\mathbf{preorder}_k(g)$ are projected spectrahedra.

For instance, in the case of two variables ($n = 2$), $\mathbf{qmodule}_1(1 - x_1^2 - x_2^2)$ admits the semidefinite representation with one lifting variable λ:

$$\left\{ a + 2b^T x + x^T C x \;\middle|\; \begin{bmatrix} a - \lambda & b \\ b^T & C + \lambda I_2 \end{bmatrix} \succeq 0 \,, \lambda \geq 0 \right\}.$$

6.3.2 Necessary Conditions

The geometry of the boundary is very important in investigating semidefinite representability of convex sets. The notion of *curvature* plays a crucial role.

Let f be a polynomial in $\mathbb{R}[x]$. Consider its real variety

$$V_{\mathbb{R}}(f) = \{x \in \mathbb{R}^n : f(x) = 0\}$$

and a point $u \in V_{\mathbb{R}}(f)$. We say f is nonsingular at u if $\nabla f(u) \neq 0$. If f is nonsingular at $u \in V_{\mathbb{R}}(f)$, we say $V_{\mathbb{R}}(f)$ has *positive curvature* at u if for either $s = 1$ or $s = -1$

$$s \cdot v^T \nabla^2 f(u) v > 0 \quad \text{for all } 0 \neq v \in \nabla f(u)^\perp. \tag{6.9}$$

Here $\nabla f(u)^\perp$ denotes the orthogonal complement of the subspace spanned by $\nabla f(u)$. When $V_{\mathbb{R}}(f)$ has positive curvature at u and $s = -1$ in (6.9) (respectively, $s = 1$), the intersection $\{f(x) \geq 0\} \cap B(u, \delta)$ (respectively, $\{f(x) \leq 0\} \cap B(u, \delta)$) is convex for a small $\delta > 0$. The definition of positive curvature of a nonsingular hypersurface Z is independent of the choice of its defining functions (cf. [13, Section 3]). Geometrically, when f is nonsingular at a point $u \in V_{\mathbb{R}}(f)$, the variety $V_{\mathbb{R}}(f)$ has positive curvature at u if and only if there exists a neighborhood $\mathcal{O}$ of u such that $V_{\mathbb{R}}(f) \cap \mathcal{O}$ is the graph of a strictly convex function (here strict convexity means the Hessian is positive definite). For a subset $V \subset V_{\mathbb{R}}(f)$, we say $V_{\mathbb{R}}(f)$ has *positive curvature* on V if $f(x)$ is nonsingular everywhere on V and $V_{\mathbb{R}}(f)$ has positive curvature at every $u \in V$. When $>$ is replaced by $\geq$ in (6.9), we similarly say $V_{\mathbb{R}}(f)$ has *nonnegative curvature* at u. We refer to [39] for more properties of curvature.

Example 6.14. Consider the TV screen $1 - x_1^4 - x_2^4 \geq 0$. Note that

$$-v^T\left(\nabla^2(1 - x_1^4 - x_2^4)\right)v = 12(x_1^2 v_1^2 + x_2^2 v_2^2) \geq 0 \quad \text{for all } v \in \mathbb{R}^2.$$

Its boundary has zero curvature on four points $(\pm 1, 0), (0, \pm 1)$ and has positive curvature everywhere else. ■

A polynomial function $f(x)$ is said to be *strictly quasi-concave* at u if the condition (6.9) holds for $s = -1$. For a subset $V \subset \mathbb{R}^n$, we say $f(x)$ is strictly quasi-concave on V if $f(x)$ is strictly quasi-concave on every point of V. When $>$ is replaced by $\geq$ in (6.9) for $s = -1$, we can similarly define $f(x)$ to be *quasi-concave.* Similarly, *quasi-convexity* and *strict quasi-convexity* are defined by requiring $s = 1$ in (6.9). Our definitions of quasi-convexity and quasi-concavity are slightly less demanding than the ones in the existing literature (e.g., [8, Section 3.4.3]).

Example 6.15. Consider the two-dimensional hyperboloid

$$H := \{x \in \mathbb{R}^2_+ : x_1 x_2 - 1 \geq 0\}.$$

We see that

$$-v^T\big(\nabla^2(x_1 x_2 - 1)\big)v = -2v_1 v_2 > 0$$

whenever $0 \neq v \perp x$ and $x_1 x_2 = 1$. Hence the boundary ∂H has positive curvature. The defining polynomial is not convex anywhere, but it is strictly quasi-concave on the boundary of H. ■

Now we present some necessary conditions for a set to be a projected spectrahedron. We are interested in closed semialgebraic sets:

$$S = \bigcup_{k=1}^{m} T_k, \quad T_k = \{x \in \mathbb{R}^n : g_1^k(x) \geq 0, \ldots, g_{m_k}^k(x) \geq 0\}.$$

Each g_i^k is a polynomial and the sets T_k are called *basic closed semialgebraic.* Denote by ∂T_k the boundary of T_k in the standard Euclidean topology. For any $u \in \partial T_k$, the active set $I_k(u) := \{1 \leq i \leq m_k : g_i^k(u) = 0\}$ is nonempty.

The description of a semialgebraic set by polynomials is usually not unique, and its boundary might have singularities. We say u is a *nonsingular point* of ∂T_k if $|I_k(u)| = 1$ and $\nabla g_i^k(u) \neq 0$ for $i \in I_k(u)$; otherwise, we say u is a *singular* point of ∂T_k. A point u on ∂T_k is called a *corner point* of T_k if $|I_k(u)| > 1$. For $u \in \partial S$ and $i \in I_k(u) \neq \emptyset$, we say g_i^k is *irredundant* at u with respect to ∂S (or just *irredundant* at u if the set S is clear from the context) if there exists a sequence of nonsingular points $\{u_N\} \subset V(g_i^k) \cap \partial S$ of ∂T_k such that $u_N \to u$; otherwise, we say g_i^k is *redundant* at u. We say g_i^k is at u if $\nabla g_i^k(u) \neq 0$. Geometrically, when g_i^k is nonsingular at $u \in \partial S$, g_i^k being redundant at u means that the inequality $g_i^k(x) \geq 0$ is not necessary for describing S in a small neighborhood of u.

Example 6.16. Consider the convex set that is drawn in the shaded area of Figure 6.3. It is the union of the following two basic closed semialgebraic sets:

$$T_1 = \{g_1^1(x) := x_2 \geq 0,\ g_2^1(x) := 1 - x_2 \geq 0,\ g_3^1(x) := x_2^4 - x_1^6 \geq 0\},$$
$$T_2 = \{g_1^2(x) := x_1 \geq 0,\ g_2^2(x) := 1 - x_2 \geq 0,\ g_3^2(x) := 10x_2^3 - x_1^5 \geq 0\}.$$

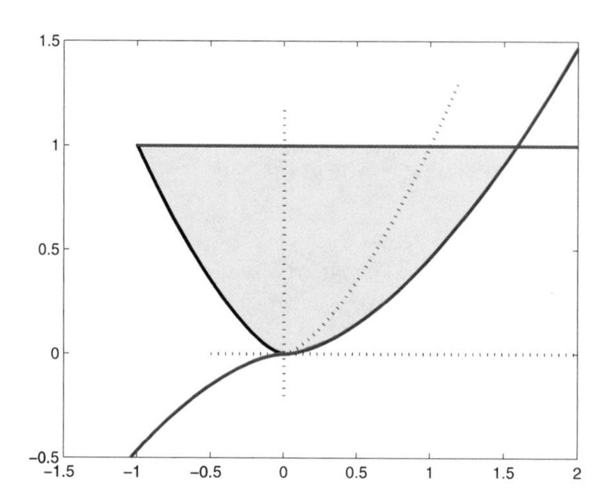

Figure 6.3. *The shaded area is the union of* T_1 *and* T_2 *in Example* 6.16.

The corner points of T_1 are $(-1,1),(0,0),(1,1)$. The polynomial g_3^1 is irredundant at $(-1,1)$ and $(0,0)$ but redundant at $(1,1)$. The polynomials g_3^1 in nonsingular at $(-1,1)$ but singular at $(0,0)$. The polynomial g_1^1 is redundant at $(0,0)$. The corner points of T_2 are $(0,0),(0,1),(\sqrt[5]{10},1)$. The polynomial g_3^2 is irredundant at both $(0,0)$ and $(\sqrt[5]{10},1)$. It is nonsingular at $(\sqrt[5]{10},1)$ but singular at $(0,0)$. The polynomial g_1^2 is redundant at $(0,1)$ and $(0,0)$. Both g_2^1 and g_2^2 are irredundant on the section $x_2 = 1$ of the boundary. ∎

Now we present necessary conditions for semidefinite representability.

Theorem 6.17 ([13]). *Let $S \subset \mathbb{R}^n$ be a projected spectrahedron. Then S is convex and has the following additional properties:*

(a) *The interior* $\operatorname{int}(S)$ *of S is a finite union of basic open semialgebraic sets, i.e.,*

$$\operatorname{int}(S) = \bigcup_{k=1}^{m} T_k, \quad T_k = \{x \in \mathbb{R}^n : g_1^k(x) > 0, \dots, g_{m_k}^k(x) > 0\}.$$

(b) *The closure $\overline{S}$ of S is a finite union of basic closed semialgebraic sets:*

$$\overline{S} = \bigcup_{k=1}^{m} T_k, \quad T_k = \{x \in \mathbb{R}^n : g_1^k(x) \geq 0, \dots, g_{m_k}^k(x) \geq 0\}.$$

(The polynomials g_i^k may be different from those in (a).)

(c) *For each $u \in \partial\overline{S}$ and $i \in I_k(u) \neq \emptyset$, if g_i^k from (b) is irredundant and nonsingular at u, then g_i^k is quasi-concave at u.*

Theorem 6.17 says that a projected spectrahedron must be convex and semialgebraic, and its boundary must have nonnegative curvature at smooth points. In particular, the first two parts establish the necessary algebraic structure of projected spectrahedra, while nonnegativity of curvature follows from convexity. In other words, convexity and being semialgebraic are necessary conditions for semidefinite representability. It is not clear whether they are also sufficient. Indeed, it was conjectured in [13] that *every convex semialgebraic set in $\mathbb{R}^n$ is semidefinite representable.*

***Proof of Theorem* 6.17.** The convexity of S is obvious. Parts (a) and (b) immediately follow from the Tarski–Seidenberg quantifier elimination [6].

(c) Let $u \in \partial\overline{S} \cap \partial T_k$. Note that $\overline{S}$ is a convex set and has the same boundary as S. (If a set is not closed, then its boundary is defined to be the boundary of its closure.)

First, consider the case that u is a smooth point. Since $\overline{S}$ is convex, $\partial\overline{S}$ has a supporting hyperplane $u + w^\perp = \{u + x : w^T x = 0\}$. $\overline{S}$ lies on one side of $u + w^\perp$ and so does T_k, since T_k is contained in $\overline{S}$. Since u is a smooth point, $I_k(u) = \{i\}$ has cardinality one. For some $\delta > 0$ sufficiently small, we have

$$T_k \cap B(u, \delta) = \{x \in \mathbb{R}^n : g_i^k(x) \geq 0, \delta^2 - \|x - u\|^2 > 0\}.$$

Note $u + w^\perp$ is also a supporting hyperplane of T_k passing through u. So, the gradient $\nabla g_i^k(u)$ must be parallel to w, i.e., $\nabla g_i^k(u) = \alpha_i^k w$ for some nonzero scalar $\alpha_i^k \neq 0$. Thus, for all $0 \neq v \in w^\perp$ and $\epsilon > 0$ small enough, the point $u + \frac{\epsilon}{\|v\|} v$ is not in the interior of $T_k \cap B(u, \delta)$, which implies

$$g_i^k\left(u + \frac{\epsilon}{\|v\|} v\right) \leq 0 \text{ for all } 0 \neq v \in w^\perp = \nabla g_i^k(u)^\perp.$$

By the second order Taylor expansion, we have

$$-v^T \nabla^2 g_i^k(u) v \geq 0 \text{ for all } 0 \neq v \in \nabla g_i^k(u)^\perp;$$

that is, g_i^k is quasi-concave at u.

Second, consider the case that $u \in \partial\overline{S}$ is a corner point. By assumption that g_i^k is irredundant and nonsingular at u, there exists a sequence of smooth points $\{u_N\} \subset Z(g_i^k) \cap \partial\overline{S}$ such that $u_N \to u$ and $\nabla g_i^k(u) \neq 0$.

So $\nabla g_i^k(u_N) \neq 0$ for N sufficiently large. From the above, we know that

$$-v^T \nabla^2 g_i^k(u_N) v \geq 0 \text{ for all } 0 \neq v \in \nabla g_i^k(u_N)^\perp.$$

Note that the subspace $\nabla g_i^k(u_N)^\perp$ equals the range space of the matrix $R(u_N)$ where

$$R(v) := I_n - \|\nabla(g_i^k(v)\|_2^{-2} \cdot \nabla g_i^k(v) \nabla g_i^k(v)^T.$$

So the quasi-concavity of g_i^k at u_N is equivalent to

$$-R(u_N)^T \nabla^2 g_i^k(u_N) R(u_N) \succeq 0.$$

Since $\nabla g_i^k(u) \neq 0$, we have $R(u_N) \to R(u)$. Therefore, letting $N \to \infty$, we get

$$-R(u)^T \nabla^2 g_i^k(u) R(u) \succeq 0,$$

which implies

$$-v^T \nabla^2 g_i^k(u) v \geq 0 \text{ for all } 0 \neq v \in \nabla g_i^k(u)^\perp;$$

that is, g_i^k is quasi-concave at u. $\square$

In part (c) of Theorem 6.17, the condition that g_i^k is irredundant cannot be dropped. We show this in the following example.

Example 6.18. Consider the basic closed semialgebraic set

$$\left\{ x \in \mathbb{R}^2 \;\middle|\; \begin{array}{c} g_1(x) := x_1^3 - x_1 - x_2^2 - (x_1^2 - x_1)^2 \geq 0 \\ g_2(x) := x_1^2 + x_2^2 - 1 \geq 0 \end{array} \right\}.$$

It is shaded in Figure 6.4. The point $u = (1, 0)$ lies on the boundary of the set. The real variety of g_1 is not connected and has two components, so the inequality $g_2(x) \geq 0$ cannot be dropped in the description of this semialgebraic set. The polynomial g_2 is redundant at u, and it is not quasi-concave at u. Indeed, g_2 is strictly convex since its Hessian is always positive definite. ■

Given a semidefinite representation of a projected spectrahedron S, finding polynomials g_i^k as in Theorem 6.17 is generally very difficult. However, for some

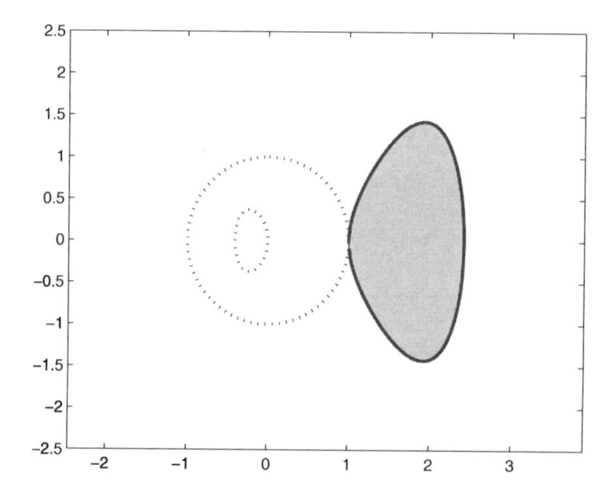

Figure 6.4. *The semialgebraic set of Example* 6.18.

simple cases, they could be obtained by eliminating lifting variables. Techniques for doing so were presented in Chapter 5, and we also refer the readers to Tarski–Seidenberg quantifier elimination [6]. We show how to do this in the following example.

Example 6.19. Let S be the projected spectrahedron defined by

$$\begin{bmatrix} 1 & 0 & x_1 \\ 0 & 1 & x_2 - y \\ x_1 & x_2 - y & y \end{bmatrix} \succeq 0.$$

Its picture is shown in the shaded area of Figure 6.5. The above linear matrix inequality is equivalent to

$$f(x, y) := x_1^2 + (x_2 - y)^2 - y \leq 0,$$

where $f(x, y)$ is the determinant of the defining linear pencil. If a point x lies on the boundary of S, then there exists y such that $f(x, y) = 0$ and y is a local maximizer of the function $y \mapsto f(x, y)$, which implies

$$f_y = -2x_2 + 2y - 1 = 0.$$

Eliminating y from $f(x, y) = f_y(x, y) = 0$ gives the equation

$$g(x) := 1 + 4(x_2 - x_1^2) = 0.$$

On the other hand, for every x satisfying $g(x) \geq 0$, the equation $f(x, y) = 0$ has a real solution y and the pair (x, y) satisfies $f(x, y) \leq 0$. Therefore, we get an equivalent description for S as

$$S = \{(x_1, x_2) \, : \, 1 + 4(x_2 - x_1^2) \geq 0\}.$$

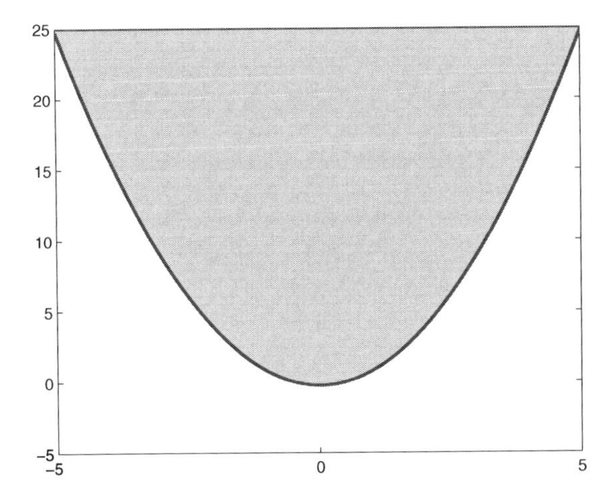

Figure 6.5. *Projected spectrahedron defined in Example* 6.19.

The defining polynomial $g(x)$ is concave. The boundary of S has positive curvature everywhere. ∎

6.3.3 Exercises

Exercise 6.20. Are the following sets projected spectrahedra?

(a) $\{X \in \mathcal{S}^n : 0 \preceq X^3 + 2X^2 - 2X \preceq I_n\}$.

(b) $\{X \in \mathbb{R}^{m\times n} : \|X\|_{p,q} \leq 1\}$ ($p, q \geq 1$ being integers).

(c) $\{(X, Y) \in \mathcal{S}_+^2 \times \mathcal{S}_+^2 : XY + YX \succeq I_2\}$.

(d) $\{(X, Y) \in \mathcal{S}^2 \times \mathcal{S}^2 : X^4 + Y^4 \preceq I_2\}$.

If yes, find a semidefinite representation for it; if no, give reasons.

Exercise 6.21. Describe the following projected spectrahedra in (x_1, x_2)-space in terms of polynomials in (x_1, x_2).

$$(a): \begin{bmatrix} 1 & x_1 & y \\ x_1 & 1 & x_2 \\ y & x_2 & 1 \end{bmatrix} \succeq 0; \quad (b): \begin{bmatrix} x_1 & y_1 & 1 \\ y_1 & 1 & y_2 \\ 1 & y_2 & x_2 \end{bmatrix} \succeq 0; \quad (c): \begin{bmatrix} 1 & y_1 & y_2 \\ y_1 & y_2 & x_1 \\ y_2 & x_1 & x_2 \end{bmatrix} \succeq 0.$$

Verify your description by drawing the above projected spectrahedra in Matlab.

Exercise 6.22. For each integer $m \geq n \geq 1$, find a semidefinite representation for the hyperboloid:

$$\{(x, t) \in \mathbb{R}_+^n \times \mathbb{R}_+ : x_1 \cdots x_n \geq t^m\}.$$

Exercise 6.23. If a convex cone $K \subset \mathbb{R}^n$ is a projected spectrahedron, show that its dual cone K^* is also a projected spectrahedron.

Exercise 6.24. Let P be the convex cone in the space $\mathbb{R}^5$:

$$P = \Big\{(f_0, f_1, f_2, f_3, f_4) : f_0 + f_1 x + f_2 x^2 + f_3 x^3 + f_4 x^4 \geq 0 \text{ for all } x \in [-1, 1]\Big\}.$$

Find a semidefinite representation for P and its dual cone P^*.

Exercise 6.25. Let Q be the convex cone:

$$Q = \Big\{(A, b, c) \in \mathcal{S}^n \times \mathbb{R}^n \times \mathbb{R} : x^T A x + 2b^T x + c \geq 0 \text{ for all } \|x\|_2 \leq 1\Big\}.$$

Find a semidefinite representation for Q and its dual cone Q^*.

Exercise 6.26. A symmetric matrix $A \in \mathcal{S}^n$ is called *copositive* if $x^T A x \geq 0$ for all $x \in \mathbb{R}_+^n$. Find a semidefinite representation for the cone $\mathcal{C}_3$ of all 3×3 copositive matrices and its dual cone $\mathcal{C}_3^*$. Repeat this for the 4×4 case.

Exercise 6.27. Is the convex set ($n > 1$)

$$\left\{(A, B, C) \in (\mathcal{S}^n)^3 : A + 2xB + x^2C \succeq 0 \text{ for all } x \in \mathbb{R}\right\}$$

a projected spectrahedron? If yes, find a semidefinite representation for it; if no, give reasons.

6.4 Constructing Semidefinite Representations

A general approach for constructing explicit semidefinite representations is to use *moments*. This was originally proposed in [19, 33]. We first describe a basic moment construction of a possible semidefinite representation, and then present tighter moment constructions. The basic construction produces a semidefinite representation when the convex set is *sos-convex*. The tighter constructions produce semidefinite representations for convex sets whose boundaries have positive curvature. In many applications, a convex set is defined by rational function inequalities, or polynomial matrix inequalities. In these cases semidefinite representations can also be constructed by using moments in a similar way. We describe these constructions and the conditions under which they work.

6.4.1 A Basic Moment Construction

To illustrate the basic idea of moment constructions, we begin with a simple example of a one-dimensional convex set defined by a single quartic inequality

$$a_0 + a_1x + a_2x^2 + a_3x^3 + a_4x^4 \geq 0.$$

We introduce a new variable y_i for each monomial x^i and convert the defining quartic inequality to the following system:

$$a_0y_0 + a_1y_1 + a_2y_2 + a_3y_3 + a_4y_4 \geq 0,$$

$$\begin{bmatrix} y_0 & y_1 & y_2 \\ y_1 & y_2 & y_3 \\ y_2 & y_3 & y_4 \end{bmatrix} = \begin{bmatrix} 1 & x & x^2 \\ x & x^2 & x^3 \\ x^2 & x^3 & x^4 \end{bmatrix}.$$

The matrix

$$\begin{bmatrix} 1 & x & x^2 \\ x & x^2 & x^3 \\ x^2 & x^3 & x^4 \end{bmatrix}$$

is always positive semidefinite. Therefore we can relax the above system to

$$a_0 + a_1x + a_2y_2 + a_3y_3 + a_4y_4 \geq 0, \qquad \begin{bmatrix} 1 & x & y_2 \\ x & y_2 & y_3 \\ y_2 & y_3 & y_4 \end{bmatrix} \succeq 0,$$

which yields a projected spectrahedron with lifting variables y_2, y_3, y_4.

This construction can be readily applied in higher-dimensional cases. Let S be a convex basic closed semialgebraic set given by

$$S = \{x \in \mathbb{R}^n : g_1(x) \geq 0, \ldots, g_m(x) \geq 0\}. \tag{6.10}$$

Let

$$d = \max\{\lceil \deg(g_1)/2 \rceil, \ldots, \lceil \deg(g_m)/2 \rceil\}. \tag{6.11}$$

Write every g_i as

$$g_i(x) = \sum_{|\alpha| \leq 2d} g_\alpha^{(i)} x^\alpha.$$

If we let $y_\alpha = x^\alpha$ for every α, then S is equal to the set

$$\left\{x \in \mathbb{R}^n : L_{g_1}(y) \geq 0, \ldots, L_{g_m}(y) \geq 0,\ y_\alpha = x^\alpha \text{ for all } |\alpha| \leq 2d\right\}.$$

The linear functionals $L_{g_i}(y)$ are linearizations of the polynomials g_i:

$$L_{g_i}(y) = \sum_{|\alpha| \leq 2d} g_\alpha^{(i)} y_\alpha.$$

The vector y is called a *truncated moment vector*, indexed by $\alpha \in \mathbb{N}^n$ with $|\alpha| \leq 2d$. Now we define a linear pencil $M_d(y)$ by substituting $y_\alpha = x^\alpha$ into $[x]_d[x]_d^T$ and call $M_d(y)$ a *moment matrix* of order d in n variables. Since the matrix inequality

$$[x]_d[x]_d^T \succeq 0$$

holds, we also get that $M_d(y) \succeq 0$. Thus, the set S defined in (6.10) can be equivalently described as

$$S = \left\{ x \in \mathbb{R}^n \,\middle|\, \begin{array}{c} L_{g_1}(y) \geq 0, \ldots, L_{g_m}(y) \geq 0, \\ M_d(y) \succeq 0,\ y_\alpha = x^\alpha \text{ for all } |\alpha| \leq 2d \end{array} \right\}.$$

Note that $y_0 = x^0 = 1$. As before, we can obtain a relaxation of S by dropping the constraints $y_\alpha = x^\alpha$ with $|\alpha| > 1$ and get the projected spectrahedron

$$R = \left\{ x \in \mathbb{R}^n \,\middle|\, \begin{array}{c} L_{g_1}(y) \geq 0, \ldots, L_{g_m}(y) \geq 0, \\ y_0 = 1, \quad M_d(y) \succeq 0, \\ x_1 = y_{e_1}, \ldots, x_n = y_{e_n} \end{array} \right\}. \tag{6.12}$$

The lifting variables in R are y_α, where $|\alpha| \geq 2$.

Example 6.28. Consider the set $S = \{(x_1, x_2) \in \mathbb{R}^2 : 1 - x_1^4 - x_2^4 - x_1^2 x_2^2 \geq 0\}$. The construction (6.12) gives a semidefinite relaxation R of S defined by

$$1 - y_{40} - y_{04} - y_{22} \geq 0, \quad \begin{bmatrix} 1 & x_1 & x_2 & y_{20} & y_{11} & y_{02} \\ x_1 & y_{20} & y_{11} & y_{30} & y_{21} & y_{12} \\ x_2 & y_{11} & y_{02} & y_{21} & y_{12} & y_{03} \\ y_{20} & y_{30} & y_{21} & y_{40} & y_{31} & y_{22} \\ y_{11} & y_{21} & y_{12} & y_{31} & y_{22} & y_{13} \\ y_{02} & y_{12} & y_{03} & y_{22} & y_{13} & y_{04} \end{bmatrix} \succeq 0.$$

There are 12 lifting variables y_{ij}, and the equality $R = S$ holds for this example, as will be shown in Section 6.4.3. ∎

Example 6.29 ([14]). Consider the set

$$S = \{x \in \mathbb{R}^n : 1 - (x^d)^T B x^d \geq 0\},$$

where B is a symmetric matrix and $x^d := \begin{bmatrix} x_1^d & x_2^d & \cdots & x_n^d \end{bmatrix}^T$. The basic moment relaxation R is

$$\left\{ x \in \mathbb{R}^n \;\middle|\; \begin{array}{c} 1 - \sum_{i,j=1}^n B_{ij} y_{d(e_i+e_j)} \geq 0, \\ y_0 = 1, M_d(y) \succeq 0, x_1 = y_{e_1}, \ldots, x_n = y_{e_n} \end{array} \right\}.$$

When B is positive semidefinite with nonnegative entries and d is even, the equality $S = R$ holds, which will be shown in Section 6.4.3. ∎

6.4.2 Tighter moment constructions

In general, the semidefinite relaxation R given by (6.12) does not equal S, except in the special case of sos-convex sets (defined in the next subsection). Hence tighter constructions by using higher order moments are necessary. We describe two basic types of refined moment constructions: *Putinar* and *Schmüdgen* semidefinite relaxations.

To describe them, we need to define *localizing matrices.* Let p be a polynomial with $\deg(p) \leq 2N$. Write

$$p(x)[x]_{N-k}[x]_{N-k}^T = \sum_{|\alpha| \leq 2N} A_\alpha^{(N)} x^\alpha \quad (k = \lceil \deg(p)/2 \rceil);$$

then define a linear pencil $L_p^{(N)}(y)$ by linearizing $y_\alpha = x^\alpha$ as before,

$$L_p^{(N)}(y) = \sum_{|\alpha| \leq 2N} A_\alpha^{(N)} y_\alpha.$$

The pencil $L_p^{(N)}(y)$ is called the *Nth order localizing matrix* of p. If p is nonnegative on S, then for every $x \in S$ we have

$$L_p^{(N)}(y) \succeq 0 \quad \text{if every } y_\alpha = x^\alpha.$$

Note that $g_0 = 1$ and $L_{g_0}^{(d)} = M_d(y)$ as before. Since all $g_0, g_1, \ldots, g_m$ are nonnegative on S, for every N the set S is contained in the projected spectrahedron

$$\tilde{S}_N = \left\{ x \in \mathbb{R}^n \;\middle|\; \begin{array}{c} L_{g_i}^{(N)}(y) \succeq 0,\ i = 0, 1, \ldots, m \\ y_0 = 1,\ x_1 = y_{e_1}, \ldots, x_n = y_{e_n} \end{array} \right\}. \quad (6.13)$$

The set $\tilde{S}_N$ is called a *Putinar semidefinite relaxation* of S.

The product of polynomials from any subset of $g_1, \ldots, g_m$ is also nonnegative on S. For every $\nu \in \{0,1\}^m$, define $g_\nu := g_1^{\nu_1} \cdots g_m^{\nu_m}$. Each g_ν is nonnegative on S. So every $x \in S$ satisfies

$$y_0 = 1, \quad L_{g_\nu}^{(N)}(y) \succeq 0 \quad \text{for all } \nu \in \{0,1\}^m \quad \text{if every } y_\alpha = x^\alpha.$$

This implies that for every N the set S is contained in the projected spectrahedron

$$\hat{S}_N = \left\{ x \in \mathbb{R}^n \;\middle|\; \begin{array}{l} L_{g_\nu}^{(N)}(y) \succeq 0 \text{ for all } \nu \in \{0,1\}^m, \\ y_0 = 1,\, x_1 = y_{e_1}, \ldots, x_n = y_{e_n} \end{array} \right\}. \tag{6.14}$$

The set $\hat{S}_N$ is called a *Schmüdgen semidefinite relaxation* of S. Clearly, for every N, $\hat{S}_N \subseteq \tilde{S}_N$ because (6.14) has extra conditions in addition to those in (6.13). We have the nesting relation

$$\begin{array}{ccccccccc} \tilde{S}_1 & \supseteq & \cdots & \supseteq & \tilde{S}_N & \supseteq & \cdots & \supseteq & S \\ \cup & & \cdots & & \cup & & \cdots & & \parallel \\ \hat{S}_1 & \supseteq & \cdots & \supseteq & \hat{S}_N & \supseteq & \cdots & \supseteq & S \end{array}.$$

Later we will see that both $\hat{S}_N$ and $\tilde{S}_N$ are equal to S for N large enough, under some general conditions. Typically, it is very difficult to get explicit bounds on N for which $\hat{S}_N = S$ or $\tilde{S}_N = S$. In some special cases, such bounds can be estimated, e.g., in [29, Section 3].

Example 6.30. Consider the convex set S defined by

$$g_1(x) := x_2 - x_1^3 \geq 0, \quad g_2(x) := x_2 + x_1^3 \geq 0.$$

The relaxation $\tilde{S}_3$ is given by

$$L_{g_1}^{(3)}(y) = \begin{bmatrix} y_{01} - y_{30} & y_{11} - y_{40} & y_{02} - y_{31} \\ y_{11} - y_{40} & y_{21} - y_{50} & y_{12} - y_{41} \\ y_{02} - y_{31} & y_{12} - y_{41} & y_{03} - y_{32} \end{bmatrix} \succeq 0, \quad x_1 = y_{10}, \quad x_2 = y_{01},$$

$$L_{g_2}^{(3)}(y) = \begin{bmatrix} y_{01} + y_{30} & y_{11} + y_{40} & y_{02} + y_{31} \\ y_{11} + y_{40} & y_{21} + y_{50} & y_{12} + y_{41} \\ y_{02} + y_{31} & y_{12} + y_{41} & y_{03} + y_{32} \end{bmatrix} \succeq 0, \quad y_{00} = 1, \quad M_3(y) \succeq 0.$$

In addition to the above, the relaxation $\hat{S}_3$ has the extra inequality

$$L_{g_{12}}^{(3)}(y) = y_{02} - y_{60} \geq 0.$$

Higher order relaxations $\tilde{S}_N$ and $\hat{S}_N$ can be constructed in a similar way. ∎

6.4.3 Sos-convex Sets

A symmetric matrix polynomial $P \in \mathbb{R}[x]^{r \times r}$ is a sum of squares if there exists a matrix polynomial W such that $P(x) = W(x)^T W(x)$. A polynomial f is called *sos-convex* if the matrix polynomial given by its Hessian $\nabla^2 f$ is a sum of squares. Similarly, f is called *sos-concave* if $-f$ is sos-convex. If for the set S defined in (6.10) every g_i is sos-concave, then we say that S is *sos-convex*.

Theorem 6.31 ([12]). *Let S be defined as in (6.10) with nonempty interior. If every defining polynomial g_i is sos-concave, then the projected spectrahedron R given by (6.12) is a semidefinite representation of S.*

In the rest of this subsection we present the proof of this result. It gives a general framework for proving that moment relaxations provide semidefinite representations. A typical approach for proving equality of two convex sets is to use duality theory via separating hyperplanes. Let S be as in Theorem 6.31. Suppose $a^T x + b = 0\,(a \neq 0)$ is a supporting hyperplane of S, then

$$a^T x + b \geq 0 \text{ for all } x \in S, \quad a^T u + b = 0 \text{ for some } u \in S.$$

The point u is a minimizer of $a^T x + b$ over S and belongs to the boundary ∂S. Since S has nonempty interior, there exists a point $v \in \mathbb{R}^n$ such that every $g_i(v) > 0$ (*Slater's condition*) and every g_i is concave. So, the first order optimality condition holds at u (cf. [5, Proposition 5.3.5]); i.e., there exist Lagrange multipliers $\lambda_1 \geq 0, \dots,$ $\lambda_m \geq 0$ such that

$$a = \lambda_1 \nabla g_1(u) + \cdots + \lambda_m \nabla g_m(u), \quad \lambda_i g_i(u) = 0 \ \ (1 \leq i \leq m).$$

Clearly, the Lagrangian

$$\mathcal{L}(x) := a^T x + b - \lambda_1 g_1(x) - \cdots - \lambda_m g_m(x) \tag{6.15}$$

is convex and nonnegative everywhere and vanishes at u, and the gradient $\nabla \mathcal{L}(u) = 0$. Interestingly, the Lagrangian $\mathcal{L}(x)$ is sos if every g_i is sos-concave. This is the key point in proving Theorem 6.31.

Lemma 6.32 ([12]). *If a symmetric matrix polynomial $P \in \mathbb{R}[x]^{r \times r}$ is sos, then for any $u \in \mathbb{R}^n$, the symmetric matrix polynomial*

$$\int_0^1 \int_0^t P(u + s(x-u))\, ds\, dt$$

is sos. In case of $r = 1$, the above integral is an sos polynomial.

Proof. This is left as an exercise. $\square$

Lemma 6.33 ([12]). *Let p be a polynomial such that $p(u) = 0$ and $\nabla p(u) = 0$ for some $u \in \mathbb{R}^n$. If its Hessian $\nabla^2 p$ is sos, then p is also sos.*

Proof. Let $q(t) = p(u + t(x-u))$ be a univariate polynomial in t. Then

$$q''(t) = (x-u)^T \nabla^2 p(u + t(x-u))(x-u).$$

Thus, $p(x) = q(1)$ has the expansion

$$(x-u)^T \left(\int_0^1 \int_0^t \nabla^2 p(u + s(x-u))\, ds\, dt \right) (x-u).$$

Since $\nabla^2 p(x)$ is sos, the double integral above is sos by Lemma 6.32. Thus $p(x)$ is also sos. $\square$

Using the above lemmas we now prove Theorem 6.31.

Proof. We have already seen that $S \subseteq R$. If $S \neq R$, there must exist $\hat{y}$ satisfying (6.12) and $\hat{x} = (\hat{y}_{e_1}, \dots, \hat{y}_{e_n}) \notin S$. Since S is closed, there exists a supporting hyperplane $a^T x + b = 0$ of S separating $\hat{x}$ strictly from S:

$$a^T x + b \geq 0 \quad \text{for all } x \in S, \quad a^T \hat{x} + b < 0, \quad a^T u + b = 0 \quad \text{for some } u \in \partial S.$$

The point u minimizes $a^T x + b$ over S. Since $\mathrm{int}(S) \neq \emptyset$ and each g_i is concave, the first order optimality condition holds (cf. [5]) and there must exist $(\lambda_1, \dots, \lambda_m) \geq 0$ such that the Lagrangian $\mathcal{L}(x)$ in (6.15) is a convex nonnegative polynomial satisfying $\mathcal{L}(u) = 0$ and $\nabla \mathcal{L}(u) = 0$. Furthermore, its Hessian

$$\nabla^2 \mathcal{L}(x) = \sum_{i=1}^{m} \lambda_i (-\nabla^2 g_i(x))$$

is sos, and Lemma 6.33 implies $\mathcal{L}(x)$ is sos. The degree of $\mathcal{L}(x)$ is at most $2d$. So there exists a symmetric matrix $W \succeq 0$ such that

$$a^T x + b = \sum_{i=1}^{m} \lambda_i g_i(x) + [x]_d^T W [x]_d.$$

The above is an identity in x. Replacing each x^α by $\hat{y}_\alpha$ results in

$$a^T \hat{x} + b = \sum_{i=1}^{m} \lambda_i L_{g_i}(\hat{y}) + Tr\big(W \cdot M_d(\hat{y})\big) \geq 0,$$

which contradicts the previous assertion that $a^T \hat{x} + b < 0$. Thus $S = R$. □

Example 6.34. Consider the set in Example 6.28. The defining polynomial there is sos-concave, because the Hessian $\nabla^2(-1 + x_1^4 + x_2^4 + x_1^2 x_2^2)$ has the sos decomposition

$$4 \begin{bmatrix} x_1 \\ x_2 \end{bmatrix} \begin{bmatrix} x_1 \\ x_2 \end{bmatrix}^T + 2 \begin{bmatrix} 2x_1 & \\ & x_1 \end{bmatrix}^2 + 2 \begin{bmatrix} x_2 & \\ & 2x_2 \end{bmatrix}^2.$$

By Theorem 6.31, the projected spectrahedron R given by (6.12) is a semidefinite representation for S. ■

Example 6.35. The set in Example 6.29 is sos-convex because the Hessian of $(x^d)^T B x^d$ has the decomposition

$$\mathrm{Diag}(x^{d-1}) \cdot W \cdot \mathrm{Diag}(x^{d-1}) + \mathrm{Diag}(a_1(x), \dots, a_n(x)),$$

where W and each $a_i(x)$ are given as

$$W = 2d^2 B + 2d(d-1)\mathrm{Diag}(B), \quad a_i(x) = 2d(d-1) \sum_{j \neq i} B_{ij} x_i^{d-2} x_j^d.$$

If $B \succeq 0$ and $d \geq 1$, then $W \succeq 0$ and must be sos; if each $B_{ij} \geq 0$ and $d > 0$ is even, then all $a_i(x)$ are sos. Therefore, when $B \succeq 0$, every $B_{ij} \geq 0$ and $d > 0$ is even, the form $(x^d)^T B x^d$ is sos-convex, and by Theorem 6.31 the projected spectrahedron R given by (6.12) is a semidefinite representation for S. ∎

Sos-convexity is a very strong condition, and not all convex polynomials are sos-convex. An explicit example is given in [1]. More generally, a nonnegative convex polynomial need not be a sum of squares (cf. Chapter 4). Generally, the projected spectrahedron R given by (6.12) does not equal S if g_i are not sos-concave. On the other hand, sos-convexity can be verified by semidefinite programming. A polynomial f is sos-convex if and only if its Hessian $\nabla^2 f$ is sos. This can be checked numerically by solving a single SDP feasibility problem, and therefore, sos-convexity is a favorable condition in practice.

6.4.4 Strictly Convex Sets

When S is not sos-convex, the basic moment relaxation R given by (6.12) might not be a semidefinite representation of S. The projected spectrahedra $\tilde{S}_N$ in (6.13) and $\hat{S}_N$ in (6.14) are better candidates for a semidefinite representation of S. We now examine weaker conditions than sos-convexity that guarantee that $\tilde{S}_N = S$ (or $\hat{S}_N = S$) for some finite N.

A sufficient condition for $\tilde{S}_N = S$ or $\hat{S}_N = S$ is the *bounded degree representation (BDR)* introduced by Lasserre in [19]. BDR is typically very difficult to check. More easily checkable conditions are strict convexity and strict quasi-convexity. We now discuss these cases.

Bounded Degree Representation Condition

A general approach for showing that a moment relaxation produces a semidefinite representation is given in the proof of Theorem 6.31. The key point is to prove a weighted sos representation with uniform degree bounds for *all* linear functionals nonnegative on S. If a linear functional $a^T x + b$ is positive on S, then Putinar's Positivstellensatz [37] says that

$$a^T x + b = \sigma_0 + \sigma_1 g_1 + \cdots + \sigma_m g_m, \tag{6.16}$$

where each σ_i is an sos polynomial. To make sure that (6.16) holds, we require that the presentation of S satisfies the *archimedean condition*: there exist sos polynomials $s_0, s_1, \ldots, s_m$ and a number $M > 0$ such that

$$M - \|x\|_2^2 = s_0 + s_1 g_1 + \cdots + s_m g_m.$$

The archimedean condition implies that S is compact, but the reverse is not necessarily true. However, the presentation of any compact set S can be strengthened to satisfy the archimedean condition by adding a "redundant" ball constraint $M - \|x\|_2^2 \geq 0$ for a sufficiently large M. Generally, the degrees of the polynomials σ_i in (6.16) go to infinity as the minimum value of $a^T x + b$ on S tends to zero.

Moment relaxations present the dual side of sum of squares relaxations. If we have $\tilde{S}_N = S$ for some N then any linear functional positive on S is also positive on $\tilde{S}_N$. Linear functionals positive on $\tilde{S}_N$ have a weighted sum of squares representation with bounded degree $2N$:

$$a^T x + b > 0 \text{ on } S \quad \Rightarrow \quad a^T x + b = \sigma_0 + \sigma_1 g_1 + \cdots + \sigma_m g_m$$

with σ_i sos and $\deg(\sigma_i g_i) \leq 2N$ for all i. If almost all positive linear functionals on S have such a representation, then we say that the presentation of S admits a *Putinar–Prestel bounded degree representation (PP-BDR)* of order N (cf. [19]).

For the Schmüdgen moment relaxation $\hat{S}_N$ in (6.14), to guarantee that $\hat{S}_N = S$ for some order N, we need a *Schmüdgen bounded degree representation (S-BDR)* of order N (cf. [19]): for *almost every* pair $(a, b) \in \mathbb{R}^n \times \mathbb{R}$

$$a^T x + b > 0 \text{ on } S \quad \Rightarrow \quad a^T x + b = \sum_{\nu \in \{0,1\}^m} \sigma_\nu g_\nu,$$

with every σ_ν being sos and $\deg(\sigma_\nu g_\nu) \leq 2N$.

Theorem 6.36 ([19]). *Suppose S is defined by* (6.10) *and is compact.*

(a) *If PP-BDR of order N holds for S, then conv(S) equals $\tilde{S}_N$.*

(b) *If S-BDR of order N holds for S, then conv(S) equals $\hat{S}_N$.*

Proof. We sketch the proof given by Lasserre in [19].

(a) We have already seen in the construction of (6.13) that $S \subseteq \tilde{S}_N$, regardless of whether S is convex. Therefore, $\text{conv}(S) \subseteq \tilde{S}_N$. Now we prove the reverse containment by contradiction. Suppose there exists a $\hat{y}$ satisfying the linear matrix inequalities in (6.13) and $\hat{x} = (\hat{y}_{e_1}, \ldots, \hat{y}_{e_n}) \notin \text{conv}(S)$. Since conv$(S)$ is closed, there exists a hyperplane strictly separating $\hat{x}$ from conv(S): there exist a nonzero $a \in \mathbb{R}^n$ and $b \in \mathbb{R}$ such that

$$a^T x + b > 0 \text{ for all } x \in S, \qquad a^T \hat{x} + b < 0.$$

Since conv(S) is compact, we can choose the above (a, b) generically. Since PP-BDR of order N holds for the presentation of S, there exist sos polynomials $\sigma_0, \ldots, \sigma_m$ such that (6.16) is true and $\deg(\sigma_i g_i) \leq 2N$. For each i, we can find a symmetric $W_i \succeq 0$ such that $\sigma_i(x) = [x]_{N-d_i}^T W_i [x]_{N-d_i}$ with $d_i = \lceil \deg(g_i)/2 \rceil$. Replacing each monomial x^α by $\hat{y}_\alpha$, we get

$$a^T \hat{x} + b = Tr\left(L_{g_0}^{(N)}(\hat{y}) W_0\right) + \cdots + Tr\left(L_{g_m}^{(N)}(\hat{y}) W_m\right) \geq 0,$$

which contradicts the previous assertion that $a^T \hat{x} + b < 0$. Therefore, conv$(S) = \tilde{S}_N$.

Part (b) is proved in almost exactly the same way. □

Generally, it is quite difficult to check PP-BDR for a given semialgebraic set. In the definition of PP-BDR, we require that *almost every* linear polynomial $a^T x + b$ *positive* on S admits a representation like (6.16) with uniform degree bounds on the sos polynomials σ_i. This is almost impossible to check in practice. Indeed, to check PP-BDR, one often needs to prove that the BDR holds for *every* linear polynomial $a^T x + b$ that is *nonnegative* on S. Interestingly, this stronger version of PP-BDR is satisfied under some general conditions (cf. [12]). The situation for S-BDR is similar.

Theorem 6.37 ([12]). *Suppose* $S = \{x \in \mathbb{R}^n : g_1(x) \geq 0, \ldots, g_m(x) \geq 0\}$ *is compact and convex and has nonempty interior. Assume each* $g_i(x)$ *is concave on* S *and let* $S_i := \{x : g_i(x) \geq 0\}$.

(i) *Suppose for each* i, *either* $-\nabla^2 g_i(x)$ *is sos or* $-\nabla^2 g_i(u) \succ 0$ *for all* $u \in \partial S_i \cap \partial S$. *Then there exists* $N > 0$ *such that if* $a^T x + b$ *is nonnegative on* S, *then*

$$a^T x + b \quad = \sum_{\nu \in \{0,1\}^m} \sigma_\nu g_\nu$$

for some sos polynomials σ_ν *satisfying* $\deg(\sigma_\nu g_\nu) \leq 2N$. *Thus, the S-BDR of order* N *holds for* S.

(ii) *If, in addition,* S *satisfies the archimedean condition, then the PP-BDR of some order* N' *holds for* S.

A detailed proof of Theorem 6.37 is given in [12]. We sketch the basic ideas behind the proof. Suppose $a^T x + b$ is nonnegative on S and $a^T u + b = 0$ for some point $u \in S$. Under some general assumptions, the KKT conditions hold at u, i.e., there exist Lagrange multipliers $\lambda_1 \geq 0, \ldots, \lambda_m \geq 0$ such that

$$a = \sum_{i=1}^{m} \lambda_i \nabla g_i(u), \quad \lambda_1 g_1(u) = \cdots = \lambda_m g_m(u) = 0.$$

Let $\mathcal{L}(x)$ be the Lagrangian defined in (6.15). Note that $\mathcal{L}(u) = 0$ and $\nabla \mathcal{L}(u) = 0$. By Taylor expansion

$$\mathcal{L}(x) = (x-u)^T \left(\sum_{i=1}^{m} \lambda_i \underbrace{\int_0^1 \int_0^t \nabla^2 g_i(u + s(x-u)) ds\, dt}_{H_i(x,u)} \right) (x-u).$$

For any fixed u, $H_i(x,u)$ is a matrix polynomial in x. If each $H_i(x,u)$ is an sos matrix in x, then $\mathcal{L}(x)$ must be sos since each $\lambda_i \geq 0$. This further implies that $a^T x + b$ has the desired Putinar- or Schmüdgen-type representation. Conditions such as sos-convexity or strict convexity ensure that each matrix polynomial $H_i(x,u)$ admits a Putinar- or Schmüdgen-type weighted sos representation with uniform degree bounds that are independent of u, and hence the PP-BDR or S-BDR holds.

We close by noting that if the set S is convex, then Theorem 6.37 gives concrete conditions under which $\tilde{S}_N$ and $\hat{S}_N$ give semidefinite representations of S.

Convex sets with positively curved boundaries

When a semialgebraic set S is convex its defining polynomials are not necessarily concave. For instance, the hyperboloid $\{x \in \mathbb{R}^2_+ : x_1x_2 - 1 \geq 0\}$ is convex, while its defining polynomial is neither concave nor convex. However, because of convexity, the boundary of S must have nonnegative curvature at smooth points (see Theorem 6.17). Therefore, the defining polynomials are quasi-concave at smooth points. This observation leads to weaker conditions, such as strict quasi-concavity of the defining polynomials.

Theorem 6.38 ([15]). *Assume that the set S defined in (6.10) is compact and convex and has nonempty interior. If each $g_i(x)$ is either sos-concave or strictly quasi-concave on S, then $\hat{S}_N$ equals S for N sufficiently large. If, in addition, the archimedean condition holds, then $\tilde{S}_N$ equals S for N sufficiently large.*

The proof of Theorem 6.38 is based on Theorem 6.37. The basic idea is that we are able to find a different set of strictly concave defining polynomials for S by using strict quasi-concavity. When $g_i(x)$ is strictly quasi-concave on S, we can find a polynomial $h_i(x)$ positive on S such that $p_i(x) = g_i(x)h_i(x)$ is strictly concave on S. We refer to [15] for the details of the proof but provide an example below. Consider the set

$$S = \left\{ x \in \mathbb{R}^2 \; : \; g_1(x) := x_1x_2 - 1 \geq 0, \quad g_2(x) := \frac{1}{9} - (x_1-1)^2 - (x_2-1)^2 \geq 0 \right\}.$$

The set S is compact and convex. The polynomial g_1 is strictly quasi-concave, but not concave. However, the set S can also be equivalently described as

$$S = \left\{ x \in \mathbb{R}^2 \;\middle|\; \begin{array}{c} p_1(x) := (x_1x_2 - 1)(3 - x_1x_2) \geq 0 \\ g_2(x) := \frac{1}{9} - (x_1-1)^2 - (x_2-1)^2 \geq 0 \end{array} \right\},$$

where $p_1(x)$ is strictly concave on S.

For a convex basic closed semialgebraic set S, the Putinar moment relaxation produces a semidefinite representation of S *only if* all faces of S are exposed (cf. [25]). There are further different conditions under which $\tilde{S}_N$ or $\hat{S}_N$ gives semidefinite representations of S (cf. [20, 28]).

6.4.5 Generalizations

In many applications convex sets are naturally defined by rational function inequalities or polynomial matrix inequalities. In these cases semidefinite representations can also be constructed by using moments. We show some examples without going into the details. Further results on these topics can be found in [28, 29].

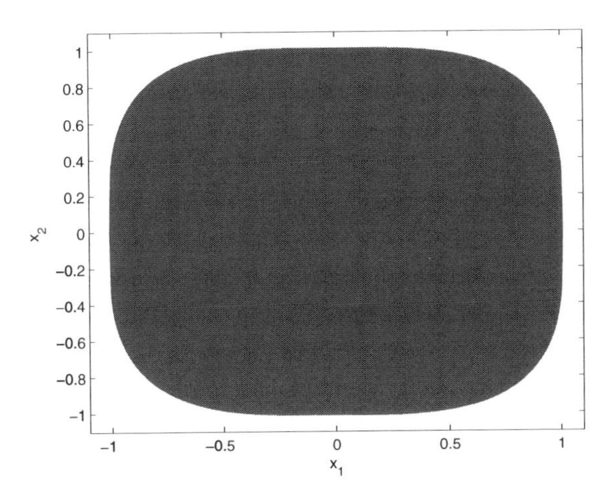

Figure 6.6. *The convex set defined by* $x_1^2 + x_2^2 \geq x_1^4 + x_1^2 x_2^2 + x_2^4$.

First, we consider the case of a convex set defined by a rational function inequality, $f(x) \geq 0$. Of course, one could describe this set by polynomial inequalities. However, by doing so, one might lose some nice properties of f (e.g., concavity of the defining inequalities could be lost). As we have seen in the preceding subsections, convexity plays a crucial role in constructing valid semidefinite representations. When f is rational, to construct a semidefinite relaxation, we need moments with fractional weights. The general constructions are described in [28]. Here is an example.

Example 6.39 ([28]). Consider the two-dimensional set S defined by the polynomial inequality $x_1^2 + x_2^2 - x_1^4 - x_1^2 x_2^2 - x_2^4 \geq 0$. Its defining polynomial is not concave in $\mathbb{R}^2$ (it is actually convex near the origin). This set is drawn in Figure 6.6. Clearly, S can also be described by the rational inequality

$$f(x) = 1 - \frac{x_1^4 + x_1^2 x_2^2 + x_2^4}{x_1^2 + x_2^2} \geq 0.$$

Interestingly, the rational function $f(x)$ is concave everywhere. It satisfies a so-called *first order sos-concavity* condition, and the set S admits the following semidefinite representation (cf. [28]):

$$\left\{ x \in \mathbb{R}^2 \;\middle|\; \begin{array}{c} \exists\, y = (y_{ij}),\, z = (z_{ij}), \text{ s.t.} \\ 1 \geq y_{20} + z_{04} \\ L_1(x, y, z) + L_2(x, y, z) \succeq 0 \end{array} \right\},$$

where L_1, L_2 are linear pencils defined as

$$L_1(x,y,z) = \begin{bmatrix} 0 & 0 & 0 & 1 & 0 & 0 \\ 0 & 1 & 0 & x_1 & x_2 & 0 \\ 0 & 0 & 0 & x_2 & 0 & 0 \\ 1 & x_1 & x_2 & y_{20} - y_{02} & y_{11} & y_{02} \\ 0 & x_2 & 0 & y_{11} & y_{02} & 0 \\ 0 & 0 & 0 & y_{02} & 0 & 0 \end{bmatrix},$$

$$L_2(x,y,z) = \begin{bmatrix} z_{00} & z_{10} & z_{01} & -z_{02} & z_{11} & z_{02} \\ z_{10} & -z_{02} & z_{11} & -z_{12} & -z_{03} & z_{12} \\ z_{01} & z_{11} & z_{02} & -z_{03} & z_{12} & z_{03} \\ -z_{02} & -z_{12} & -z_{03} & z_{04} & -z_{13} & -z_{04} \\ z_{11} & -z_{03} & z_{12} & -z_{13} & -z_{04} & z_{13} \\ z_{02} & z_{12} & z_{03} & -z_{04} & z_{13} & z_{04} \end{bmatrix}.$$

The lifting variables y_{ij} correspond to regular moments, while z_{ij} correspond to moments with the weight $(x_1^2 + x_2^2)^{-1}$, i.e., the integrals of type

$$\int \frac{x_i x_j}{x_1^2 + x_2^2} d\mu(x)$$

with respect to some positive measure μ on $\mathbb{R}^n$. The details of constructing L_1, L_2 are described in [28]. ∎

Now we consider the case of a convex set defined by a *polynomial matrix inequality*. A semidefinite relaxation as in (6.12) can be constructed by using moments. Under a *matrix sos-convexity* condition, this construction gives a semidefinite representation of the convex set (cf. [29]).

Example 6.40 ([29]). Consider the set S defined by the polynomial matrix inequality:

$$\begin{bmatrix} 2 - x_1^2 - 2x_3^2 & 1 + x_1x_2 & x_1x_3 \\ 1 + x_1x_2 & 2 - x_2^2 - 2x_1^2 & 1 + x_2x_3 \\ x_1x_3 & 1 + x_2x_3 & 2 - x_3^2 - 2x_2^2 \end{bmatrix} \succeq 0.$$

The above quadratic matrix polynomial is matrix sos-concave (cf. [29]). A picture of this set is drawn in Figure 6.7. As in (6.12), a basic moment semidefinite relaxation of S is

$$\left\{ x \in \mathbb{R}^3 \;\middle|\; \begin{array}{c} \begin{bmatrix} 2 - y_{200} - 2y_{002} & 1 + y_{110} & y_{101} \\ 1 + y_{110} & 2 - y_{020} - 2y_{200} & 1 + y_{011} \\ y_{101} & 1 + y_{011} & 2 - y_{002} - 2y_{020} \end{bmatrix} \succeq 0 \\ \exists y_{ijk} \ s.t. \\ \begin{bmatrix} 1 & x_1 & x_2 & x_3 \\ x_1 & y_{200} & y_{110} & y_{101} \\ x_2 & y_{110} & y_{020} & y_{011} \\ x_3 & y_{101} & y_{011} & y_{002} \end{bmatrix} \succeq 0 \end{array} \right\}.$$

Indeed, the above is a semidefinite representation of S, as shown in [29]. Therefore, S is a projected spectrahedron. ∎

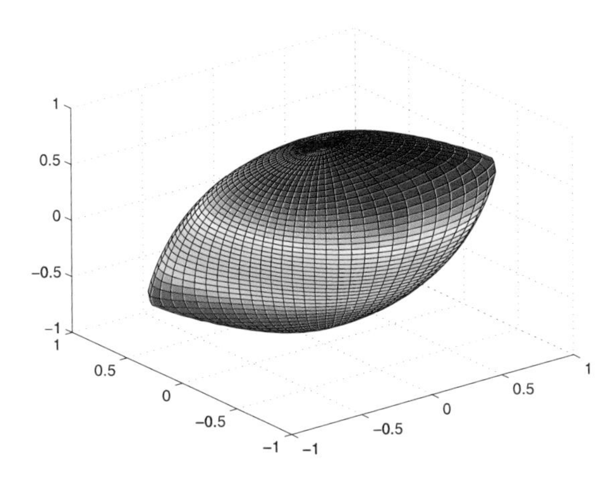

Figure 6.7. *The convex set in Example* 6.40.

6.4.6 Convex Hulls of Unions

Suppose we can divide a convex set S into several parts and find a semidefinite representation for each piece. Then a natural question is whether these representations can be glued together to provide a semidefinite representation of S. This brings us to the main question of this section: Is the convex hull of a union of projected spectrahedra a projected spectrahedron? If so, how can we construct a semidefinite representation of it? Interestingly, there exist positive answers to these questions.

A simple implementation of the above idea is to cover the compact set by finitely many balls. If the intersection of each ball with the convex set is a projected spectrahedron, then we can glue them together to get a uniform semidefinite representation for the whole convex set. This approach is called *localization*. The necessary tool is building a single semidefinite representation for the convex hull of several projected spectrahedra. Since balls (ellipsoids) are spectrahedra, the question of semidefinite representability of a convex set reduces to the representability of the intersections of balls with the boundary of the set. Thus we can focus on local properties of the boundary.

Let $W_1, \ldots, W_m \subset \mathbb{R}^n$ be convex sets. Their *Minkowski sum* is the convex set defined as

$$W_1 + \cdots + W_m := \left\{ \sum_{k=1}^{m} x_k \,\middle|\, x_k \in W_k,\ k = 1, \ldots, m \right\}.$$

If each W_k is a projected spectrahedron described by

$$L_k\left(x, y^{(k)}\right) := A^{(k)} + \sum_{i=1}^{n} x_i B_i^{(k)} + \sum_{j=1}^{N_k} y_j^{(k)} C_j^{(k)} \succeq 0, \qquad (6.17)$$

then a semidefinite representation for the Minkowski sum $W_1 + \cdots + W_m$ is

$$\left\{ \sum_{k=1}^{m} x^{(k)} \,\middle|\, L_k\left(x^{(k)}, y^{(k)}\right) \succeq 0 \text{ for pairs } \left(x^{(1)}, y^{(1)}\right), \ldots, \left(x^{(m)}, y^{(m)}\right) \right\}.$$

Lemma 6.41. *If $W_1, \ldots, W_m$ are nonempty convex sets, then*

$$\text{conv}\left(\bigcup_{k=1}^{m} W_k\right) = \bigcup_{\lambda \in \Delta_m} \left(\lambda_1 W_1 + \cdots + \lambda_m W_m\right), \tag{6.18}$$

where $\Delta_m = \{\lambda \in \mathbb{R}_+^m : \lambda_1 + \cdots + \lambda_m = 1\}$ is the standard simplex.

Proof. The proof follows readily from the definitions of convex hull and Minkowski sum. See, for instance, [13]. □

Using Lemma 6.41, we can get a single semidefinite representation for the convex hull $\text{conv}(\cup_{k=1}^m W_k)$ from those of the individual W_k.

Theorem 6.42 ([13]). *Let $W_1, \ldots, W_m$ be nonempty projected spectrahedra defined in (6.17), and $W := \text{conv}(\cup_{k=1}^m W_k)$ be the convex hull of their union. Define*

$$\mathcal{C} := \left\{ \sum_{k=1}^{m} x^{(k)} \,\middle|\, \begin{array}{c} \exists\, \lambda_k, u^{(k)}\, (k = 1, \ldots, m) \\ \lambda_1, \ldots, \lambda_m \geq 0,\ \lambda_1 + \cdots + \lambda_m = 1, \\ \lambda_k A^{(k)} + \sum_{i=1}^n x_i^{(k)} B_i^{(k)} + \sum_{j=1}^{N_k} u_j^{(k)} C_j^{(k)} \succeq 0 \end{array} \right\}. \tag{6.19}$$

Then $W \subseteq \mathcal{C}$ and $\overline{\mathcal{C}} = \overline{W}$. If, in addition, every W_k is bounded, then $\mathcal{C} = W$.

Proof. This is left as an exercise. □

Example 6.43. Let W_1, W_2 be the spectrahedra defined by

$$\begin{bmatrix} 2 + x_1 & x_2 + 1 \\ x_2 + 1 & -x_1 \end{bmatrix} \succeq 0, \quad \begin{bmatrix} x_1 & x_2 - 1 \\ x_2 - 1 & 2 - x_1 \end{bmatrix} \succeq 0.$$

They are unit balls centered at $\pm(1, 1)$. The convex hull of their union has the semidefinite representation

$$\left\{ x = x^{(1)} + x^{(2)} \,\middle|\, \begin{array}{c} \begin{bmatrix} 2\lambda_1 + x_1^{(1)} & x_2^{(1)} + \lambda_1 \\ x_2^{(1)} + \lambda_1 & -x_1^{(1)} \end{bmatrix} \succeq 0 \\ \begin{bmatrix} x_1^{(2)} & x_2^{(2)} - \lambda_2 \\ x_2^{(2)} - \lambda_2 & 2\lambda_2 - x_1^{(2)} \end{bmatrix} \succeq 0 \\ \lambda_1 + \lambda_2 = 1, \lambda_1, \lambda_2 \geq 0 \end{array} \right\}.$$

Setting $x^{(2)} = (u_1, u_2)$, we get a projected spectrahedron with three lifting variables:

$$\left\{ x \in \mathbb{R}^2 \;\middle|\; \begin{array}{c} \begin{bmatrix} 2\lambda_1 + x_1 - u_1 & x_2 - u_2 + \lambda_1 \\ x_2 - u_2 + \lambda_1 & u_1 - x_1 \end{bmatrix} \succeq 0 \\[2ex] \begin{bmatrix} u_1 & u_2 + \lambda_1 - 1 \\ u_2 + \lambda_1 - 1 & 2 - 2\lambda_1 - u_1 \end{bmatrix} \succeq 0 \\[2ex] \lambda_1 \geq 0, \quad 1 - \lambda_1 \geq 0 \end{array} \right\}. \quad \blacksquare$$

When some W_k are unbounded, $\mathcal{C}$ and the convex hull W may not be equal, but they have the same closure and interior. Note that both $\mathcal{C}$ and W are not necessarily closed even when all W_i are.

Example 6.44 ([13]). (i) Consider the following spectrahedra:

$$W_1 = \left\{ x \in \mathbb{R}^2 : \begin{bmatrix} x_1 & 1 \\ 1 & x_2 \end{bmatrix} \succeq 0 \right\}, \quad W_2 = \{0\}.$$

Their convex hull is

$$\mathrm{conv}(W_1 \cup W_2) = \{x \in \mathbb{R}^2_+ : x_1 = x_2 = 0 \text{ or } x_1 x_2 > 0\}.$$

However, the set $\mathcal{C}$ in (6.19) is

$$\left\{ x \in \mathbb{R}^2 : \exists \;\; 0 \leq \lambda_1 \leq 1, \begin{bmatrix} x_1 & \lambda_1 \\ \lambda_1 & x_2 \end{bmatrix} \succeq 0 \right\} = \mathbb{R}^2_+.$$

So, $\mathcal{C} \neq \mathrm{conv}(W_1 \cup W_2)$, but they have same interior. Both W_1 and W_2 are closed while $\mathrm{conv}(W_1 \cup W_2)$ is not.

(ii) Consider the projected spectrahedra

$$W_1 = \left\{ x \in \mathbb{R}^2 : \exists u \geq 0, \begin{bmatrix} x_1 & 1 + x_2 \\ 1 + x_2 & 1 + u \end{bmatrix} \succeq 0 \right\}$$

and $W_2 = \{0\}$. We have

$$\mathrm{conv}(W_1 \cup W_2) = \{x \in \mathbb{R}^2 : x_1 > 0, \text{ or } x_1 = 0 \text{ and } -1 \leq x_2 \leq 0\},$$

$$\overline{\mathrm{conv}(W_1 \cup W_2)} = \{x \in \mathbb{R}^2 : x_1 \geq 0\},$$

and hence, $\mathrm{conv}(W_1 \cup W_2)$ is not closed. However, one can verify that the projected spectrahedron $\mathcal{C}$ is equal to $\mathrm{conv}(W_1 \cup W_2)$. $\blacksquare$

As seen above, the equality $\mathcal{C} = W$ is possible even if $W_1, \ldots, W_m$ are not all bounded. In particular, we always have $\mathcal{C} = W$ if every W_k is homogeneous (i.e., $A^{(k)} = 0$ in the semidefinite representation of W_k). This fact is implied by Lemma 6.41.

Example 6.45 ([13]). Consider the two spectrahedra in $\mathbb{R}^2$ defined by

$$\begin{bmatrix} -x_1 & 1 \\ 1 & x_2 \end{bmatrix} \succeq 0, \quad \begin{bmatrix} x_1 & 1 \\ 1 & x_2 \end{bmatrix} \succeq 0.$$

The convex hull of their union is the open half space $\{x : x_2 > 0\}$, which is precisely equal to the projected spectrahedron $\mathcal{C}$. The description of $\mathcal{C}$ can be simplified to

$$\left\{ x \in \mathbb{R}^2 \;\middle|\; \begin{array}{c} \begin{bmatrix} u_1 - x_1 & \lambda_1 \\ \lambda_1 & x_2 - u_2 \end{bmatrix} \succeq 0, \begin{bmatrix} u_1 & 1-\lambda_1 \\ 1-\lambda_1 & u_2 \end{bmatrix} \succeq 0 \\ u \in \mathbb{R}^2,\ 0 \le \lambda_1 \le 1 \end{array} \right\}.$$

This is a semidefinite representation with three lifting variables. ∎

Putting all of the above together we obtain the following result.

Theorem 6.46 ([13]). *Let $S \subseteq \mathbb{R}^n$ be a compact convex set. Then S is a projected spectrahedron if and only if for every $u \in \partial S$ there exists $\delta > 0$ such that the intersection $S \cap \overline{B(u,\delta)}$ is a projected spectrahedron.*

Proof. The "only if" part is trivial, because the closed ball

$$\overline{B(u,\delta)} = \{x : \|x-u\|_2 \le \delta\}$$

is a spectrahedron. For the "if" part, suppose for every $u \in \partial S$ and some $\delta_u > 0$, the set $S \cap \overline{B(u,\delta_u)}$ is a projected spectrahedron. Note that $\{B(u,\delta_u) : u \in \partial S\}$ is an open cover for the compact set ∂S. So there are a finite number of balls, say, $B(u_1,\delta_1),\dots,B(u_L,\delta_L)$, to cover ∂S. Note that

$$S = \operatorname{conv}(\partial S) = \operatorname{conv}\left(\bigcup_{k=1}^{L} (\partial S \cap \overline{B(u_k,\delta_k)})\right) \subseteq \operatorname{conv}\left(\bigcup_{k=1}^{L} (S \cap \overline{B(u_k,\delta_k)})\right) \subseteq S.$$

The sets $S \cap \overline{B(u_k,\delta_k)}$ are all bounded. By Theorem 6.42, we see that

$$S = \operatorname{conv}\left(\bigcup_{k=1}^{L} S \cap \overline{B(u_k,\delta_k)}\right)$$

is a projected spectrahedron. □

6.4.7 Sufficient Conditions for Semidefinite Representability

We now have all the tools to present a sufficient condition for a compact convex semialgebraic set $S \subset \mathbb{R}^n$ to be a projected spectrahedron. The condition essentially requires that the boundary of S has positive curvature.

Theorem 6.47 ([13]). *Suppose $S \subset \mathbb{R}^n$ is a compact convex set defined by*

$$S \quad = \quad \bigcup_{k=1}^{m} T_k := \{x \in \mathbb{R}^n : g_1^k(x) \ge 0, \dots, g_{m_k}^k(x) \ge 0\},$$

where g_i^k are polynomials. If for every $u \in \partial S$ and every g_i^k satisfying $g_i^k(u) = 0$, T_k has interior near u (i.e., for any $\delta > 0$, the ball $B(u, \delta)$ intersects the interior of T_k) and $g_i^k(x)$ is strictly quasi-concave at u, then S is a projected spectrahedron.

Theorem 6.47 is proved by applying Theorem 6.46. It is enough to show that for every $u \in \partial S$, there exists a ball $B(u, \delta)$ so that $S \cap B(u, \delta)$ is a projected spectrahedron. Note that $S \cap B(u, \delta)$ is a finite union of intersections $T_k \cap B(u, \delta)$. By Theorem 6.38, every $T_k \cap B(u, \delta)$ is a projected spectrahedron, under the assumption of strict quasi-concavity of the defining polynomials. A complete proof of this result can be found in Theorem 4.5 of Helton and Nie [13].

In Theorem 6.47, if the set S is not convex, but the other conditions are satisfied, then we can conclude that the convex hull of S is a projected spectrahedron.

Here we give some remarks on the conditions in Theorems 6.31, 6.37, and 6.47. Theorem 6.31 assumes that all g_i are sos-concave, which is the strongest assumption, but its conclusion is also the strongest: (6.12) is an explicit representation of S as a projected spectrahedron. Theorem 6.37 assumes that g_i are either sos-concave or strictly quasi-concave, which is weaker than Theorem 6.31, and its conclusion is also weaker: $\hat{S}_N$ or $\tilde{S}_N$ provides a representation of S as a projected spectrahedron for some large enough N. Theorem 6.47 assumes the weakest condition, but its conclusion is also the weakest: there *exists* a semidefinite representation of S (an explicit description is typically quite complicated).

By comparing Theorems 6.17 and 6.47, we can see that the presented necessary and sufficient conditions for semidefinite representability are not too far away from each other. The difference between them is nonnegative versus positive curvature and singularity versus nonsingularity.

6.4.8 Exercises

Exercise 6.48. Prove Lemma 6.32.

Exercise 6.49. Show that a polynomial $f(x)$ is sos-convex if and only if the following polynomial in (x, y) is sos:

$$f(x) - f(y) - \nabla f(y)^T (x - y).$$

Exercise 6.50. Show that a univariate polynomial $f(x)$ is sos-convex if and only if it is convex everywhere. Show that this is also true if $f(x)$ is a bivariate quartic polynomial.

Exercise 6.51. Let $f(x)$ be a cubic polynomial that is concave over $\mathbb{R}_+^n$ and

$$S = \{x \in \mathbb{R}^n : x_1 \geq 0, \ldots, x_n \geq 0, f(x) \geq 0\}.$$

Show that the equality $\tilde{S}_N = S$ holds for some order N, where $\tilde{S}_N$ is given by (6.13). What is the smallest value of such an N?

Exercise 6.52. Consider the following convex set

$$S = \{x \in \mathbb{R}^2 : x_1 \geq 0, x_2 \geq 0, x_1^5 + x_2^5 \leq 1\}.$$

Find a semidefinite representation for S with the smallest number of lifting variables.

Exercise 6.53. Consider the following basic closed semialgebraic set:

$$S = \{x \in \mathbb{R}^2 : x_1^2 - x_2^2 - (x_1^2 + x_2^2)^2 \geq 0, x_1 \geq 0\}.$$

Does there exist an order $N > 0$ such that $\tilde{S}_N = S$, where $\tilde{S}_N$ is given by (6.13)? If so, what is the smallest N making the equality occur? If no, give reasons. How about $\hat{S}_N$ given by (6.14)?

Exercise 6.54. For each of the following cases, find a semidefinite representation with the *smallest* number of lifting variables for the convex hull of the union of the given convex sets:

(a) Two balls $B(-\mathbf{1}, 1), B(\mathbf{1}, 1/2)$ in $\mathbb{R}^n$ ($\mathbf{1}$ is the vector of all ones).

(b) Three pairwise touching balls in $\mathbb{R}^2$:

$$(x_1 + 1)^2 + x_2^2 \leq 1, \quad (x_1 - 1)^2 + x_2^2 \leq 1, \quad x_1^2 + (x_2 - 2)^2 \leq (\sqrt{5} - 1)^2.$$

(c) Two elliptopes in $\mathbb{R}^3$:

$$\begin{bmatrix} 1 & x_1 & x_2 \\ x_1 & 1 & x_3 \\ x_2 & x_3 & 1 \end{bmatrix} \succeq 0, \quad \begin{bmatrix} 1 & x_1 - 1 & x_2 - 1 \\ x_1 - 1 & 1 & x_3 - 1 \\ x_2 - 1 & x_3 - 1 & 1 \end{bmatrix} \succeq 0.$$

(d) The semidefinite cone and nonnegative orthant embed in $\mathbb{R}^3$:

$$\begin{bmatrix} x_1 & x_2 \\ x_2 & x_3 \end{bmatrix} \succeq 0, \quad \begin{bmatrix} x_1 \\ x_2 \\ x_3 \end{bmatrix} \geq 0.$$

Exercise 6.55. Let P be the set of univariate quadratic polynomials that are either nonnegative on $[-1, 0]$ or nonnegative on $[0, 1]$. Find a semidefinite representation for the convex hull of P with the smallest number of lifting variables.

Exercise 6.56. Prove Theorem 6.42. (Hint: use Lemma 6.41.)

Exercise 6.57. Let T be a compact nonconvex set in $\mathbb{R}^n$. Its *convex boundary* is defined as $\partial_c T := \partial T \cap \partial \text{conv}(T)$. Show that $\text{conv}(\partial_c T) = \text{conv}(T)$. Is this also true if T is not compact?

Bibliography

[1] A. A. Ahmadi and P. A. Parrilo. A convex polynomial that is not sos-convex. *Math. Program.*, 135:275–292, 2012.

[2] F. Alizadeh and D. Goldfarb. Second-order cone programming. *Math. Program.*, 95:3–51, 2003.

[3] A. Ben-Tal and A. Nemirovski. *Lectures on Modern Convex Optimization: Analysis, Algorithms, and Engineering Applications*, MPS/SIAM Ser. Optim. SIAM, Philadelphia, 2001.

[4] G. Blekherman. Convex forms that are not sums of squares. Preprint, 2009. http://arxiv.org/abs/0910.0656.

[5] D. P. Bertsekas. *Convex Optimization Theory.* Athena Scientific, Belmont, MA, 2009.

[6] J. Bochnak, M. Coste, and M.-F. Roy. *Real Algebraic Geometry*, Springer, Berlin, 1998.

[7] S. Boyd, L. El Ghaoui, E. Feron, and V. Balakrishnan. *Linear Matrix Inequalities in System and Control Theory*, SIAM Stud. Appl. Math. 15, SIAM, Philadelphia, 1994.

[8] S. Boyd and L. Vandenberghe. *Convex Optimization.* Cambridge University Press, Cambridge, UK, 2004.

[9] P. Brändén. Obstructions to determinantal representability. *Adv. Math.*, 226:1202–1212, 2011.

[10] J. B. Conway. *A Course in Functional Analysis.* Grad. Texts in Math. Springer, Berlin, 1985.

[11] D. Cox, J. Little, and D. O'Shea. *Ideals, Varieties, and Algorithms. An Introduction to Computational Algebraic Geometry and Commutative Algebra*, 3rd edition, Undergrad. Texts in Math. Springer, New York, 2007.

[12] J. W. Helton and J. Nie. Semidefinite representation of convex sets. *Math. Program.*, 122:21–64, 2010.

[13] J. W. Helton and J. Nie. Sufficient and necessary conditions for semidefinite representability of convex hulls and sets. *SIAM J. Optim.*, 20:759–791, 2009.

[14] J. W. Helton and J. Nie. Structured semidefinite representation of some convex sets. *Proceedings of 47th IEEE Conference on Decision and Control (CDC)*, Cancun, Mexico, Dec. 9–11, 2008, pp. 4797–4800.

[15] J. W. Helton and J. Nie. Semidefinite representation of convex sets and convex hulls. In M. Anjos and J. Lasserre, editors, *Handbook on Semidefinite, Cone and Polynomial Optimization: Theory, Algorithms, Software and Applications*, to appear.

[16] J. W. Helton, S. McCullough, and V. Vinnikov. Noncommutative convexity arises from linear matrix inequalities. *J. Funct. Anal.*, 240:105–191, 2006.

[17] J. W. Helton and V. Vinnikov. Linear matrix inequality representation of sets. *Comm. Pure Appl. Math.*, 60:654–674, 2007.

[18] J. Lasserre. Global optimization with polynomials and the problem of moments. *SIAM J. Optim.*, 11:796–817, 2001.

[19] J. Lasserre. Convex sets with semidefinite representation. *Math. Program.*, 120:457–477, 2009.

[20] J. Lasserre. Convexity in semialgebraic geometry and polynomial optimization. *SIAM J. Optim.*, 19:1995–2014, 2009.

[21] P. Lax. Differential equations, difference equations and matrix theory. *Comm. Pure Appl. Math.*, 6:175–194, 1958.

[22] M. Marshall. Representation of non-negative polynomials, degree bounds and applications to optimization. *Canad. J. Math.*, 61:205–221, 2009.

[23] Y. Nesterov and A. Nemirovski. *Interior-Point Polynomial Algorithms in Convex Programming*, SIAM Stud. Appl. Math. 13. SIAM, Philadelphia, 1994.

[24] A. Nemirovskii. Advances in convex optimization: conic programming. *Plenary Lecture, International Congress of Mathematicians (ICM)*, Madrid, Spain, 2006.

[25] T. Netzer, D. Plaumann, and M. Schweighofer. Exposed faces of semidefinitely representable sets. *SIAM J. Optim.*, 20:1944–1955, 2010.

[26] T. Netzer and A. Thom. Polynomials with and without determinantal representations. *Linear Algebra Appl.*, 437:1579–1595, 2012.

[27] J. Nie and M. Schweighofer. On the complexity of Putinar's Positivstellensatz. *J. Complexity*, 23:135–150, 2007.

[28] J. Nie. First order conditions for semidefinite representations of convex sets defined by rational or singular polynomials. *Math. Program. Ser. A*, 131:1–36, 2012.

[29] J. Nie. Polynomial matrix inequality and semidefinite representation. *Math. Oper. Res.*, 36:398–415, 2011.

[30] J. Nie and B. Sturmfels. Matrix cubes parametrized by eigenvalues. *SIAM J. Matrix Anal. Appl.*, 31:755–766, 2009.

[31] J. Nie, P. A. Parrilo, and B. Sturmfels. Semidefinite representation of the k-ellipse. In A. Dickenstein, F.-O. Schreyer, and A. Sommese, editors, *Algorithms in Algebraic Geometry*. Springer, New York, 2008, pp. 117–132.

[32] J. Nie and J. Demmel. Minimum ellipsoid bounds for solutions of polynomial systems via sum of squares. *J. Global Optim.*, 33:511–525, 2005.

[33] P. A. Parrilo. Exact semidefinite representation for genus zero curves. *Talk at the Banff Workshop "Positive Polynomials and Optimization"*, Banff, Canada, October 8–12, 2006.

[34] P. A. Parrilo and B. Sturmfels. Minimizing polynomial functions. In S. Basu and L. Gonzalez-Vega, editors, *Proceedings of the DIMACS Workshop on Algorithmic and Quantitative Aspects of Real Algebraic Geometry in Mathematics and Computer Science* (*March 2001*), American Mathematical Society, Providence, RI, 2003, pp. 83–100.

[35] D. Plaumann, B. Strumfels, and C. Vinzant. Computing linear matrix representations of Helton-Vinnikov curves. In H. Dym, M. de Oliveira, and M. Putinar, editors, *Mathematical Methods in Systems, Optimization, and Control*, Oper. Theory Adv. Appl., Birkhauser, Basel, 2011.

[36] S. Prajna, A. Papachristodoulou, P. Seiler, and P. Parrilo. *SOSTOOLS User's Guide.* Website: http://www.mit.edu/~parrilo/sosTOOLS/.

[37] M. Putinar. Positive polynomials on compact semi-algebraic sets, *Indiana Univ. Math. J.*, 42:969–984, 1993.

[38] K. Schmüdgen. The K-moment problem for compact semialgebraic sets. *Math. Ann.*, 289:203–206, 1991.

[39] M. Spivak. *A Comprehensive Introduction to Differential Geometry.* Vol. II, 2nd edition. Publish or Perish, Inc., Wilmington, DE, 1979.

[40] H. Wolkowicz, R. Saigal, and L. Vandenberghe, editors. *Handbook of Semidefinite Programming.* Kluwer, Amsterdam, 2000.

Chapter 7

Spectrahedral Approximations of Convex Hulls of Algebraic Sets

João Gouveia[†] *and Rekha R. Thomas*

This chapter describes a method for finding spectrahedral approximations of the convex hull of a real algebraic variety (the set of real solutions to a finite system of polynomial equations). The procedure creates a nested sequence of convex approximations of the convex hull of the variety. Computations can be done modulo the ideal generated by the polynomials which has several advantages. We examine conditions under which the sequence of approximations converges to the closure of the convex hull of the real variety, either asymptotically or in finitely many steps, with special attention to the case in which the very first approximation yields a semidefinite representation of the convex hull. These methods allow optimization, or approximation of the optimal value, of a linear function over a real algebraic variety via semidefinite programming.

7.1 Introduction

A central problem in optimization is to find the maximum (or minimum) value of a linear function over a set S in $\mathbb{R}^n$. For example, in a *linear program*

$$\text{maximize}\,\{\langle c, x\rangle : Ax \le b\}$$

with $c \in \mathbb{R}^n, A \in \mathbb{R}^{m\times n}$, and $b \in \mathbb{R}^m$, the set $S = \{x \in \mathbb{R}^n : Ax \le b\}$ is a polyhedron, while in a *semidefinite program*,

$$\text{maximize}\left\{\langle c, x\rangle \;:\; A_0 + \sum_{i=1}^{n} A_i x_i \succeq 0\right\}$$

[†]João Gouveia was partially supported by NSF grant DMS-0757371 and by Fundação para a Ciência e Technologia.

with $c \in \mathbb{R}^n$ and symmetric matrices $A_0, A_1, \ldots, A_n$, the feasible region is the set $S = \{x \in \mathbb{R}^n : A_0 + \sum_{i=1}^n A_i x_i \succeq 0\}$ which is a *spectrahedron*. In both cases, S is a convex *semialgebraic set* as it is convex and can be defined by a finite list of polynomial inequalities. A *real algebraic variety*, which is the set of all real solutions to a finite list of polynomial equations, is a special case of a semialgebraic set. Optimizing a linear function over any set $S \subset \mathbb{R}^n$, in particular, a real algebraic variety, is equivalent to optimizing the linear function over the closure of $\text{conv}(S)$, the convex hull of S. In this chapter we describe a method to construct semidefinite approximations of the closure of the convex hull of a real algebraic variety.

Representing the convex hull of a real algebraic variety is a multifaceted problem that arises in many contexts in both theory and practice. In Chapter 5 we saw a method using dual projective varieties for explicitly finding the polynomials that describe the boundary of the convex hull of a real variety. These bounding polynomials use the same variables as those describing the variety and can be highly complicated. Their computation boils down to eliminating variables from a larger polynomial system and can be challenging in practice, although they can be computed using existing computer algebra packages in examples with a small number of variables. If one is allowed to use more variables than those describing the variety, then there is more freedom in finding representations and approximations and the key idea then is to express the convex hull implicitly as the projection of a higher-dimensional object. This approach is more flexible than the former and has the potential to yield a representation of a complicated set as the projection of a simple set in higher dimensions. The method we will describe adopts this philosophy for finding approximations and representations of the convex hull of a real algebraic variety.

We present a procedure for finding a sequence of approximations of the convex hull of a real algebraic variety (sometimes just called an *algebraic set*) in the form of *projected spectrahedra*. While the convex hull of a real algebraic variety is a convex semialgebraic set, recall from Chapter 6 that it is not known which convex semialgebraic sets are projected spectrahedra. Regardless, we will develop an automatic method that finds semidefinite representations (as projected spectrahedra) for a sequence of outer approximations of $\text{conv}(S)$, when S is an algebraic set. In many cases, these approximations will converge to $\text{conv}(S)$. If our procedure yields an exact representation of $\text{conv}(S)$ as a projected spectrahedron, then as a by product we can optimize a linear function over S by solving a semidefinite program. In the nice cases where the representation uses spectrahedra of small size (relative to the size of S), semidefinite programming becomes an efficient method for optimizing a linear function over S. In fact, there are several families of algebraic sets where this spectrahedral approach yields polynomial time algorithms for linear optimization. Similarly, the spectrahedral approach can, in some cases, yield efficient algorithms for finding good approximations of the optimal value of a linear function over S.

While we will see many examples of real algebraic varieties (and their defining ideals) for which our method yields an exact representation of its convex hull in a few iterations of our procedure, many open questions remain. For instance, there is no complete understanding of when the method is guaranteed to converge to the convex hull of the variety in finitely many steps of the procedure. Even in the cases where finite convergence is guaranteed, good upper bounds on the number of iterations required by the procedure are lacking. The work presented in this chapter was inspired by a question posed by Lovász in [19] that asked for a characterization of ideals for which the first approximation in our hierarchy will yield a semidefinite representation of the convex hull of the variety of the ideal. In Section 7.3 we answer this question for finite varieties. The case of infinite varieties is far less understood. We identify conditions that prevent finite convergence of these approximations to the closure of the convex hull of the variety. However, again a full characterization is missing. Thus, the material in this chapter offers both advances in spectrahedral representations of algebraic sets as well as many avenues for further research.

This chapter is organized as follows. In Section 7.2 we explain the procedure for finding spectrahedral approximations of the convex hull of an algebraic set. These techniques were developed in [8], coauthored with Parrilo. One of the key theorems needed in this section (Theorem 7.6) was strengthened in this presentation with the help of Greg Blekherman. We illustrate the method with various examples and explain the underlying computations. In Section 7.3 we discuss the situations in which this method converges, either asymptotically or finitely, to an exact semidefinite representation of the convex hull of the variety. The most useful scenario is when the first approximation yields an exact semidefinite representation of the convex hull of the variety. We characterize all finite varieties for which this happens. We conclude in Section 7.4 with examples from combinatorial optimization where the underlying varieties are all finite. The methods we describe have algorithmic impact on certain classes of combinatorial optimization problems and the algebra becomes endowed with rich combinatorics in these cases.

7.2 The Method

Let $f_1, \ldots, f_m \in \mathbb{R}[x_1, \ldots, x_n] =: \mathbb{R}[x]$ be polynomials and

$$V_{\mathbb{R}}(f_1, \ldots, f_m) := \{x \in \mathbb{R}^n \,:\, f_1(x) = f_2(x) = \cdots = f_m(x) = 0\}$$

be their set of real zeros. We are interested in representing $\operatorname{conv}(V_{\mathbb{R}}(f_1, \ldots, f_m))$, the convex hull of $V_{\mathbb{R}}(f_1, \ldots, f_m)$ in $\mathbb{R}^n$ as projected spectrahedra.

Recall that the *ideal* generated by $f_1, \ldots, f_m$ in $\mathbb{R}[x]$ is the set

$$I = \langle f_1, \ldots, f_m \rangle = \left\{ \sum_{i=1}^{m} g_i f_i \,:\, g_i \in \mathbb{R}[x],\ m \in \mathbb{N} \right\} \subset \mathbb{R}[x].$$

The *real variety* of I is the set $V_{\mathbb{R}}(I) := \{x \in \mathbb{R}^n \,:\, h(x) = 0 \text{ for all } h \in I\}$ of real zeros of all polynomials in I. Note that if $s \in V_{\mathbb{R}}(f_1, \ldots, f_m)$, then $s \in V_{\mathbb{R}}(I)$ since $f_i(s) = 0$ implies that $h(s) = \sum_{i=1}^{m} g_i(s) f_i(s) = 0$ for all $h \in I$. Conversely, if $s \in V_{\mathbb{R}}(I)$, then for all $i = 1, \ldots, m$, $f_i(s) = 0$ since $f_i \in I$. Therefore, $V_{\mathbb{R}}(f_1, \ldots, f_m) = V_{\mathbb{R}}(I)$, and our goal can be viewed more generally as wanting to find semidefinite representations of the convex hull of the real variety of an ideal in $\mathbb{R}[x]$, or approximations of it.

For any set $S \subseteq \mathbb{R}^n$, the closure of $\operatorname{conv}(S)$ is exactly the intersection of all closed half spaces $\{x \in \mathbb{R}^n \,:\, l(x) \geq 0\}$ as l varies over all *linear* polynomials that are nonnegative on S. Throughout this chapter, linear polynomials include *affine* linear polynomials (those with a constant term). In particular, given an ideal I,

$$\mathrm{cl}(\mathrm{conv}(V_{\mathbb{R}}(I))) = \bigcap_{l \text{ linear},\ l|_{V_{\mathbb{R}}(I)} \geq 0} \{x : l(x) \geq 0\}.$$

It is not so clear how to work with this description. Even for a single linear polynomial l, checking whether $l(x)$ is nonnegative on $V_{\mathbb{R}}(I)$ is a difficult task. A natural idea is to relax the condition $l|_{V_{\mathbb{R}}(I)} \geq 0$ to something easier to check, at the risk of losing some of the $l(x)$ in the above intersection, and obtaining a superset of $\mathrm{cl}(\mathrm{conv}(V_{\mathbb{R}}(I)))$. As seen already in Chapters 3 and 4, the classical method to certify the nonnegativity of a polynomial on all of $\mathbb{R}^n$ is to write it as a *sum of squares* (sos) of other polynomials. In our case, we just need to certify that $l(x)$ is nonnegative on $V_{\mathbb{R}}(I)$, a subset of $\mathbb{R}^n$.

Let Σ denote the set of all sos polynomials in $\mathbb{R}[x]$, $\mathbb{R}[x]_k$ the set of all polynomials in $\mathbb{R}[x]$ of degree at most k, and Σ_{2k} the set of all sos polynomials $\sum h_j^2$, where $h_j \in \mathbb{R}[x]_k$. Nonnegativity of $l(x)$ on $V_{\mathbb{R}}(I)$ is guaranteed if

$$l(x) = \sigma(x) + \sum_{i=1}^{m} g_i(x) f_i(x) \tag{7.1}$$

for $\sigma(x) \in \Sigma$ and $g_i \in \mathbb{R}[x]$, since then for all $s \in V_{\mathbb{R}}(I)$, $l(s) = \sigma(s) \geq 0$. In Chapter 3 we saw that semidefinite programming can be used to check whether a polynomial is sos. In (7.1) we need to find both $\sigma(x)$ and the polynomials g_i to write $l(x)$ as *sos mod I*. Therefore, to check (7.1) in practice, we impose degree restrictions and proceed in one of two possible ways.

(i) In the first method, we ask that $\sigma \in \Sigma_{2k}$ and $g_i f_i \in \mathbb{R}[x]_{2k}$ for a fixed positive integer k and, if so, say that $l(x)$ is *k-sos mod* $\{f_1, \ldots, f_m\}$. This is the basic idea that underlies Lasserre's moment method for approximating the convex hull of a semialgebraic set described in Chapter 6.

(ii) In the second method, we ask only that $\sigma \in \Sigma_{2k}$ for a fixed positive integer k which reduces (7.1) to $l(x) = \sigma(x) + h(x)$ where $h(x) \in I$. If this is the case, we say that $l(x)$ is *k-sos mod I*. This method is more natural if one is interested in the geometry of $V_{\mathbb{R}}(I)$ and $\mathrm{conv}(V_{\mathbb{R}}(I))$ as it removes the dependence of the method on the choice of a particular generating set of I. The only issue is if the computation can be done in practice at the level of the ideal I and not the input $f_1, \ldots, f_m$.

Both methods yield a hierarchy of convex relaxations of $\mathrm{conv}(V_{\mathbb{R}}(I))$ obtained as the intersection of all half spaces $\{x : l(x) \geq 0\}$ as $l(x)$ ranges over the linear polynomials that are k-sos in the sense of the method. Since if $l(x)$ is k-sos mod $\{f_1, \ldots, f_m\}$ then it is also k-sos mod I, method (ii) yields a relaxation that is no worse than that from method (i) for each value of k. On the other hand, method (ii) requires the knowledge of a basis of $\mathbb{R}[x]/I$ as we will see below, which for some problems may be hard to compute in practice. To see the computational differences that can occur between the two methods, consult Remark 7.14.

In this chapter we focus on method (ii). The kth iteration of (ii) yields a closed convex set, called the kth *theta body* of I, defined as

$$\mathrm{TH}_k(I) := \{x \in \mathbb{R}^n \, : \, l(x) \geq 0 \text{ for all } l \text{ linear and } k\text{-sos mod } I\}.$$

Clearly $V_{\mathbb{R}}(I)$, and hence $\mathrm{cl}(\mathrm{conv}(V_{\mathbb{R}}(I)))$, is contained in $\mathrm{TH}_k(I)$ for all k. Thus the theta bodies of I form a hierarchy of closed convex approximations of $\mathrm{conv}(V_{\mathbb{R}}(I))$ as follows:

$$\mathrm{TH}_1(I) \supseteq \mathrm{TH}_2(I) \supseteq \cdots \supseteq \mathrm{TH}_k(I) \supseteq \mathrm{TH}_{k+1}(I) \supseteq \cdots \supseteq \mathrm{cl}(\mathrm{conv}(V_{\mathbb{R}}(I))).$$

An immediate question is when this hierarchy converges to $\mathrm{cl}(\mathrm{conv}(V_{\mathbb{R}}(I)))$ either finitely or asymptotically. Finite convergence allows an exact representation of $\mathrm{cl}(\mathrm{conv}(V_{\mathbb{R}}(I)))$ as a theta body which would be extremely useful if we can represent and optimize over a theta body efficiently. We will show in Section 7.2.2 that each $\mathrm{TH}_k(I)$ is the closure of a projected spectrahedron. This enables optimization over a real variety using semidefinite programming. In Section 7.4, we will learn the motivation for the name "theta bodies." We begin with some background on working modulo a polynomial ideal.

7.2.1 Sum of Squares Modulo an Ideal

Let $I \subseteq \mathbb{R}[x]$ be an ideal and $V_{\mathbb{R}}(I)$ be its real variety. For two polynomials $f, g \in \mathbb{R}[x]$, if $f - g \in I$, then $f(s) = g(s)$ for all $s \in V_{\mathbb{R}}(I)$. If $f - g \in I$, then f and g are said to be *congruent mod* I, written as $f \equiv g \bmod I$. Congruence mod I is an equivalence relation on $\mathbb{R}[x]$. The equivalence class of f is denoted as $f + I$, and the set of equivalence classes is denoted as $\mathbb{R}[x]/I$. The set $\mathbb{R}[x]/I$ is both an $\mathbb{R}$-vector space and a ring over $\mathbb{R}$ where addition, scalar multiplication, and multiplication are defined as follows. Given $f, g \in \mathbb{R}[x]$ and $\lambda \in \mathbb{R}$, $(f + I) + (g + I) = (f + g) + I$, $\lambda(f + I) = \lambda f + I$, and $(f + I)(g + I) = fg + I$. We will denote vector space bases of $\mathbb{R}[x]/I$ by $\mathcal{B}$ in this chapter. By the *degree* of an equivalence class $f + I$, we mean the smallest degree of an element in the class. With this definition, we may assume that the elements of $\mathcal{B}$ are listed in order of increasing degree. Further, for each $k \in \mathbb{N}$, the set $\mathcal{B}_k$ of all elements in $\mathcal{B}$ of degree at most k is then well-defined.

Computations in $\mathbb{R}[x]/I$ can be done via *Gröbner bases* of I. Recall that if G is any reduced Gröbner basis of I, then a polynomial h lies in I if and only if the *normal form* of h with respect to G is zero. Therefore, $f \equiv g \bmod I$ if and only if the normal form of $f - g$ with respect to G is zero, or equivalently, f and g have the same normal form with respect to G. This provides an algorithm to check whether two polynomials are congruent mod I. The unique normal form of all polynomials in the same equivalence class serves as a canonical representative for this class given G. If M is the *initial ideal* of I corresponding to the reduced Gröbner basis G, then recall that the *standard monomials* of M form an $\mathbb{R}$-vector space basis for $\mathbb{R}[x]/I$. Therefore, the normal form of a polynomial with respect to G can be written as an $\mathbb{R}$-linear combination of the standard monomials of the initial ideal M. The vector space $\mathbb{R}[x]/I$ has many other bases, some of which may be better suited for computations than the standard monomial bases coming from an initial ideal of I. See Chapter 3 for a discussion of alternative bases of $\mathbb{R}[x]$ and hence $\mathbb{R}[x]/I$. In this chapter we will use only a standard monomial basis of $\mathbb{R}[x]/I$. A quick tour of the algebraic notions needed in this chapter can be found in the appendix. For a thorough introduction to the theory of Gröbner bases and related notions, we refer the reader to [6].

We now come to sum of squares polynomials modulo an ideal I, and the question of how to check whether a polynomial $f \in \mathbb{R}[x]$ is k-sos mod I. A polynomial $f \in \mathbb{R}[x]$ is *sos mod I* if $f \equiv \sum h_j^2$ mod I for some $h_j \in \mathbb{R}[x]$, and *k-sos mod I* if $h_j \in \mathbb{R}[x]_k$ for all j. Hence, the equivalence classes of polynomials that are sos mod I (respectively, k-sos mod I) are precisely those in

$$\Sigma/I := \{\sigma + I \,:\, \sigma \in \Sigma\}$$

(respectively, Σ_{2k}/I). It is worthwhile to note that many polynomials that are not sos in $\mathbb{R}[x]$ can become sos mod an ideal I. For instance, the univariate linear polynomial x is congruent to x^2 mod the ideal $\langle x - x^2 \rangle \subset \mathbb{R}[x]$.

Let $[x]_k$ denote the vector of all monomials in $\mathbb{R}[x]_k$ in a fixed order, say degree lexicographic. Recall from Chapter 3 that a polynomial $f \in \Sigma_{2k}$ if and only if there exists a positive semidefinite matrix A, denoted $A \succeq 0$, such that $f = [x]_k^T A [x]_k$. The matrix A can be solved for using semidefinite programming and a Cholesky factorization of it as $A = V^T V$ yields an sos expression $\sum h_j^2$ for f, where $h_j(x)$ is the inner product of the jth row of V and the vector of monomials $[x]_k$. This method can be adapted to check whether f is k-sos mod I as follows. The vector $[x]_k$ can be replaced by the vector of monomials from $\mathcal{B}_k$, denoted as $[x]_{\mathcal{B}_k}$, since $\mathbb{R}[x]_k/I$ is spanned by $\mathcal{B}_k$. Since the size of $\mathcal{B}_k$ is no larger than the size of a basis of $\mathbb{R}[x]_k$, this can decrease the size of the unknown matrix A considerably, making the final SDP much smaller than before. Setting up A as a symmetric matrix of indeterminates A_{ij} and multiplying out $[x]_{\mathcal{B}_k}^T A [x]_{\mathcal{B}_k}$, we get a polynomial $g \in \mathbb{R}[x]_{2k}$. Let the normal forms of f and g with respect to a reduced Gröbner basis G of I be f' and g', respectively. Then since $f \equiv f'$ and $g \equiv g'$ mod I and f' and g' are fully reduced with respect to G, we have that $f \equiv g$ mod I if and only if $f' = g'$. Therefore, to check if f is k-sos mod I, we equate the coefficients of f' and g' for like monomials and check whether the resulting linear system in the A_{ij}'s has a solution with $A \succeq 0$.

Example 7.1. Consider the polynomial $f(x, y) = x^4 + y^4 + 2x^2y^2 - x^2 + y^2$ and the principal ideal $I = \langle f \rangle \subset \mathbb{R}[x, y]$. The real variety $V_{\mathbb{R}}(I)$, which is the set of real zeros of f, is a *Bernoulli lemniscate* (shown in Figure 7.1) with foci $(\pm\frac{1}{\sqrt{2}}, 0)$.

It is easy to check that the horizontal line $y = \frac{1}{\sqrt{8}}$ is a bitangent to $V_{\mathbb{R}}(I)$ and that $l(x, y) := -y + \frac{1}{\sqrt{8}}$ is nonnegative on $V_{\mathbb{R}}(I)$. Since f has degree 4 and l has degree 1, l cannot be 1-sos mod I but has a chance to be 2-sos mod I. We apply the method described above to verify this.

The set $\{f\}$ is a reduced Gröbner basis of I with respect to every term order. The initial ideal of I under the total degree order with ties broken lexicographically with $x > y$, is generated by x^4. Hence a basis $\mathcal{B}$ for $\mathbb{R}[x, y]/I$ is given by the infinite set of standard monomials of $\langle x^4 \rangle \subseteq \mathbb{R}[x, y]$ which are all the monomials in x and y

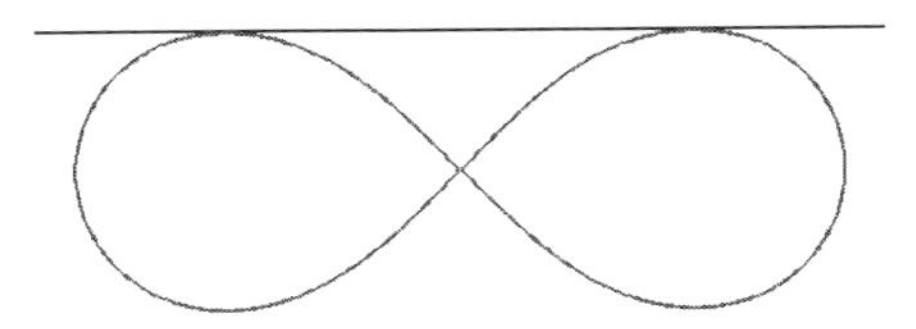

Figure 7.1. *The lemniscate $x^4 + y^4 + 2x^2y^2 - x^2 + y^2 = 0$ with a bitangent.*

that are not divisible by x^4. In particular, $\mathcal{B}_1 = \{1, x, y\}$, $\mathcal{B}_2 = \{1, x, y, x^2, xy, y^2\}$, and $[x]_{\mathcal{B}_2} = (1\ x\ y\ x^2\ xy\ y^2)$.

The general 2-sos polynomial mod I is therefore of the form

$$g = \begin{bmatrix} 1 \\ x \\ y \\ x^2 \\ xy \\ y^2 \end{bmatrix}^T \begin{bmatrix} a_{11} & a_{12} & a_{13} & a_{14} & a_{15} & a_{16} \\ a_{12} & a_{22} & a_{23} & a_{24} & a_{25} & a_{26} \\ a_{13} & a_{23} & a_{33} & a_{34} & a_{35} & a_{36} \\ a_{14} & a_{24} & a_{34} & a_{44} & a_{45} & a_{46} \\ a_{15} & a_{25} & a_{35} & a_{45} & a_{55} & a_{56} \\ a_{16} & a_{26} & a_{36} & a_{46} & a_{56} & a_{66} \end{bmatrix} \begin{bmatrix} 1 \\ x \\ y \\ x^2 \\ xy \\ y^2 \end{bmatrix},$$

where $A = (a_{ij}) \succeq 0$. Multiplying out the above expression we get that

$$g := a_{11} + 2a_{12}x + 2a_{13}y + (2a_{14} + a_{22})x^2 + (2a_{23} + 2a_{15})xy + (2a_{16} + a_{33})y^2 + 2a_{24}x^3 + (2a_{34} + 2a_{25})x^2y + (2a_{26} + 2a_{35})xy^2 + 2a_{36}y^3 + a_{44}x^4 + 2a_{45}x^3y + (a_{55} + 2a_{46})x^2y^2 + 2a_{56}xy^3 + a_{66}y^4.$$

We now reduce g by the Gröbner basis $\{f\}$, which means replacing every occurrence of x^4 with

$$-y^4 - 2x^2y^2 + x^2 - y^2,$$

and obtain the normal form of g, which is

$$g' := a_{11} + 2a_{12}x + 2a_{13}y + (2a_{14} + a_{22} + a_{44})x^2 + (2a_{23} + 2a_{15})xy + (2a_{16} + a_{33} - a_{44})y^2 + 2a_{24}x^3 + (2a_{34} + 2a_{25})x^2y + (2a_{26} + 2a_{35})xy^2 + 2a_{36}y^3 + 2a_{45}x^3y + (a_{55} + 2a_{46} - 2a_{44})x^2y^2 + 2a_{56}xy^3 + (a_{66} - a_{44})y^4.$$

Since $l(x, y) = -y + \frac{1}{\sqrt{8}}$ is already reduced with respect to $\{f\}$, if l is 2-sos mod I, then $l = g'$, and hence to verify this, we need to check whether there exists $A \succeq 0$ such that $a_{11} = \frac{1}{\sqrt{8}}, 2a_{13} = -1$, and all other coefficients of g' equal zero. Writing out all the linear conditions, we need to check whether there exists a positive semidefinite matrix of the form

$$\begin{bmatrix} \frac{1}{\sqrt{8}} & 0 & -\frac{1}{2} & a_{14} & a_{15} & a_{16} \\ 0 & a_{22} & -a_{15} & 0 & a_{25} & a_{26} \\ -\frac{1}{2} & -a_{15} & a_{33} & -a_{25} & -a_{26} & 0 \\ a_{14} & 0 & -a_{25} & a_{44} & 0 & a_{46} \\ a_{15} & a_{25} & -a_{26} & 0 & a_{55} & 0 \\ a_{16} & a_{26} & 0 & a_{46} & 0 & a_{44} \end{bmatrix}$$

that satisfies the conditions

$$2a_{14} + a_{22} + a_{44} = 0, \quad 2a_{16} + a_{33} - a_{44} = 0, \quad a_{55} + 2a_{46} - 2a_{44} = 0.$$

Check that the matrix

$$A = \begin{bmatrix} 2^{-3/2} & 0 & -1/2 & -2^{-3/2} & 0 & -2^{-3/2} \\ 0 & 0 & 0 & 0 & 0 & 0 \\ -1/2 & 0 & 2^{1/2} & 0 & 0 & 0 \\ -2^{-3/2} & 0 & 0 & 2^{-1/2} & 0 & 2^{-1/2} \\ 0 & 0 & 0 & 0 & 0 & 0 \\ -2^{-3/2} & 0 & 0 & 2^{-1/2} & 0 & 2^{-1/2} \end{bmatrix}$$

is positive semidefinite and satisfies the conditions given above. This matrix A factors as $A = V^T V$ with

$$V = \begin{bmatrix} -2^{-5/4} & 0 & 0 & 2^{-1/4} & 0 & 2^{-1/4} \\ -2^{-5/4} & 0 & 2^{1/4} & 0 & 0 & 0 \end{bmatrix},$$

and hence,

$$\left(\frac{1}{\sqrt{8}} - y\right) \equiv \frac{1}{4\sqrt{2}}\left(2x^2 + 2y^2 - 1\right)^2 + \sqrt{2}\left(y - \frac{1}{\sqrt{8}}\right)^2 \mod I.$$

In general, finding exact sos expressions, as above, is difficult. This particular sos decomposition was found by Bruce Reznick using a series of tricks. He showed that

$$\begin{aligned} &(\tfrac{1}{\sqrt{8}} - y) + \tfrac{1}{\sqrt{2}}((x^2+y^2)^2 - (x^2 - y^2)) \\ &= \tfrac{1}{4\sqrt{2}}\left(2x^2 + 2y^2 - 1\right)^2 + \sqrt{2}\left(y - \tfrac{1}{\sqrt{8}}\right)^2. \end{aligned}$$

In practice, one can use an SDP solver to find A. Using MATLAB, to do this computation in YALMIP [17] we input the following code:

```
sdpvar a14 a15 a16 a22 a25 a26 a33 a44 a46 a55
A=[ 1/sqrt(8) 0    -1/2  a14   a15 a16;
    0          a22 -a15  0     a25 a26;
   -1/2       -a15  a33 -a25 -a26 0   ;
    a14        0   -a25  a44   0   a46;
    a15        a25 -a26  0     a55 0   ;
    a16        a26  0    a46   0   a44];
```

```
l1=2*a14 + a22 + a44;
l2=2*a16 + a33 - a44;
l3=a55 + 2*a46 -2*a44;
solvesdp([A>0,l1==0,l2==0,l3==0],0);
```

We ran this code with SeDuMi 1.1 as the underlying SDP solver in YALMIP. The matrix can now be recovered by simply typing `double(A)` and we obtain

$$A = \begin{bmatrix} 0.3536 & 0.0000 & -0.5000 & -0.4052 & 0.0000 & -0.1985 \\ 0.0000 & 0.1034 & 0.0000 & 0.0000 & -0.2924 & 0.0000 \\ -0.5000 & 0.0000 & 1.1041 & 0.2924 & 0.0000 & 0.0000 \\ -0.4052 & 0.0000 & 0.2924 & 0.7071 & 0.0000 & 0.2936 \\ 0.0000 & -0.2924 & 0.0000 & 0.0000 & 0.8270 & 0.0000 \\ -0.1985 & 0.0000 & 0.0000 & 0.2936 & 0.0000 & 0.7071 \end{bmatrix},$$

in which the entries are shown up to four digits of precision. After factorizing A as $V^T V$ we obtain the sos decomposition:

$$\begin{aligned} & \left(0.5946427499 - 0.8408409925\, y - 0.6814175403\, x^2 - 0.3338138740\, y^2\right)^2 \\ + \; & (0.3215587038\, x - 0.9093207446\, xy)^2 \\ + \; & \left(0.6301479392\, y - 0.4452348146\, x^2 - 0.4454261796\, y^2\right)^2 \\ + \; & \left(0.2110357686\, x^2 - 0.6263671431\, y^2\right)^2 \\ + \; & 0.0001357833655\, x^2y^2 \\ + \; & 0.004928018144\, y^4, \end{aligned}$$

which simplifies to

$$\begin{aligned} & 0.3536000000 - \;y \\ + \; & 0.707(x^4 + 2x^2y^2 + y^4 - x^2 + y^2) \\ + \; & 10^{-11}(8.089965190\, x^2y - 3.247827064\, y^3). \end{aligned}$$

This provides fairly strong computational evidence that $l = \frac{1}{\sqrt{8}} - y$ is 2-sos mod I even though it is not an exact 2-sos representation of l mod I.

The above approach becomes cumbersome as we search for higher and higher degree sums of squares modulo an ideal. Luckily there are ways of using the existing software to simplify our input. In our example, checking whether l is 2-sos modulo I is the same as checking if there exists some $\lambda \in \mathbb{R}$ such that $l(x, y) + \lambda f(x, y)$ is sos, which can be done via YALMIP with the following commands:

```
sdpvar x y lambda
f=x^4+y^4+2*x^2*y^2-x^2+y^2;
l=1/sqrt(8)-y;
F=sos(l+lambda*f);
solvesos(F,0,[],lambda);
sdisplay(sosd(F))
```

The last command will actually display a list of polynomials whose squares sum up to (approximately) $l(x, y) + \lambda f(x, y)$. In our example, the following output is obtained

```
'-0.5919274724+0.8880*y+0.6222*x^2+0.3571*y^2'
'-0.03240303655-0.5699*y+0.4037*x^2+0.6602*y^2'
'-0.3036*x+0.8587*x*y'
'-0.0461010126-0.1559*y+0.3963*x^2-0.3792*y^2'
'9.2958e-05*x+3.2868e-05*x*y'
'3.789017278e-05+1.3396e-05*y+1.4209e-05*x^2+4.7355e-06*y^2'
```

which should be interpreted as saying that $l(x,y)$ is the sum of squares of the polynomials shown on each line. Note that the last two polynomials in the list above again point to the fact that the software only provided reasonable evidence that $l(x,y)$ is 2-sos mod I. ∎

The above computations also give a glimpse into the intertwining of algebraic and numerical methods that is prevalent in convex algebraic geometry. The question of whether a polynomial is a sum of squares modulo an ideal is purely algebraic. However, the search for an sos expression is done via semidefinite programming which is solved using numerical methods. The answer provided by these numerical solvers is often not exact. Massaging the numerical information into a certifiable answer can sometimes be an art.

Example 7.2. Consider the polynomial $g(x,y) := y^2(1-x^2) - (x^2+2y-1)^2$ and the ideal $I = \langle g(x,y) \rangle$ defining the *bicorn curve* shown in Figure 7.2. It is clear that $y \geq 0$ over the curve. Instead of checking if y is k-sos mod I for some k (which is never the case as we will see in the next section), it is in general more useful to search for the smallest μ such that $y+\mu$ is k-sos mod I. That way, if y is not sos mod I, we will at least obtain a valid inequality $y+\mu \geq 0$ on $V_{\mathbb{R}}(I)$ which will then be valid for $\mathrm{TH}_k(I)$. In general, $y+\mu$ is k-sos mod I if there exists some polynomial $h(x,y)$ of degree $2k-4$ such that $(y+\mu)+h(x,y)g(x,y)$ is sos. As before, this can be checked easily using YALMIP.

```
k=2;
sdpvar x y mu
[h,c]=polynomial([x y],2*k-4);
g=y^2*(1-x^2)-(x^2+2*y-1)^2;
F=sos(y+mu-h*g);
solvesos(F,mu,[],[mu;c]);
```

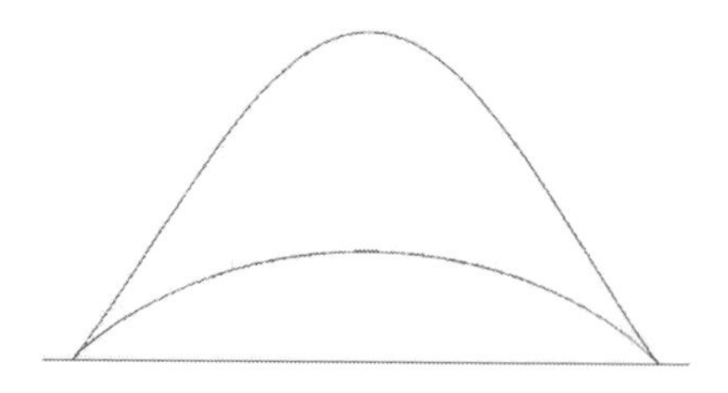

Figure 7.2. *A bicorn curve.*

By successively setting k to be 2, 3, and 4, we get that the minimum value of μ (recovered using `double(mu)`) is 0.1776, 0.0370, and 0.0161, respectively. So while μ is approaching 0, it seems that y is at least not 4-sos mod I. ∎

7.2.2 Theta Bodies

We now come back to theta bodies of the ideal I and their representations. Recall that the kth theta body of I is

$$\mathrm{TH}_k(I) := \{x \in \mathbb{R}^n \ : \ l(x) \geq 0 \text{ for all } \ l \text{ linear and } k\text{-sos mod } I\}.$$

Given any polynomial, it is possible to check whether it is k-sos mod I using Gröbner bases and semidefinite programming as seen in Section 7.2.1. The bottleneck in using the definition of $\mathrm{TH}_k(I)$ in practice is that it requires knowledge of all the linear polynomials (infinitely many) that are k-sos mod I. To overcome this difficulty we will now derive an alternative description of $\mathrm{TH}_k(I)$ as a projected spectrahedron (up to closure) which enables computations via semidefinite programming.

We may assume that there are no linear polynomials in the ideal I since otherwise, some variable x_i is congruent to a linear combination of other variables mod I, and we may work in a smaller polynomial ring. Therefore, $\mathbb{R}[x]_1/I \cong \mathbb{R}[x]_1$ and $\{1 + I, x_1 + I, \ldots, x_n + I\}$ can be completed to a basis $\mathcal{B}$ of $\mathbb{R}[x]/I$. Recall the definition of degree of $f + I$. We will assume that each element in a basis $\mathcal{B} = \{f_i + I\}$ of $\mathbb{R}[x]/I$ is represented by a polynomial whose degree equals the degree of its equivalence class, and that $\mathcal{B}$ is ordered so that $\deg(f_i + I) \leq \deg(f_{i+1} + I)$. Further, $\mathcal{B}_k$ denotes the ordered subset of $\mathcal{B}$ of degree at most k.

Definition 7.3. *Let $I \subseteq \mathbb{R}[x]$ be an ideal. A basis $\mathcal{B} = \{f_0 + I, f_1 + I, \ldots\}$ of $\mathbb{R}[x]/I$ is a θ-basis if it has the following properties:*

1. *$\mathcal{B}_1 = \{1 + I, x_1 + I, \ldots, x_n + I\}$.*
2. *If $\deg(f_i + I), \deg(f_j + I) \leq k$, then $f_i f_j + I$ is in the $\mathbb{R}$-span of $\mathcal{B}_{2k}$.*

Our goal will be to first express the kth theta body $\mathrm{TH}_k(I)$ as the closure of a certain set of linear functionals on the k-sos polynomials mod I. This will be achieved in Theorem 7.6. In the case where I contains the polynomials $x_i^2 - x_i$ for all $i = 1, \ldots, n$, the closure can be removed (Theorem 7.8). Such ideals appear in combinatorial optimization and hence this result will have an important role in Section 7.4. After this, we use a θ-basis of the quotient ring $\mathbb{R}[x]/I$ to turn the description of $\mathrm{TH}_k(I)$ in Theorem 7.6 to an explicit semidefinite representation. This allows concrete computations and examples. We proceed toward Theorem 7.6.

In what follows, we identify a linear polynomial $\alpha + \langle a, x\rangle \in \mathbb{R}[x]_1$ with the vector $(\alpha, a) \in \mathbb{R}^{n+1}$. Let $\Sigma_1^k(I) := \{f + I \ : \ f \in \mathbb{R}[x]_1, f \ k\text{-sos mod } I\}$. Then $\Sigma_1^k(I)$ is a cone in the vector space $\mathbb{R}[x]_1/I \cong \mathbb{R}[x]_1$, and its dual cone $\Sigma_1^k(I)^*$ lives in $(\mathbb{R}[x]_1/I)^* \cong \mathbb{R}[x]_1^* \cong \mathbb{R}^{n+1}$. Thus,

$$\Sigma_1^k(I)^* = \{(t, x) \in \mathbb{R} \times \mathbb{R}^n \ : \ \alpha t + \langle a, x\rangle \geq 0 \text{ for all } (\alpha, a) \in \Sigma_1^k(I)\}.$$

Consider the hyperplane $H := \{(1,x) : x \in \mathbb{R}^n\}$ in $\mathbb{R}^{n+1}$. We may think of H also as $H = \{L \in (\mathbb{R}[x]_1/I)^* : L(1+I) = 1\}$. It then follows immediately that

$$\{1\} \times \mathrm{TH}_k(I) = \Sigma_1^k(I)^* \cap H. \tag{7.2}$$

Lemma 7.4. *The hyperplane H intersects the relative interior of $\Sigma_1^k(I)^*$.*

Proof. A sufficient condition for a hyperplane L to intersect the relative interior of a closed convex cone P is that $\mathrm{cl}(\mathrm{cone}(\mathrm{relint}(P \cap L))) = P$. If L does not intersect the relative interior of P, then $P \cap L$ is contained in some proper face F of P (possibly the empty face). Therefore, $\mathrm{cl}(\mathrm{cone}(\mathrm{relint}(P \cap L)))$ is also contained in this face which is a proper subset of P.

By (7.2), $C := \{(\lambda, \lambda x) : \lambda \geq 0, x \in \mathrm{relint}(\mathrm{TH}_k(I))\}$ is the cone over the relative interior of $\Sigma_1^k(I)^* \cap H$. We will show that $\mathrm{cl}(C) = \Sigma_1^k(I)^*$. Let $(\alpha, a) \in \Sigma_1^k(I)$ and $x \in \mathrm{relint}(\mathrm{TH}_k(I))$. Then since $x \in \mathrm{TH}_k(I)$, $0 \leq \alpha + \langle a, x\rangle = \langle(\alpha, a), (1, x)\rangle$ which implies that $0 \leq \langle(\alpha, a), (\lambda, \lambda x)\rangle$ for all $\lambda \geq 0$. Hence $C \subseteq \Sigma_1^k(I)^*$, and since $\Sigma_1^k(I)^*$ is closed, $\mathrm{cl}(C) \subseteq \Sigma_1^k(I)^*$.

Suppose $\Sigma_1^k(I)^* \not\subseteq \mathrm{cl}(C)$. Then there exists $(t, x) \in \Sigma_1^k(I)^* \backslash \mathrm{cl}(C)$. Since the constant polynomial 1 lies in $\Sigma_1^k(I)$ and $(t, x) \in \Sigma_1^k(I)^*$, $t \geq 0$. Also, since $\mathrm{cl}(C)$ is closed and there exists $(s, y) \in C$ with $s > 0$, we can find a small enough $\epsilon > 0$ such that $(t, x) + \epsilon(s, y) \in \Sigma_1^k(I)^* \backslash \mathrm{cl}(C)$, and the first coordinate of $(t, x) + \epsilon(s, y)$ is positive. Scaling this element, we may assume that there is an element $(1, x) \in \Sigma_1^k(I)^* \backslash \mathrm{cl}(C)$. Since $(1, x) \in \Sigma_1^k(I)^*$, $\alpha + \langle a, x\rangle \geq 0$ for all $(\alpha, a) \in \Sigma_1^k(I)$, which implies that $x \in \mathrm{TH}_k(I)$ and hence $(1, x) \in \mathrm{cl}(C)$, which is a contradiction. □

We will also need the following lemma which can be proved using standard tools of convex geometry.

Lemma 7.5. *Let P be a closed convex cone and Q be a convex subcone of P such that $\mathrm{cl}(Q) = P$. Then $\mathrm{relint}(P) \subseteq Q$, and for any affine hyperplane H passing through the relative interior of P, $P \cap H = \mathrm{cl}(Q \cap H)$.*

We now examine the cone $\Sigma_1^k(I)^*$ more closely. Let $\Sigma^k(I)$ denote the set of all $f + I$ such that f is k-sos mod I. Then $\Sigma^k(I) = \Sigma_{2k}/I$ is a cone in $\mathbb{R}[x]_{2k}/I$, and $\Sigma_1^k(I) = \Sigma^k(I) \cap \mathbb{R}[x]_1/I$. Therefore, the dual cone of $\Sigma_1^k(I)$ in $(\mathbb{R}[x]/I)^*$ is the closure of the projection of $\Sigma^k(I)^*$ into $(\mathbb{R}[x]_1/I)^*$ as explained in Section 2.1 of Chapter 5. Hence we may identify $\Sigma_1^k(I)^*$ with the closure of the set

$$S_k(I) := \{(L(1+I), L(x_1+I), \ldots, L(x_n+I)) : L \in \Sigma^k(I)^*\}.$$

Further, define $Q_k(I) := \big\{(L(x_1+I), \ldots, L(x_n+I)) : L \in \Sigma^k(I)^*, L(1+I) = 1\big\}$. We will see shortly that $Q_k(I)$ is a projected spectrahedron, but first we establish the connection between $\mathrm{TH}_k(I)$ and $Q_k(I)$.

Theorem 7.6. $\mathrm{TH}_k(I) = \mathrm{cl}(Q_k(I))$.

Proof. Since $\{1\} \times Q_k(I) = S_k(I) \cap H$, we have $\{1\} \times \mathrm{cl}(Q_k(I)) = \mathrm{cl}(S_k(I) \cap H)$. Since $\mathrm{cl}(S_k(I)) = \Sigma_1^k(I)^*$, it follows from (7.2) that $\{1\} \times \mathrm{TH}_k(I) = \mathrm{cl}(S_k(I)) \cap H$. Therefore, the theorem will follow if we can show that

$$\mathrm{cl}(S_k(I)) \cap H = \mathrm{cl}(S_k(I) \cap H).$$

By Lemma 7.5, this equality holds if H intersects $S_k(I)$ in its relative interior. Again, by Lemma 7.5, $\mathrm{relint}(\Sigma_1^k(I)^*) \subseteq S_k(I)$. Lemma 7.4 showed that H intersects the relative interior of $\Sigma_1^k(I)^*$ and hence the relative interior of $S_k(I)$. □

We now focus on an important situation where the closure is not needed in Theorem 7.6. In many cases in practice, we are interested in finding the convex hull of a set $S \subseteq \mathbb{R}^n$ that may not be presented as the real variety of an ideal. However, the approximation $\mathrm{TH}_k(I)$ of $\mathrm{conv}(S)$ is defined with respect to an ideal I whose real variety is S. In this case, the canonical choice for such an ideal is the *vanishing ideal* of S, denoted as $I(S)$, which consists of all polynomials in $\mathbb{R}[x]$ that vanish on S. The *real radical* of an ideal $I \subseteq \mathbb{R}[x]$ is the ideal

$$\sqrt[\mathbb{R}]{I} = \left\{ f \in \mathbb{R}[x] : f^{2m} + \sum g_i^2 \in I, m \in \mathbb{N}, g_i \in \mathbb{R}[x] \right\},$$

and the ideal I is said to be real radical if $I = \sqrt[\mathbb{R}]{I}$. The *real Nullstellensatz* [21] states that I is real radical if and only if $I = I(V_{\mathbb{R}}(I))$. This is the analogue of Hilbert's Nullstellensatz for real algebraic varieties. Computing any ideal I such that $V_{\mathbb{R}}(I) = S$ might be hard, and in general, computing $I(S)$, given S, might also be hard. However, in many cases of practical interest, $I(S)$ is available. A large source of such examples is combinatorial optimization, where S is usually a finite set of 0/1 points for which a generating set for $I(S)$ can be computed using combinatorial arguments. We will see several such examples in Section 7.4. If S is a subset of $\{0,1\}^n$ and $I = I(S)$, then Theorem 7.6 can be improved to Theorem 7.8. We first prove a lemma.

Lemma 7.7. *Let J be any ideal that contains $x_i^2 - x_i$ for all $i = 1, \ldots, n$. Then $1 + J$ is in the relative interior of* $\Sigma^k(J) = \{f + J : f \text{ is } k\text{-sos mod } J\rangle$.

Proof. Let $\mathcal{I} := \langle x_i^2 - x_i \text{ for all } i = 1, \ldots, n\rangle$. We will first show that $1 + \mathcal{I}$ is in the relative interior of $\Sigma^k(\mathcal{I}) \subseteq \mathbb{R}[x]_{2k}/\mathcal{I}$. The cone $\Sigma^k(J)$ is a projection of $\Sigma^k(\mathcal{I})$ since $\mathcal{I} \subseteq J$, and hence, if $1 + \mathcal{I} \in \mathrm{relint}(\Sigma^k(\mathcal{I}))$, then $1 + J \in \mathrm{relint}(\Sigma^k(J))$. $1 + \mathcal{I}$ is in the relative interior of $\Sigma^k(\mathcal{I})$, which is a cone in the vector space $\mathbb{R}[x]_{2k}/\mathcal{I}$.

We will show that for any polynomial $p \in \mathbb{R}[x]_{2k}$, $(1 + \epsilon p) + \mathcal{I} \in \Sigma^k(\mathcal{I})$ for some $\epsilon > 0$. Since we are working modulo $\mathcal{I}$, we may assume that every monomial in p is square-free. Further, since every monomial is a square modulo $\mathcal{I}$, it suffices to show that $(1 - \epsilon q) + \mathcal{I} \in \Sigma^k(\mathcal{I})$ for any square-free monomial q of degree at most $2k$ and some $\epsilon > 0$. Write $q = q_1 q_2$ for some square-free monomials q_1, q_2 of degree at most k. Now note that

$$(1 - q_2)^2 = 1 - 2q_2 + q_2^2 \equiv 1 - q_2 \bmod \mathcal{I}, \text{ and}$$
$$(1 - q_1 + q_2)^2 = 1 + q_1^2 + q_2^2 - 2q_1 + 2q_2 - 2q_1q_2 \equiv 1 - q_1 + 3q_2 - 2q_1q_2 \bmod \mathcal{I}.$$

Therefore, $(1 - q_1 + q_2)^2 + 3(1 - q_2)^2 + q_1^2 \equiv 4 - 2q_1q_2 = 4 - 2q \bmod \mathcal{I}$. Since $q_1, q_2 \in \mathbb{R}[x]_k$, it follows that $(4-2q) + \mathcal{I} \in \Sigma^k(\mathcal{I})$, which implies that $(1 - \frac{q}{2}) + \mathcal{I} \in \Sigma^k(\mathcal{I})$. □

Theorem 7.8. *If $S \subseteq \{0,1\}^n$ and $I = I(S)$, then* $\mathrm{TH}_k(I) = Q_k(I)$.

Proof. Since $S \subseteq \{0,1\}^n$, its vanishing ideal $I = I(S)$ contains $x_i^2 - x_i$ for all $i = 1, \dots, n$, and so by Lemma 7.7, $1 + I$ is in the relative interior of $\Sigma^k(I)$. Hence, $\Sigma_1^k(I)^* = S_k(I)$. (No closure operation is needed by [24, Corollary 16.4.2].) Therefore,

$$\{1\} \times \mathrm{TH}_k(I) = \Sigma_1^k(I)^* \cap H = S_k(I) \cap H = \{1\} \times Q_k(I),$$

and the result follows. □

We have thus far seen that the kth theta body $\mathrm{TH}_k(I)$ is the closure of $Q_k(I)$. However, this description is still abstract and in order to work with theta bodies in practice, we now give an explicit (coordinate based) description of $Q_k(I)$ using a basis of $\mathbb{R}[x]/I$ which will make it transparent that $Q_k(I)$ is the projection of a spectrahedron. This involves the theory of *moments* and *moment matrices* as explained below.

Fix a θ-basis $\mathcal{B} = \{f_i + I\}$ of $\mathbb{R}[x]/I$ and define $[x]_{\mathcal{B}_k}$ to be the column vector formed by all the elements of $\mathcal{B}_k$ in order. Then $[x]_{\mathcal{B}_k}[x]_{\mathcal{B}_k}^T$ is a square matrix indexed by $\mathcal{B}_k$ and its (i,j)-entry is equal to $f_if_j + I$. By hypothesis, the entries of $[x]_{\mathcal{B}_k}[x]_{\mathcal{B}_k}^T$ lie in the $\mathbb{R}$-span of $\mathcal{B}_{2k}$. Let $\{\ \lambda_{i,j}^l\ \}$ be the unique set of real numbers such that $f_if_j + I = \sum_{f_l + I \in \mathcal{B}_{2k}} \lambda_{i,j}^l (f_l + I)$.

Definition 7.9. *Let I, $\mathcal{B}$, and $\{\ \lambda_{i,j}^l\ \}$ be as above. Let y be a real vector indexed by $\mathcal{B}_{2k}$ with $y_0 = 1$, where y_0 is the first entry of y, indexed by the basis element $1 + I$. The* kth reduced moment matrix *$M_{\mathcal{B}_k}(y)$ of I is the real matrix indexed by $\mathcal{B}_k$ whose (i,j)-entry is $[M_{\mathcal{B}_k}(y)]_{i,j} = \sum_{f_l + I \in \mathcal{B}_{2k}} \lambda_{i,j}^l y_l$.*

We now give examples of reduced moment matrices. For simplicity, we often write f for $f + I$. Also, in this chapter we consider only monomial bases of $\mathbb{R}[x]/I$ (i.e., f_i is a monomial for all $f_i + I \in \mathcal{B}$) which we can obtain via Gröbner basis theory. In this case, $[x]_{\mathcal{B}_k}$ is a vector of monomials and we identify the vector $[x]_{\mathcal{B}_k}$ with the vector of monomials that represent the elements of $\mathcal{B}_k$. The method is to compute a reduced Gröbner basis of I and take $\mathcal{B}$ to be the equivalence classes of the standard monomials of the corresponding initial ideal. If the reduced Gröbner basis is with respect to a *total degree ordering*, then the second condition in the definition of a θ-basis is satisfied by $\mathcal{B}$.

Example 7.10. Consider the ideal I generated by $f := (x+1)x(x-1)^2$. Clearly, $V_{\mathbb{R}}(I) = \{-1, 0, 1\}$ with a double root at 1, and $\mathrm{conv}(V_{\mathbb{R}}(I)) = [-1, 1]$. The polynomial $f = x^4 - x^3 - x^2 + x$ is the unique element in every reduced Gröbner basis of I with $\langle x^4 \rangle$ as initial ideal. The standard monomials of this initial ideal are

$1, x, x^2, x^3$, and hence $\mathcal{B} = \{1 + I, x + I, x^2 + I, x^3 + I\}$ is a θ-basis for $\mathbb{R}[x]/I$. The biggest reduced moment matrix we could construct is $M_{\mathcal{B}_3}(y)$, whose rows and columns are indexed by $\mathcal{B}_3 = \mathcal{B}$.

We have $[x]_{\mathcal{B}_3} = (1\ x\ x^2\ x^3)$ and

$$[x]_{\mathcal{B}_3}[x]_{\mathcal{B}_3}^T = \begin{bmatrix} 1 & x & x^2 & x^3 \\ x & x^2 & x^3 & x^4 \\ x^2 & x^3 & x^4 & x^5 \\ x^3 & x^4 & x^5 & x^6 \end{bmatrix},$$

which is entrywise equivalent mod I to

$$\begin{bmatrix} 1 & x & x^2 & x^3 \\ x & x^2 & x^3 & x^3 + x^2 - x \\ x^2 & x^3 & x^3 + x^2 - x & 2x^3 - x \\ x^3 & x^3 + x^2 - x & 2x^3 - x & 2x^3 + x^2 - 2x \end{bmatrix}.$$

We now linearize using $y = (1, y_1, y_2, y_3)$ and obtain

$$M_{\mathcal{B}_3}(y) = \begin{bmatrix} 1 & y_1 & y_2 & y_3 \\ y_1 & y_2 & y_3 & y_3 + y_2 - y_1 \\ y_2 & y_3 & y_3 + y_2 - y_1 & 2y_3 - y_1 \\ y_3 & y_3 + y_2 - y_1 & 2y_3 - y_1 & 2y_3 + y_2 - 2y_1 \end{bmatrix}.$$

The reduced moment matrices $M_{\mathcal{B}_1}(y)$ and $M_{\mathcal{B}_2}(y)$ are the upper left 2×2 and 3×3 principal submatrices of $M_{\mathcal{B}_3}(y)$. ■

Example 7.11. Consider the ideal $I = \langle x^4 - y^2 - z^2, x^4 + x^2 + y^2 - 1 \rangle$. Using a computer algebra package such as Macaulay2 [10] one can calculate a total degree reduced Gröbner basis of I as follows:

```
Macaulay2, version 1.3

i1 : R = QQ[x,y,z,Weights => {1,1,1}];
i2 : I = ideal(x^4-y^2-z^2, x^4+x^2+y^2-1);
i3 : G = gens gb I
o3 = | x2+2y2+z2-1 4y4+4y2z2+z4-5y2-3z2+1 |
```

which says that this Gröbner basis consists of the two polynomials

$$x^2 + 2y^2 + z^2 - 1 \quad \text{and} \quad 4y^4 + 4y^2z^2 + z^4 - 5y^2 - 3z^2 + 1.$$

A basis for the quotient ring $\mathbb{R}[x, y, z]/I$ is given by the standard monomials of the initial ideal $\langle x^2, y^4 \rangle$, which gives the following partial bases:

$$\begin{aligned} \mathcal{B}_1 &= \{1, x, y, z\}, \\ \mathcal{B}_2 &= \mathcal{B}_1 \cup \{xy, y^2, xz, yz, z^2\}, \\ \mathcal{B}_3 &= \mathcal{B}_2 \cup \{xy^2, y^3, xyz, y^2z, xz^2, yz^2, z^3\}, \\ \mathcal{B}_4 &= \mathcal{B}_3 \cup \{xy^3, xy^2z, y^3z, xyz^2, y^2z^2, xz^3, yz^3, z^4\}. \end{aligned}$$

Linearizing the elements of $\mathcal{B}_4$, we get the following table:

1	x	y	z	xy	y^2	xz	yz	z^2	xy^2	y^3	xyz	y^2z	xz^2	yz^2	z^3
1	y_1	y_2	y_3	y_4	y_5	y_6	y_7	y_8	y_9	y_{10}	y_{11}	y_{12}	y_{13}	y_{14}	y_{15}

xy^3	xy^2z	y^3z	xyz^2	y^2z^2	xz^3	yz^3	z^4
y_{16}	y_{17}	y_{18}	y_{19}	y_{20}	y_{21}	y_{22}	y_{23}.

We can now calculate various reduced moment matrices. For instance,

$$M_{\mathcal{B}_2}(y) = \begin{bmatrix} 1 & y_1 & y_2 & y_3 & y_4 & y_5 & y_6 & y_7 & y_8 \\ & T_1 & y_4 & y_6 & T_2 & y_9 & T_3 & y_{11} & y_{13} \\ & & y_5 & y_7 & y_9 & y_{10} & y_{11} & y_{12} & y_{14} \\ & & & y_8 & y_{11} & y_{12} & y_{13} & y_{14} & y_{15} \\ & & & & T_4 & y_{16} & T_5 & y_{17} & y_{19} \\ & & & & & T_6 & y_{17} & y_{18} & y_{20} \\ & & & & & & T_7 & y_{19} & y_{21} \\ & & & & & & & y_{20} & y_{22} \\ & & & & & & & & y_{23} \end{bmatrix},$$

where we have filled in only the upper triangular region. The unknowns $T_1, T_2, \ldots$ stand for the following expressions:

$$\begin{aligned} T_1 &= -2y_5 - y_8 + 1, \\ T_2 &= -2y_{10} - y_{14} + y_2, \\ T_3 &= -2y_{12} - y_{15} + y_3, \\ T_4 &= y_{20} + \tfrac{y_{23}}{2} - \tfrac{3y_5}{2} - \tfrac{3y_8}{2} + \tfrac{1}{2}, \\ T_5 &= -2y_{18} - y_{22} + 1, \\ T_6 &= -y_{20} - \tfrac{y_{23}}{4} + \tfrac{5y_5}{4} + \tfrac{3y_8}{4} - \tfrac{1}{4}, \\ T_7 &= -2y_{20} - y_{23} + y_8. \end{aligned}$$

The T_i's can be calculated using Macaulay2 by first finding the normal form of the needed monomial with respect to the Gröbner basis that was calculated and then linearizing using the y_i's. For instance, T_2 is the linearization of the normal form of x^2y, which by the calculation below, is $-2y^3 - yz^2 + y$.

```
i6 : x^2*y%G
            3         2
o6 = - 2y   - y*z   + y
```

The reduced moment matrix $M_{\mathcal{B}_k}(y)$ can also be defined in terms of linear functionals on $\mathbb{R}[x]_{2k}/I$. For a vector $y = (y_b) \in \mathbb{R}^{\mathcal{B}_{2k}}$, define $L_y \in (\mathbb{R}[x]_{2k}/I)^*$ as $L_y(b) := y_b$ for all $b \in \mathcal{B}_{2k}$. Then every $L \in (\mathbb{R}[x]_{2k}/I)^*$ is equal to L_y for $y = (L(b) : b \in \mathcal{B}_{2k}) \in \mathbb{R}^{\mathcal{B}_{2k}}$. If $y \in \mathbb{R}^{\mathcal{B}_{2k}}$, let $y_0 := y_{1+I}$, $y_i := y_{x_i+I}$ for $i = 1, \ldots, n$. Further, let $\pi_{\mathbb{R}^n}$ be the projection map that sends $y \in \mathbb{R}^{\mathcal{B}_{2k}}$ to $(y_1, \ldots, y_n) \in \mathbb{R}^n$.

Lemma 7.12.

1. *For a vector* $y \in \mathbb{R}^{\mathcal{B}_{2k}}$ *with* $y_0 = 1$, *the entry of* $M_{\mathcal{B}_k}(y)$ *indexed by* $b_i, b_j \in \mathcal{B}_k$ *is* $L_y(b_i b_j)$.
2. $M_{\mathcal{B}_k}(y) \succeq 0 \Leftrightarrow L_y(f^2 + I) \geq 0$ *for all* $f + I \in \mathbb{R}[x]_k/I$.

Proof. The first part follows from the definition of $M_{\mathcal{B}_k}(y)$ and L_y. For $f + I \in \mathbb{R}[x]_k/I$, let $\hat{f}$ be the unique vector in $\mathbb{R}^{\mathcal{B}_k}$ such that $f + I = \sum_{b_i \in \mathcal{B}_k} \hat{f}_i b_i$. Therefore, $f^2 + I = \sum_{b_i, b_j \in \mathcal{B}_k} \hat{f}_i \hat{f}_j (b_i b_j)$ which implies that

$$L_y(f^2 + I) = \sum_{b_i, b_j \in \mathcal{B}_k} \hat{f}_i \hat{f}_j L_y(b_i b_j) = \hat{f}^T M_{\mathcal{B}_k}(y) \hat{f}.$$

Therefore, $M_{\mathcal{B}_k}(y) \succeq 0 \Leftrightarrow L_y(f^2 + I) \geq 0$ for all $f + I \in \mathbb{R}[x]_k/I$. □

Putting all this together, we obtain the following specific semidefinite representation of $Q_k(I)$, and hence $\mathrm{TH}_k(I)$ up to closure. We will use this explicit coordinate based description of $\mathrm{TH}_k(I)$ in the the calculations below.

Theorem 7.13. *The kth theta body of* I, $\mathrm{TH}_k(I)$, *is the closure of*

$$Q_k(I) = \pi_{\mathbb{R}^n} \left\{ y \in \mathbb{R}^{\mathcal{B}_{2k}} \; : \; M_{\mathcal{B}_k}(y) \succeq 0, \; y_0 = 1 \right\}.$$

Proof. Recall that $Q_k(I)$ is the set

$$\left\{ (L(x_1 + I), \ldots, L(x_n + I)) \; : \; \begin{array}{l} L(g + I) \geq 0 \text{ for all } g + I \in \Sigma_{2k}/I, \\ L(1 + I) = 1 \end{array} \right\}.$$

Equivalently, $Q_k(I)$ is the set

$$\left\{ (L(b) \; : \; b \in \mathcal{B}_1 \backslash \{1 + I\}) \; : \; \begin{array}{l} L(f^2 + I) \geq 0 \text{ for all } f + I \in \mathbb{R}[x]_k/I, \\ L(1 + I) = 1 \end{array} \right\}.$$

By Lemma 7.12 (2), it then follows that

$$Q_k(I) = \pi_{\mathbb{R}^n} \left\{ y \in \mathbb{R}^{\mathcal{B}_{2k}} \; : \; M_{\mathcal{B}_k}(y) \succeq 0, \; y_0 = 1 \right\} =: Q_{\mathcal{B}_k}(I). \quad \square$$

When working with a specific basis $\mathcal{B}$, we use $Q_{\mathcal{B}_k}(I)$ instead of $Q_k(I)$ to make the choice of basis clear. In the examples that follow, please bear in mind that this abuse of notation is simply to keep track of which θ-basis of $\mathbb{R}[x]/I$ was used in the explicit semidefinite representation of $Q_k(I)$. The proof of Theorem 7.13 shows that any θ-basis of $\mathbb{R}[x]/I$ can be used to coordinatize $Q_k(I)$.

Example 7.10 continued. We write down $Q_{\mathcal{B}_k}(I)$ for $k = 1, 2, 3$ for the ideal $I = \langle (x+1)x(x-1)^2 \rangle$ from Example 7.10. Using the matrix $M_{\mathcal{B}_3}(y)$ (with $y_0 = 1$) that was already computed we see that

$$Q_{\mathcal{B}_1}(I) = \{ y_1 \; : \; \exists (y_1, y_2) \in \mathbb{R}^2 \text{ s.t. } y_2 \geq y_1^2 \},$$

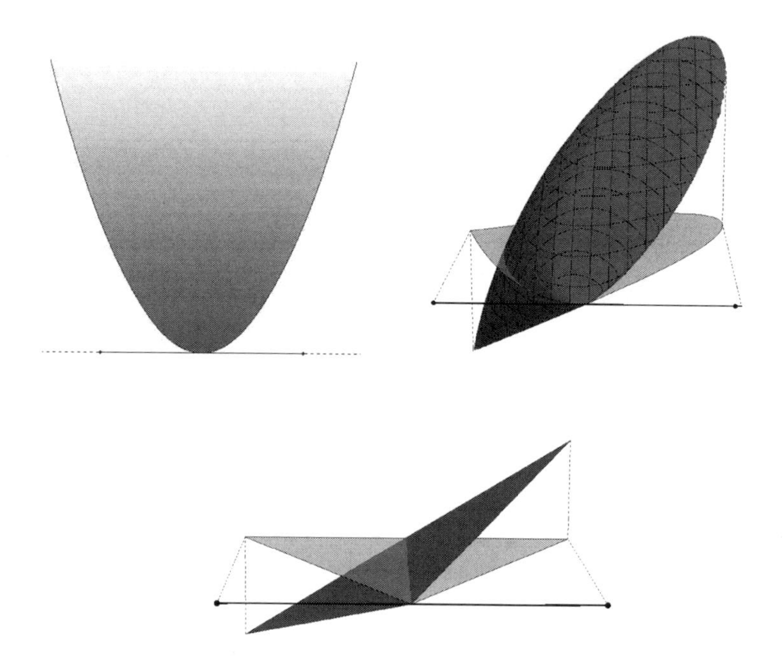

Figure 7.3. *The spectrahedra* $\{y \in \mathbb{R}^{\mathcal{B}_{2k}} : y_0 = 1, M_{\mathcal{B}_k}(y) \succeq 0\}$ *for* $k = 1, 2, 3$ *for* $I = \langle (x+1)x(x-1)^2 \rangle$ *and their projections to the* y_1*-axis.*

which is the projection onto the y_1-axis of the convex hull of the parabola $y_2 = y_1^2$. Therefore, $Q_{\mathcal{B}_1}(I) = \mathbb{R}$ and hence $\mathrm{TH}_1(I) = \mathbb{R}$, which is a trivial relaxation of $\mathrm{conv}(V_{\mathbb{R}}(I)) = [-1, 1]$.

The body $Q_{\mathcal{B}_2}(I) = \{y_1 : \exists y \in \mathbb{R}^3 \text{ s.t. } M_{\mathcal{B}_2}(y) \succeq 0\}$. We know the exact form of the moment matrices so we can use YALMIP to find $\mathrm{cl}(Q_{\mathcal{B}_2}(I))$, by minimizing x and $-x$ over that body.

```
sdpvar y1 y2 y3
M=[1  y1 y2;
   y1 y2 y3;
   y2 y3 y3+y2-y1];
solvesdp(M>0,y1);
double(y1)
solvesdp(M>0,-y1);
double(y1)
```

We then get $\mathrm{cl}(Q_{\mathcal{B}_2}(I)) \approx [-1.0000, 1.0417]$, and we will later see that it is actually exactly $[-1, \frac{25}{24}]$.

To finish, we compute $Q_{\mathcal{B}_3}(I) = \{y_1 : \exists y \in \mathbb{R}^3 \text{ s.t. } M_{\mathcal{B}_3}(y) \succeq 0\}$. This is the projection onto the y_1-coordinate of the spectrahedron in $\mathbb{R}^3$ described by all the

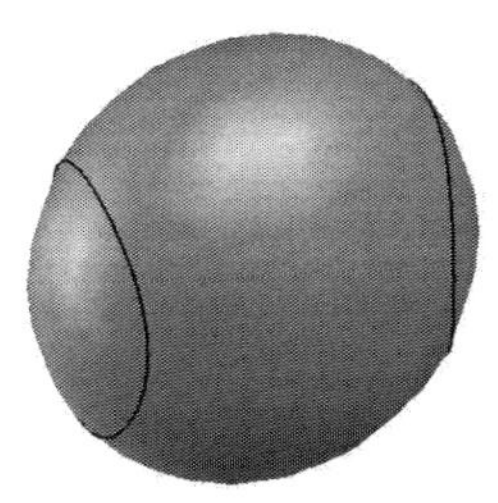

Figure 7.4. *The variety of Example* 7.11 *and its first theta body.*

Figure 7.5. *The second theta body from Example* 7.11.

inequalities obtained from the condition $M_{\mathcal{B}_3}(y) \succeq 0$. This body is the convex hull of the moment vectors (x, x^2, x^3) evaluated at $x = -1, 0, 1$, which is the triangle with vertices $(-1, 1, -1), (0, 0, 0), (1, 1, 1)$. Projecting onto the y_1-coordinate, we get $\text{cl}(Q_{\mathcal{B}_3}(I)) = [-1, 1]$. See Figure 7.3 for $Q_{\mathcal{B}_i}(I)$, $i = 1, 2, 3$, and their spectrahedral preimages.

Example 7.11 continued. We now draw a few theta bodies of the ideal

$$I = \langle x^4 - y^2 - z^2, x^4 + x^2 + y^2 - 1 \rangle$$

from Example 7.11, where we calculated the second reduced moment matrix $M_{\mathcal{B}_2}(y)$. This allows us to write down $Q_{\mathcal{B}_1}(I)$ and $Q_{\mathcal{B}_2}(I)$.

From the Gröbner basis of I that we computed, we see that the polynomial $x^2 + 2y^2 + z^2 - 1$ is in I. We will see in Example 7.36 that the first theta body of I is the ellipsoid $\{(x, y, z) \in \mathbb{R}^3 : x^2 + 2y^2 + z^2 \leq 1\}$. This ellipsoid along with $V_{\mathbb{R}}(I)$ (the two black rings) is shown in Figure 7.4. The second theta body is shown in Figure 7.5 and it appears to equal $\text{conv}(V_{\mathbb{R}}(I))$.

Remark 7.14. *This example shows the difference between Lasserre's method to convexify* $V_{\mathbb{R}}(I)$ *and the reduced moment method that underlies theta bodies. Recall that in step* k *of Lasserre's method, the relaxation of* $\text{conv}(V_{\mathbb{R}}(I))$ *that is computed is the common intersection of all half spaces* $l(x) \geq 0$ *containing* $V_{\mathbb{R}}(I)$ *and*

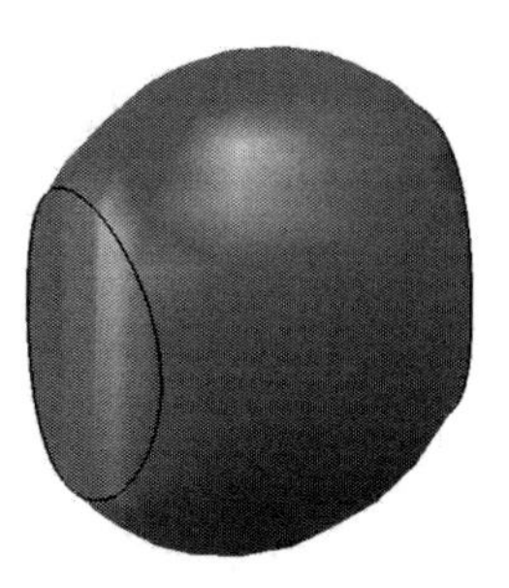

Figure 7.6. *The second Lasserre relaxation for Example* 7.11.

$l(x) = \sigma(x) + \sum_{i=1}^{m} g_i(x) f_i(x)$, where $\sigma(x)$ is a k-sos polynomial and $g_i(x) f_i(x) \in \mathbb{R}[x]_{2k}$. Using the software package Bermeja [25] *we can draw the second relaxation in Lasserre's method which is shown in Figure* 7.6.

Now that we have seen several examples of theta bodies of ideals, we give a few comments and examples to point out some of the subtleties involved. We start with an example to show that $Q_{\mathcal{B}_k}(I)$ may not be closed, which emphasizes the need to take its closure to get $\mathrm{TH}_k(I)$.

Example 7.15. Consider the principal ideal $I = \langle x_1^2 x_2 - 1 \rangle \subset \mathbb{R}[x_1, x_2]$. Then $\mathrm{conv}(V_{\mathbb{R}}(I)) = \{(s_1, s_2) \in \mathbb{R}^2 : s_2 > 0\}$, which is not a closed set. Any linear polynomial that is nonnegative over $V_{\mathbb{R}}(I)$ is of the form $\alpha x_2 + \beta$, where $\alpha, \beta \geq 0$. Since $\alpha x_2 + \beta \equiv (\sqrt{\alpha} x_1 x_2)^2 + (\sqrt{\beta})^2 \bmod I$, $\mathrm{TH}_2(I) = \mathrm{cl}(\mathrm{conv}(V_{\mathbb{R}}(I)))$.

The set $\mathcal{B} = \bigcup_{k \in \mathbb{N}} \{x_1^k + I, x_2^k + I, x_1 x_2^k + I\}$ is a θ-basis for $\mathbb{R}[x_1, x_2]/I$ for which

$$\mathcal{B}_4 = \{1, x_1, x_2, x_1^2, x_1 x_2, x_2^2, x_1 x_2^2, x_1^3, x_2^3, x_1 x_2^3, x_1^4, x_2^4\} + I.$$

The reduced moment matrix $M_{\mathcal{B}_2}(y)$ for $y = (1, y_1, \ldots, y_{11}) \in \mathbb{R}^{\mathcal{B}_4}$ is

$$\begin{array}{c} \\ 1 \\ x_1 \\ x_2 \\ x_1^2 \\ x_1 x_2 \\ x_2^2 \end{array} \begin{array}{c} \begin{array}{cccccc} 1 & x_1 & x_2 & x_1^2 & x_1 x_2 & x_2^2 \end{array} \\ \left(\begin{array}{cccccc} 1 & y_1 & y_2 & y_3 & y_4 & y_5 \\ y_1 & y_3 & y_4 & y_6 & 1 & y_7 \\ y_2 & y_4 & y_5 & 1 & y_7 & y_8 \\ y_3 & y_6 & 1 & y_9 & y_1 & y_2 \\ y_4 & 1 & y_7 & y_1 & y_2 & y_{10} \\ y_5 & y_7 & y_8 & y_2 & y_{10} & y_{11} \end{array} \right) \end{array}.$$

If $M_{\mathcal{B}_2}(y) \succeq 0$, then the principal minor indexed by x_1 and $x_1 x_2$ implies that $y_2 y_3 \geq 1$, and so in particular, $y_2 \neq 0$ for all $y \in Q_{\mathcal{B}_2}(I)$. However, since $Q_{\mathcal{B}_2}(I) \supseteq \mathrm{conv}(V_{\mathbb{R}}(I)) = \{(s_1, s_2) \in \mathbb{R}^2 : s_2 > 0\}$, it must be that $Q_{\mathcal{B}_2}(I) = \mathrm{conv}(V_{\mathbb{R}}(I))$, which shows that $Q_{\mathcal{B}_2}(I)$ is not closed. ∎

We will see in the next section that when S is a finite set of points in $\mathbb{R}^n$, the ideal $I = I(S)$ of all polynomials that vanish on S, has the property that

$\mathrm{TH}_l(I) = \mathrm{conv}(V_{\mathbb{R}}(I)) = \mathrm{conv}(S)$ for a finite l that depends on I. However, since $\mathrm{conv}(S) \subseteq Q_{\mathcal{B}_l}(I) \subseteq \mathrm{TH}_l(I)$, we also get that $Q_{\mathcal{B}_l}(I)$ is closed. Even in this case, $Q_{\mathcal{B}_k}(I)$ may not be closed for some $k < l$.

Example 7.16. Consider the finite set of points $S = \{(\pm t, 1/t^2) : t = 1, \ldots, 7\}$ lying on the curve $x_1^2 x_2 = 1$. Then

$$I(S) = \langle x_1^2 x_2 - 1, (x_1^2 - 1)(x_1^2 - 4)(x_1^2 - 9)(x_1^2 - 16)(x_1^2 - 25)(x_1^2 - 36)(x_1^2 - 49)\rangle.$$

This is a zero-dimensional ideal, and a basis for $\mathbb{R}[x_1, x_2]/I(S)$ is given by

$$\mathcal{B} = \{1, x_1, x_2, x_1^2, x_1 x_2, x_2^2, x_1 x_2^2, x_1^3, x_2^3, x_1 x_2^3, x_1^4, x_2^4, x_1^5, x_1 x_2^4\} + I.$$

In particular, $\mathcal{B}_4$ is the same as the $\mathcal{B}_4$ in Example 7.15 and the initial ideal of $I(S)$ whose standard monomials are the monomials in $\mathcal{B}$ is generated by $\{x_1^2 x_2, x_2^5, x_1^6\}$. Therefore, $M_{\mathcal{B}_2}(I(S))$ and $Q_{\mathcal{B}_2}(I(S))$ agree with those in Example 7.15, which implies that $Q_{\mathcal{B}_2}(I(S))$ is not closed. ∎

Another natural question is whether the theta bodies of different ideals with the same real variety can have drastically different behaviors, especially with respect to convergence to the convex hull of the variety. For instance, an ideal I and its real radical $\sqrt[\mathbb{R}]{I}$ have the same real variety and $I \subseteq \sqrt[\mathbb{R}]{I}$, $\mathrm{TH}_k(\sqrt[\mathbb{R}]{I}) \subseteq \mathrm{TH}_k(I)$ for all k.

Theorem 7.17. *Fix an ideal I. Then there exists a function $\Psi : \mathbb{N} \to \mathbb{N}$ such that $\mathrm{TH}_{\Psi(k)}(I) \subseteq \mathrm{TH}_k(\sqrt[\mathbb{R}]{I})$ for all k.*

We refer the reader to [9, Section 2.2] for a proof. The main message to take away from this result is that whether or not the theta body hierarchy of an ideal converges to $\mathrm{cl}(\mathrm{conv}(V_{\mathbb{R}}(I)))$ is determined by the real variety of I. In particular, whether the theta body sequence of an ideal converges to $\mathrm{cl}(\mathrm{conv}(V_{\mathbb{R}}(I)))$ in finitely many steps, or not, is determined by $\sqrt[\mathbb{R}]{I}$.

7.2.3 Possible Extensions

The focus of this chapter is on polynomial equations, and sums of squares relaxations. However, all this theory can potentially be adapted to work in some more complicated cases. In this section we give examples of some constructions that give a flavor of possible extensions. Similar constructions were also seen in Chapter 6, and we refer to [22] for a more systematic study of the types of techniques we will see below (in a slightly different setting).

Example 7.18. The theta body sequence can be modified to deal with polynomial inequalities, using Lasserre's ideas. Given an ideal I and some polynomials $g_1, \ldots, g_t$, we might want to find the convex hull of the semialgebraic set $S = \{x \in V_{\mathbb{R}}(I) : g_1(x) \geq 0, \ldots, g_t(x) \geq 0\}$. To do this we use *shifted reduced moment matrices* in addition to the reduced moment matrices of I.

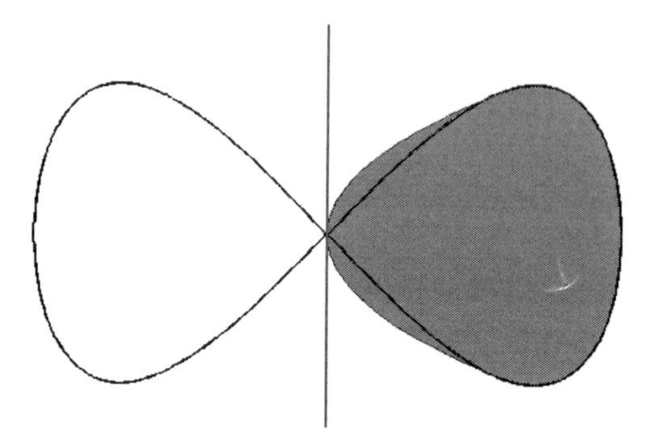

Figure 7.7. *Sum of squares approximation to the half-lemniscate of Gerono.*

Recall that to obtain the kth reduced moment matrix $M_{\mathcal{B}_k}(y)$ of I, we would take the matrix $[x]_{\mathcal{B}_k}[x]_{\mathcal{B}_k}^T$, write it in terms of a basis $\mathcal{B}$ of $\mathbb{R}[x]/I$, and linearize using the new variables y with $y_0 = 1$. To define the shifted reduced moment matrix $M_{\mathcal{B}_k}(g * y)$ (with respect to g), we take the matrix $g(x)[x]_{\mathcal{B}_k}[x]_{\mathcal{B}_k}^T$ and do precisely as before.

Consider for example the ideal $I = \langle x^4 - x^2 + y^2 \rangle$ of the *lemniscate of Gerono*, together with the inequality $x \geq 0$. The semialgebraic set S in this case is the right half-lemniscate shown in Figure 7.7. The second reduced moment matrix of I is given by

$$\begin{pmatrix} 1 & x & y & w_2^0 & w_1^1 & w_0^2 \\ x & w_2^0 & w_1^1 & w_3^0 & w_2^1 & w_1^2 \\ y & w_1^1 & w_0^2 & w_2^1 & w_1^2 & w_0^3 \\ w_2^0 & w_3^0 & w_2^1 & w_2^0 - w_0^2 & w_3^1 & w_2^2 \\ w_1^1 & w_2^1 & w_1^2 & w_3^1 & w_2^2 & w_1^3 \\ w_0^2 & w_1^2 & w_0^3 & w_2^2 & w_1^3 & w_0^4 \end{pmatrix},$$

where w_i^j is the linearization of $x^i y^j$. The combinatorial moment matrix shifted by x and truncated at $k = 1$ is

$$\begin{pmatrix} x & w_2^0 & w_1^1 \\ w_2^0 & w_3^0 & w_2^1 \\ w_1^1 & w_2^1 & w_1^2 \end{pmatrix}.$$

If we force both matrices to be positive semidefinite and project over the x, y coordinates, we get an approximation of the convex hull of the right half of the lemniscate, as shown in Figure 7.7. By increasing the truncation parameter of the reduced moment matrix and the shifted moment matrix we get better approximations to the convex hull.

Note that in this example we are essentially searching for certificates of nonnegativity of the form $l(x, y) \equiv \sigma_0(x, y) + x\sigma_1(x, y) \mod I$, where σ_0 and σ_1 are 2-sos and 1-sos, respectively. ∎

Example 7.19. Consider the *teardrop curve* given by $p(x,y) := x^4 - x^3 + y^2 = 0$. We will see in Corollary 7.45 that the singularity at the origin will prevent the theta bodies of $\langle p \rangle$ from converging in a finite number of steps to the convex hull of the curve. We can, however, get rid of that problem by strengthening the hierarchy in a simple way. Recall that the second theta body in this case will be obtained as the closure of the set of all points $(x,y) \in \mathbb{R}^2$ for which there exists a positive definite matrix of the form

$$\begin{pmatrix} 1 & x & y & w_2^0 & w_1^1 & w_0^2 \\ x & w_2^0 & w_1^1 & w_3^0 & w_2^1 & w_1^2 \\ y & w_1^1 & w_0^2 & w_2^1 & w_1^2 & w_0^3 \\ w_2^0 & w_3^0 & w_2^1 & w_3^0 - w_0^2 & w_3^1 & w_2^2 \\ w_1^1 & w_2^1 & w_1^2 & w_3^1 & w_2^2 & w_1^3 \\ w_0^2 & w_1^2 & w_0^3 & w_2^2 & w_1^3 & w_0^4 \end{pmatrix},$$

where w_i^j is a variable that linearizes the monomial $x^i y^j$, and so the rows and columns are indexed by $\{1, x, y, x^2, xy, y^2\}$. One can in this case strengthen the condition by adding a new row and column to the matrix, indexed not by a monomial but by the fraction $\frac{y}{x}$ that we linearize as w_{-1}^1. We then use the same strategy as before, of linearizing all resulting products modulo the relation $x^4 = x^3 - y^2$ (which allows us to get rid of $w_{4,0}$) and the relations $\frac{y^2}{x} = x^2 - x^3$ and $\frac{y^2}{x^2} = x - x^2$ (which eliminates two more variables). This new *pseudomoment matrix* is given by

$$M(x,y,w) = \begin{pmatrix} 1 & x & y & w_2^0 & w_1^1 & w_0^2 & w_{-1}^1 \\ x & w_2^0 & w_1^1 & w_3^0 & w_2^1 & w_1^2 & y \\ y & w_1^1 & w_0^2 & w_2^1 & w_1^2 & w_0^3 & w_2^0 - w_3^0 \\ w_2^0 & w_3^0 & w_2^1 & w_3^0 - w_0^2 & w_3^1 & w_2^2 & w_1^1 \\ w_1^1 & w_2^1 & w_1^2 & w_3^1 & w_2^2 & w_1^3 & w_0^2 \\ w_0^2 & w_1^2 & w_0^3 & w_2^2 & w_1^3 & w_0^4 & w_{-1}^3 \\ w_{-1}^1 & y & w_2^0 - w_3^0 & w_1^1 & w_0^2 & w_{-1}^3 & x - w_2^0 \end{pmatrix}.$$

Since the original moment matrix is a submatrix of $M(x,y,w)$, the body $Q = \{(x,y) : \exists w \text{ s.t. } M(x,y,w) \succeq 0\}$ must be contained in $\mathrm{TH}_2(\langle p \rangle)$, and a simple numeric computation seems to show that Q actually matches the convex hull of the real variety $V_{\mathbb{R}}(p)$, as we can see in Figure 7.8. In this figure we see a comparison of the second theta body and Q, drawn numerically using YALMIP. The fact that Q seems to be exact is related to the fact that we can now use the term $\frac{x}{y}$ to get sos certificates. For example, $x = x^2 + (\frac{x}{y})^2$ modulo the new identities that we introduced. ∎

Exercise 7.20. Let $I = \langle x^2 \rangle$.

1. Show that x is not k-sos mod I for any k.
2. Show that for any $\varepsilon > 0$, the polynomial $x + \varepsilon$ is 1-sos mod I.
3. Describe $\mathrm{TH}_1(I)$.

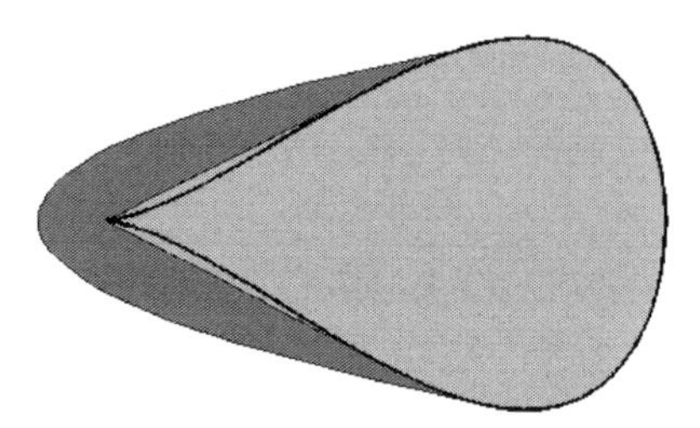

Figure 7.8. *In the darker color we see* $\mathrm{TH}_2(\langle p\rangle)$*, while in the lighter color we see the strengthening* Q *as defined in Example* 7.19*. In black we see the variety itself.*

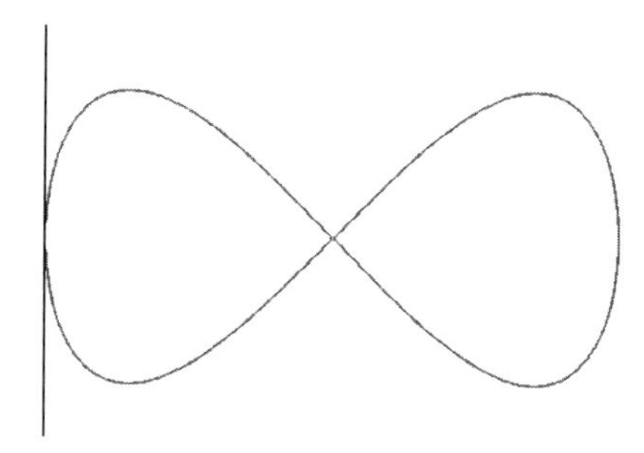

Figure 7.9. *Lemniscate of Gerono.*

Exercise 7.21. Using YALMIP or other software, find the smallest ϵ such that $x+\epsilon$ is 2-sos modulo the ideal $I = \langle x^4 - x^3 + y^2\rangle$. What about 3-sos? What about 4-sos?

Exercise 7.22. The lemniscate of Gerono is given by the equation $x^4 - x^2 + y^2 = 0$ shown in Figure 7.9. Using YALMIP give an approximate 2-sos decomposition of $x+1$ modulo the equation of the curve. Can you find an exact one?

Exercise 7.23. Using reduced moment matrices, give semidefinite descriptions of the following bodies:

1. $Q_{\mathcal{B}_2}(I)$ for the ideal of the lemniscate of Gerono.

2. $Q_{\mathcal{B}_1}(I)$ and $Q_{\mathcal{B}_2}(I)$ where $I = \langle y^2 - x - 1, x^2 - y - 1\rangle$.

3. $Q_{\mathcal{B}_1}(I)$ where I is the vanishing ideal of the vertices of the 0/1 cube in $\mathbb{R}^3$.

Exercise 7.24. Let I be the vanishing ideal of a finite set of points in $\mathbb{R}^n$.

1. Prove that $p(x)$ is nonnegative over $V_{\mathbb{R}}(I)$ if and only if it is a sum of squares modulo the ideal I.

2. Using the above fact, prove that for $\mathcal{B}$, a θ-basis of $\mathbb{R}[x]/I$, the spectrahedron $\{y \in \mathbb{R}^{\mathcal{B}} : M_{\mathcal{B}}(y) \succeq 0, y_0 = 1\}$ is the simplex whose vertices are the vectors $(f_i(s) : f_i + I \in \mathcal{B})$ as s varies over the finitely many points in $V_{\mathbb{R}}(I)$.

7.3 Convergence of Theta Bodies

One of the main questions after defining a sequence of approximations to a convex set is if they actually approximate the set, and further, if some approximation in the sequence is guaranteed to coincide with the set. In this section we examine conditions under which the sequence of theta bodies of an ideal I converges, either finitely or asymptotically, to $\operatorname{conv}(V_{\mathbb{R}}(I))$.

Definition 7.25. *Let $I \subset \mathbb{R}[x]$ be an ideal.*

1. *The theta body sequence of I converges to $\operatorname{cl}(\operatorname{conv}(V_{\mathbb{R}}(I)))$ if*

$$\bigcap_{k=1}^{\infty} \mathrm{TH}_k(I) = \operatorname{cl}(\operatorname{conv}(V_{\mathbb{R}}(I))).$$

2. *For a finite integer k, the ideal I is* TH_k-exact *if* $\mathrm{TH}_k(I) = \operatorname{cl}(\operatorname{conv}(V_{\mathbb{R}}(I)))$.

3. *If I is TH_k-exact for a finite integer k, then we say that the theta body sequence of I converges to $\operatorname{cl}(\operatorname{conv}(V_{\mathbb{R}}(I)))$ in finitely many steps. If the theta body sequence of I converges to $\operatorname{cl}(\operatorname{conv}(V_{\mathbb{R}}(I)))$ but there is no finite k for which I is TH_k-exact, then we say that the theta body sequence of I converges asymptotically to $\operatorname{cl}(\operatorname{conv}(V_{\mathbb{R}}(I)))$.*

We will see in Section 7.3.1 that if $V_{\mathbb{R}}(I)$ is finite, then there is always some finite k for which I is TH_k-exact. However, tight bounds on k for which I is TH_k-exact are not known in general. The best scenario is when I is TH_1-exact. We characterize finite varieties whose real radical ideal is TH_1-exact. Recall from the discussion following Theorem 7.17 that there is no loss of generality in passing to the real radical of I in discussing convergence.

When $V_{\mathbb{R}}(I)$ is infinite, much less is understood about the convergence of the theta body sequence of I. In Section 7.3.2 we explain what we know about this case. The best general result is that when $V_{\mathbb{R}}(I)$ is compact, the theta body sequence is guaranteed to converge to $\operatorname{cl}(\operatorname{conv}(V_{\mathbb{R}}(I)))$ asymptotically. However, finite convergence, and even convergence in the first step are sometimes possible for infinite varieties, although no characterization is known in either case. We show that certain singularities can prevent finite convergence when the variety is compact.

7.3.1 Finite Real Varieties

Theorem 7.26. *Let I be an ideal such that $V_{\mathbb{R}}(I)$ is finite; then there exists some k such that* $\mathrm{TH}_k(I) = \mathrm{conv}(V_{\mathbb{R}}(I))$.

Proof. First note that by Theorem 7.17 we just need to prove the existence of such a k for $J = \sqrt[\mathbb{R}]{I}$. Let $V_{\mathbb{R}}(I) := \{P_1, \ldots, P_m\} \subset \mathbb{R}^n$ and, for each P_i, let q_i be a polynomial such that $q_i(P_i) = 1$ and $q_i(P_j) = 0$ for $j \neq i$. Then given any polynomial $f(x)$ that is nonnegative on $V_{\mathbb{R}}(I)$ we have that

$$f(x) - \sum_{j=1}^{m} \left(\sqrt{f(P_j)} q_j(x) \right)^2$$

vanishes at all P_i, and hence it belongs to J, and f is sos modulo J. So all nonnegative polynomials on $V_{\mathbb{R}}(J)$ are sos modulo J, which in particular implies that each of them is nonnegative over some $\mathrm{TH}_k(J)$. Since the convex hull of $V_{\mathbb{R}}(I)$ is a polytope, it is cut out by a finite number of linear inequalities. Pick k large enough for all these linear inequalities to be valid on $\mathrm{TH}_k(J)$ simultaneously. Then $\mathrm{conv}(V_{\mathbb{R}}(I)) = \mathrm{TH}_k(J)$. □

Clearly, Theorem 7.26 implies that when $V_{\mathbb{C}}(I)$ is finite, the ideal I is TH_k-exact for some finite k. When the ideal I is also radical, finite convergence of its theta body sequence to the convex hull of the variety was proved by Parrilo (see Theorem 2.4 in [16]). Having established finite convergence of the theta body sequence of I when $V_{\mathbb{R}}(I)$ is finite, one can ask the more ambitious question of when such an I is TH_1-exact. This is the most useful and computationally practical case of finite convergence. If the ideal defining a finite set of points is always assumed to be the vanishing ideal of the variety (and hence real radical), we can give a complete geometric characterization of when they are TH_1-exact. We will need the following fact about real radical ideals.

Lemma 7.27 ([8]). *If $I \subset \mathbb{R}[x]$ is a real radical ideal, then a linear inequality $l(x) \geq 0$ is valid for $\mathrm{TH}_k(I)$ if and only if $l(x)$ is k-sos modulo I.*

In order to characterize real radical ideals with finite real varieties, we need a new definition.

Definition 7.28. *Given a polytope P, we say that P is* 2-level *if for each facet F of P and its affine span H_F, all vertices of P are either in F or in a unique translate of H_F.*

Example 7.29. In $\mathbb{R}^3$, up to affine equivalence there are five three-dimensional 2-level polytopes, shown in the upper part of Figure 7.10. It is easy to see that a 2-level polytope must be affinely equivalent to a 0/1-polytope. In the bottom of Figure 7.10 we show the three remaining 0/1-polytopes (up to affine equivalence) with a face that fails to verify the 2-level condition highlighted. ■

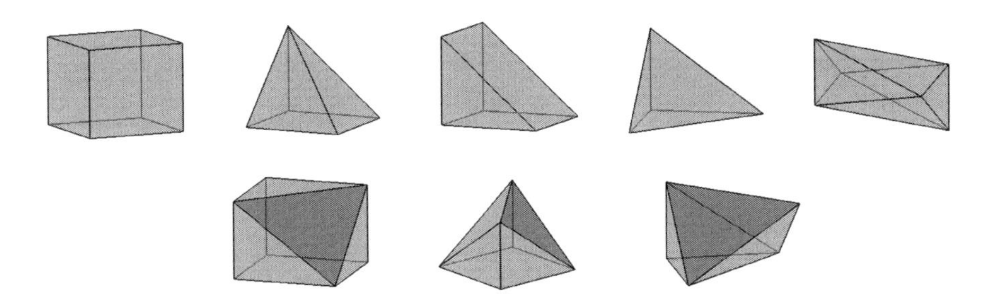

Figure 7.10. *The top row contains all $0/1$ three-dimensional 2-level polytopes (up to affine equivalence). The bottom row contains all $0/1$ three-dimensional polytopes (up to affine equivalence) that are not 2-level.*

Theorem 7.30. *Let I be real radical with $S := V_{\mathbb{R}}(I)$ finite. Then I is* TH_1*-exact if and only if S is the set of vertices of a 2-level polytope.*

Proof. Assume without loss of generality that S spans the entire space and let $f_1(x) \geq 0, \ldots, f_m(x) \geq 0$ be a minimal list of linear inequalities describing $P := \mathrm{conv}(S)$, i.e., each f_i corresponds to a facet F_i of P and is zero on that facet. By Lemma 7.27, I is TH_1-exact if and only if all f_i are 1-sos mod I, since every affine linear polynomial that is nonnegative on S is a nonnegative linear combination of the f_i's.

If I is TH_1-exact, for each $i = 1, \ldots, m$, we have $f_i(x) \equiv \sum (h_k(x))^2 \mod I$, where all h_k are linear. But since f_i vanishes on $S \cap F_i$ so must all h_k and therefore, since they are linear, they must vanish on the affine space generated by F_i. This means that they are actually just scalar multiples of f_i and we have $f_i(x) \equiv \lambda (f_i(x))^2 \mod I$, for some nonnegative λ. In particular, all points $P \in S$ must satisfy either $f_i(P) = 0$ or $f_i(P) = 1/\lambda$ proving the 2-level condition.

Suppose now that P is 2-level. Then for each f_i, all points $P \in S$ must satisfy $f_i(P) = 0$ or $f_i(P) = \lambda_i$, for some fixed $\lambda_i > 0$. But then $f_i(f_i - \lambda_i)$ vanishes on S, and therefore belongs to I. This implies $f_i \equiv (1/\lambda_i) f_i^2 \mod I$ and f_i is 1-sos modulo I. $\square$

Theorem 7.30 will turn out to be very useful in the context of combinatorial optimization as we will see in the next section. Polytopes with integer vertices that are 2-level are called *compressed polytopes* in the literature [34, 35] and play an important role in other research areas. Being 2-level is a highly restrictive condition that immediately gives us much information on the polytope. Since all the vertices of a 2-level polytope in $\mathbb{R}^n$ can be assumed to be $0/1$ vectors, it is clear that they have at most 2^n vertices. It was shown in [8] that they also have at most 2^n facets which is not obvious. There are many infinite families of 2-level polytopes such as simplices, hypercubes, cross polytopes, and hypersimplices.

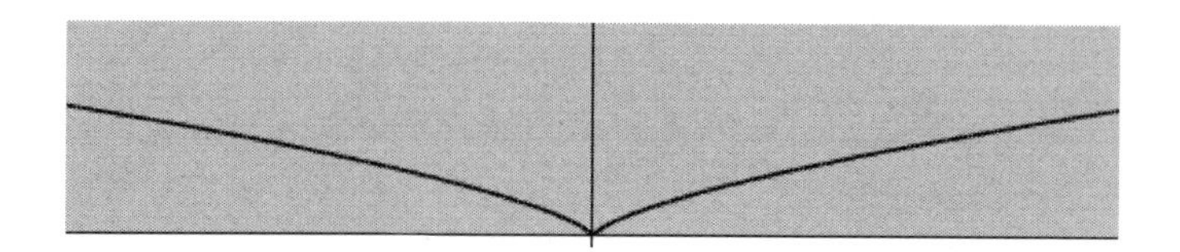

Figure 7.11. *Cusp and its convex hull.*

7.3.2 Infinite Real Varieties

We begin by showing that unlike for finite varieties, the theta body approximations can fail drastically when $V_{\mathbb{R}}(I)$ is infinite. The following simple example is adapted from Example 1.3.2 in [21].

Example 7.31. Consider the ideal $I = \langle x^2 - y^3 \rangle$ defining the cusp in Figure 7.11. The closure of the convex hull of this curve is the upper half-plane, so the only linear inequalities valid on the curve are of the form $l_\varepsilon(x, y) = y + \varepsilon$, where $\varepsilon \geq 0$. Suppose there exists some l_ε with an sos certificate modulo I, then $l_\varepsilon(x, y) \equiv \sum p_i(x, y)^2 \mod I$ for some polynomials p_i. Note that any polynomial p has a unique standard form of the type $a(y) + xb(y)$ modulo this ideal, which we can obtain by reducing all multiples of x^2, using the fact that $x^2 \equiv y^3 \mod I$. Two polynomials are the same modulo the ideal if they have the same standard form. Since $l_\varepsilon(x, y)$ is already in this form, we can simply reduce the right-hand side in the congruence relation to its standard form too. Suppose each $p_i = a_i(y) + xb_i(y)$. Then it is easy to check that

$$\sum p_i(x, y)^2 \equiv \sum (a_i(y)^2 + y^3 b_i(y)^2) + \sum (2x a_i(y) b_i(y)) \mod I.$$

Since the right-hand side is in standard form, to be congruent to l_ε it must be the same as l_ε. Looking at the maximum degree of y in the first sum on the right, we see that it is smaller than two only if the a_i's are all constants and the b_i's are all zero, since the highest degree terms cannot all cancel. In particular we get $y + \varepsilon$ is a constant, which is clearly a contradiction. This proves that $\mathrm{TH}_k(I) = \mathbb{R}^2$ for all k, and the theta bodies are completely ineffective in approximating $\mathrm{conv}(V_{\mathbb{R}}(I))$. In fact, the same proof would work for any curve of the form $x^2 - p(y)$ where p has odd degree. ∎

However, despite the existence of "badly behaved" varieties such as the one presented above, there is a large, very interesting class of infinite real varieties where such behavior never occurs, namely, compact varieties.

Theorem 7.32. *Let I be an ideal such that $V_{\mathbb{R}}(I)$ is compact. Then the theta body sequence of I converges to the convex hull of the variety $V_{\mathbb{R}}(I)$ in the sense that*

$$\bigcap_{k=1}^{\infty} \mathrm{TH}_k(I) = \mathrm{conv}(V_{\mathbb{R}}(I)).$$

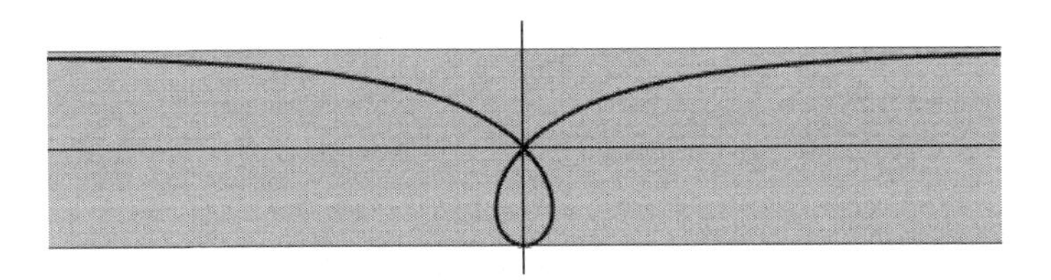

Figure 7.12. *Strophoid curve and its convex hull.*

This is an immediate consequence of Schmudgën's *Positivstellensatz* (see Chapter 3). To see the connection, just consider any set of generators $\{g_1, \ldots, g_t\}$ for I and the semialgebraic set $S = \{x \in \mathbb{R}^n : \pm g_1 \geq 0, \ldots \pm g_t \geq 0\} = V_{\mathbb{R}}(I)$. When applied to S, Schmudgën's Positivstellensatz guarantees that every linear polynomial that is strictly positive over $V_{\mathbb{R}}(I)$ is sos modulo I.

Example 7.33. The existence of varieties as in Example 7.31 does not imply that for all unbounded varieties we have problems with the theta body sequence. Consider the *strophoid* curve given by $p(x, y) := (1 - y)x^2 - (1 + y)y^2 = 0$, shown in Figure 7.12. The closure of the convex hull of this variety is the band B defined by $-1 \leq y \leq 1$. We claim that $\mathrm{TH}_2(I) = B$. To show this it is enough to prove that both $1 - y$ and $1 + y$ are 2-sos modulo I, which is true since

$$1 \pm y = \left(1 \pm \frac{1}{2}y - \frac{1}{2}y^2\right)^2 + \frac{1}{4}\left(\mp y - y^2\right)^2 + \frac{1}{2}\left(xy - x\right)^2 + \frac{1}{2}(y - 1)p(x, y). \qquad \blacksquare$$

In what follows we concentrate our efforts on the compact case, where asymptotic convergence of the theta body sequence is guaranteed. The next natural question when $V_{\mathbb{R}}(I)$ is infinite but compact is whether we can understand when the theta body sequence converges in finitely many steps to $\mathrm{cl}(\mathrm{conv}(V_{\mathbb{R}}(I)))$. Finite convergence would prove that $\mathrm{conv}(V_{\mathbb{R}}(I))$ is the projection of a spectrahedron, which is an important feature of a convex semialgebraic set as seen in Chapter 6. There is no complete understanding of this situation, but in the remainder of this section, we discuss the known results.

TH_1-exactness. We begin by discussing the strongest scenario within finite convergence, namely TH_1-exactness of an ideal. In spite of the strength of this property, there are surprisingly many interesting examples of such ideals with infinite real varieties. We begin by taking a general look at the notion of TH_1-exactness for all ideals. Roughly speaking, TH_1-exact ideals are those whose quadratic elements are enough to describe their convex geometry, a statement that will be made precise shortly. We start with a small lemma concerning convex quadrics.

Lemma 7.34. *If $p \in \mathbb{R}[x]$ is a convex quadric polynomial, then $\langle p \rangle$ is* TH_1*-exact.*

Proof. This result will follow from Proposition 7.41, where we will show that the first theta body of any quadric is simply the convex hull of its graph intersected with the x-plane. This intersection is precisely $\text{conv}(p)$ if p is convex. $\square$

We now give an alternative characterization of $\text{TH}_1(I)$ for any ideal I.

Proposition 7.35. *For any ideal* $I \subseteq \mathbb{R}[x]$, $\text{TH}_1(I)$ *equals the intersection of* $\text{conv}(V_{\mathbb{R}}(p))$ *as* p *varies over all convex quadrics in* I.

Proof. The inclusion $\text{TH}_1(I) \subseteq \text{conv}(V_{\mathbb{R}}(p))$ for all convex quadrics $p \in I$ is easy, since a linear inequality is valid over the second set if and only if it is 1-sos modulo $\langle p \rangle$, which immediately implies that it is 1-sos modulo I and therefore valid on $\text{TH}_1(I)$. For the second inclusion note that if $l(x)$ is 1-sos mod I, then

$$l(x) = \sigma(x) + g(x),$$

where σ is a sum of squares and g is a quadric in I. But note that $-\nabla^2 g = \nabla^2 \sigma \succeq 0$ which implies $-g$ is a convex quadric in I, and $l(x)$ is 1-sos modulo $\langle -g \rangle$. Therefore, $l(x) \geq 0$ is valid on $\text{conv}(V_{\mathbb{R}}(-g))$ and hence also valid on the intersection of $\text{conv}(V_{\mathbb{R}}(p))$ as p varies over all convex quadrics in I. $\square$

Example 7.36. Consider the ideal $I = \langle x^4 - y^2 - z^2, x^4 + x^2 + y^2 - 1 \rangle$ that we introduced in Example 7.11. This is the intersection of two quartic surfaces in $\mathbb{R}^3$. The Gröbner basis computation we did then shows that there exists a single quadric in this ideal (up to scalar multiplication), which is the polynomial $-1+x^2+2y^2+z^2$. Therefore, $\text{TH}_1(I)$ equals the ellipsoid $\{(x, y, z) \in \mathbb{R}^3 : x^2 + 2y^2 + z^2 \leq 1\}$, as seen in Figure 7.4. ■

Proposition 7.35 can sometimes be used to prove TH_1-exactness.

Example 7.37. Consider the ideal $I = \langle x^2 + y^2 + z^2 - 4, (x-1)^2 + y^2 - 1 \rangle$, from Example 7.47. Note that the quadratic polynomials $p_1 = (x-1)^2 + y^2 - 1$ and $p_2 = 2x + z^2 - 4$ belong to I. Write $I_1 = \langle p_1 \rangle$ and $I_2 = \langle p_2 \rangle$. Then we claim that

$$\text{conv}(V_{\mathbb{R}}(I)) = \text{conv}(V_{\mathbb{R}}(I_1)) \cap \text{conv}(V_{\mathbb{R}}(I_2)),$$

and therefore I is TH_1-exact. To see this note that the variety $V_{\mathbb{R}}(I)$ can be written as

$$\{(x, \pm\sqrt{1-(x-1)^2}, \pm\sqrt{4-2x}) \, : \, 0 \leq x \leq 2\}.$$

In particular for each fixed x we get four points, and the rectangle they form must be contained in the convex hull of $V_{\mathbb{R}}(I)$. This means

$$\{(x,y,z) \in \mathbb{R}^3 \, : \, |y| \leq \sqrt{1-(x-1)^2}, |z| \leq \sqrt{4-2x}, 0 \leq x \leq 2\} \subseteq \text{conv}(V_{\mathbb{R}}(I)),$$

but it is clear that this set can be rewritten as

$$\{(x,y,z) \in \mathbb{R}^3 \, : \, y^2 \leq 1-(x-1)^2, z^2 \leq 4-2x\} = \text{conv}(V_{\mathbb{R}}(I_1)) \cap \text{conv}(V_{\mathbb{R}}(I_2)),$$

which contains $\text{conv}(V_{\mathbb{R}}(I))$, so we get the intended equality. ■

An important open question concerning TH_1-exactness of varieties comes from oriented Grassmannians and illustrates that the TH_1 relaxation can be surprisingly powerful. For the purposes of this discussion, we define the oriented Grassmannian $G_{k,n}$ to be the set of all oriented k-subspaces of $\mathbb{R}^n$, embedded in $\mathbb{R}^{\binom{n}{k}}$ by taking Plücker coordinates, i.e., by picking an oriented basis of the space, writing the vectors as an $n \times k$ matrix, and taking all $k \times k$ minors and scaling them by a positive scalar to a point on the sphere $S^{\binom{n}{k}-1}$.

The ideal $I_{k,n}$, generated by all the quadratic relations among the $k \times k$ minors of an $n \times k$ matrix, is called the *Plücker* ideal. The Grassmann variety is then the compact real variety of the ideal $I = I_{n,k} + \langle 1 - \|x\|^2 \rangle$, so it makes sense to approximate it with theta bodies. It is unknown whether all Grassmann varieties are TH_1-exact, in fact even the $G_{3,6}$ case is unknown, but numerical simulations seem to say it is, at least for the relatively small examples for which numerical computations are doable. Unpublished work by Sanyal and Rostalski [26] makes connections between TH_1-exactness of these ideals and some classical open questions of Harvey and Lawson on calibrated geometries [12].

Exactness in one step for principal ideals. Principal ideals are the simplest ideals with infinite varieties. However, even in this case, TH_1-exactness is not to be expected. In fact, if p has degree d and $2k < d$, $\mathrm{TH}_k(p)$ is the full ambient space $\mathbb{R}^n$, since any k-sos linear inequality would verify $l(x) = \sigma(x) + g(x)$ with degree of the sums of squares σ less than or equal to $2k$. But the degree of $g \in I$ must be at least d so there would be no cancellation of the highest degree and the sum could never be a linear polynomial. An interesting question in this case is whether and when the first meaningful theta body would equal $\mathrm{conv}(V_{\mathbb{R}}(p))$ when $I = \langle p \rangle$. We will focus on the following problem: given a polynomial p of degree $2k$, decide if $\langle p \rangle$ is TH_k-exact. In this generality there is a simple necessary criterion, but we have to introduce a few definitions in order to state it.

Definition 7.38. *Consider a polynomial $p \in \mathbb{R}[x_1, \dots, x_n]$ and define $\tilde{p} = x_0 - p(x_1, \dots, x_n) \in \mathbb{R}[x_0, x_1, \dots, x_n]$. Consider the convex set $C = \mathrm{conv}(V_{\mathbb{R}}(\tilde{p}))$, which is simply the convex hull of the graph of p, and define the* shadow area *of p, denoted by $\mathrm{sh}(p)$, as the intersection of C with the plane $x_0 = 0$.*

This shadow area clearly contains $\mathrm{conv}(V_{\mathbb{R}}(p))$ since it is convex and contains the variety. However we can easily establish a more interesting inclusion.

Proposition 7.39. *For $p \in \mathbb{R}[x]$ of degree $2k$, $\mathrm{sh}(p) \subseteq \mathrm{TH}_k(\langle p \rangle)$. In particular if $\mathrm{sh}(p)$ strictly contains the closure of the convex hull of $V_{\mathbb{R}}(p)$, then $\langle p \rangle$ is not TH_k-exact.*

Proof. Let $l(x)$ be k-sos modulo $\langle p \rangle$, i.e., $l(x) = \sigma(x) + \lambda p(x)$ where σ is a sum of squares of degree at most $2k$ and $\lambda \in \mathbb{R}$. Then $l(x) - \lambda p(x) = \sigma(x)$ implies $l(x) - \lambda p(x) \geq 0$ everywhere and therefore $\tilde{l}(x_0, x) := l(x) - \lambda x_0$ is valid over $V_{\mathbb{R}}(\langle \tilde{p} \rangle)$ and hence over its convex hull too. But by intersecting with $x_0 = 0$ we

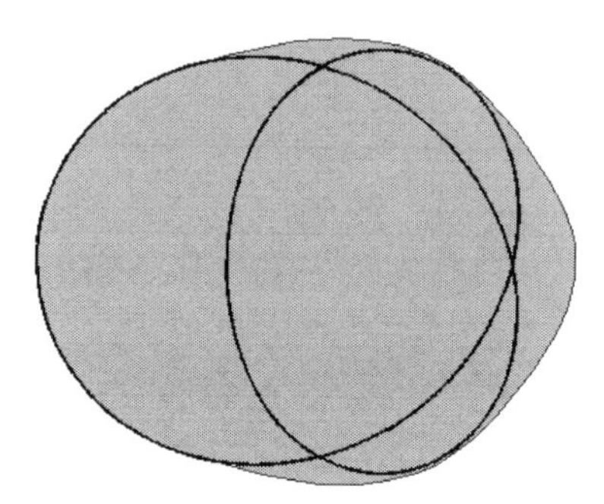

Figure 7.13. *Scarabaeus curve and its third theta body.*

get that $l(x) \geq 0$ must be valid on $\mathrm{sh}(p)$. From the definition of $\mathrm{TH}_k(I)$ it follows immediately that $\mathrm{sh}(p) \subseteq \mathrm{TH}_k(I)$ as intended. □

Despite the simplicity of the criterion, it is a handy tool to prove that a principal ideal is not exact at the first step, without relying on numerical approximations.

Example 7.40. Consider the *scarabaeus* curve given by

$$p(x,y) := (x^2+y^2)(x^2+y^2+4x)^2 - (x^2-y^2)^2 = 0.$$

A simple numerical computation with an SDP solver shows us that $\mathrm{TH}_3(\langle p \rangle)$ does not match the convex hull of the curve, as can be seen in Figure 7.13. To provide a short exact proof, one just has to point out that $p(-4,0) = 256$ and $p(1,0) = 24$, and since the point $(\frac{4}{7}, 0, 0)$ lies in the segment between $(-4, 0, 256)$ and $(1, 0, 24)$, the point $\xi = (\frac{4}{7}, 0)$ must be contained in $\mathrm{sh}(p)$ and therefore in $\mathrm{TH}_3(\langle p \rangle)$. It is, however, easy to calculate that the maximum value that x attains on the curve is $(-50 + 11\sqrt{22})/27 \approx 0.06$, which implies that the convex hull must not contain ξ. ■

In some very special cases we can actually say a bit more about the first meaningful theta body.

Proposition 7.41. *Let p be a polynomial in n variables and degree $2d$. Then*

1. *if $n = 1$, $\mathrm{sh}(p) = \mathrm{TH}_d(\langle p \rangle)$;*
2. *if $d = 1$, $\mathrm{sh}(p) = \mathrm{TH}_1(\langle p \rangle)$;*
3. *if $n = 2$ and $d = 2$, $\mathrm{sh}(p) = \mathrm{TH}_2(\langle p \rangle)$.*

Proof. We just have to prove that in these cases $\mathrm{sh}(p) \supseteq \mathrm{TH}_d(\langle p \rangle)$. To do this let $l(x) > 0$ be a valid linear inequality over $\mathrm{sh}(p)$. This means that the line $L = \{(x_0, x) : x_0 = 0, l(x) = 0\}$ does not intersect $C = \mathrm{conv}(V_{\mathbb{R}}(\langle \tilde{p} \rangle))$. By the

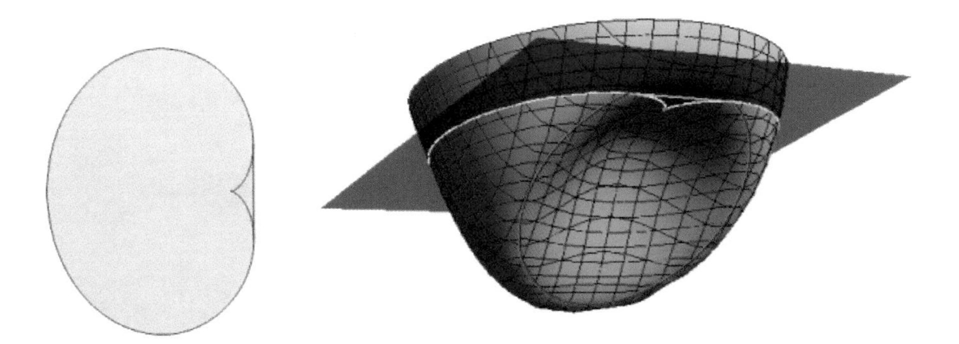

Figure 7.14. *On the left we see the cardioid $p(x) = 0$ and its convex hull. On the right we see the graph of p, its intersection with the plane $z = 0$ and the ellipsoidal region where the graph and the boundary of its convex hull differ.*

separation theorem for convex sets we can therefore take a hyperplane H that strictly separates L and C. Since H does not touch the graph of p, it depends on x_0, and since it does not touch L, it must be parallel to it. Therefore we have a hyperplane of the form $l'(x_0, x) := x_0 + \lambda(l(x) - \varepsilon) = 0$, with $\lambda \neq 0$, $\varepsilon > 0$. Since $\tilde{p}(x_0, x) = x_0 - p(x)$, this means that $\sigma(x) := p(x) + \lambda(l(x) - \varepsilon)$ is always nonnegative or always nonpositive. Without loss of generality assume it is always nonnegative (which implies $\lambda > 0$). Since the degree and number of variables of this polynomial fall under Hilbert's result (see Chapter 4), $\sigma(x)$ is a sum of squares. Hence, $l(x) = \sigma(x)/\lambda + \varepsilon - p(x)/\lambda$ is d-sos modulo the ideal, which implies that $l(x) \geq 0$ is valid over $\mathrm{TH}_d(\langle p \rangle)$, proving the inclusion. ◻

Example 7.42. We use the above result to prove TH_2-exactness of the following principal ideal. Consider

$$p(x, y) = (x^2 + y^2 + 2x)^2 - 4(x^2 + y^2)$$

defining a *cardioid*, and the function

$$q(x, y) = \begin{cases} p(x, y) & \text{if } (x+1)^2 + y^2 \geq 3, \\ 8x - 4 & \text{if } (x+1)^2 + y^2 < 3. \end{cases}$$

One can check that q is smooth and convex by noticing that $p(x, y) = ((x+1)^2 + y^2 - 3)^2 + 8x - 4$ and by looking at its Hessian. Furthermore, the convex hull of the graph of p is just the region above the graph of q. Therefore $\mathrm{sh}(p) = \{(x, y) : q(x, y) \leq 0\}$, and we can see in Figure 7.14 that $\mathrm{sh}(p)$ is the convex hull of the cardioid. ■

Even for one-variable polynomials this result is interesting.

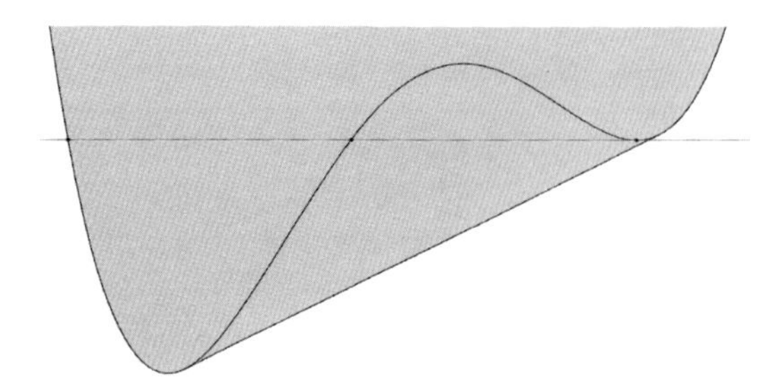

Figure 7.15. *Graph of the polynomial $x - x^2 - x^3 + x^4$, its convex hull, and intersection with the x-axis.*

Example 7.43. Consider the polynomial $p(x) = x - x^2 - x^3 + x^4$. In Figure 7.15 we can see that this polynomial is not TH_2-exact, and why that happens. The double root at $x = 1$ forces the convex hull of the graph to include some points to the right of $x = 1$. In fact one can compute precisely the double tangent that defines the boundary of the convex hull and show that $\mathrm{TH}_2(\langle p \rangle) = \left[-1, \frac{25}{24}\right]$. ■

Singularities and convergence. We now return to the more general question of finite convergence of the theta body sequence for an ideal with an infinite real variety. There is no complete understanding of the obstructions to finite convergence, but we now show that if $V_{\mathbb{R}}(I)$ has certain types of *singularities*, then finite convergence is not possible.

Given an ideal I and a point P on the real variety of I, we define the *normal space* $N_P(I)$ to be the linear space $\{\nabla f(P) \,:\, f \in I\}$.

Proposition 7.44. *Let $l(x)$ be an affine polynomial such that $l(P) = 0$ for some P in $V_{\mathbb{R}}(I)$. If $\nabla l \notin N_P(I)$, then l is not a sum of squares modulo I.*

Proof. Suppose l is a sum of squares. Then

$$l(x) = \sigma(x) + g(x) \tag{7.3}$$

for some sum of squares σ and some polynomial $g \in I$. By evaluating at P we get that $\sigma(P) = 0$, which immediately implies $\nabla\sigma(P) = 0$. By differentiating (7.3) we get

$$\nabla l = \nabla\sigma(x) + \nabla g(x), \tag{7.4}$$

and by evaluating at P we get that $\nabla l = \nabla g(P) \in N_P(I)$. □

If I is real radical we can say even more.

Corollary 7.45. *If I is real radical and $l(x) \geq 0$ is a linear inequality valid on $V_{\mathbb{R}}(I)$ with $l(P) = 0$ at a point $P \in V_{\mathbb{R}}(I)$ such that $\nabla l \notin N_P(I)$, then I is not TH_k-exact for any k.*

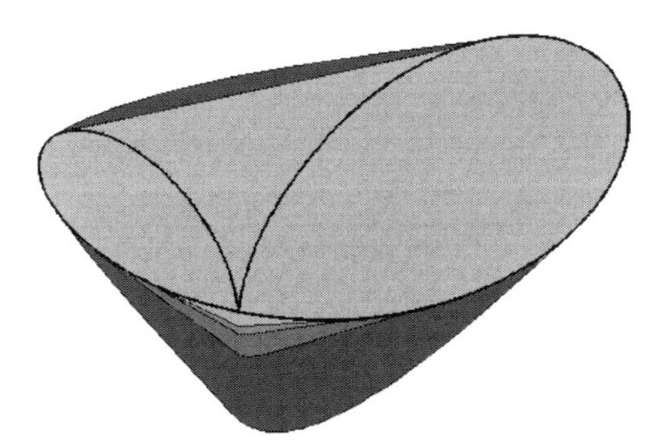

Figure 7.16. $\mathrm{TH}_2(I)$, $\mathrm{TH}_3(I)$, $\mathrm{TH}_4(I)$, *and* $\mathrm{TH}_5(I)$*: all contain the origin in their interior.*

Proof. This follows from the previous proposition and Lemma 7.27. □

Example 7.46. Let $p(x,y) = (x^2+y^2)^2 - (x+5y)x^2$ and $I = \langle p \rangle$. This ideal defines a *bifolium* with a singularity at the origin, which implies $N_{(0,0)}(I) = \{(0,0)\}$. Furthermore the linear inequality $x + 5y \geq 0$ is valid on the variety and holds with equality at the origin. Since $(1,5) \notin N_{(0,0)}(I)$ we immediately have that this inequality does not hold for any theta body relaxation of this ideal. In Figure 7.16 we can see $\mathrm{TH}_k(I)$ for $k = 2,3,4,5$, and see that in fact the inequality does not hold for any of them. ■

Corollary 7.45 essentially tells us that certain singularities of the ideal I that are in the boundary of the convex hull of $V_{\mathbb{R}}(I)$ affect the convergence of the theta bodies of I. For a point $P \in V_{\mathbb{R}}(I)$, the expected dimension of the normal space $N_P(I)$ is the codimension of $V_{\mathbb{R}}(I)$. A reasonable notion of a singularity of I is a point $P \in V_{\mathbb{R}}(I)$ for which $N_P(I)$ has smaller dimension than expected. The next example will show that just the existence of singularities of I on the boundary of $\mathrm{conv}(V_{\mathbb{R}}(I))$ is not enough for Corollary 7.45 to apply.

Example 7.47. Consider the variety $V_{\mathbb{R}}(I)$ in $\mathbb{R}^3$ defined by the ideal

$$I = \langle x^2 + y^2 + z^2 - 4, (x-1)^2 + y^2 - 1 \rangle.$$

As seen in Figure 7.17, this variety looks like a curved figure-eight and has a singularity at the point $p = (2,0,0)$, which belongs to the boundary of $\mathrm{conv}(V_{\mathbb{R}}(I))$. This happens since $N_P(I) = \mathbb{R}\{(1,0,0)\}$ has dimension one, smaller than the codimension of the variety, which is two. However, $(2,0,0)$ does not cause problems for the convergence of theta bodies since the only linear polynomial that is zero at p and nonnegative on $V_{\mathbb{R}}(I)$ is the polynomial $2 - x$, whose gradient is in $N_P(I)$. Indeed, the first theta body of I already equals $\mathrm{conv}(V_{\mathbb{R}}(I))$, as we will see in Example 7.37. ■

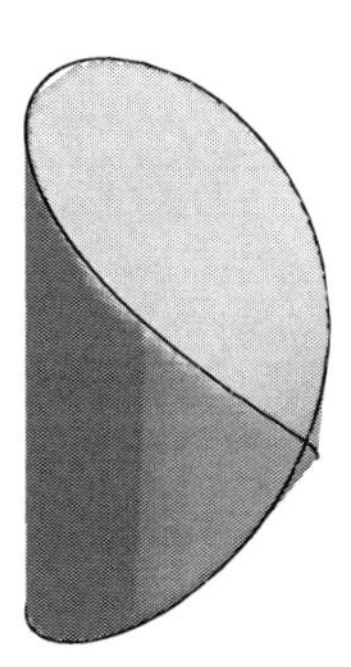

Figure 7.17. *The curved eight variety and its convex hull.*

A better, more refined, way of looking at singularities was introduced by Omar and Osserman in [23]. They introduce a stronger notion of nonnegativity over varieties that yields a stronger necessary condition for finite convergence of the theta body hierarchy. As a byproduct they prove the following result.

Theorem 7.48. *Let $f(x)$ be a polynomial such that there exists some positive integer n and an $\mathbb{R}$-algebra homomorphism $\varphi : \mathbb{R}[x]/I \to \mathbb{R}[\varepsilon]/\langle \varepsilon^n \rangle$ for which $\varphi(f) = a_0 + a_1\varepsilon + \cdots + a_{n-1}\varepsilon^{n-1}$. If the first nonzero (leading) coefficient a_i is negative, then f is not a sum of squares modulo I.*

Proof. Just note that homomorphisms send sums of squares to sums of squares, and sums of squares in $\mathbb{R}[\varepsilon]/\langle \varepsilon^n \rangle$ always have their leading coefficient nonnegative. □

Again this immediately gives us a new criterion.

Corollary 7.49. *Let I be a real radical ideal and $l(x) \geq 0$ a linear inequality valid on $V_{\mathbb{R}}(I)$. If there exists an $\mathbb{R}$-algebra homomorphism $\varphi : \mathbb{R}[x]/I \to \mathbb{R}[\varepsilon]/\langle \varepsilon^n \rangle$ for which $\varphi(l)$ has negative leading coefficient, then I is not* TH_k*-exact for any k.*

This corollary is much stronger than Corollary 7.45, and examples showing the difference are presented in [23]. In our next example we just show that we can recover Corollary 7.45 from Corollary 7.49 for the variety in Example 7.46 but, in fact, we can do so for any variety just by considering maps to $\mathbb{R}[\varepsilon]/\langle \varepsilon^2 \rangle$.

Example 7.50. Let $p(x, y) = (x^2+y^2)^2-(x+5y)x^2$ and $I = \langle p \rangle$ as in Example 7.46. Then the map $\varphi : \mathbb{R}[x, y]/I \to \mathbb{R}[\varepsilon]/\langle \varepsilon^2 \rangle$ defined by $\varphi(x) = \varphi(y) = -\varepsilon$ is well defined, since $\varphi(p) = 0$. However, $\varphi(x+5y) = -6\varepsilon$ has a negative leading coefficient despite $x + 5y \geq 0$ being valid on the variety. Hence, $\langle p \rangle$ is not TH_k-exact for any k. ■

One should keep in mind that singularities are not necessarily the only things that prevent finite convergence of the theta body sequence to $\mathrm{cl}(\mathrm{conv}(V_{\mathbb{R}}(I)))$. For compact smooth curves and surfaces, Scheiderer proved that nonnegativity and

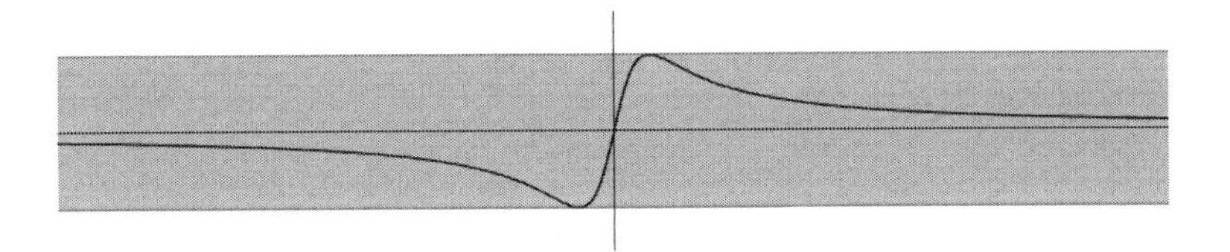

Figure 7.18. *Serpentine curve and the closure of its convex hull.*

sums of squares modulo the ideal are equivalent [28, 29]. However, even in these cases, it is an open question if one can bound the degree needed to represent every nonnegative affine polynomial as a sum of squares modulo the ideal. Thus there might be examples of smooth curves and surfaces with no finite convergence of the theta body hierarchy to $\mathrm{conv}(V_{\mathbb{R}}(I))$. The only cases where we know a little more is when the genus of the curve is one.

Proposition 7.51 (Theorem 2.1 [30]). *If $V_{\mathbb{R}}(I)$ is a smooth curve of genus 1 with at least one nonreal point at infinity, then I is TH_k-exact for some k.*

Genus zero curves can be rationally parametrized which allows semidefinite representations of their convex hulls by means of sums of squares, as seen in [13]. However such constructions do not automatically translate to finite convergence of the theta body sequence to the convex hull of the curve, even in the smooth case.

For varieties of dimension greater than two, there always exist nonnegative polynomials that are not sums of squares modulo any ideal that defines them, even in the smooth compact case, as seen in [27]. It is therefore very natural to expect examples of smooth compact varieties with no finite convergence of the theta body hierarchy, but we do not know a concrete example at this point.

Exercise 7.52. Consider the serpentine curve given by $p(x) := y(x^2+1) - x = 0$, depicted in Figure 7.18. The closure of its convex hull is the band cut out by the inequalities $-1/2 \leq y \leq 1/2$. Show that the ideal $I = \langle p \rangle$ is TH_2-exact by giving an exact expression of $1 - 2y$ and $1 + 2y$ as 2-sos polynomials modulo I.

Exercise 7.53. Using Proposition 7.35 show that the first theta body of the vanishing ideal of the points $\{(0,0),(1,0),(0,1),(2,2)\}$ is cut out by precisely two polynomial inequalities, and write them explicitly.

Exercise 7.54. Consider the ideal $I = \langle y^2 - x^5, z - x^3 \rangle$. The inequality $z \geq 0$ is valid on the variety $V_{\mathbb{R}}(I)$.

1. Can we use Proposition 7.44 to prove that z is not k-sos modulo I for any k?

2. Use Theorem 7.48 to prove that z is not k-sos modulo I for any k.

Exercise 7.55. Similarly to our definition of 2-level polytope, we can define a k-level polytope to be one where given a facet F, and the affine plane H_F that it spans, all vertices of the polytope are contained either in H_F or in one of $k-1$ parallel translates of H_F. Prove that if S is the set of vertices of a $(k+1)$-level polytope then the vanishing ideal of S, $I(S)$, is TH_k-exact.

Exercise 7.56. Consider the univariate quartic polynomial $p(x) = x^4 - 3x^3 + 3x^2 - 3x + 2$ which has two real roots, 1 and 2. Compute $\mathrm{TH}_2(\langle p \rangle)$ exactly. Is the ideal TH_2-exact?

Exercise 7.57. Consider the bifolium given by $p(x,y) := (x^2+y^2)^2 - yx^2 = 0$. This curve has a singularity at the origin, which is also on the boundary of its convex hull and satisfies the conditions of Corollary 7.45, and hence we know that its theta body hierarchy does not converge. Using the same ideas as in Example 7.19, add to the second moment matrix of $I = \langle p \rangle$ a row and a column indexed by $\frac{y^2}{x}$. Plot the resulting approximation and compare it with the convex hull of the curve.

7.4 Combinatorial Optimization

In this final section, we focus on combinatorial optimization where a typical problem involves optimizing a linear function over all combinatorial objects of a certain kind. Many of these problems are modeled using graphs and can sometimes be studied combinatorially. However, a more systematic approach is to model these problems as integer or linear programs, which puts an emphasis on the underlying geometry. These models work as follows. The combinatorial objects of interest are typically defined as subsets of the ground set $[n] := \{1, 2, \ldots, n\}$ and the object $T \subseteq [n]$ is recorded via its *characteristic vector* $\chi^T \in \{0,1\}^n$ defined as $\chi_i^T = 1$ if $i \in T$ and $\chi_i^T = 0$ otherwise. This creates a simple bijection between the objects and certain elements of $\{0,1\}^n$. Then, for a vector $c \in \mathbb{R}^n$, maximizing $\sum_{i \in T} c_i$ over all the objects $\{T\}$ is equivalent to maximizing $\sum c_i x_i$ over the characteristic vectors $\{\chi^T\}$ which in turn is equivalent to maximizing $\sum c_i x_i$ over $\mathrm{conv}(\{\chi^T\})$ which is a $0/1$ polytope by construction. (Recall that a $0/1$ polytope in $\mathbb{R}^n$ is the convex hull of vectors in $\{0,1\}^n$.) In principle this is a linear program but the difficulty is that no description of $\mathrm{conv}(\{\chi^T\})$ is usually known, and one resorts to relaxations of $\mathrm{conv}(\{\chi^T\})$ over which $\sum c_i x_i$ is maximized to obtain an upper bound on the value of $\max\{\langle c, x \rangle : x \in \mathrm{conv}(\{\chi^T\})\}$.

The theory of integer programming offers general methods to construct polyhedral relaxations of $\mathrm{conv}(\{\chi^T\})$ by first finding a polytope whose integer points are precisely $\{\chi^T\}$. See [31, Chapter 23] for linear programming-based methods. Polyhedral relaxations can sometimes be found using combinatorial arguments that depend explicitly on the structure of the problem. Automatic methods for constructing relaxations have also come about from *lift-and-project* methods that find a sequence of polyhedral or spectrahedral relaxations of $\mathrm{conv}(\{\chi^T\})$. Some examples of lift-and-project methods besides, the theta body method described in this chapter, can be found in [2, 14, 20, 33] (see also [15]). Theta bodies construct relaxations of $\text{conv}(V_{\mathbb{R}}(I))$ for an ideal I. In the special case of the combinatorial optimization model described above, the starting point is the finite set $\{\chi^T\}$ which is a finite algebraic variety, and we typically take its vanishing ideal as the ideal whose theta bodies are to be computed. As we saw in Section 7.3.1, these real radical ideals are always TH_k-exact for some finite k. We take a closer look at some combinatorial optimization problems whose theta bodies have been explored.

7.4.1 Stable Sets in a Graph

An example that is at the heart of the history of theta bodies is the *maximum stable set problem* in an undirected graph $G = ([n], E)$ with vertex set $[n]$ and edge set E. A *stable set* in G is a set $U \subseteq [n]$ such that for all $i, j \in U$, $\{i, j\} \notin E$. The maximum stable set problem seeks the stable set of largest cardinality in G, the size of which is the *stability number*, $\alpha(G)$, of G.

The maximum stable set problem can be modeled as follows. For each stable set $U \subseteq [n]$, let $\chi^U \in \{0,1\}^n$ be its characteristic vector defined as $\chi_i^U = 1$ if $i \in U$ and $\chi_i^U = 0$ otherwise. Let $S_G \subseteq \{0,1\}^n$ be the set of characteristic vectors of all stable sets in G. Then $\text{STAB}(G) := \text{conv}(S_G)$ is called the *stable set polytope* of G and the maximum stable set problem is, in theory, the linear program $\max\{\sum_{i=1}^n x_i \,:\, x \in \text{STAB}(G)\}$ with optimal value $\alpha(G)$. However, $\text{STAB}(G)$ is not known a priori, and so one resorts to relaxations of it over which to optimize $\sum_{i=1}^n x_i$.

Polyhedral relaxations of $\text{STAB}(G)$ can be constructed from combinatorial arguments. For instance, a well-known relaxation is the polytope

$$\text{FRAC}(G) := \{x \in \mathbb{R}^n \,:\, x_i + x_j \leq 1 \text{ for all } \{i,j\} \in E,\ \ x_i \geq 0 \text{ for all } i \in [n]\},$$

where the constraint $x_i + x_j \leq 1$ for $\{i, j\} \in E$ comes from the fact that both endpoints of an edge cannot be in a stable set. It can be checked that $\text{STAB}(G)$ is exactly the convex hull of the integer points in $\text{FRAC}(G)$. The polytope $\text{FRAC}(G)$ and several tighter polyhedral relaxations of $\text{STAB}(G)$ have been studied extensively in the literature; see [11, Chapter 9].

Since the set S_G is an algebraic variety, the theta bodies of its vanishing ideal offer convex relaxations of $\text{STAB}(G)$. This vanishing ideal is:

$$I_G := \langle x_i^2 - x_i \text{ for all } i \in [n],\ \ x_i x_j \text{ for all } \{i,j\} \in E\rangle \subset \mathbb{R}[x_1, \ldots, x_n].$$

For $U \subseteq [n]$, let $x^U := \prod_{i \in U} x_i$. From the generators of I_G it follows that if $f \in \mathbb{R}[x]$, then $f \equiv g \bmod I_G$ where g is in the $\mathbb{R}$-span of the set of monomials $\{x^U \,:\, U \text{ is a stable set in } G\}$. In particular,

$$\mathcal{B} := \{x^U + I_G \,:\, U \text{ stable set in } G\}$$

is a θ-basis of $\mathbb{R}[x]/I_G$ (containing $1 + I_G, x_1 + I_G, \ldots, x_n + I_G$). This implies that $\mathcal{B}_k = \{x^U + I_G \,:\, U \text{ stable set in } G,\ |U| \leq k\}$, and for $x^{U_i} + I_G, x^{U_j} + I_G \in \mathcal{B}_k$, their product is $x^{U_i \cup U_j} + I_G$, which is $0 + I_G$ if $U_i \cup U_j$ is not a stable set in G. This product formula allows us to compute $M_{\mathcal{B}_k}(y)$, where we index the element

$x^U + I_G \in \mathcal{B}_k$ by the set U. Since $S_G \subseteq \{0,1\}^n$ and $I(G)$ is the vanishing ideal of S_G, by Theorems 7.8, we have that

$$\mathrm{TH}_k(I_G) = \left\{ y \in \mathbb{R}^n : \begin{array}{l} \exists\, M \succeq 0,\ M \in \mathbb{R}^{|\mathcal{B}_k| \times |\mathcal{B}_k|} \text{ such that} \\ M_{\emptyset\emptyset} = 1, \\ M_{\emptyset\{i\}} = M_{\{i\}\emptyset} = M_{\{i\}\{i\}} = y_i \\ M_{UU'} = 0 \text{ if } U \cup U' \text{ is not stable in } G \\ M_{UU'} = M_{WW'} \text{ if } U \cup U' = W \cup W' \end{array} \right\}.$$

In particular, indexing the one-element stable sets by the vertices of G,

$$\mathrm{TH}_1(I_G) = \left\{ y \in \mathbb{R}^n : \begin{array}{l} \exists\, M \succeq 0, M \in \mathbb{R}^{(n+1)\times(n+1)} \text{ such that} \\ M_{00} = 1, \\ M_{0i} = M_{i0} = M_{ii} = y_i\ \forall\, i \in [n] \\ M_{ij} = 0 \text{ for all } \{i,j\} \in E \end{array} \right\}.$$

Example 7.58. Let $G = ([5], \{\{1,2\}, \{2,3\}, \{3,4\}, \{4,5\}, \{1,5\}\})$ be a 5-cycle. The vanishing ideal of the characteristic vectors of stable sets in G is

$$I_G = \langle x_1x_2, x_2x_3, x_3x_4, x_4x_5, x_1x_5, x_i^2 - x_i \text{ for all } i = 1, \ldots, 5\rangle,$$

and a θ-basis for $\mathbb{R}[x]/I_G$ is given by

$$\mathcal{B} = \{1, x_1, x_2, x_3, x_4, x_5, x_1x_3, x_1x_4, x_2x_4, x_2x_5, x_3x_5\} + I_G.$$

Let $y \in \mathbb{R}^{10}$ be the vector of variables whose coordinates are indexed by $\mathcal{B}$ in the given order and with $y_0 = 1$. Then

$$\mathrm{TH}_1(I_G) = \left\{ y \in \mathbb{R}^5 : \exists y_6, \ldots, y_{10} \text{ s.t. } M_{\mathcal{B}_1}(y) \succeq 0 \right\},$$

where

$$M_{\mathcal{B}_1}(y) = \begin{pmatrix} 1 & y_1 & y_2 & y_3 & y_4 & y_5 \\ y_1 & y_1 & 0 & y_6 & y_7 & 0 \\ y_2 & 0 & y_2 & 0 & y_8 & y_9 \\ y_3 & y_6 & 0 & y_3 & 0 & y_{10} \\ y_4 & y_7 & y_8 & 0 & y_4 & 0 \\ y_5 & 0 & y_9 & y_{10} & 0 & y_5 \end{pmatrix}.$$

Note that $x_i \equiv x_i^2$ and $1 - x_i \equiv (1 - x_i)^2 \bmod I_G$ for any graph G, so $\mathrm{TH}_1(I_G)$ is always contained in the $[0,1]$ cube. ■

The first example of an SDP relaxation of a combinatorial optimization problem was the *theta body* of a graph $G = ([n], E)$ constructed by Lovász in [18] while studying the Shannon capacity of graphs. The theta body of G, denoted as $\mathrm{TH}(G)$, is a relaxation of $\mathrm{STAB}(G)$ that was originally defined as the intersection of the infinitely many half spaces that arise from the *orthonormal representations* of G. Several equivalent definitions can be found in [18] and [11, Chapter 9]. However, none of them point to an obvious generalization of the construction to other discrete optimization problems. In [20], Lovász and Schrijver observe that $\mathrm{TH}(G)$ can be formulated via semidefinite programming exactly as the formulation for $\mathrm{TH}_1(I_G)$ shown above. This is still specialized to the stable set problem. Then in [19], Lovász observes that, in fact, $\mathrm{TH}(G)$ is cut out by all linear polynomials that are 1-sos mod the ideal I_G. For the stable set problem, this fact can be proven without all the machinery introduced in this paper. This connection leads naturally to the definition of $\mathrm{TH}_k(I_G)$ for any positive integer k and more generally $\mathrm{TH}_k(I)$ for any ideal $I \subseteq \mathbb{R}[x]$ and any k. Problem 8.3 in [19] (roughly) asks to characterize all ideals $I \subseteq \mathbb{R}[x]$ such that $\mathrm{cl}(\mathrm{conv}(V_{\mathbb{R}}(I)))$ equals $\mathrm{TH}_1(I)$ or more generally, $\mathrm{TH}_k(I)$. It was this problem that motivated us to study theta bodies in general and develop the methods in this chapter.

Example 7.59. Let us return to the example Example 7.58. When Lovász introduced the theta body of a graph G, he also introduced the concept of *theta number* of a graph, $\vartheta(G)$ (c.f. Chapter 2). This is just the number

$$\max\left\{\sum_{i=1}^{n} x_i : x \in \mathrm{TH}(G) = \mathrm{TH}_1(I_G)\right\},$$

which is an upper bound (and approximation) for the stability number $\alpha(G)$ of a graph. We can now easily compute $\vartheta(C_5)$, the theta number of the 5-cycle, numerically using YALMIP, since we have the precise structure of the reduced moment matrix.

```
y=sdpvar(1,10);
M=[1    y(1) y(2) y(3)  y(4) y(5) ;
   y(1) y(1) 0    y(6)  y(7) 0    ;
   y(2) 0    y(2) 0     y(8) y(9) ;
   y(3) y(6) 0    y(3)  0    y(10);
   y(4) y(7) y(8) 0     y(4) 0    ;
   y(5) 0    y(9) y(10) 0    y(5) ];
obj=y(1)+y(2)+y(3)+y(4)+y(5);
solvesdp(M>=0,-obj);
double(obj)
```

This will return the answer $\vartheta(C_5) \approx 2.361$. Note that $\alpha(C_5) = 2$, so we do get an upper approximation as expected, but it is clear that I_{C_5} is not TH_1-exact. ■

A particular reason for Lovász's interest in [19, Problem 8.3] was due to the fact that $\mathrm{STAB}(G) = \mathrm{TH}(G)$ if and only if G is a *perfect graph* [11, Corollary 9.3.27]. Recall that a graph is perfect if and only if it has no induced odd cycle of length at least five or its complement [4]. Since $\mathrm{TH}(G) = \mathrm{TH}_1(I_G)$ for all graphs G, it follows that I_G is TH_1-exact if and only if G is perfect. The pentagon in Example 7.58 is not perfect, which justifies our observation that its ideal I_G is not TH_1-exact. Chvátal and Fulkerson had shown that $\mathrm{STAB}(G) = \mathrm{QSTAB}(G)$ if and only if G is a perfect graph where

$$\mathrm{QSTAB}(G) := \left\{x \in \mathbb{R}^n \,:\, x_i \geq 0 \text{ for all } i \in [n],\ \sum_{i \in K} x_i \leq 1 \text{ for all cliques } K \text{ in } G\right\}.$$

A clique in G is a complete subgraph in G. Since every edge in G is a clique, $\mathrm{FRAC}(G) \supseteq \mathrm{QSTAB}(G) \supseteq \mathrm{STAB}(G)$ in general. A hexagon is perfect, in which case, $\mathrm{FRAC}(G) = \mathrm{QSTAB}(G)$ since the only cliques in G are its edges. Therefore, for the hexagon, $\mathrm{STAB}(G) = \mathrm{TH}(G) = \mathrm{TH}_1(I_G) = \mathrm{QSTAB}(G) = \mathrm{FRAC}(G)$. Since I_G is TH_1-exact if and only if G is perfect, by Theorem 7.30, we also have that $\mathrm{STAB}(G)$ is 2-level if and only if G is perfect.

The above discussion leads naturally to the question of which graphs G have the property that I_G is TH_2-exact, or more generally, TH_k-exact. These problems are open at the moment, although isolated examples of TH_k-exact ideals are known for specific values of $k > 1$. In practice it is quite difficult to find examples of graphs G for which I_G is not TH_2-exact although such graphs have to exist unless $P = NP$. Recent results of Au and Tunçel prove that if G is the line graph of the complete graph on $2n + 1$ vertices, then the smallest k for which I_G is TH_k-exact grows linearly with n [1].

7.4.2 A General Framework

The stable set problem and many others in combinatorial optimization can be modeled as arising from a simplicial complex. A *simplicial complex* or *independence system*, Δ, with vertex set $[n]$, is a collection of subsets of $[n]$, called the *faces* of the Δ, such that whenever $S \in \Delta$ and $T \subset S$, then $T \in \Delta$. The *Stanley–Reisner* ideal of Δ is the ideal J_Δ generated by the square-free monomials $x_{i_1} x_{i_2} \cdots x_{i_k}$ such that $\{i_1, i_2, \ldots, i_k\} \subseteq [n]$ is not a face of Δ. If $I_\Delta := J_\Delta + \langle x_i^2 - x_i \,:\, i \in [n]\rangle$, then $V_{\mathbb{R}}(I_\Delta) = \{s \in \{0,1\}^n \,:\, \mathrm{support}(s) \in \Delta\}$. The support of a vector $v \in \mathbb{R}^n$ is the set $\{i \in [n] \,:\, v_i \neq 0\}$. Further, for $T \subseteq [n]$, if $x^T := \prod_{i \in T} x_i$, then $\mathcal{B} := \{x^T + I_\Delta \,:\, T \in \Delta\}$ is a θ-basis of $\mathbb{R}[x]/I_\Delta$. This implies that the kth theta body of I_Δ is

$$\mathrm{TH}_k(I_\Delta) = \pi_{\mathbb{R}^n}\{y \in \mathbb{R}^{\mathcal{B}_{2k}} \,:\, M_{\mathcal{B}_k}(y) \succeq 0,\ y_0 = 1\}.$$

Since $\mathcal{B}$ is in bijection with the faces of Δ and $x_i^2 - x_i \in I_\Delta$ for all $i \in [n]$, the theta body can be written explicitly as

$$\mathrm{TH}_k(I_\Delta) = \left\{ y \in \mathbb{R}^n \,:\, \begin{array}{l} \exists\, M \succeq 0,\ M \in \mathbb{R}^{|\mathcal{B}_k| \times |\mathcal{B}_k|} \text{ such that} \\ M_{\emptyset\emptyset} = 1, \\ M_{\emptyset\{i\}} = M_{\{i\}\emptyset} = M_{\{i\}\{i\}} = y_i, \\ M_{UU'} = 0 \text{ if } U \cup U' \notin \Delta, \\ M_{UU'} = M_{WW'} \text{ if } U \cup U' = W \cup W' \end{array} \right\}.$$

If the dimension of Δ is $d - 1$ (i.e., the largest faces in Δ have size d), then I_Δ is TH_d-exact since all elements of $\mathcal{B}$ have degree at most d and hence the last possible theta body $\mathrm{TH}_d(I_\Delta)$ must coincide with $\mathrm{conv}(V_{\mathbb{R}}(I_\Delta))$ as $V_{\mathbb{R}}(I_\Delta)$ is finite. However, in many examples, I_Δ could be TH_k-exact for a k much smaller than d.

In the case of the stable set problem on $G = ([n], E)$, Δ is the set of all stable sets in G. This is a simplicial complex with vertex set $[n]$ whose nonfaces are the sets $T \subseteq [n]$ containing a pair $i, j \in [n]$ such that $\{i, j\} \in E$. Hence the minimal nonfaces (by set inclusion) are precisely the edges of G and so $J_\Delta = \langle x_i x_j \,:\, \{i, j\} \in E\rangle$.

Then $I_\Delta = J_\Delta + \langle x_i^2 - x_i \,:\, i \in [n]\rangle$, which is precisely the ideal I_G from Section 7.4.1, and the remaining facts about the θ-basis $\mathcal{B}$ used in Section 7.4.1 and the structure of the theta bodies of I_G follow from the general set up described above.

An example from combinatorial optimization that does not follow the simplicial complex framework is the *maximum cut problem* of finding the largest size *cut* in a graph. Recall that a cut in G is the collection of edges that go between the two parts of a partition of the vertices of G. Note that a subset of a cut is not necessarily a cut and hence the set of cuts in a graph do not form a simplicial complex. In [7] the theta body hierarchy for the maximum cut problem, and more generally for *binary matroids*, is studied. In this case, a θ-basis for the ideal in question is not obvious as in the simplicial complex model.

7.4.3 Triangle-free Subgraphs in a Graph

We finish the chapter with a second example from combinatorial optimization that fits the simplicial complex model. A subgraph H of a graph $G = ([n], E)$ is *triangle-free* if it does not contain a triangle (K_3, the complete graph on 3 vertices). Given weights on the edges of G, the *triangle-free subgraph problem* in G asks for a triangle-free subgraph of G of maximum weight. If all the edge weights are one, then the problem seeks a triangle-free subgraph in G with the most number of edges. The triangle-free subgraph problem is known to be NP-hard [36] and is relevant in various contexts within optimization.

The integer programming formulation of the triangle-free subgraph problem optimizes the linear function $\sum_{e\in E} w_e x_e$, where w_e is the weight on edge $e \in E$, over the characteristic vectors $\{\chi^H \,:\, H \text{ is triangle-free in } G\}$. This is equivalent to maximizing $\sum_{e\in E} w_e x_e$ over

$$P_{\text{tf}}(G) := \operatorname{conv}\{\chi^H \,:\, H \text{ is triangle-free in } G\},$$

the *triangle-free subgraph polytope* of G. Note that $P_{\text{tf}}(G)$ is a full-dimensional 0/1 polytope in $\mathbb{R}^E$. The triangle-free subgraph polytope of a graph has been studied by various authors (see, for instance, [3, 5]), and a number of facet defining inequalities of the polytope are known, although a full inequality description is not known or expected.

Taking Δ to be the simplicial complex on E consisting of all triangle-free subgraphs in G, and $I_{\text{tf}}(G) := I_\Delta$, we have that

$$V_{\mathbb{R}}(I_{\text{tf}}(G)) = \{\chi^H \,:\, H \text{ is triangle-free in } G\}.$$

Hence the theta bodies of $I_{\text{tf}}(G)$ provide convex relaxations of the triangle-free subgraph polytope $P_{\text{tf}}(G)$. From the general framework in Section 7.4.2, $\mathcal{B} = \{x^H + I_{\text{tf}}(G) \,:\, H \text{ triangle-free in } G\}$ is a θ-basis of $\mathbb{R}[x]/I_{\text{tf}}(G)$. Therefore, the rows and columns of $M_{\mathcal{B}_k}(y)$ are indexed by the triangle-free subgraphs in G with at most k edges. For ease of exposition, let us denote the entry of $M_{\mathcal{B}_k}(y)$ corresponding to row indexed by x^{H_1} and column indexed by x^{H_2} by $M_{\mathcal{B}_k}(y)_{H_1 H_2}$, let $H_1 \cup H_2$ denote the subgraph of G whose edge set is the union of the edge sets of H_1 and H_2, and y_H denote the entry of $y \in \mathbb{R}^{\mathcal{B}}$ corresponding to the basis element $x^H + I_{\text{tf}}(G)$. Then

$$\text{TH}_k(I_{\text{tf}}(G)) = \left\{ y \in \mathbb{R}^E : \begin{array}{l} \exists\, M \succeq 0,\ M \in \mathbb{R}^{|\mathcal{B}_k| \times |\mathcal{B}_k|} \text{ such that} \\ M_{\emptyset\emptyset} = 1, \\ M_{H_1 H_2} = \begin{cases} 0 \text{ if } H_1 \cup H_2 \text{ has a triangle} \\ y_{H_1 \cup H_2} \text{ otherwise} \end{cases} \end{array} \right\}.$$

Since all subgraphs of G with at most two edges are triangle-free, and $\mathcal{B}_1 = \{1 + I_{\text{tf}}(G)\} \cup \{x_e + I_{\text{tf}}(G) : e \in E\}$, $\text{TH}_1(I_{\text{tf}}(G))$ is exactly the same as the first theta body of the ideal $\langle x_e^2 - x_e : e \in E\rangle$ which is TH_1-exact by Theorem 7.30. Hence $\text{TH}_1(I_{\text{tf}}(G)) = [0,1]^E$, and $I_{\text{tf}}(G)$ is TH_1-exact if and only if every subgraph of G is triangle-free, or equivalently, G is triangle-free.

For graphs G that contain triangles, the second theta body of $I_{\text{tf}}(G)$ is more interesting as triples and quadruples of edges in G can contain triangles which forces some of the entries in $M_{\mathcal{B}_2}(y)$ to be zero.

Example 7.60. Suppose $G = K_3$ with edges labeled $1, 2, 3$. Then $P_{\text{tf}}(G)$ is the convex hull of all $0/1$ vectors in $\mathbb{R}^3$ except $(1,1,1)$ which is the first polytope shown in the second row of polytopes in Figure 7.10. This polytope is TH_2-exact since

$$\mathcal{B}_2 = \{1, x_1, x_2, x_3, x_1x_2, x_1x_3, x_2x_3\} + I_{\text{tf}}(G) = \mathcal{B}.$$

Denoting $y \in \mathbb{R}^{\mathcal{B}_2}$, with first entry one, to be $y = (1, y_1, y_2, y_3, y_{12}, y_{13}, y_{23})$, we have that

$$M_{\mathcal{B}_2}(y) = \begin{pmatrix} 1 & y_1 & y_2 & y_3 & y_{12} & y_{13} & y_{23} \\ y_1 & y_1 & y_{12} & y_{13} & y_{12} & y_{13} & 0 \\ y_2 & y_{12} & y_2 & y_{23} & y_{12} & 0 & y_{23} \\ y_3 & y_{13} & y_{23} & y_3 & 0 & y_{13} & y_{23} \\ y_{12} & y_{12} & y_{12} & 0 & y_{12} & 0 & 0 \\ y_{13} & y_{13} & 0 & y_{13} & 0 & y_{13} & 0 \\ y_{23} & 0 & y_{23} & y_{23} & 0 & 0 & y_{23} \end{pmatrix}.$$

Hence the triangle-free subgraph polytope of K_3 has the spectrahedral description $P_{\text{tf}}(G) = \{(y_1, y_2, y_3) : M_{\mathcal{B}_2}(y) \succeq 0\}$. ■

Several families of facet inequalities for the triangle-free subgraph polytope of a graph can be found in the literature, and a complete facet description of $P_{\text{tf}}(G)$ for an arbitrary graph is unknown. An easy class of facets of $P_{\text{tf}}(G)$ come from the obvious fact that in any triangle in G at most two edges can be in a triangle-free subgraph. Mathematically, if $a, b, c \in E$ induce a triangle in G, then $2 - x_a - x_b - x_c \geq 0$ is a valid inequality for $P_{\text{tf}}(G)$. We now show that this inequality is valid for $\text{TH}_2(I_{\text{tf}}(G))$. First check that

$$(1 - x_c - x_a x_b) \equiv (1 - x_c - x_a x_b)^2 \bmod I_{\text{tf}}(G)$$

and also

$$(1 - x_a - x_b + x_a x_b) \equiv (1 - x_a - x_b + x_a x_b)^2 \bmod I_{\text{tf}}(G).$$

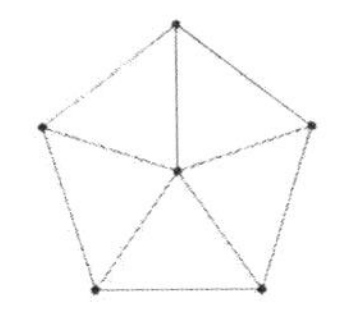 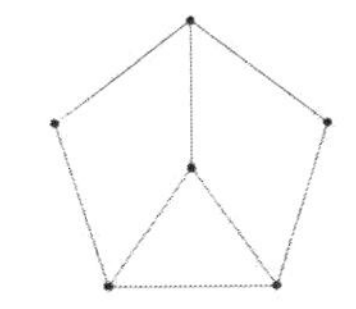 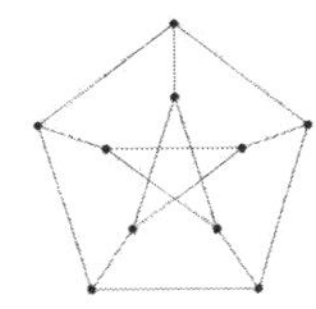

Figure 7.19. 5-*wheel, partial* 5-*wheel, and Petersen graph.*

This implies that $2 - x_a - x_b - x_c = (1 - x_a - x_b + x_a x_b) + (1 - x_c - x_a x_b)$ is 2-sos mod $I_{\mathrm{tf}}(G)$ and hence $2 - x_a - x_b - x_c \geq 0$ is valid for $\mathrm{TH}_2(I_{\mathrm{tf}}(G))$.

Exercise 7.61. We saw in Example 7.59 how to compute $\vartheta(G)$ numerically for a graph G. Find $\vartheta(G)$ for the graphs in Figure 7.19.

1. G a 5-wheel;

2. G the 5-wheel with two missing nonconsecutive rays;

3. G the Petersen graph.

Exercise 7.62. Compute the value of $\vartheta(G)$ for the 5-cycle exactly. (Hint: take advantage of the symmetries of the graph.)

Exercise 7.63. Prove that for any graph G, $\mathrm{TH}_1(I_G) \subseteq \mathrm{QSTAB}(G)$. Note that it is enough to prove that x_i and $1 - \sum_{i \in C} x_i$ are 1-sos mod I_G for all vertices i and all cliques C.

Exercise 7.64. It is known that the stable set polytope of C_{2k+1}, the odd cycle of $2k+1$ nodes, is defined by the inequalities $x_i \geq 0$ for all $i \in [2k+1]$, $x_i + x_j \leq 1$ for all $\{i, j\} \in E$, which by the previous exercise are 1-sos mod I_G, and the single *odd cycle inequality* $\sum x_i \leq k$ [32, Corollary 65.12a].

1. Show that C_5 is TH_2-exact.

2. Show that C_{2k+1} is TH_2-exact for all k.

Exercise 7.65. In Exercise 7.55 we have shown that the vanishing ideal of the set of vertices of a $(k+1)$-level polytope is TH_k-exact. We also have seen in Theorem 7.30 that the reverse implication is true for $k = 1$: if a real radical ideal is TH_1-exact, then its variety must be the set of vertices of a 2-level polytope. Using what we know of the theta body approximations to the stable set polytope, show that the reverse implication (TH_k-exact $\Rightarrow$ k-level) fails for $k \geq 2$.

Exercise 7.66. The triangle-free subgraph problem is closely related to another important problem in combinatorial optimization, the K_3-cover subgraph problem.

A subgraph of G is said to be a K_3-cover if it contains at least an edge of every triangle of G. What is the relation between a maximum triangle-free subgraph and a minimum K_3-cover? How is that reflected in the polytopes underlying those combinatorial problems?

Exercise 7.67. A $(2k+1)$-odd wheel is the graph on $2k+2$ vertices with $2k+1$ of the vertices forming a $2k+1$-cycle and the last vertex connected to each of the vertices of the cycle. Such a wheel yields the inequality $\sum_{e \in EW} x_e \leq 3k+1$ that is valid for the triangle-free subgraph polytope of G. For example, an induced 5-wheel in a graph gives the inequality

$$x_{12} + x_{23} + x_{34} + x_{45} + x_{15} + x_{16} + x_{26} + x_{36} + x_{46} + x_{56} \leq 7,$$

which is valid for the triangle-free subgraph polytope of the graph.

1. Use YALMIP to see that the 5-wheel and 7-wheel inequalities appear to be 2-sos mod $I_{\mathrm{tf}}(G)$, where G is the corresponding wheel.

2. Can you express them exactly as 2-sos modulo the ideals?

3. Can you prove that all odd wheel inequalities are 2-sos modulo its ideal?

Exercise 7.68. Another version of the triangle-free subgraph problem is vertex-based. Given a subset of nodes of G we say it is triangle-free if its induced subgraph is triangle-free. This also falls into the simplicial complex model, so we know how to construct reduced moment matrices. Using the first theta body, compute an approximation for the maximum triangle-free subset of nodes of the 4-wheel.

Bibliography

[1] Y. H. Au and L. Tunçel. Complexity analyses of Bienstock-Zuckerberg and Lasserre relaxations on the matching and stable set polytopes. In *Integer Programming and Combinatorial Optimization*, Lecture Notes in Comput. Sci. 6655, Springer, Heidelberg, 2011, pp. 14–26.

[2] E. Balas, S. Ceria, and G. Cornuéjols. A lift-and-project cutting plane algorithm for mixed 0-1 programs. *Math. Program.*, 58:295–324, 1993.

[3] F. Bendali, A. R. Mahjoub, and J. Mailfert. Composition of graphs and the triangle-free subgraph polytope. *J. Comb. Optim.*, 6:359–381, 2002.

[4] M. Chudnovsky, N. Robertson, P. Seymour, and R. R. Thomas. The strong perfect graph theorem. *Ann. of Math.* (2), 164:51–229, 2006.

[5] M. Conforti, D. G. Corneil, and A. R. Mahjoub. K_i-covers. I. Complexity and polytopes. *Discrete Math.*, 58:121–142, 1986.

[6] D. Cox, J. Little, and D. O'Shea. *Ideals, Varieties and Algorithms.* Springer-Verlag, New York, 1992.

[7] J. Gouveia, M. Laurent, P. A. Parrilo, and R. R. Thomas. A new hierarchy of semidefinite programming relaxations for cycles in binary matroids and cuts in graphs. *Math. Program., Ser. A*, 2010, to appear.

[8] J. Gouveia, P. A. Parrilo, and R. R. Thomas. Theta bodies for polynomial ideals. *SIAM J. Optim.*, 20:2097–2118, 2010.

[9] J. Gouveia and R. R. Thomas. Convex hulls of algebraic sets. In M. Anjos and J.-B. Lasserre, editors, *Handbook of Semidefinite, Cone and Polynomial Optimization: Theory, Algorithms, Software and Applications*, to appear.

[10] D. Grayson and M. Stillman. Macaulay 2, a software system for research in algebraic geometry. Available at http://www.math.uiuc.edu/Macaulay2.

[11] M. Grötschel, L. Lovász, and A. Schrijver. *Geometric Algorithms and Combinatorial Optimization*, 2nd edition, Algorithms Combin. Springer-Verlag, Berlin, 1993.

[12] R. Harvey and H. B. Lawson, Jr. Calibrated geometries. *Acta Math.*, 148:47–157, 1982.

[13] D. Henrion. Semidefinite representation of convex hulls of rational varieties. LAAS-CNRS Research Report 09001, 2009.

[14] J. B. Lasserre. Global optimization with polynomials and the problem of moments. *SIAM J. Optim.*, 11:796–817, 2001.

[15] M. Laurent. A comparison of the Sherali-Adams, Lovász-Schrijver, and Lasserre relaxations for 0-1 programming. *Math. Oper. Res.*, 28:470–496, 2003.

[16] M. Laurent. Sums of squares, moment matrices and optimization over polynomials. In *Emerging Applications of Algebraic Geometry*, IMA Vol. Math. Appl. 149. Springer, Berlin, 2009.

[17] J. Löfberg. YALMIP: A toolbox for modeling and optimization in MATLAB. In *Proceedings of the CACSD Conference*, Taipei, Taiwan, 2004.

[18] L. Lovász. On the Shannon capacity of a graph. *IEEE Trans. Inform. Theory*, 25:1–7, 1979.

[19] L. Lovász. Semidefinite programs and combinatorial optimization. In *Recent Advances in Algorithms and Combinatorics*, CMS Books Math./Ouvrages Math. SMC 11. Springer, New York, 2003, pp. 137–194.

[20] L. Lovász and A. Schrijver. Cones of matrices and set-functions and 0-1 optimization. *SIAM J. Optim.*, 1:166–190, 1991.

[21] M. Marshall. *Positive Polynomials and Sums of Squares*, Math. Surveys Monogr. 146. American Mathematical Society, Providence, RI, 2008.

[22] J. Nie. First order conditions for semidefinite representations of convex sets defined by rational or singular polynomials. *Math. Program.*, 131:1–36, 2012.

[23] M. Omar and B. Osserman. Strong nonnegativity and sums of squares on real varieties. arXiv:1101.0826.

[24] R. T. Rockafellar. *Convex Analysis*, Princeton Landmarks in Mathematics and Physics. Princeton University Press, Princeton, NJ, 1996.

[25] P. Rostalski. Bermeja, Software for Convex Algebraic Geometry. Available at http://math.berkeley.edu/~philipp/Software/Software.

[26] R. Sanyal. Orbitopes and theta bodies. Talk at IPAM Workshop on Convex Optimization and Algebraic Geometry, slides available at http://math.berkeley.edu/~bernd/raman.pdf, 2010.

[27] C. Scheiderer. Sums of squares of regular functions on real algebraic varieties. *Trans. Amer. Math. Soc.*, 352:1039–1069, 2000.

[28] C. Scheiderer. Sums of squares on real algebraic curves. *Math. Z.*, 245:725–760, 2003.

[29] C. Scheiderer. Sums of squares on real algebraic surfaces. *Manuscripta Math.*, 119:395–410, 2006.

[30] C. Scheiderer. Convex hulls of curves of genus one. *Adv. Math.*, 228:2606–2622, 2011.

[31] A. Schrijver. *Theory of Linear and Integer Programming*, Wiley-Interscience Series in Discrete Mathematics and Optimization. Wiley, New York, 1986.

[32] A. Schrijver. *Combinatorial Optimization. Polyhedra and Efficiency. Vol. B*, Algorithms Combin. 24. Springer-Verlag, Berlin, 2003.

[33] H. D. Sherali and W. P. Adams. A hierarchy of relaxations between the continuous and convex hull representations for zero-one programming problems. *SIAM J. Discrete Math.*, 3:411–430, 1990.

[34] R. P. Stanley. Decompositions of rational convex polytopes. *Ann. Discrete Math.*, 6:333–342, 1980.

[35] S. Sullivant. Compressed polytopes and statistical disclosure limitation. *Tohoku Math. J.* (2), 58:433–445, 2006.

[36] M. Yannakakis. Edge-deletion problems. *SIAM J. Comput.*, 10:297–309, 1981.

Chapter 8

Free Convex Algebraic Geometry

J. William Helton[†], *Igor Klep*[‡], *and Scott McCullough*[§]

A new development is extension of the algebraic certificates of real algebraic geometry to noncommutative polynomials, thereby giving a theory of noncommutative polynomial inequalities. Here we shall focus on convexity aspects of noncommutative real algebraic geometry, and we shall see this leads to a very rigid structure. Our subject pertains to optimization problems where the unknowns are matrices.

8.1 Introduction

This chapter is a tutorial on techniques and results in *free convex algebraic geometry* and *free positivity*. As such it also serves as a point of entry into the larger field of *free real algebraic geometry* and makes contact with noncommutative real algebraic geometry [27, 30, 32, 33, 38, 47, 48, 53, 59, 62, 63], free analysis and free probability (lying at the origins of free analysis; cf. [64]), and free analytic function theory and free harmonic analysis [28, 29, 34, 54, 60, 69, 70, 46].

The term free here refers to the central role played by algebras of noncommuting polynomials $\mathbb{R}<x>$ in free (freely noncommuting) variables $x = (x_1, \ldots, x_g)$. A striking difference between the free and classical settings is the following Positivstellensatz.

[†]J. William Helton was partially supported by NSF grants DMS-0700758, DMS-0757212, and DMS-1160802 and by the Ford Motor Company.

[‡]Igor Klep was supported by the Faculty Research Development Fund (FRDF) of The University of Auckland (project 3701119) and was partially supported by the Slovenian Research Agency (program P1-0222).

[§]Scott McCullough was supported by NSF grant DMS-1101137.

Theorem 8.1 (Helton [27]). *A nonnegative (suitably defined) free polynomial is a sum of squares.*

The subject of free real algebraic geometry flows in two branches. One, free positivity is an analogue of classical real algebraic geometry, a theory of polynomial inequalities embodied in Positivstellensätze. As is the case with the sum of squares result above (Theorem 8.1), generally free Positivstellensätze have cleaner statements than do their commutative counterparts; see, e.g., [53, 27, 39, 33] for a sample. Free convexity, the second branch of free real algebraic geometry, arose in an effort to unify a torrent of ad hoc techniques which came on the linear systems engineering scene in the mid 1990s. We will soon give a quick sketch of the engineering motivation, based on the slightly more complete sketch given in the survey article [13]. Mathematically, much as in the commutative case, free convexity is connected with free positivity through the second derivative: A free polynomial is convex if and only if its Hessian is positive.

The tutorial proper starts with Section 8.2. In the remainder of this introduction, motivation for the study of free positivity and convexity arising in *linear systems engineering*, *quantum phenomena*, and other subjects such as *free probability* is provided, as are some suggestions for further reading.

8.1.1 Motivation

While the theory is both mathematically pleasing and natural, much of the excitement of free convexity and positivity stems from its applications. Indeed, the fact that a large class of linear systems engineering problems naturally lead to free inequalities provided the main force behind the development of the subject. In this motivational section, we describe in some detail the linear systems point of view. We also give a brief introduction to other applications.

Linear systems engineering

The layout of a linear systems problem is typically specified by a signal flow diagram. Signals go into boxes and other signals come out. The boxes in a linear system contain constant coefficient linear differential equations which are specified entirely by matrices (the coefficients of the differential equations). Often many boxes appear and many signals transmit between them. In a typical problem some boxes are given, and some we get to design subject to the condition that the L^2-norm of various signals must compare in a prescribed way; e.g., the input to the system has L^2-norm bigger than the output. The signal flow diagram itself and corresponding problems do not specify the size of matrices involved. So ideally any algorithms derived apply to matrices of all sizes. Hence the problems are called *dimension free.*

An empirical observation is that system problems of this type convert to inequalities on polynomials in matrices, the form of the polynomials being determined entirely by the signal flow layout (and independent of the matrices involved). Thus the systems problem naturally leads to free polynomials and free positivity conditions.

For yet a more detailed discussion of this example, see [13, Section 4.1]. Those who read Chapter 2 saw a basic example of this in Section 2.2.1. Next we give more of an idea of how the correspondence between linear systems and noncommutative polynomials occurs. This is done primarily with an example.

Linear systems

A *linear system* $\mathfrak{F}$ is given by the constant coefficient linear differential equations

$$\frac{dx}{dt} = Ax + Bu,$$
$$y = Cx,$$

with the vector

- $x(t)$ at each time t being in the vector space $\mathcal{X}$ called the *state space*,
- $u(t)$ at each time t being in the vector space $\mathcal{U}$ called the *input space*,
- $y(t)$ at each time t being in the vector space $\mathcal{Y}$ called the *output space*,

and A, B, C being linear maps on the corresponding vector spaces.

Connecting linear systems

Systems can be connected in incredibly complicated configurations. We describe a simple connection and this goes a long way toward illustrating the general idea. Given two linear systems $\mathfrak{F}$, $\mathfrak{G}$, we describe the formulas for connecting them in feedback.

One basic feedback connection is described by the diagram

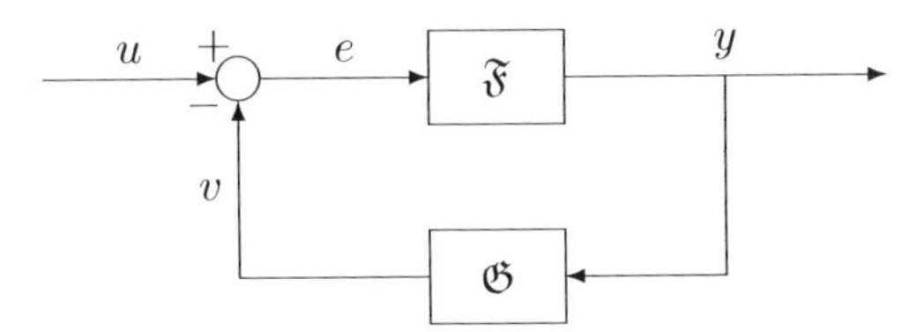

called a *signal flow diagram*. Here u is a signal going into the *closed loop system* and y is the signal coming out. The signal flow diagram is equivalent to a collection of equations. The systems $\mathfrak{F}$ and $\mathfrak{G}$ themselves are, respectively, given by the linear differential equations

$$\frac{dx}{dt} = Ax + Be, \qquad \frac{d\xi}{dt} = Q\,\xi + R\,w,$$
$$y = Cx, \qquad v = S\,\xi.$$

The feedback connection is described algebraically by

$$w = y \qquad \text{and} \qquad e = u - v.$$

Putting these relations together gives that the closed loop system is described by differential equations

$$\frac{dx}{dt} = Ax - BS\xi + Bu,$$
$$\frac{d\xi}{dt} = Q\,\xi + R\,y = Q\,\xi + R\,Cx,$$
$$y = Cx,$$

which is conveniently described in matrix form as

$$\frac{d}{dt}\begin{bmatrix} x \\ \xi \end{bmatrix} = \begin{bmatrix} A & -BS \\ RC & Q \end{bmatrix}\begin{bmatrix} x \\ \xi \end{bmatrix} + \begin{bmatrix} B \\ 0 \end{bmatrix} u, \qquad y = \begin{bmatrix} C & 0 \end{bmatrix}\begin{bmatrix} x \\ \xi \end{bmatrix}, \tag{8.1}$$

where the state space of the closed loop systems is the direct sum $\mathcal{X} \oplus \mathcal{Y}$ of the state spaces $\mathcal{X}$ of $\mathfrak{F}$ and $\mathcal{Y}$ of $\mathfrak{G}$. From (8.1), the coefficients of the ODE are (block) matrices whose entries are (in this case simple) polynomials in the matrices A, B, C, Q, R, S.

This illustrates the moral of the general story:

System connections produce a new system whose coefficients are matrices with entries which are noncommutative polynomials (or at worst "rational expressions") in the coefficient matrices of the component systems.

Complicated signal flow diagrams give complicated matrices of noncommutative polynomials or rationals. Note that in what was said the dimensions of vector spaces and matrices A, B, C, Q, R, S never entered explicitly; the algebraic form of (8.1) is completely determined by the flow diagram. Thus, such linear systems lead to *dimension free* problems.

Next we turn to how "noncommutative inequalities" arise. The main constraint producing them can be thought of as energy dissipation, a special case of which are the Lyapunov functions already seen in Section 2.2.1.

Energy dissipation

We have a system $\mathfrak{F}$ and want a condition which checks whether

$$\int_0^\infty |u|^2 dt \geq \int_0^\infty |\mathfrak{F}u|^2 dt, \qquad x(0) = 0,$$

holds for all input functions u, where $\mathfrak{F}u = y$ in the above notation. If this holds $\mathfrak{F}$ is called a *dissipative system*.

$L^2[0,\infty]$ → $\mathfrak{F}$ → $L^2[0,\infty]$

The energy dissipative condition is formulated in the language of analysis, but it converts to algebra (or at least an algebraic inequality) because of the following construction, which assumes the existence of a "potential energy"-like function V on the state space. A function V which satisfies $V \geq 0$, $V(0) = 0$, and

$$V(x(t_1)) + \int_{t_1}^{t_2} |u(t)|^2 dt \;\geq\; V(x(t_2)) + \int_{t_1}^{t_2} |y(t)|^2 dt$$

for all input functions u and initial states x_1 is called a *storage function*. The displayed inequality is interpreted physically as

potential energy now + energy in $\geq$ potential energy then + energy out.

Assuming enough smoothness of V, we can differentiate this integral condition and use $\frac{d}{dt}x(t_1) = Ax(t_1) + Bu(t_1)$ to obtain a differential inequality

$$0 \geq \nabla V(x)(Ax + Bu) + |Cx|^2 - |u|^2 \tag{8.2}$$

on what is called the "reachable set" (which we do not need to define here).

In the case of linear systems, V can be chosen to be a quadratic. So it has the form $V(x) = \langle Ex, x\rangle$ with $E \succeq 0$ and $\nabla V(x) = 2Ex$.

Theorem 8.2. *The linear system A, B, C is dissipative if inequality* (8.2) *holds for all $u \in \mathcal{U}, x \in \mathcal{X}$. Conversely, if A, B, C is "reachable,"*[1] *then dissipativity implies that inequality* (8.2) *holds for all $u \in \mathcal{U}$, $x \in \mathcal{X}$.*

In the linear case, we may substitute $\nabla V(x) = 2Ex$ in (8.2) to obtain

$$0 \geq 2(Ex)^{\mathsf{T}}(Ax + Bu) + |Cx|^2 - |u|^2$$

for all u, x. Then maximize in x to get

$$0 \geq x^{\mathsf{T}}[EA + A^{\mathsf{T}}E + EBB^{\mathsf{T}}E + C^{\mathsf{T}}C]x.$$

Thus the classical *Riccati matrix inequality*

$$0 \succeq EA + A^{\mathsf{T}}E + EBB^{\mathsf{T}}E + C^{\mathsf{T}}C \quad \text{with} \quad E \succeq 0 \tag{8.3}$$

ensures dissipativity of the system and, it turns out, is also implied by dissipativity when the system is reachable.

It is inequality (8.3), applied in many many contexts, which leads to positive semidefinite inequalities throughout all of linear systems theory.

As an aside we return to the very special case of dissipativity, namely Lyapunov stability, described in Section 2.2.1. Our discussion starts with the "miracle of inequality (8.3)": when $B = 0$ it becomes the Lyapunov inequality. However, this is merely magic (no miracle whatsoever); the trick being that the if input u is identically zero, then dissipativity implies stability. The converse is less intuitive, but true: stability of $\dot{x} = Ax$ implies existence of a "virtual" potential energy $V(x) = \langle Ex, x\rangle$ and output C making the "virtual" system dissipative.

Schur complements and linear matrix inequalities

Using Schur complements, the Riccati inequality of (8.3) is equivalent to the inequality

$$L(E) := \begin{bmatrix} EA + A^{\mathsf{T}}E + C^{\mathsf{T}}C & EB \\ B^{\mathsf{T}}E & -I \end{bmatrix} \preceq 0.$$

[1] A mild technical condition.

Here A, B, C describe the system and E is an unknown matrix. If the system is reachable, then A, B, C is dissipative if and only if $L(E) \preceq 0$ and $E \succeq 0$.

The key feature in this reformulation of the Riccati inequality is that $L(E)$ is linear in E, so the inequality $L(E) \preceq 0$ is a *linear matrix inequality* in E.

Putting it together

We have shown two ingredients of linear system theory, connection laws (algebraic) and dissipation (inequalities), but have yet to put them together. It is in fact a very mechanical procedure. After going through the procedure one sees that the problem a software toolbox designer faces is this:

> (GRAIL) Given a symmetric matrix of noncommutative polynomials
>
> $$p(a, x) = \Big[p_{ij}(a, x)\Big]_{i,j=1}^{k},$$
>
> and a tuple of matrices A, provide an algorithm for finding X making $p(A, X) \succeq 0$ or, better yet, as large as possible.

Algorithms for doing this are based on numerical optimization or a close relative, so even if they find a local solution there is no guarantee that it is global. If p is convex in X, then these problems disappear.

Thus, systems problems described by signal flow diagrams produce a mess of matrix inequalities with some matrices known and some unknown and the constraints that some polynomials are positive semidefinite. The inequalities can get very complicated as one might guess, since signal flow diagrams get complicated. These considerations thus naturally lead to the emerging subject of free real algebraic geometry, the study of noncommutative (free) polynomial inequalities, and free semialgebraic sets. Indeed, much of what is known about this very new subject is touched on in this chapter.

The engineer would like for these polynomial inequalities to be convex in the unknowns. Convexity guarantees that local optima are global optima (finding global optima is often of paramount importance) and facilitates numerics.

Hence the major issues in linear systems theory are as follows:

1. *Which problems convert to a convex matrix inequality? How does one do the conversion?*

2. *Find numerics which will solve large convex problems. How do you use special structure, such as most unknowns are matrices and the formulas are all built of noncommutative rational functions?*

3. *Are convex matrix inequalities more general than linear matrix inequalities?*

The mathematics here can be motivated by the problem of writing a toolbox for engineers to use in designing linear systems. What goes in such toolboxes are algebraic formulas with matrices A, B, C unspecified and reliable numerics for solving them when a user does specify A, B, C as matrices. A user who designs a controller for a helicopter puts in the mathematical systems model for his helicopter and puts in matrices, for example, A is a particular 8×8 real matrix etc. Another user who designs a satellite controller might have a 50-dimensional state space and of course would pick completely different A, B, C. Essentially any matrices of any compatible dimensions can occur. Any claim we make about our formulas must be valid regardless of the size of the matrices plugged in.

The toolbox designer faces two completely different tasks. One is manipulation of algebraic inequalities; the other is numerical solutions. Often the first is far more daunting since the numerics is handled by some standard package (although for numerics problem size is a demon). Thus there is a great need for algebraic theory. Most of this chapter bears on questions like (3) above, where the unknowns are matrices. *The first two questions will not be addressed.* Here we treat (3) when there are no a variables. When there are a variables, see [26, 1]. Thus we shall consider polynomials $p(x)$ in free noncommutative variables x and focus on their convexity on free semialgebraic sets.

What are the implications of our study for engineering? Herein you will see strong results on free convexity but what do they say to an engineer? We foreshadow the forthcoming answer by saying it is fairly negative, but postpone further disclosure till the final page of these writings not so much to promote suspense but for the conclusion to arrive after you have absorbed the theory.

Quantum phenomena

Free Positivstellensätze—algebraic certificates for positivity—of which Theorem 8.1 is the grandfather, have physical applications. Applications to quantum physics are explained by Pironio, Navascués, and Acín [59], who also consider computational aspects related to noncommutative sum of squares. How this pertains to operator algebras is discussed by Klep and Schweighofer in [47]. The important Bessis–Moussa–Villani conjecture (BMV) from quantum statistical mechanics is tackled in [48, 7]. Doherty et al. [12] employ noncommutative positivity and the Positivstellensatz [37] of the first and the third author to consider the quantum moment problem and multiprover games.

A particularly elegant recent development, independent of the line of history containing the work in this chapter, was initiated by Effros. The classic "perspective" transformation carries a function on $\mathbb{R}^n$ to a function on $\mathbb{R}^{n+1}$. It is used for various purposes, one being in algebraic geometry to produce "blowups" of singularities, thereby removing them. It has the property that convex functions map to convex functions. What about convex functions on free variables? This question was asked by Effros and settled affirmatively in [18] for natural cases as a way to show that quantum relative entropy is convex. Subsequently, [19] showed that the perspective transformation in free variables always maps convex functions to convex functions.

Miscellaneous applications

A number of other scientific disciplines use free analysis, though less systematically than in free real algebraic geometry.

Free probability. Voiculescu developed it to attack one of the purest of mathematical questions regarding von Neumann algebras. From the outset (about 20 years ago) it was elegant and it came to have great depth. Subsequently, it was discovered to bear forcefully and effectively on random matrices. The area is vast, so we do not dive in but refer the reader to an introduction [64, 71].

Nonlinear engineering systems. A classical technique in nonlinear systems theory developed by Fliess is based on manipulation of power series with noncommutative variables (the Chen series). The area has a new impetus coming from the problem of data compression, so now is a time when these correspondences are being worked out; cf. [21, 22, 52].

8.1.2 Further Reading

We pause here to offer some suggestions for further reading. For further engineering motivation we recommend the paper [65] or the longer version [66] for related new directions. Descriptions of Positivstellensätze are in the surveys [31, 13, 43, 63], with the first three also briskly touring free convexity. The survey article [40] is aimed at engineers.

Noncommutative is a broad term, encompassing essentially all algebras. In between the extremes of commutative and free lie many important topics, such as Lie algebras, Hopf algebras, quantum groups, C^*-algebras, von Neumann algebras, etc. For instance, there are elegant noncommutative real algebraic geometry results for the Weyl algebra [62]; cf. [63].

8.1.3 Guide to the Chapter

The goal of this tutorial is to introduce the reader to the main results and techniques used to study free convexity. Fortunately, the subject is new and the techniques not too numerous so that one can quickly become an expert.

The basics of free, or noncommutative, polynomials and their evaluations are developed in Section 8.2. The key notions are positivity and convexity for free polynomials. The principal fact is that the second directional derivative (in direction h) of a free convex polynomial is a positive quadratic polynomial in h (just like in the commutative case). Free quadratic (in h) polynomials have a Gram-type representation which thus figures prominently in studying convexity. The nuts and bolts of this Gram representation and some of its consequences, including Theorem 8.1, are the subjects of Sections 8.4 and 8.5, respectively.

The Gram representation techniques actually require only a small amount of convexity, and thus there is a theory of geometry on free varieties having signed (e.g., positive) curvature. Some details are in Section 8.6.

A couple of free real algebraic geometry results which have a heavy convexity component are described in the last section, Section 8.7. The first is an optimal free convex Positivstellensatz which generalizes Theorem 8.1. The second says that free convex semialgebraic sets are free spectrahedra, giving another example of the much more rigid structure in the free setting.

Section 8.3 introduces software which handles free noncommutative computations. You may find it useful in your free studies.

In what follows, mildly incorrectly but in keeping with the usage in the literature, the terms noncommutative and free are used synonymously.

8.2 Basics of Noncommutative Polynomials and Their Convexity

This section treats the basics of polynomials in noncommutative variables, noncommutative differential calculus, and noncommutative inequalities. There is also a brief introduction to noncommutative rational functions and inequalities.

8.2.1 Noncommutative Polynomials

Before turning to the formalities, we give, by examples, an informal introduction to noncommutative polynomials.

A noncommutative polynomial p is a polynomial in a finite set $x = (x_1, \dots, x_g)$ of relation free variables. A canonical example, in the case of two variables $x = (x_1, x_2)$, is the *commutator*

$$c(x_1, x_2) = x_1 x_2 - x_2 x_1. \tag{8.4}$$

It is precisely the fact that x_1 and x_2 do not commute that makes c nonzero.

While a commutative polynomial $q \in \mathbb{R}[t_1, t_2]$ is naturally evaluated at points $t \in \mathbb{R}^2$, noncommutative polynomials are naturally evaluated on tuples of square matrices. For instance, with

$$X_1 = \begin{bmatrix} 0 & 1 \\ 1 & 0 \end{bmatrix}, \quad X_2 = \begin{bmatrix} 1 & 0 \\ 0 & 0 \end{bmatrix},$$

and $X = (X_1, X_2)$, one finds

$$c(X) = \begin{bmatrix} 0 & 1 \\ -1 & 0 \end{bmatrix}.$$

Importantly, c can be evaluated on any pair (X, Y) of symmetric matrices of the same size. (Later in the section we will also consider evaluations involving not necessarily symmetric matrices.) Note that if X and Y are $n \times n$, then $c(X, Y)$ is itself an $n \times n$ matrix. In the case of $c(x, y) = xy - yx$, the matrix $c(X, Y) = 0$ if and only if X and Y commute. In particular, c is zero on $\mathbb{R}^2$ (2-tuples of 1×1 matrices).

For another example, if $d(x_1, x_2) = 1 + x_1 x_2 x_1$, then with X_1 and X_2 as above, we find

$$d(X) = I_2 + X_1 X_2 X_1 = \begin{bmatrix} 1 & 0 \\ 0 & 2 \end{bmatrix}.$$

Note that although X is a tuple of symmetric matrices, it need not be the case that $p(X)$ is symmetric. Indeed, the matrix $c(X)$ above is not. In the present context, we say that p is *symmetric* if $p(X)$ is symmetric whenever $X = (X_1, \ldots, X_g)$ is a tuple of symmetric matrices. Another more algebraic definition of symmetric for noncommutative polynomials appears in Section 8.2.2.

Noncommutative convexity for polynomials

Many standard notions for polynomials, and even functions, on $\mathbb{R}^g$ extend to the noncommutative setting, though often with unexpected ramifications. For example, the commutative polynomial $q \in \mathbb{R}[t_1, t_2]$ is convex if, given $s, t \in \mathbb{R}^2$,

$$\frac{1}{2}\big(q(s) + q(t)\big) \geq q\Big(\frac{s+t}{2}\Big).$$

There is a natural ordering on symmetric $n \times n$ matrices defined by $X \succeq Y$ if the symmetric matrix $X - Y$ is positive semidefinite, i.e., if its eigenvalues are all nonnegative. Similarly, $X \succ Y$ if $X - Y$ is positive definite, i.e., all its eigenvalues are positive. This order yields a canonical notion of convex noncommutative polynomial. Namely, a symmetric polynomial p is *convex* if for each n and each pair of g tuples of $n \times n$ symmetric matrices $X = (X_1, \ldots, X_g)$ and $Y = (Y_1, \ldots, Y_g)$, we have

$$\frac{1}{2}\big(p(X) + p(Y)\big) \succeq p\Big(\frac{X+Y}{2}\Big).$$

Equivalently,

$$\frac{p(X) + p(Y)}{2} - p\Big(\frac{X+Y}{2}\Big) \succeq 0. \tag{8.5}$$

Even in one variable, convexity for a noncommutative polynomial is a serious constraint. For instance, consider the polynomial x^4. It is symmetric, but with

$$X = \begin{bmatrix} 4 & 2 \\ 2 & 2 \end{bmatrix} \text{ and } Y = \begin{bmatrix} 2 & 0 \\ 0 & 0 \end{bmatrix}$$

it follows that

$$\frac{X^4 + Y^4}{2} - \Big(\frac{1}{2}X + \frac{1}{2}Y\Big)^4 = \begin{bmatrix} 164 & 120 \\ 120 & 84 \end{bmatrix}$$

is not positive semidefinite. Thus x^4 is not convex.

Noncommutative polynomial inequalities and convexity

The study of polynomial inequalities, real algebraic geometry or semialgebraic geometry, has a noncommutative version. A *basic open semialgebraic set* is a subset of $\mathbb{R}^g$ defined by a list of polynomial inequalities; i.e., a set S is a basic open semialgebraic set if

$$S = \{t \in \mathbb{R}^g : p_1(t) > 0, \ldots, p_k(t) > 0\}$$

for some polynomials $p_1, \ldots, p_k \in \mathbb{R}[t_1, \ldots, t_g]$.

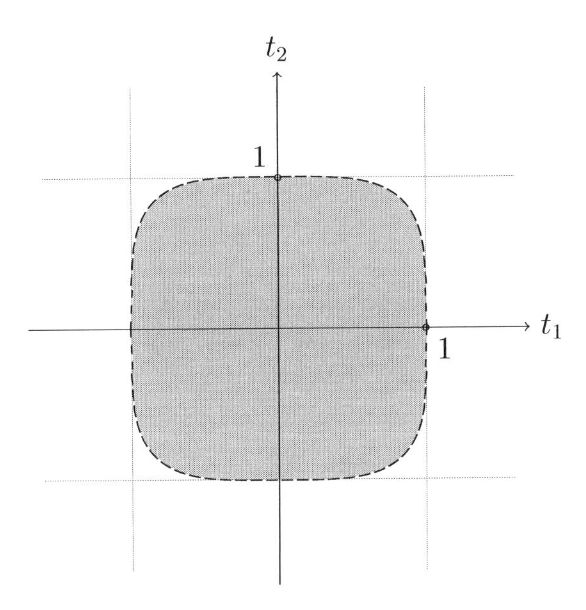

$$\mathrm{ncTV}(1) = \{(t_1, t_2) \in \mathbb{R}^2 \colon 1 - t_1^4 - t_2^4 > 0\}.$$

Because noncommutative polynomials are evaluated on tuples of matrices, a noncommutative (free) basic open semialgebraic set is a sequence. For positive integers n, let $(\mathbb{S}^{n\times n})^g$ denote the set of g-tuples of $n \times n$ symmetric matrices. Given symmetric noncommutative polynomials $p_1, \ldots, p_k$, let

$$\mathcal{P}(n) = \{X \in (\mathbb{S}^{n\times n})^g \colon p_1(X) \succ 0, \ldots, p_k(X) \succ 0\}.$$

The sequence $\mathcal{P} = (\mathcal{P}(n))$ is then a *noncommutative (free) basic open semialgebraic set.* The sequence

$$\mathrm{ncTV}(n) = \{X \in (\mathbb{S}^{n\times n})^2 \colon I_n - X_1^4 - X_2^4 \succ 0\}$$

is an entertaining example. When $n = 1$, ncTV(1) is a subset of $\mathbb{R}^2$ often called the *TV screen.* Numerically it can be verified, though it is rather tricky to do so (see Exercise 8.23) that the set ncTV(2) is not a convex set. An analytic proof that ncTV(n) is not a convex set for some n can be found in [15]. It also follows by combining results in [38] and [44]. For properties of the classical commutative TV screen, see Chapters 5 and 6 of this book.

Example 8.3. Let $p_\epsilon := \epsilon^2 - \sum_{j=1}^{g} x_j^2$. Then the *$\epsilon$-neighborhood of* 0,

$$\mathcal{N}_\epsilon := \bigcup_{n\in\mathbb{N}} \{X \in (\mathbb{S}^{n\times n})^g \colon p_\epsilon(X) \succ 0\},$$

is an important example of a noncommutative basic open semialgebraic set. ∎

8.2.2 Noncommutative Polynomials: The Formalities

We now take up the formalities of noncommutative polynomials, their evaluations, convexity, and positivity.

Let $x = (x_1, \ldots, x_g)$ denote a g-tuple of free noncommuting variables and let $\mathbb{R}<x>$ denote the associative $\mathbb{R}$-algebra freely generated by x, i.e., the elements of $\mathbb{R}<x>$ are polynomials in the noncommuting variables x with coefficients in $\mathbb{R}$. Its elements are called (*noncommutative*) *polynomials*. An element of the form aw, where $0 \neq a \in \mathbb{R}$ and w is a *word* in the variables x, is called a *monomial* and a its *coefficient*. Hence words are monomials whose coefficient is 1. Note that the empty word $\emptyset$ plays the role of the multiplicative identity for $\mathbb{R}<x>$.

There is a natural *involution* $^\intercal$ on $\mathbb{R}<x>$ that reverses words. For example, $(2 - 3x_1^2x_2x_3)^\intercal = 2 - 3x_3x_2x_1^2$. A polynomial p is a *symmetric polynomial* if $p^\intercal = p$. Later we will see that this notion of symmetric is equivalent to that in the previous subsection. For now we note that of

$$c(x) = x_1x_2 - x_2x_1,$$
$$\jmath(x) = x_1x_2 + x_2x_1,$$

$\jmath$ is symmetric, but c is not. Indeed, $c^\intercal = -c$. Because $x_j^\intercal = x_j$ we refer to the variables as *symmetric variables*. Occasionally we emphasize this point by writing $\mathbb{R}<x = x^\intercal>$ for $\mathbb{R}<x>$.

The *degree* of a noncommutative polynomial p, denoted $\deg(p)$, is the length of the longest word appearing in p. For instance the polynomials c and $\jmath$ above both have degree 2 and the degree of

$$r(x) = 1 - 3x_1x_2 - 3x_2x_1 - 2x_1^2x_2^4x_1^2$$

is 8. Let $\mathbb{R}<x>_k$ denote the polynomials of degree at most k.

Noncommutative matrix polynomials

Given positive integers d, d', let $\mathbb{R}^{d\times d'}<x>$ denote the $d \times d'$ matrices with entries from $\mathbb{R}<x>$. Thus elements of $\mathbb{R}^{d\times d'}<x>$ are *matrix-valued noncommutative polynomials*. The involution on $\mathbb{R}<x>$ naturally extends to a mapping $\intercal : \mathbb{R}^{d\times d'}<x> \to \mathbb{R}^{d'\times d}<x>$. In particular, if

$$P = \left[p_{i,j}\right]_{i,j=1}^{d,d'} \in \mathbb{R}^{d\times d'}<x>,$$

then

$$P^\intercal = \left[p_{j,i}^\intercal\right]_{i,j=1}^{d,d'} \in \mathbb{R}^{d'\times d}<x>.$$

In the case that $d = d'$, such a P is symmetric if $P^\intercal = P$.

Linear pencils

Given a positive integer n, let $\mathbb{S}^{n\times n}$ denote the real symmetric $n \times n$ matrices. For $A_0, A_1, \ldots, A_g \in \mathbb{S}^{d\times d}$, the expression

$$L(x) = A_0 + \sum_{j=1}^{g} A_jx_j \in \mathbb{S}^{d\times d}<x> \tag{8.6}$$

in the noncommuting variables x is a *symmetric affine linear pencil.* In other words, these are precisely the symmetric degree one matrix-valued noncommutative polynomials. If $A_0 = I$, then L is *monic.* If $A_0 = 0$, then L is a *linear pencil.* The homogeneous linear part $\sum_{j=1}^{g} A_j x_j$ of a linear pencil L as in (8.6) will be denoted by $L^{(1)}$.

Example 8.4. Let

$$A_1 = \begin{bmatrix} 0 & 1 & 0 & 0 \\ 1 & 0 & 0 & 0 \\ 0 & 0 & 0 & 0 \\ 0 & 0 & 0 & 0 \end{bmatrix}, \quad A_2 = \begin{bmatrix} 0 & 0 & 0 & 0 \\ 0 & 0 & 1 & 0 \\ 0 & 1 & 0 & 0 \\ 0 & 0 & 0 & 0 \end{bmatrix}, \quad A_3 = \begin{bmatrix} 0 & 0 & 0 & 0 \\ 0 & 0 & 0 & 0 \\ 0 & 0 & 0 & 1 \\ 0 & 0 & 1 & 0 \end{bmatrix}.$$

Then

$$I + \sum A_j x_j = \begin{bmatrix} 1 & x_1 & 0 & 0 \\ x_1 & 1 & x_2 & 0 \\ 0 & x_2 & 1 & x_3 \\ 0 & 0 & x_3 & 1 \end{bmatrix}$$

is the corresponding monic affine linear pencil. ■

Polynomial evaluations

If $p \in \mathbb{R}^{d \times d'}<x>$ is a noncommutative polynomial and $X \in (\mathbb{S}^{n \times n})^g$, the *evaluation* $p(X) \in \mathbb{R}^{dn \times d'n}$ is defined by simply replacing x_i by X_i. Throughout we use lowercase letters for variables and the corresponding capital letter for matrices substituted for that variable.

Example 8.5. Suppose $p(x) = Ax_1x_2$ where $A = \left[\begin{smallmatrix} -4 & 2 \\ 3 & 0 \end{smallmatrix}\right]$. That is,

$$p(x) = \begin{bmatrix} -4x_1x_2 & 2x_1x_2 \\ 3x_1x_2 & 0 \end{bmatrix}.$$

Thus $p \in \mathbb{R}^{2\times 2}<x>$ and one example of an evaluation is

$$p\left(\begin{bmatrix} 0 & 1 \\ 1 & 0 \end{bmatrix}, \begin{bmatrix} 1 & 0 \\ 0 & -1 \end{bmatrix}\right) = A \otimes \left(\begin{bmatrix} 0 & 1 \\ 1 & 0 \end{bmatrix} \begin{bmatrix} 1 & 0 \\ 0 & -1 \end{bmatrix}\right) = A \otimes \left(\begin{bmatrix} 0 & -1 \\ 1 & 0 \end{bmatrix}\right)$$

$$= \begin{bmatrix} 0 & 4 & 0 & -2 \\ -4 & 0 & 2 & 0 \\ 0 & -3 & 0 & 0 \\ 3 & 0 & 0 & 0 \end{bmatrix}.$$

Similarly, if p is a constant matrix-valued noncommutative polynomial, $p(x) = A$, and $X \in (\mathbb{S}^{n \times n})^g$, then $p(X) = A \otimes I_n$. Here we have taken advantage of the usual *tensor* (or *Kronecker*) *product* of matrices. Given an $\ell \times \ell'$ matrix $A = (A_{i,j})$ and an $n \times n'$ matrix B, by definition, $A \otimes B$ is the $n \times n'$ *block* matrix

$$A \otimes B = \left[A_{i,j} B\right],$$

with $\ell \times \ell'$ matrix entries. We have reserved the tensor product notation for the tensor product of matrices and have eschewed the strong temptation of using $A \otimes x_\ell$ in place of Ax_ℓ when x_ℓ is one of the variables. ■

Proposition 8.6. *Suppose $p \in \mathbb{R}<x>$. In increasing levels of generality,*

1. *if $p(X) = 0$ for all n and all $X \in (\mathbb{S}^{n\times n})^g$, then $p = 0$;*
2. *if there is a nonempty noncommutative basic open semialgebraic set $\mathcal{O}$ such that $p(X) = 0$ on $\mathcal{O}$ (meaning for every n and $X \in \mathcal{O}(n)$, $p(X) = 0$), then $p = 0$;*
3. *there is an N, depending only upon the degree of p, so that for any $n \geq N$ if there is an open subset $O \subseteq (\mathbb{S}^{n\times n})^g$ with $p(X) = 0$ for all $X \in O$, then $p = 0$.*

Proof. See Exercises 8.28, 8.31, and 8.34. □

Exercise 8.7. Use Proposition 8.6 to prove the following statement.

Proposition 8.8. *Suppose $p \in \mathbb{R}<x>$. Show $p(X)$ is symmetric for every n and every $X \in (\mathbb{S}^{n\times n})^g$ if and only if $p^\intercal = p$.*

8.2.3 Noncommutative Convexity Revisited and Noncommutative Positivity

Now we return with a bit more detail to our main theme, convexity. A symmetric polynomial p is *matrix convex* if, for each positive integer n, each pair of g-tuples $X = (X_1, \ldots, X_g)$ and $Y = (Y_1, \ldots, Y_g)$ in $(\mathbb{S}^{n\times n})^g$, and each $0 \leq t \leq 1$,

$$tp(X) + (1-t)p(Y) - p\big(tX + (1-t)Y\big) \succeq 0,$$

where, for an $n \times n$ matrix $A \in \mathbb{R}^{n\times n}$, the notation $A \succeq 0$ means A is positive semidefinite. Synonyms for matrix convex include both *noncommutative convex* and simply *convex*.

Exercise 8.9. Show that the definition here of (matrix) convex is equivalent to that given in (8.5) in the informal introduction to noncommutative polynomials.

As we have already seen in the informal introduction to noncommutative polynomials, even in one variable, convexity in the noncommutative setting differs from convexity in the commutative case because here Y need not commute with X. Thus, although the polynomial x^4 is a convex function of one real variable, it is not matrix convex. On the other hand, to verify that x^2 is a matrix convex polynomial, observe that

$$\begin{aligned} &tX^2 + (1-t)Y^2 - (tX + (1-t)Y)^2 \\ &\qquad = t(1-t)(X^2 - XY - YX + Y^2) = t(1-t)(X-Y)^2 \succeq 0. \end{aligned}$$

A polynomial $p \in \mathbb{R}<x>$ is *matrix positive*, synonymously *noncommutative positive* or simply *positive*, if $p(X) \succeq 0$ for all tuples $X = (X_1, \ldots, X_g) \in (\mathbb{S}^{n\times n})^g$.

A polynomial p is a *sum of squares* if there exists $k \in \mathbb{N}$ and polynomials $h_1, \ldots, h_k$ such that

$$p = \sum_{j=1}^{k} h_j^{\intercal} h_j.$$

Because, for a matrix A, the matrix $A^{\intercal}A$ is positive semidefinite, if p is a sum of squares, then p is positive. Though we will not discuss its proof in this chapter, we mention that, in contrast with the commutative case, the converse is true [27, 53].

Theorem 8.10. *If $p \in \mathbb{R}<x>$ is positive, then p is a sum of squares.*

As for convexity, note that $p(x)$ is convex if and only if the polynomial $q(x, y)$ in $2g$ noncommutative variables given by

$$q(x,y) = \frac{1}{2}\big(p(x) + p(y)\big) - p\Big(\frac{x+y}{2}\Big)$$

is positive.

8.2.4 Directional Derivatives Versus Noncommutative Convexity and Positivity

Matrix convexity can be formulated in terms of positivity of the Hessian, just as in the case of a real variable. Thus we take a few moments to develop a very useful noncommutative calculus.

Given a polynomial $p \in \mathbb{R}<x>$, the *ℓth directional derivative* of p in the "direction" h is

$$p^{(\ell)}(x)[h] := \left.\frac{d^{\ell} p(x+th)}{dt^{\ell}}\right|_{t=0}.$$

Thus $p^{(\ell)}(x)[h]$ is the polynomial that evaluates to

$$\left.\frac{d^{\ell} p(X+tH)}{dt^{\ell}}\right|_{t=0} \quad \text{for every choice of} \quad X,\, H \in (\mathbb{S}^{n\times n})^g.$$

We let $p'(x)[h]$ denote the first derivative, and the *Hessian*, denoted $p''(x)[h]$ of $p(x)$, is the second directional derivative of p in the direction h.

Equivalently, the Hessian of $p(x)$ can also be defined as the part of the polynomial

$$r(x)[h] := 2\big(p(x+h) - p(x)\big)$$

in

$$\mathbb{R}<x>[h] := \mathbb{R}<x_1, \ldots, x_g,\, h_1, \ldots, h_g>$$

that is homogeneous of degree two in h.

If $p'' \neq 0$, that is, if $p = p(x)$ is a noncommutative polynomial of degree two or more, then the polynomial $p''(x)[h]$ in the $2g$ variables $x_1, \ldots, x_g, h_1 \ldots, h_g$ is homogeneous of degree 2 in h and has degree equal to the degree of p.

Example 8.11.

(1) The Hessian of the polynomial $p = x_1^2 x_2$ is

$$p''(x)[h] = 2(h_1^2 x_2 + h_1 x_1 h_2 + x_1 h_1 h_2).$$

(2) The Hessian of the polynomial $f(x) = x^4$ (just one variable) is

$$f''(x)[h] = 2(h^2x^2 + hxhx + hx^2h + xhxh + xh^2x + x^2h^2). \quad \blacksquare$$

Noncommutative convexity is neatly described in terms of the Hessian.

Lemma 8.12. *$p \in \mathbb{R}<x>$ is noncommutative convex if and only if $p''(x)[h]$ is noncommutative positive.*

Proof. See Exercise 8.26. $\square$

8.2.5 Symmetric, Free, Mixed, and Classes of Variables

To this point, our variables x have been *symmetric* in the sense that, under the involution, $x_j^\intercal = x_j$. The corresponding polynomials, elements of $\mathbb{R}<x>$ are then the noncommutative analogue of polynomials in real variables, with evaluations at tuples in $\mathbb{S}^{n\times n}$. In various applications and settings it is natural to consider noncommutative polynomials in other types of variables.

Free variables

The noncommutative analogue of polynomials in complex variables is obtained by allowing evaluations on tuples X of not necessarily symmetric matrices. In this case, the involution must be interpreted differently, and the variables are called *free.*

In this setting, given the noncommutative variables $x = (x_1, \dots, x_g)$, let $x^\intercal = (x_1^\intercal, \dots, x_g^\intercal)$ denote another collection of noncommutative variables. On the ring $\mathbb{R}<x, x^\intercal>$ define the involution $^\intercal$ by requiring $x_j \mapsto x_j^\intercal$; $x_j^\intercal \mapsto x_j$; $^\intercal$ reverses the order of words; and linearity. For instance, for

$$q(x) = 1 + x_1^\intercal x_2 - x_2^\intercal x_1 \in \mathbb{R}<x, x^\intercal>,$$

we have

$$q^\intercal(x) = 1 + x_2^\intercal x_1 - x_1^\intercal x_2.$$

Elements of $\mathbb{R}<x, x^\intercal>$ are polynomials in *free variables*, and in this setting the variables themselves are *free.*

A polynomial $p \in \mathbb{R}<x, x^\intercal>$ is *symmetric* provided $p^\intercal = p$. In particular, q above is not symmetric, but

$$p = 1 + x_1^\intercal x_2 + x_2^\intercal x_1 \tag{8.7}$$

is.

A polynomial $p \in \mathbb{R}{<}x, x^\intercal{>}$ is *analytic* if there are no transposes, i.e., if p is a polynomial in x alone.

Elements of $\mathbb{R}{<}x, x^\intercal{>}$ are naturally evaluated on tuples $X = (X_1, \ldots, X_g) \in (\mathbb{R}^{\ell \times \ell})^g$. For instance, if p is the polynomial in (8.7) and $X = (X_1, X_2) \in (\mathbb{R}^{2\times 2})^2$, where

$$X_1 = \begin{bmatrix} 0 & 0 \\ 1 & 0 \end{bmatrix} = X_2,$$

then

$$p(X) = \begin{bmatrix} 3 & 0 \\ 0 & 1 \end{bmatrix}.$$

The space $\mathbb{R}^{d\times d'}{<}x, x^\intercal{>}$ is defined by analogy with $\mathbb{R}^{d\times d'}{<}x{>}$, and evaluation of elements in $\mathbb{R}^{d\times d'}{<}x, x^\intercal{>}$ at a tuple $X \in (\mathbb{R}^{\ell\times\ell})^g$ is defined in the obvious way.

Exercise 8.13. State and prove analogues of Propositions 8.6 and 8.8 for $\mathbb{R}{<}x, x^\intercal{>}$ and evaluations from $(\mathbb{R}^{\ell\times\ell})^g$.

Mixed variables

At times it is desirable to mix free and symmetric variables. We won't introduce notation for this situation, as it will generally be understood from the context. Here are some examples:

Example 8.14.

$$p(x) = x_1^\intercal x_1 + x_2 + \frac{3}{4} x_1 x_2 x_1^\intercal, \qquad x_2 = x_2^\intercal; \tag{8.8}$$

$$\mathrm{ric}(a_1, a_2, x) = a_1 x + x a_1^\intercal - x a_2 a_2^\intercal x, \qquad x = x^\intercal.$$

In the first case x_1 is free, but x_2 is symmetric; and in the second a_1 and a_2 are free, but x is symmetric. Two additional remarks are in order about the second polynomial. First, it is a *Riccati polynomial* ubiquitous in control theory. Second, we have separated the variables into *two classes* of variables, the a variables and the x variable(s); thus $p \in \mathbb{R}{<}a, x = x^\intercal{>}$. In applications, the a variables can be chosen to represent known (system parameters), while the x variables are unknown(s). Of course, it could be that some of the a variables are symmetric and some free and ditto for the x variables. ∎

Example 8.15. Various directional derivatives of p in (8.8) are

$$D_{x_1} p(x)[h_1] = h_1^\intercal x_1 + x_1^\intercal h_1 + \frac{3}{4} h_1 x_2 x_1^\intercal + \frac{3}{4} x_1 x_2 h_1^\intercal, \qquad D_{x_2} p(x)[h_2] = h_2 + \frac{3}{4} x_1 h_2 x_1^\intercal,$$

$$D_x p(x)[h] = h_1^\intercal x_1 + x_1^\intercal h_1 + h_2 + \frac{3}{4} h_1 x_2 x_1^\intercal + \frac{3}{4} x_1 x_2 h_1^\intercal + \frac{3}{4} x_1 h_2 x_1^\intercal, \qquad ∎$$

Continuing with the variable class warfare, consider the following matrix-valued example.

Example 8.16. Let

$$L(a_1, a_2, x) = \begin{bmatrix} a_1 x + x a_1^\intercal & a_2^\intercal x \\ x a_2 & 1 \end{bmatrix}.$$

We consider $L \in \mathbb{R}^{2\times 2}<a, x = x^\intercal>$; i.e., the a variables are free, and the x-variables symmetric. Note that L is linear in x if we consider a_1, a_2 fixed. Of course, if a_1, a_2, and x are all scalars, then using Schur complements tells us there is a close relation between L in this example and the Riccati of the previous example. ∎

8.2.6 Noncommutative Rational Functions

While it is possible to define noncommutative functions [67, 64, 69, 70, 60, 61, 46, 28, 29], in this section we content ourselves with a relatively informal discussion of noncommutative rational functions [10, 11, 41, 45].

Rational functions, a gentle introduction

Noncommutative *rational expressions* are obtained by allowing inverses of polynomials. An example is the discrete time algebraic Riccati equation

$$r(a, x) = a_1^\intercal x a_1 - (a_1^\intercal x a_2) a_1 (a_3 + a_2^\intercal x a_2)^{-1} (a_2^\intercal x a_1) + a_4, \qquad x = x^\intercal.$$

It is a rational expression in the free variables a and the symmetric variable x, as is r^{-1}. An example, in free variables, which arises in operator theory is

$$s(x) = x^\intercal (1 - x x^\intercal)^{-1}. \tag{8.9}$$

Thus, we define (scalar) *noncommutative rational expressions* for free noncommutative variables x by starting with noncommutative polynomials and then applying successive arithmetic operations—addition, multiplication, and inversion. We emphasize that an expression includes the order in which it is composed, and no two distinct expressions are identified, e.g., $(x_1) + (-x_1)$, $(-1) + (((x_1)^{-1})(x_1))$, and 0 are different noncommutative rational expressions.

Evaluation on polynomials naturally extends to rational expressions. If r is a rational expression in free variables and $X \in (\mathbb{R}^{\ell\times\ell})^g$, then $r(X)$ is defined—in the obvious way—as long as any inverses appearing actually exist. Indeed, our main interest is in the evaluation of a rational expression. For instance, for the polynomial s above in one free variable, $s(X)$ is defined as long as $I - XX^\intercal$ is invertible and in this case,

$$s(X) = X^\intercal (I - XX^\intercal)^{-1}.$$

Generally, a noncommutative rational expression r can be evaluated on a g-tuple X of $n \times n$ matrices in its *domain of regularity*, $\operatorname{dom} r$, which is defined as the set of all g-tuples of square matrices of all sizes such that all the inverses involved in the calculation of $r(X)$ exist. For example, if $r = (x_1 x_2 - x_2 x_1)^{-1}$, then $\operatorname{dom} r = \{X = (X_1, X_2)\colon \det(X_1X_2 - X_2X_1) \neq 0\}$. We assume that $\operatorname{dom} r \neq \emptyset$. In other words, when forming noncommutative rational expressions we never invert an expression that is nowhere invertible.

Two rational expressions r_1 and r_2 are *equivalent* if $r_1(X) = r_2(X)$ at any X where both are defined. For instance, for the rational expression t in one free variable,

$$t(x) = (1 - x^\mathsf{T} x)^{-1} x^\mathsf{T},$$

and s from (8.9), it is an exercise to check that $s(X)$ is defined if and only if $t(X)$ is and moreover in this case $s(X) = t(X)$. Thus s and t are equivalent rational expressions. We call an equivalence class of rational expressions a *rational function.* The set of all rational functions will be denoted by $\mathbb{R}\langle\!\langle x \rangle\!\rangle$.

Here is an interesting example of a noncommutative rational function with nested inverses. It is taken from [2, Theorem 6.3].

Example 8.17. Consider two free variables x, y. For any $r \in \mathbb{R}\langle\!\langle x, y \rangle\!\rangle$ let

$$W(r) := c\big(x, c(x,r)^2\big) \cdot c\big(x, c(x,r)^{-1}\big)^{-1} \in \mathbb{R}\langle\!\langle x, y \rangle\!\rangle. \tag{8.10}$$

Recall that c denotes the commutator (8.4). Bergman's noncommutative rational function is given by

$$\begin{aligned} b := W(y) \cdot W\big(c(x,y)\big) \cdot W\Big(c\big(x, c(x,y)\big)^{-1}\Big) \cdot W\Big(c\big(x, c(x, c(x,y))\big)^{-1}\Big) \\ \in \mathbb{R}\langle\!\langle x, y \rangle\!\rangle. \quad \blacksquare \end{aligned} \tag{8.11}$$

Exercise 8.18. Consider the function W from (8.10). Let R, X be $n \times n$ matrices and assume $c\big(X, c(X,R)^{-1}\big)$ exists and is invertible. Prove the following:

(1) If $n = 2$, then $W(R) = 0$.

(2) If $n = 3$, then $W(R) = \det(c(X,R))$.

Exercise 8.19. Consider Bergman's rational function (8.11).

(1) Show that on a dense set of 2×2 matrices (X, Y), $b(X,Y) = 0$.

(2) Prove that on a dense set of 3×3 matrices (X, Y), $b(X,Y) = 1$.

The moral of Exercise 8.19 is that, unlike in the case of polynomial identities, a noncommutative rational function that vanishes on (a dense set of) 3×3 matrices need not vanish on (a dense set of) 2×2 matrices.

Matrices of rational functions; LDL^T

One of the main ways noncommutative rational functions occur in systems engineering is in the manipulation of matrices of polynomials. Extremely important is the LDL^T *decomposition.* Consider the 2×2 matrix with noncommutative entries

$$M = \begin{bmatrix} a & b^\mathsf{T} \\ b & c \end{bmatrix},$$

where $a = a^\intercal$. The entries themselves could be noncommutative polynomials or even rational functions. If a is not zero, then M has the following decomposition:

$$M = LDL^\intercal = \begin{bmatrix} I & 0 \\ ba^{-1} & I \end{bmatrix} \begin{bmatrix} a & 0 \\ 0 & c - ba^{-1}b^\intercal \end{bmatrix} \begin{bmatrix} I & a^{-1}b^\intercal \\ 0 & I \end{bmatrix}.$$

Note that this formula holds in the case that c is itself a (square) matrix noncommutative rational function and b (and thus $b^\intercal$) are vector-valued noncommutative rational functions. On the other hand, if both $a = c = 0$, then M is the block matrix

$$M = \begin{bmatrix} 0 & b \\ b^\intercal & 0 \end{bmatrix}.$$

If M is a $k \times k$ matrix, then iterating this procedure produces a decomposition of a permutation $\Pi M \Pi^\intercal$ of M of the form $\Pi M \Pi^\intercal = LDL^\intercal$, where D and L have the form

$$D = \begin{bmatrix} d_1 & 0 & 0 & 0 & 0 & 0 & 0 \\ \vdots & \ddots & 0 & 0 & \cdots & 0 & 0 \\ 0 & \cdots & d_k & 0 & \cdots & 0 & 0 \\ 0 & \cdots & 0 & D_{k+1} & \cdots & 0 & 0 \\ \vdots & \cdots & \vdots & \vdots & \ddots & 0 & 0 \\ 0 & \cdots & 0 & 0 & \cdots & D_\ell & 0 \\ 0 & \cdots & 0 & 0 & \cdots & 0 & E \end{bmatrix} \tag{8.12}$$

and

$$L = \begin{bmatrix} 1 & 0 & 0 & 0 & 0 & 0 & 0 \\ * & \ddots & 0 & 0 & 0 & 0 & \\ * & * & 1 & 0 & 0 & 0 & 0 \\ * & * & * & I_2 & 0 & 0 & 0 \\ * & * & * & * & \ddots & 0 & 0 \\ * & * & * & * & * & I_2 & 0 \\ * & * & * & * & * & * & I_a \end{bmatrix}, \tag{8.13}$$

where d_j are symmetric rational functions, and the D_j are nonzero 2×2 matrices of the form

$$D_j = \begin{bmatrix} 0 & b_j \\ b_j^\intercal & 0 \end{bmatrix}.$$

E is a square 0 matrix (possibly of size 0×0 and thus absent), and I_2 is the 2×2 identity and the $*$'s represent possibly nonzero rational expressions (in some cases matrices of rational functions), some of the 0's are zero matrices (of the appropriate sizes), and a is the dimension of the space that E acts upon. The permutation Π is necessary in cases where the procedure hits a 0 on the diagonal, necessitating a permutation to bring a nonzero diagonal entry into the "pivot" position.

Theorem 8.20. *Suppose $M(x) \in \mathbb{R}\langle x \rangle^{\ell \times \ell}$ is symmetric, and $\Pi M \Pi^\intercal = LDL^\intercal$ where L, D are $\ell \times \ell$ matrices with noncommutative rational entries as in* (8.13) *and* (8.12) *and L, respectively. If n is a positive integer and $X \in (\mathbb{S}^{n \times n})^g$ is in the domains of both L and D, then $M(X)$ is positive semidefinite if and only if $D(X)$ is positive semidefinite.*

Proof. The proof is an easy exercise based on the fact that a square block lower triangular matrix whose diagonal blocks are invertible is itself invertible. In this case, $L(X)$ is block lower triangular, with the $n \times n$ identity I_n as each diagonal entry. Thus $M(X)$ and $D(X)$ are congruent and thus have the same number of negative eigenvalues. ☐

Remark 8.21. Note that if D has any 2×2 blocks D_j, then $D(X) \succeq 0$ if and only if each $D_j(X) = 0$. Thus, if D has any 2×2 blocks, generically $D(X)$, and hence $M(X)$, is not positive semidefinite. (Recall that we assume, without loss of generality, that D_j are not zero.)

More on rational functions

The matrix positivity and convexity properties of noncommutative rational functions go just like those for polynomials. One only tests a rational function r on matrices X in its domain of regularity. The definition of directional derivatives goes as before and it is easy to compute them formally. There are issues of equivalences which we avoid here, instead referring the reader to [10, 45] or our treatment in [41].

We emphasize that proving the assertions above takes considerable effort, because of dealing with the equivalence relation. In practice one works with rational expressions, and calculations with noncommutative rational expressions themselves are straightforward. For instance, computing the derivative of a symmetric noncommutative rational function r leads to an expression of the form

$$Dr(x)[h] = \text{symmetrize}\left[\sum_{\ell=1}^{k} a_\ell(x) h b_\ell(x)\right],$$

where a_ℓ, b_ℓ are noncommutative rational functions of x, and the symmetrization of a (not necessarily symmetric) rational expression s is $\frac{s+s^\intercal}{2}$.

8.2.7 Exercises

Section 8.3 gives a very brief introduction on noncommutative computer algebra and some might enjoy playing with computer algebra in working some of these exercises.

Define for use in later exercises the noncommutative polynomials

$$\begin{aligned}
p &= x_1^2 x_2^2 - x_1 x_2 x_1 x_2 - x_2 x_1 x_2 x_1 - x_2^2 x_1^2, \\
q &= x_1 x_2 x_3 + x_2 x_3 x_1 + x_3 x_1 x_2 - x_1 x_3 x_2 - x_2 x_1 x_3 - x_3 x_2 x_1, \\
s &= x_1 x_3 x_2 - x_2 x_3 x_1.
\end{aligned}$$

Exercise 8.22.

(a) What is the derivative with respect to x_1 in direction h_1 of q and s?

(b) Concerning the formal derivative with respect to x_1 in direction h_1,

 (i) show the derivative of $r(x_1) = x_1{}^{-1}$ is $-x_1^{-1}h_1x_1^{-1}$;

 (ii) what is the derivative of $u(x_1, x_2) = x_2(1 + 2x_1)^{-1}$?

Exercise 8.23. Consider the polynomials p, q, s and rational functions r, u from above.

(a) Evaluate the polynomials p, q, s on some matrices of size 1×1, 2×2, and 3×3.

(b) Redo part (a) for the rational functions r, u.

Try to use Mathematica or MATLAB.

Exercise 8.24. Show that $c = x_1x_2 - x_2x_1$ is not symmetric by finding n and $X = (X_1, X_2)$ such that $c(X)$ is not a symmetric matrix.

Exercise 8.25. Consider the following polynomials in two and three variables, respectively:

$$\begin{aligned} h_1 &= c^2 = (x_1x_2)^2 - x_1x_2^2x_1 - x_2x_1^2x_2 + (x_2x_1)^2, \\ h_2 &= h_1x_3 - x_3h_1. \end{aligned}$$

(a) Compute $h_1(X_1, X_2)$ and $h_2(X_1, X_2, X_3)$ for several choices of 2×2 matrices X_j. What do you find? Can you formulate and prove a statement?

(b) What happens if you plug in 3×3 matrices into h_1 and h_2?

Exercise 8.26. Prove that a symmetric noncommutative polynomial p is matrix convex if and only if the Hessian $p''(x)[h]$ is matrix positive by completing the following exercise.

Fix n, suppose ℓ is a positive linear functional on $\mathbb{S}^{n\times n}$, and consider

$$f = \ell \circ p : (\mathbb{S}^{n\times n})^g \to \mathbb{R}.$$

(a) Show f is convex if and only if $\frac{d^2 f(X+tH)}{dt^2} \geq 0$ at $t = 0$ for all $X, H \in (\mathbb{S}^{n\times n})^g$.

Given $v \in \mathbb{R}^n$, consider the linear functional $\ell(M) := v^\intercal Mv$ and let $f_v = \ell \circ p$.

(b) *Geometric*: Fix n. Show, each f_v satisfies the convexity inequality if and only if p satisfies the convexity inequality on $(\mathbb{S}^{n\times n})^g$.

(c) *Analytic*: Show, for each $v \in \mathbb{R}^n$, $f_v''(X)[H] \geq 0$ for every $X, H \in (\mathbb{S}^{n\times n})^g$ if and only if $p''(X)[H] \succeq 0$ for every $X, H \in (\mathbb{S}^{n\times n})^g$.

Exercise 8.27. For $n \in \mathbb{N}$ let

$$s_n = \sum_{\tau \in \mathrm{Sym}_n} \mathrm{sign}(\tau) x_{\tau(1)} \cdots x_{\tau(n)}$$

be a polynomial of degree n in n variables. Here Sym_n denotes the symmetric group on n elements.

(a) Prove that s_4 is a polynomial identity for 2×2 matrices. That is, for any choice of 2×2 matrices $X_1, \ldots, X_4$, we have

$$s_4(X_1, \ldots, X_4) = 0.$$

(b) Fix $d \in \mathbb{N}$. Prove that there exists a nonzero polynomial p vanishing on all tuples of $d \times d$ matrices.

Several of the next exercises use a version of the shift operators on Fock space. With g fixed, the corresponding *Fock space*, $\mathcal{F} = \mathcal{F}_g$, is the Hilbert space obtained from $\mathbb{R}<x>$ by declaring the words to be an orthonormal basis; i.e., if v, w are words, then

$$\langle v, w \rangle = \delta_{v,w},$$

where $\delta_{v,w} = 1$ if $v = w$ and is 0 otherwise. Thus $\mathcal{F}_g$ is the closure of $\mathbb{R}<x>$ in this inner product. For each j, the operator S_j on $\mathcal{F}_g$ densely defined by $S_j p = x_j p$, for $p \in \mathbb{R}<x>$ is an isometry (preserves the inner product) and hence extends to an isometry on all of $\mathcal{F}_g$. Of course, S_j acts on an infinite-dimensional Hilbert space and thus is not a matrix.

Exercise 8.28. Given a natural number k, note that $\mathbb{R}<x>_k$ is a finite dimensional (and hence closed) subspace of $\mathcal{F} = \mathcal{F}_g$. The dimension of $\mathbb{R}<x>_k$ is

$$\sigma(k) = \sum_{j=0}^{k} g^j. \tag{8.14}$$

Let $V : \mathbb{R}<x>_k \to \mathcal{F}$ denote the inclusion and

$$T_j = V^\intercal S_j V.$$

Thus T_j does act on a finite-dimensional space, and $T = (T_1, \ldots, T_g) \in (\mathbb{R}^{n \times n})^g$ for $n = \sigma(k)$.

(a) Show that if v is a word of length at most $k - 1$, then

$$T_j v = x_j v,$$

and $T_j v = 0$ if the length of v is k.

(b) Determine $T_j^\intercal$.

(c) Show that if p is a nonzero polynomial of degree at most k and $Y_j = T_j + T_j^\intercal$, then $p(Y)\emptyset \neq 0$.

(d) Conclude that if, for every n and $X \in (\mathbb{S}^{n\times n})^g$, $p(X) = 0$, then p is 0.

Exercise 8.28 shows there are no noncommutative polynomials vanishing on all tuples of (symmetric) matrices of all sizes. The next exercise will lead the reader through an alternative proof inspired by standard methods of polynomial identities.

Exercise 8.29. Let $p \in \mathbb{R}\langle x\rangle_n$ be an analytic polynomial that vanishes on $(\mathbb{R}^{n\times n})^g$ (same fixed n). Write $p = p_0 + p_1 + \cdots + p_n$, where p_j is the homogeneous part of p of degree j.

(a) Show that p_j also vanishes on $(\mathbb{R}^{n\times n})^g$.

(b) A polynomial q is called multilinear if it is homogeneous of degree one with respect to all of its variables. Equivalently, each of its monomials contains all variables exactly once, i.e.,

$$q = \sum_{\pi \in S_n} \alpha_\pi X_{\pi(1)} \cdots X_{\pi(n)}.$$

Using the staircase matrices $E_{11}, E_{12}, E_{22}, E_{23}, \ldots, E_{n-1\,n}, E_{nn}$ show that a nonzero multilinear polynomial q of degree n cannot vanish on all $n\times n$ matrices.

(c) By (a) we may assume p is homogeneous. By induction on the biggest degree a variable in p can have, prove that $p = 0$. *Hint:* What are the degrees of the variables appearing in

$$p(x_1 + \hat{x}_1, x_2, \ldots, x_g) - p(x_1, x_2, \ldots, x_g) - p(\hat{x}_1, x_2, \ldots, x_g)?$$

Exercise 8.30. Redo Exercise 8.29 for a polynomial

(a) $p \in \mathbb{R}\langle x, x^\intercal\rangle$, not necessarily analytic, vanishing on all tuples of matrices;

(b) $p \in \mathbb{R}\langle x\rangle$ vanishing on all tuples of symmetric matrices.

Exercise 8.31. Show that if $p \in \mathbb{R}\langle x\rangle$ vanishes on a nonempty basic open semialgebraic set, then $p = 0$.

Exercise 8.32. Suppose $p \in \mathbb{R}\langle x\rangle$, n is a positive integer, and $O \subseteq (\mathbb{S}^{n\times n})^g$ is an open set. Show that if $p(X) = 0$ for each $X \in O$, then $P(X) = 0$ for each $X \in (\mathbb{S}^{n\times n})^g$. *Hint:* Given $X_0 \in O$ and $X \in (\mathbb{S}^{n\times n})^g$, consider the matrix valued polynomial,

$$q(t) = p(X_0 + tX).$$

Exercise 8.33. Suppose $r \in \mathbb{R}\langle\!\langle x\rangle\!\rangle$ is a rational function and there is a nonempty noncommutative basic open semialgebraic set $\mathcal{O} \subseteq \operatorname{dom}(r)$ with $r|_{\mathcal{O}} = 0$. Show that $r = 0$.

Exercise 8.34. Prove item (3) of Proposition 8.6. You may wish to use Exercises 8.32 and 8.28.

Exercise 8.35. Prove the following proposition.

Proposition 8.36. *If $\pi : \mathbb{R}{<}x{>} \to \mathbb{R}^{n\times n}$ is an involution preserving homomorphism, then there is an $X \in (\mathbb{S}^{n\times n})^g$ such that $\pi(p) = p(X)$; i.e., all finite dimensional representations of $\mathbb{R}{<}x{>}$ are evaluations.*

Exercise 8.37. Do the algebra to show

$$x^\intercal(1 - xx^\intercal)^{-1} = (1 - x^\intercal x)^{-1}x^\intercal.$$

(This is a key fact used in the model theory for contractions [55].)

Exercise 8.38. Give an example of symmetric 2×2 matrices X, Y such that $X \succeq Y \succeq 0$ but $X^2 \not\succeq Y^2$.

This failure of a basic order property of $\mathbb{R}$ for $\mathbb{S}^{n\times n}$ is closely related to the rigid nature of positivity and convexity in the noncommutative setting.

Exercise 8.39. Antiderivatives.

(a) Is $q(x)[h] = xh + hx$ the derivative of any noncommutative polynomial p? If so, what is p?

(b) Is $q(x)[h] = hhx + hxh + xhh$ the second derivative of any noncommutative polynomial p? If so, what is p?

(c) Describe in general which polynomials $q(x)[h]$ are the derivative of some noncommutative polynomial $p(x)$.

(d) Check you answer against the theory in [23].

Exercise 8.40. (Requires background in algebra) Show that $\mathbb{R}\langle\!\langle x\rangle\!\rangle$ is a division ring; i.e., the noncommutative rational functions form a ring in which every nonzero element is invertible.

Exercise 8.41. In this exercise we will establish that it is possible to embed the free algebra $\mathbb{R}{<}x_1, \ldots, x_g{>}$ into $\mathbb{R}{<}x, y{>}$ for any $g \in \mathbb{N}$.

(a) Show that the subalgebra of $\mathbb{R}{<}x, y{>}$ generated by xy^n, $n \in \mathbb{N}_0$, is free.

(b) Ditto for the subalgebra generated by

$$x_1 = x, \quad x_2 = c(x_1, y), \quad x_3 = c(x_2, y), \quad \ldots, \quad x_n = c(x_{n-1}, y), \ldots.$$

Here, as before, c is the commutator, $c(a, b) = ab - ba$.

A comprehensive study of free algebras and noncommutative rational functions from an algebraic viewpoint is developed in [10, 11].

Exercise 8.42. As a hard exercise, numerically verify that the set

$$\mathrm{ncTV}(2) = \{X \in (\mathbb{S}^{2\times 2})^2 : 1 - X_1^4 - X_2^4 \succ 0\}$$

is not convex. That is, find $X = (X_1, X_2)$ and $Y = (Y_1, Y_2)$, where X_1, X_2, Y_1, Y_2 are 2×2 symmetric matrices such that both

$$1 - X_1^4 - X_2^4 \succ 0 \quad \text{and} \quad 1 - Y_1^4 - Y_2^4 \succ 0$$

but

$$1 - \left(\frac{X_1 + Y_1}{2}\right)^4 - \left(\frac{X_2 + Y_2}{2}\right)^4 \not\succ 0.$$

You may wish to write a numerical search routine.

8.3 Computer Algebra Support

There are several computer algebra packages available to ease the first contact with free convexity and positivity. In this section we briefly describe two of them:

(1) `NCAlgebra` running under Mathematica;

(2) `NCSOStools` running under MATLAB.

The former is more universal in that it implements manipulation with noncommutative variables, including noncommutative rationals, and several algorithms pertaining to convexity. The latter is focused on noncommutative positivity and numerics.

8.3.1 NCAlgebra

NCAlgebra [42] runs under Mathematica and gives it the capability of manipulating noncommuting algebraic expressions. An important part of the package (which we shall not go into here) is NCGB, which computes noncommutative Groebner bases and has extensive sorting and display features as well as algorithms for automatically discarding "redundant" polynomials.

We recommend that the user have a look at the Mathematica notebook `NCBasicCommandsDemo` available from the NCAlgebra website

http://math.ucsd.edu/~ncalg/

for the basic commands and their usage in NCAlgebra. Here is a sample.

The basic ingredients are (symbolic) variables, which can be either noncommutative or commutative. At present, single-letter lowercase variables are noncommutative by default and all others are commutative by default. To change this one can employ

NCAlgebra Command: `SetNonCommutative[listOfVariables]` to make all the variables appearing in listOfVariables noncommutative. The converse is given by

NCAlgebra Command: `SetCommutative`.

Example 8.43. Here is a sample session in Mathematica running NCAlgebra.

```
In[1]:= a ** b - b ** a
Out[1]= a ** b - b ** a

In[2]:= A ** B - B ** A
Out[2]= 0

In[3]:= A ** b - b ** a
Out[3]= A b - b ** a

In[4]:= CommuteEverything[a ** b - b ** a]
Out[4]= 0

In[5]:= SetNonCommutative[A, B]
Out[5]= {False, False}

In[6]:= A ** B - B ** A
Out[6]= A ** B - B ** A

In[7]:= SetNonCommutative[A];SetCommutative[B]
Out[7]= {True}

In[8]:= A ** B - B ** A
Out[8]= 0
```
■

Slightly more advanced is the NCAlgebra command to generate the directional derivative of a polynomial $p(x, y)$ with respect to x, which is denoted by $D_x p(x, y)[h]$:

NCAlgebra Command: `DirectionalD[Function` p, x, h`]`, and is abbreviated

NCAlgebra Command: `DirD`.

Example 8.44. Consider

```
a = x ** x ** y - y ** x ** y
```

Then

```
DirD[a, x, h] = (h ** x + x ** h) ** y - y ** h ** y
```

or in expanded form,

```
NCExpand[DirD[a, x, h]] = h ** x ** y + x ** h ** y - y ** h ** y
```

Note that we have used

NCAlgebra Command: `NCExpand[Function p]` to expand a noncommutative expression. The command comes with a convenient abbreviation

NCAlgebra Command: `NCE`. ■

NCAlgebra is capable of much more. For instance, is a given noncommutative function "convex"? You type in a function of noncommutative variables; the command

NCAlgebra Command: `NCConvexityRegion[Func, ListOfVariables]` tells you where the (symbolic) Function is convex in the Variables. The algorithm comes from the paper of Camino et al. [9].

NCAlgebra Command: $\{L, D, U, P\}$:=`NCLDUDecomposition[Matrix]`. Computes the LDU decomposition of matrix and returns the result as a 4-tuple. The last entry is a permutation matrix which reveals which pivots were used. If matrix is symmetric, then $U = L^\intercal$.

The NCAlgebra website comes with extensive documentation. A more advanced notebook with a hands-on demonstration of applied capabilities of the package is `DemoBRL.nb`; it derives the bounded real lemma for a linear system.

Exercise 8.45. For the polynomials and rational functions defined at the beginning of Section 8.2.7, use NCAlgebra to calculate

(a) `p**q` and `NCExpand[p**q]`,

(b) `NCCollect[p**q, x1]`,

(c) `D[p,x1,h1]` and `D[u,x1,h1]`.

Warning

The Mathematica substitute commands `/.`, `/>` and `/:>` are not reliable in NCAlgebra, so a user should use NCAlgebra's Substitute command.

Example 8.46. Here is an example of unsatisfactory behavior of the built-in Mathematica function.

```
In[1]:= (x ** a ** b) /. {a ** b -> c}
Out[1]= x ** a ** b
```

On the other hand, NCAlgebra performs as desired:

```
In[2]:= Substitute[x ** a ** b, a ** b -> c]
Out[2]= x ** c
```
■

8.3.2 NCSOStools

A reader mainly interested in positivity of noncommutative polynomials might be better served by NCSOStools [8]. NCSOStools is an open source MATLAB toolbox for

(a) basic symbolic computation with polynomials in noncommuting variables;

(b) constructing and solving sum of hermitian squares (with commutators) programs for polynomials in noncommuting variables.

It is normally used in combination with standard SDP software to solve these constructed linear matrix inequalities.

The NCSOStools website http://ncsostools.fis.unm.si contains documentation and a demo notebook `NCSOStoolsdemo` to give the user a gentle introduction to its features.

Example 8.47. Although it has some ability to manipulate symbolic expressions, MATLAB cannot handle noncommuting variables. They are implemented in NCSOStools.

NCSOStools Command: `NCvars` x introduces a noncommuting variable x into the workspace. ∎

NCSOStools is well equipped to work with commutators and sums of (hermitian) squares. Recall: a *commutator* is an expression of the form $fg - gf$.

Exercise 8.48. Use NCSOStools to check whether the polynomial $x^2yx + yx^3 - 2xyx^2$ is a sum of commutators. (*Hint*: Try the `NCisCycEq` command.) If so, can you find such an expression?

Let us demonstrate an example with sums of squares.

Example 8.49. Consider

```
f = 5 + x^2 - 2*x^3 + x^4 + 2*x*y + x*y*x*y - x*y^2 + x*y^2*x
    -2*y + 2*y*x + y*x^2*y - 2*y*x*y + y*x*y*x - 3*y^2 - y^2*x + y^4
```

Is f matrix positive? By Theorem 8.10 it suffices to check whether f is a sum of squares. This is easily done using

NCSOStools Command: `NCsos`(f), which checks if the polynomial f is a sum of squares. Running NCsos(f) tells us that f is indeed a sum of squares. What NCSOStools does is transform this question into a semidefinite program and then calls a solver. NCsos comes with several options. Its full command line is

```
[IsSohs,X,base,sohs,g,SDP_data,L] = NCsos(f,params)
```

The meaning of the output is as follows:

- `IsSohs` equals 1 if the polynomial f is a sum of hermitian squares and 0 otherwise;
- `X` is the Gram matrix solution of the corresponding semidefinite program returned by the solver;
- `base` is a list of words which appear in the sums of Hermitian squares decomposition;
- `sohs` is the sums of hermitian squares decomposition of f;
- `g` is the NCpoly representing $\sum_i m_i^\intercal m_i$;
- `SDP_data` is a structure holding all the data used in the SDP solver;
- `L` is the operator representing the dual optimization problem (i.e., the dual feasible SDP matrix). ∎

Exercise 8.50. Use NCSOStools to compute the smallest eigenvalue $f(X,Y)$ can attain for a pair of symmetric matrices (X,Y). Can you also find a minimizer pair (X,Y)?

Exercise 8.51. Let $f = y^2 + (xy-1)^\intercal(xy-1)$. Show the following.

(a) $f(X,Y)$ is always positive semidefinite.

(b) For each $\epsilon > 0$ there is a pair of symmetric matrices (X,Y) so that the smallest eigenvalue of $f(X,Y)$ is ϵ.

(c) Can $f(X,Y)$ be singular?

The moral of Example 8.51 is that even if a noncommutative polynomial is bounded from below, it need not attain its minimum.

Exercise 8.52. Redo the Exercise 8.51 for $f(x) = x^\intercal x + (xx^\intercal - 1)^\intercal(xx^\intercal - 1)$.

8.4 A Gram-like Representation

The next two sections are devoted to a powerful representation of quadratic functions q in noncommutative variables which takes a strong form when q is matrix positive; we call it a *QuadratischePositivstellensatz.* Ultimately we shall apply this to $q(x)[h] = p''(x)[h]$ and show that if p is matrix convex (i.e., q is matrix positive), then p has degree 2. We begin by illustrating our grand scheme with examples.

8.4.1 Illustrating the Ideas

Example 8.53. The (symmetric) polynomial $p(x) = x_1x_2x_1 + x_2x_1x_2$ (in symmetric variables) has Hessian $q(x)[h] = p''(x)[h]$, which is homogeneous quadratic in h and is

$$q(x)[h] = 2h_1h_2x_1 + 2h_1x_2h_1 + 2h_2h_1x_2 + 2h_2x_1h_2 + 2x_1h_2h_1 + 2x_2h_1h_2.$$

We can write q in the form

$$q(x)[h] = \begin{bmatrix} h_1 & h_2 & x_2h_1 & x_1h_2 \end{bmatrix} \begin{bmatrix} 2x_2 & 0 & 0 & 2 \\ 0 & 2x_1 & 2 & 0 \\ 0 & 2 & 0 & 0 \\ 2 & 0 & 0 & 0 \end{bmatrix} \begin{bmatrix} h_1 \\ h_2 \\ h_1x_2 \\ h_2x_1 \end{bmatrix}. \quad \blacksquare$$

The representation of q displayed above is of the form

$$q(x)[h] = V(x)[h]^\intercal Z(x)V(x)[h],$$

where Z is called the *middle matrix* and V the *border vector*. The middle matrix does not contain h. The border vector is linear in h with h always on the left. In Section 8.4.2 we define this border vector–middle matrix (BV-MM) representation generally for noncommutative polynomials $q(x)[h]$ which are homogeneous of degree two in the h variables. Note that the entries of the border vector are distinct monomials.

Example 8.54. Let $p = x_2x_1x_2x_1 + x_1x_2x_1x_2$. Then

$$q = p'' = 2h_1h_2x_1x_2 + 2h_1x_2h_1x_2 + 2h_1x_2x_1h_2 + 2h_2h_1x_2x_1 + 2h_2x_1h_2x_1 + 2h_2x_1x_2h_1$$
$$+ 2x_1h_2h_1x_2 + 2x_1h_2x_1h_2 + 2x_1x_2h_1h_2 + 2x_2h_1h_2x_1 + 2x_2h_1x_2h_1 + 2x_2x_1h_2h_1.$$

The BV-MM representation for q is

$$q = \begin{bmatrix} h_1 & h_2 & x_2h_1 & x_1h_2 & x_1x_2h_1 & x_2x_1h_2 \end{bmatrix}$$
$$\times \begin{bmatrix} 0 & 2x_2x_1 & 2x_2 & 0 & 0 & 2 \\ 2x_1x_2 & 0 & 0 & 2x_1 & 2 & 0 \\ 2x_1 & 0 & 0 & 2 & 0 & 0 \\ 0 & 2x_2 & 2 & 0 & 0 & 0 \\ 0 & 2 & 0 & 0 & 0 & 0 \\ 2 & 0 & 0 & 0 & 0 & 0 \end{bmatrix} \begin{bmatrix} h_1 \\ h_2 \\ h_1x_2 \\ h_2x_1 \\ h_1x_2x_1 \\ h_2x_1x_2 \end{bmatrix}. \quad \blacksquare$$

Example 8.55. In the one variable with $h_1 = h_1^\intercal$ we abbreviate h_1 to h. Fix some noncommutative variables not necessarily symmetric $w := (a, b, d, e)$ and consider

$$q(w)[h] := hah + e^\intercal hbh + hb^\intercal he + e^\intercal hdhe, \tag{8.15}$$

which is a quadratic function of h. It can be written in the BV-MM form

$$q(w)[h] = \begin{bmatrix} h & e^\intercal h \end{bmatrix} \begin{bmatrix} a & b^\intercal \\ b & d \end{bmatrix} \begin{bmatrix} h \\ he \end{bmatrix}. \tag{8.16}$$

The representation is unique.

Observe (8.16) contrasts strongly with the commutative case wherein (8.15) takes the form

$$q(w)[h] = h(a + e^\intercal b + b^\intercal e + e^\intercal de)h. \quad \blacksquare$$

Example 8.56. The Hessian of $p(x) = x^4$ is

$$q(x)[h] := p''(x)[h] = 2(x^2h^2 + xh^2x + h^2x^2) \\ + 2(xhxh + hxhx) \\ + hx^2h, \tag{8.17}$$

a polynomial that is homogeneous of degree 2 in x and homogeneous of degree 2 in h that can be expressed as

$$q(x)[h] = 2\begin{bmatrix} h & xh & x^2h \end{bmatrix} \begin{bmatrix} x^2 & x & 1 \\ x & 1 & 0 \\ 1 & 0 & 0 \end{bmatrix} \begin{bmatrix} h \\ hx \\ hx^2 \end{bmatrix}. \quad \blacksquare$$

Notice that the contribution of the main antidiagonal of the middle matrix for q in Example 8.56 (all 1's) corresponds to the right-hand side of first line of (8.17). Indeed, each antidiagonal corresponds to a line of (8.17).

Exercise 8.57. In Example 8.56, for which symmetric matrices X is $Z(X)$ positive semidefinite?

Exercise 8.58. What is the middle matrix $Z(x)$ for $p(x) = x^3$? For which symmetric matrices X is $Z(X)$ positive semidefinite?

Exercise 8.59. Compute middle matrix representations using `NCAlgebra`. The command is

$\{lt, mq, rt\}$ =`NCMatrixOfQuadratic[`q`,` $\{h, k\}$`]`

In the output mq is the middle matrix, rt is the border vector, and lt is $(rt)^{\intercal}$. For examples, see `NCConvexityRegionDemo.nb` in the `NC/DEMOS directory`.

The positivity of q vs. positivity of the middle matrix

In this section we let $q(x)[h]$ denote a polynomial which is homogeneous of degree two in h, but which is not necessarily the Hessian of a noncommutative polynomial. While we have focused on Hessians, such a q will still have a BV-MM representation. *So what good is this representation?* After all one expects that q could have wonderful properties, such as positivity, which are not shared by its middle matrix. No, the striking thing is that positivity of q implies positivity of the middle matrix. Roughly we shall prove what we call the *QuadratischePositivstellensatz*, which is essentially Theorem 3.1 of [9].

Theorem 8.60. *If the polynomial*[2] $q(x)[h]$ *is homogeneous quadratic in* h*, then* q *is matrix positive if and only if its middle matrix* Z *is matrix positive.*

[2]This theorem is true (but not proved here) for q which are noncommutative rational in x.

More generally, suppose $\mathcal{O}$ is a nonempty noncommutative basic open semialgebraic set. If $q(X)[H]$ is positive semidefinite for all $n \in \mathbb{N}$, $X \in \mathcal{O}(n)$, and $H \in (\mathbb{S}^{n\times n})^g$, then $Z(X) \succeq 0$ for all $X \in \mathcal{O}$.

We emphasize that, in the theorem, the convention that the terms of the border vector are distinct is in force.

To foreshadow Section 8.5 and to give an idea of the proof of Theorem 8.60, we illustrate it on an example in one variable. This time we use a *free* rather than symmetric variable since proofs are a bit easier.

Consider the noncommutative quadratic function q given by

$$q(w)[h] := h^\intercal bh + e^\intercal h^\intercal ch + h^\intercal c^\intercal he + e^\intercal h^\intercal ahe, \tag{8.18}$$

where $w = (a, b, c, e)$. The border vector $V(w)[h]$ and the coefficient matrix $Z(w)$ with noncommutative entries are

$$V(w)[h] = \begin{bmatrix} h \\ he \end{bmatrix} \qquad \text{and} \qquad Z(w) = \begin{bmatrix} b & c^\intercal \\ c & a \end{bmatrix};$$

that is, q has the form

$$q(w)[h] = V(w)[h]^\intercal Z(w) V(w)[h] = \begin{bmatrix} h^\intercal & e^\intercal h^\intercal \end{bmatrix} \begin{bmatrix} b & c^\intercal \\ c & a \end{bmatrix} \begin{bmatrix} h \\ he \end{bmatrix}.$$

Now, if in (8.18) the elements a, b, c, e, h are replaced by matrices in $\mathbb{R}^{n\times n}$, then the noncommutative quadratic function $q(w)[h]$ becomes a matrix-valued function $q(W)[H]$. The matrix-valued function $q[H]$ is matrix positive if and only if $v^\intercal q(W)[H]v \geq 0$ for all vectors $v \in \mathbb{R}^n$ and all $H \in \mathbb{R}^{n\times n}$, or, equivalently, the following inequality must hold:

$$\begin{bmatrix} v^\intercal H^\intercal & v^\intercal E^\intercal H^\intercal \end{bmatrix} Z \begin{bmatrix} Hv \\ HEv \end{bmatrix} \geq 0. \tag{8.19}$$

Let

$$y^\intercal := \begin{bmatrix} v^\intercal H^\intercal & v^\intercal E^\intercal H^\intercal \end{bmatrix}.$$

Then (8.19) is equivalent to $y^\intercal Z\, y \geq 0$. Now it suffices to prove that all vectors of the form y sweep $\mathbb{R}^{2n}$. This will be completely analyzed in full generality in Section 8.5.1, but next we give the proof for our simple situation.

Suppose for a given v, with $n \geq 2$, the vectors v and Ev are linearly independent. Let $y = \left[\begin{smallmatrix} v_1 \\ v_2 \end{smallmatrix}\right]$ be any vector in $\mathbb{R}^{2n}$; then we can choose $H \in \mathbb{R}^{n\times n}$ with the property that $v_1 = Hv$ and $v_2 = HEv$. It is clear that

$$\mathcal{R}^v := \left\{ \begin{bmatrix} Hv \\ HEv \end{bmatrix} : H \in \mathbb{R}^{n\times n} \right\} \tag{8.20}$$

is all $\mathbb{R}^{2n}$ as required.

Thus we are finished unless for all v the vectors v and Ev are linearly dependent. That is for all v, $\lambda_1(v)v + \lambda_2(v)Ev = 0$ for nonzero $\lambda_1(v)$ and $\lambda_2(v)$. Note $\lambda_2(v) \neq 0$, unless $v = 0$. Set $\tau(v) := \frac{\lambda_1(v)}{\lambda_2(v)}$; then the linear dependence becomes $\tau(v)v + Ev = 0$ for all v. It turns out that this does not happen unless $E = \tau I$ for some $\tau \in \mathbb{R}$. This is a baby case of Theorem 8.92 which comes later and is a subject unto itself.

To finish the proof pick a v which makes $\mathcal{R}^v$ equal all of $\mathbb{R}^{2n}$. Then $v^\mathsf{T} q(W)[H]v \geq 0$ implies that $Z \succeq 0$ by (8.19).

8.4.2 Details of the Middle Matrix Representation

The following representation for symmetric noncommutative polynomials $q(x)[h]$ that are of degree ℓ in x and homogeneous of degree 2 in h is exploited extensively in this subject:

$$q(x)[h] = \begin{bmatrix} V_0^\mathsf{T} & V_1^\mathsf{T} & \cdots & V_{\ell-1}^\mathsf{T} & V_\ell^\mathsf{T} \end{bmatrix} \begin{bmatrix} Z_{00} & Z_{01} & \cdots & Z_{0,\ell-1} & Z_{0\ell} \\ Z_{10} & Z_{11} & \cdots & Z_{1,\ell-1} & 0 \\ \vdots & \vdots & \iddots & \vdots & \vdots \\ Z_{\ell-1,0} & Z_{\ell-2,1} & \cdots & 0 & 0 \\ Z_{\ell 0} & 0 & \cdots & 0 & 0 \end{bmatrix} \begin{bmatrix} V_0 \\ V_1 \\ \vdots \\ V_{\ell-1} \\ V_\ell \end{bmatrix}, \tag{8.21}$$

where the following hold:

1. The degree d of $q(x)[h]$ is $d = \ell + 2$.

2. $V_j = V_j(x)[h]$, $j = 0, \ldots, \ell$, is a vector of height g^{j+1} whose entries are monomials of degree j in the x variables and degree 1 in the h variables. The h always appears to the left. In particular, $V(x)[h]$ is a vector of height $g\sigma(\ell)$, where as in (8.14),

$$\sigma(\ell) = 1 + g + \cdots + g^\ell .$$

3. $Z_{ij} = Z_{ij}(x)$ is a matrix of size $g^{i+1} \times g^{j+1}$ whose entries are polynomials in the noncommuting variables $x_1, \ldots, x_g$ of degree $\leq \ell - (i + j)$. In particular, $Z_{i,\ell-i} = Z_{i,\ell-i}(x)$ is a constant matrix for $i = 0, \ldots, \ell$.

4. $Z_{ij}^\mathsf{T} = Z_{ji}$.

Usually the entries of the vectors V_j are ordered lexicographically.

We note that the vector of monomials, $V(x)[h]$, might contain monomials that are not required in the representation of the noncommutative quadratic q. Therefore, we can omit all monomials from the border vector that are not required. This gives us a *minimal length* border vector and prevents extraneous zeros from occurring in the middle matrix. The matrix Z in the representation (8.21) will be referred to as the *middle matrix of the polynomial* $q(x)[h]$, and the vectors $V_j = V_j(x)[h]$ with monomials as entries will be referred to as *border vectors*. It is easy to check that a minimal length border vector contains distinct monomials, and once the ordering of entries of V is set, the middle matrix for a given q is unique; see Lemma 8.62 below.

Example 8.61. Returning to Example 8.54, we have for the middle matrix representation of q that

$$V_0 = \begin{bmatrix} h_1 \\ h_2 \end{bmatrix}, \qquad V_1 = \begin{bmatrix} h_2 x_1 \\ h_1 x_2 \end{bmatrix}, \qquad V_2 = \begin{bmatrix} h_1 x_2 x_1 \\ h_2 x_1 x_2 \end{bmatrix},$$

and, for instance,

$$Z_{00} = \begin{bmatrix} 0 & 2x_2x_1 \\ 2x_1x_2 & 0 \end{bmatrix}, \qquad Z_{01} = \begin{bmatrix} 2x^2 & 0 \\ 0 & 2x_1 \end{bmatrix}, \qquad Z_{02} = \begin{bmatrix} 0 & 2 \\ 2 & 0 \end{bmatrix}.$$

Note that generically for a polynomial q in two variables the V_j have additional terms. For instance, usually V_1 is the column

$$\begin{bmatrix} h_1x_1 \\ h_1x_2 \\ h_2x_1 \\ h_2x_2 \end{bmatrix}.$$

Likewise generically V_2 has eight terms. As for the Z_{ij}, Z_{01}, for instance, is generically 2×4. ■

Lemma 8.62. *The entries in the middle matrix $Z(x)$ are uniquely determined by the polynomial $q(x)[h]$ and the border vector $V(x)[h]$.*

Proof. Note every monomial in $q(x)[h]$ has the form

$$m_L h_i m_M h_j m_R.$$

Define

$$\mathcal{R}_j := \{h_j m : \ m_L h_i m_M h_j m \text{ is a term in } q(x)[h]\}.$$

Given the representation $V^\intercal Z V$ for q, let E_V denote the monomials in V. Then it is clear that each monomial in E_V must occur in some term of q, so it appears in $\mathcal{R}_j$ for some j. Conversely, each term $h_j m$ in $\mathcal{R}_j$ corresponds to at least one term $m_L h_i m_M h_j m$ of q, so it must be in E_V.

Exercise 8.63. Consider (8.21) and prove the degree bound on the Z_{ij} in (3). *Hint*: Read Example 8.64 first.

Example 8.64. If $p(x)$ is a symmetric polynomial of degree $d = 4$ in g noncommuting variables, then the middle matrix $Z(x)$ in the representation of the Hessian $p''(x)[h]$ is

$$Z(x) = \begin{bmatrix} Z_{00}(x) & Z_{01}(x) & Z_{02}(x) \\ Z_{10}(x) & Z_{11}(x) & 0 \\ Z_{20}(x) & 0 & 0 \end{bmatrix},$$

where the block entries $Z_{ij} = Z_{ij}(x)$ have the following structure:

Z_{00} is a $g \times g$ matrix with noncommutative polynomial entries of degree ≤ 2,
Z_{01} is a $g \times g^2$ matrix with with noncommutative polynomial entries of degree ≤ 1,
Z_{02} is a $g \times g^3$ matrix with constant entries.

All of these are proved merely by keeping track of the degrees. For example, the contribution of Z_{02} to p'' is $V_0^\mathsf{T} Z_{02} V_2$, whose degree is

$$\deg(V_0^\mathsf{T}) + \deg(Z_{02}) + \deg(V_2) = 1 + \deg(Z_{02}) + 3 \leq 4,$$

so $\deg(Z_{02}) = 0$. ∎

8.4.3 The Middle Matrix of p''

The middle matrix $Z(x)$ of the Hessian $p''(x)[h]$ of a noncommutative symmetric polynomial $p(x)$ plays a key role. These middle matrices have a very rigid structure similar to that in Example 8.56. We illustrate with an example and then with exercises.

Example 8.65. As a warm-up we first illustrate that $Z_{02}(X) = 0$ if and only if $Z_{11}(X) = 0$ for Example 8.54. To this end, observe that the contribution of the middle matrix's extreme outer diagonal element Z_{02} to q is as follows:

$$\frac{1}{2}V_0(x)[h]^\mathsf{T} Z_{02}(x) V_2(x)[h] = \begin{bmatrix} h_1 \\ h_2 \end{bmatrix}^\mathsf{T} \begin{bmatrix} 0 & 2 \\ 2 & 0 \end{bmatrix} \begin{bmatrix} h_1 x_2 x_1 \\ h_2 x_1 x_2 \end{bmatrix} = 2h_1 h_2 x_1 x_2 + 2h_2 h_1 x_2 x_1.$$

Substitute $h_j \rightsquigarrow x_j$ and get $2x_1x_2x_1x_2 + 2x_2x_1x_2x_1$, which is $2p(x)$. That is,

$$p(x) = \frac{1}{2}V_0(x)[x]^\mathsf{T} Z_{02}(x) V_2(x)[x],$$

where $V_k(x)[h]$ is the homogeneous, in x, of degree k part of the border vector V. Obviously, $Z_{02} = 0$ implies $p = 0$. ∎

Exercise 8.66. Show $p(x)$ can also be obtained from Z_{11} in a similar fashion, i.e.,

$$p(x) = \frac{1}{2}V_1(x)[x]^\mathsf{T} Z_{11}(x) V_1(x)[x].$$

Exercise 8.67. Suppose p is homogeneous of degree d and its Hessian q has the BV-MM representation $q(x)[h] = V(x)[h]^\mathsf{T} Z(x) V(x)[h]$.

(a) Show

$$p = \frac{1}{2}V_0(x)[x]^\mathsf{T} Z_{0\ell} V_\ell(x)[x]$$

with $\ell = d - 2$. Prove this formula for $d = 2$, $d = 4$.

(b) Show that likewise

$$p = \frac{1}{2} V_1(x)[x]^\intercal Z_{1,\ell-1}(x) V_{\ell-1}(x)[x].$$

Do not cheat and look this up in [14], but do compare with Exercise 8.63.

Exercise 8.68. Let Z denote the middle matrix for the Hessian of a noncommutative polynomial p. Show, if $i + j = i' + j'$, then $Z_{ij} = 0$ if and only if $Z_{i'j'} = 0$.

8.4.4 Positivity of the Middle Matrix and the Demise of Noncommutative Convexity

This section focuses on positivity of the middle matrix of a Hessian.

Why should we focus on the case where $Z(x)$ is positive semidefinite? In [35] it was shown that a polynomial $p \in \mathbb{R}<x>$ is matrix convex if and only if its Hessian $p''(x)[h]$ is positive (see Exercise 8.26). Moreover, if $Z(x)$ is positive, then the degree of $p(x)$ is at most two [36]. The proof of this degree constraint given in Proposition 8.70 below using the more manageable bookkeeping scheme in this chapter begins with the following exercise.

Exercise 8.69. Show that

$$\begin{bmatrix} A & B \\ B^\intercal & 0 \end{bmatrix}$$

is positive semidefinite if and only if $A \succeq 0$ and $B = 0$. More refined versions of this fact appear as exercises later; see Exercise 8.76.

As we shall see, we need not require our favorite functions be positive everywhere. It is possible to work locally, namely, on an open set.

Proposition 8.70. *Let $p = p(x)$ be a symmetric polynomial of degree d in g noncommutative variables and let $Z(x)$ denote the middle matrix in the BV-MM representation of the Hessian $p''(x)[h]$. If $Z(X) \succeq 0$ for all X in some nonempty noncommutative basic open semialgebraic set $\mathcal{O}$, then d is at most 2.*

Proof. Arguing by contradiction, suppose that $d \geq 3$; then $p''(x)[h]$ is of degree $\ell = d - 2 \geq 1$ in x and its middle matrix is of the form

$$Z = \begin{bmatrix} Z_{00} & \cdots & Z_{0\ell} \\ \vdots & \cdot^{\cdot^{\cdot}} & \vdots \\ Z_{\ell 0} & \cdots & 0 \end{bmatrix}.$$

Therefore, $Z(X)$ is of the form

$$Z(X) = \begin{bmatrix} A & B \\ B^\intercal & 0 \end{bmatrix},$$

where $A = A^\intercal$ and $B^\intercal = \begin{bmatrix} Z_{0\ell}(X) & 0 & \cdots & 0 \end{bmatrix}$. From Exercise 8.67, p_d, the homogeneous degree d part of p, can be reconstructed from $Z_{0\ell}$. Now there is an $X \in \mathcal{O}$ such that $p_d(X)$ is nonzero, as otherwise p_d vanishes on a basic open semialgebraic set and is equal to 0. It follows that there is an $X \in \mathcal{O}$ such that $Z_{0\ell}(X)$ is not zero. Hence $B(X)$ is not zero which implies, by Exercise 8.69, the contradiction that $Z(X)$ is not positive semidefinite. □

We have now reached our goal of showing that convex polynomials have degree ≤ 2.

Theorem 8.71. *If $p \in \mathbb{R}\langle x\rangle$ is a symmetric polynomial which is convex on a nonempty noncommutative basic open semialgebraic set $\mathcal{O}$, then it has degree at most 2.*

There is a version of the theorem for free variables, i.e., with $p \in \mathbb{R}\langle x, x^\intercal\rangle$.

Proof. The convexity of p on $\mathcal{O}$ is equivalent to $p''(X)[H]$ being positive semidefinite for all X in $\mathcal{O}$; see Exercise 8.26. By the QuadratischePositivstellensatz the middle matrix $Z(x)$ for $p''(x)[h]$ is positive on $\mathcal{O}$; that is, $Z(X) \succeq 0$ for all $X \in \mathcal{O}$. Proposition 8.70 implies degree p is at most 2. □

8.4.5 The Signature of the Middle Matrix

This section introduces the notion of the *signature* $\mu_\pm(Z(x))$ of $Z(x)$, the middle matrix of a Hessian, or more generally a polynomial $q(x)[h]$ which is homogeneous of degree 2 in h.

The *signature of a symmetric matrix* M is a triple of integers

$$\big(\mu_-(M),\ \mu_0(M),\ \mu_+(M)\big),$$

where $\mu_-(M)$ is the number of negative eigenvalues (counted with multiplicity); $\mu_+(M)$ is the number of positive eigenvalues; and $\mu_0(M)$ is the dimension of the null space of M.

Lemma 8.72. *A noncommutative symmetric polynomial $q(x)[h]$ homogeneous of degree 2 in h has middle matrix Z of the form in (8.21), and Z being positive semidefinite implies Z is of the form*

$$\begin{bmatrix} Z_{00} & Z_{01} & \cdots & Z_{0,\lfloor\frac{\ell}{2}\rfloor} & 0 & \cdots \\ Z_{10} & Z_{11} & \cdots & Z_{1,\lfloor\frac{\ell}{2}\rfloor} & 0 & \cdots \\ \vdots & \vdots & \iddots & \vdots & \vdots & \iddots \\ Z_{\lfloor\frac{\ell}{2}\rfloor,0} & Z_{\lfloor\frac{\ell}{2}\rfloor,1} & \cdots & Z_{\lfloor\frac{\ell}{2}\rfloor,\lfloor\frac{\ell}{2}\rfloor} & 0 & \iddots \\ 0 & 0 & \cdots & 0 & 0 & \iddots \\ \vdots & \vdots & \iddots & \iddots & \iddots & \end{bmatrix}.$$

This lemma follows immediately from a much more general lemma.

Lemma 8.73. *If*

$$E = \begin{bmatrix} A & B & C \\ B^\intercal & D & 0 \\ C^\intercal & 0 & 0 \end{bmatrix}$$

is a real symmetric matrix, then

$$\mu_\pm(E) \geq \mu_\pm(D) + \operatorname{rank} C.$$

This can be proved using the $LDL^\intercal$ decomposition which we shall not do here but suggest the reader apply the $LDL^\intercal$ hammer to the following simpler exercise.

8.4.6 Exercises

Exercise 8.74. True or false? If p_d is homogeneous of degree d and we let Z denote the middle matrix of the Hessian $p''(x)[h]$, then for each $k \leq d-2$ the degree of $Z_{i,k-i}$ is independent of i.

Exercise 8.75. Redo Exercise 8.26 for convexity on a noncommutative basic open semialgebraic set.

Exercise 8.76. If $F = \left[\begin{smallmatrix} A & C \\ C^\intercal & 0 \end{smallmatrix}\right]$, then $\mu_\pm(F) \geq \operatorname{rank} C$. (If you cannot do the general case, assume A is invertible.)

Exercise 8.77. If $p(x)$ is a symmetric polynomial of degree $d = 2$ in g noncommuting variables, then the middle matrix $Z(x)$ in the representation of the Hessian $p''(x)[h]$ is equal to the $g \times g$ constant matrix Z_{00}. Substituting $X \in (\mathbb{S}^{n\times n})^g$ for x gives

$$\mu_\pm(Z(X)) \geq \mu_\pm(Z_{00}).$$

Exercise 8.78. Let $f \in \mathbb{R}<x>_{2d}$ and let $V \in <x>_d^{\sigma(d)}$ be a vector consisting of all words in x of degree $\leq d$. Prove

(a) there is a matrix $G \in \mathbb{R}^{\sigma(d)\times\sigma(d)}$ with $f = V^\intercal G V$ (any such G is called a *Gram matrix* for f);

(b) if f is symmetric, then there is a symmetric Gram matrix for f.

Exercise 8.79. Find all Gram matrices for

(a) $f = x_1^4 + x_1^2 x_2 - x_1 x_2^2 + x_2 x_1^2 - x_2^2 x_1 + x_1^2 - x_2^2 + 2x_1 - x_2 + 4$;

(b) $f = c(x_1, x_2)^2$.

Exercise 8.80. Show that if $f \in \mathbb{R}<x>$ is homogeneous of degree $2d$, then it has a unique Gram matrix $G \in \mathbb{R}^{\sigma(d)\times\sigma(d)}$.

8.4.7 A Glimpse of History

There is a theory of operator monotone and operator convex functions which overlaps with the matrix convex functions considered here in the case of one variable. However, the points of view are substantially different, diverging markedly in several variables. Löwner introduced a class of real analytic functions in one real variable called matrix monotone functions, which we shall not define here. Löwner gave integral representations and these have developed substantially over the years. The contact with convexity came when Löwner's student Kraus [49] introduced matrix convex functions f in one variable. Such a function f on $[0, \infty) \subseteq \mathbb{R}$ can be represented as $f(t) = tg(t)$ with g matrix monotone, so the representations for g produce representations for f. Hansen has extensive in-depth work on matrix convex and monotone functions whose definition in several variables is different than the one we use here; see [25] or [24]. All of this gives a beautiful integral representation characterizing matrix convex functions using techniques very different from ours. An excellent treatment of the one-variable case is [3, Chapter 5]. Interestingly, to the best of our knowledge, the one-variable version of Theorem 8.71 [36] does not seem to be explicit in this classical literature. However, it is an immediate consequence of the results of [25], where (not necessarily polynomial) operator convex functions on an interval are described. This and the papers of Hansen and [56, 68] are some of the more recent references in this line of convexity history orthogonal to ours.

8.5 Der QuadratischePositivstellensatz

In this section we present the proof of the QuadratischePositivstellensatz (Theorem 8.60) which is based on the fact that local linear dependence of noncommutative rationals (or noncommutative polynomials) implies global linear dependence, a fact itself based on the forthcoming CHSY lemma [9].

8.5.1 The Camino–Helton–Skelton–Ye (CHSY) Lemma

At the root of the CHSY lemma [9] is the following linear algebra fact.

Lemma 8.81. *Fix $n > d$. If $\{z_1, \dots, z_d\}$ is a linearly independent set in $\mathbb{R}^n$, then the codimension of*

$$\left\{ \begin{bmatrix} Hz_1 \\ Hz_2 \\ \vdots \\ Hz_d \end{bmatrix} : H \in \mathbb{S}^{n\times n} \right\} \subseteq \mathbb{R}^{nd}$$

is $\frac{d(d-1)}{2}$. It is especially important that this codimension is independent of n.

The following exercise is a variant of Lemma 8.81 which is easier to prove. Thus we suggest attempting it before launching into the proof of the lemma.

Exercise 8.82. Prove that if $\{z_1, \ldots, z_d\}$ is a linearly independent set in $\mathbb{R}^n$, then

$$\left\{ \begin{bmatrix} Hz_1 \\ Hz_2 \\ \vdots \\ Hz_d \end{bmatrix} : H \in \mathbb{R}^{n\times n} \right\} = \mathbb{R}^{nd}.$$

Hint: It proceeds like the proof of (8.20).

***Proof of Lemma* 8.81.** Consider the mapping $\Phi : \mathbb{S}^{n\times n} \to \mathbb{R}^{nd}$ given by

$$H \mapsto \begin{bmatrix} Hz_1 \\ Hz_2 \\ \vdots \\ Hz_d \end{bmatrix}.$$

Since the span of $\{z_1, \ldots, z_d\}$ has dimension d, it follows that the kernel of Φ has dimension $\kappa = \frac{(n-d)(n-d+1)}{2}$, and hence the range has dimension $\frac{n(n+1)}{2} - \kappa$. To see this assertion, it suffices to assume that the span of $\{z_1, \ldots, z_d\}$ is the span of $\{e_1, \ldots, e_d\} \subseteq \mathbb{R}^n$ (the first d standard basis vectors in $\mathbb{R}^n$). In this case (since H is symmetric) $Hz_j = 0$ for all j if and only if

$$H = \begin{bmatrix} 0 & 0 \\ 0 & H' \end{bmatrix},$$

where H' is a symmetric matrix of size $(n-d) \times (n-d)$; in other words, this is the kernel of Φ.

From this we deduce that the codimension of the range of Φ is

$$nd - \left(\frac{n(n+1)}{2} - \kappa\right) = \frac{d(d-1)}{2},$$

concluding the proof. $\square$

Next is a straightforward extension of Lemma 8.81.

Lemma 8.83 ([9]). *If $n > d$ and $\{z_1, \ldots, z_d\}$ is a linearly independent subset of $\mathbb{R}^n$, then the codimension of*

$$\left\{ \oplus_{j=1}^{g} \begin{bmatrix} H_j z_1 \\ H_j z_2 \\ \vdots \\ H_j z_d \end{bmatrix} : H = (H_1, \ldots, H_g) \in (\mathbb{S}^{n\times n})^g \right\} \subseteq \mathbb{R}^{gnd}$$

is $g\frac{d(d-1)}{2}$ and is independent of n.

Proof. See Exercise 8.94. $\square$

Finally, the form in which we generally apply the lemma is the following.

Lemma 8.84. *Let $v \in \mathbb{R}^n$, $X \in (\mathbb{S}^{n\times n})^g$. If the set $\{m(X)v\colon m \in <x>_d\}$ is linearly independent, then the codimension of*

$$\{V(X)[H]v\colon H \in (\mathbb{S}^{n\times n})^g\}$$

is $g\frac{\kappa(\kappa-1)}{2}$, where $\kappa = \sigma(d) = \sum_{j=0}^{d} g^j$ and where

$$V = \bigoplus_{i=1}^{g} \bigoplus_{m\in<x>_d} H_i m$$

is the border vector associated with $<x>_d$. Again, this codimension is independent of n as it depends only upon the number of variables g and the degree d of the polynomial.

Proof. Let $z_m = m(X)v$ for $m \in <x>_d$. There are at most κ of these. Now apply the previous lemma. □

8.5.2 Linear Dependence of Symbolic Functions

The main result in this section, Theorem 8.92, says roughly that if each evaluation of a set $G_1, \dots G_\ell$ of rational functions produces linearly dependent matrices, then they satisfy a universal linear dependence relation. We begin with a clean and easily stated consequence of Theorem 8.92.

In Section 8.2.1 we defined noncommutative basic open semialgebraic sets. Here we define a noncommutative basic semialgebraic set. Given matrix-valued symmetric noncommutative polynomials ρ and $\tilde{\rho}$, let

$$\mathcal{D}_+^{\rho}(n) = \{X \in (\mathbb{S}^{n\times n})^g\colon \rho(X) \succ 0\}$$

and

$$\mathcal{D}^{\tilde{\rho}}(n) = \{X \in (\mathbb{S}^{n\times n})^g\colon \tilde{\rho}(X) \succeq 0\}.$$

Then $\mathcal{D}$ is a *noncommutative basic semialgebraic set* if there exists $\rho_1, \dots, \rho_k$ and $\tilde{\rho}_1, \dots, \tilde{\rho}_{\tilde{k}}$ such that $\mathcal{D} = (\mathcal{D}(n))_{n\in\mathbb{N}}$, where

$$\mathcal{D}(n) = \left(\bigcap_j \mathcal{D}_+^{\rho_j}(n)\right) \cap \left(\bigcap_j \mathcal{D}^{\tilde{\rho}_{\tilde{j}}}(n)\right).$$

Theorem 8.85. *Suppose $G_1, \dots, G_\ell$ are rational expressions and $\mathcal{D}$ is a nonempty noncommutative basic semialgebraic set on which each G_j is defined. If, for each $X \in \mathcal{D}(n)$ and vector $v \in \mathbb{R}^n$, the set $\{G_j(X)v\colon j = 1, 2, \dots, \ell\}$ is linearly dependent, then the set $\{G_j(X)\colon j = 1, 2, \dots, \ell\}$ is linearly dependent on $\mathcal{D}$; i.e., there* *exists a nonzero* $\lambda \in \mathbb{R}^{\ell}$ *such that*

$$0 = \sum_{j=1}^{\ell} \lambda_j G_j(X) \qquad \text{for all } X \in \mathcal{D}.$$

If, in addition, $\mathcal{D}$ *contains an* ϵ*-neighborhood of* 0 *for some* $\epsilon > 0$*, then there exists a nonzero* $\lambda \in \mathbb{R}^{\ell}$ *such that*

$$0 = \sum_{j=1}^{\ell} \lambda_j G_j.$$

Corollary 8.86. *Suppose* $G_1, \ldots, G_\ell$ *are rational expressions. If, for each* $n \in \mathbb{N}$, $X \in (\mathbb{S}^{n\times n})^g$*, and vector* $v \in \mathbb{R}^n$*, the set* $\{G_j(X)v\colon j = 1, 2, \ldots, \ell\}$ *is linearly dependent, then the set* $\{G_j\colon j = 1, 2, \ldots, \ell\}$ *is linearly dependent; i.e., there exists a nonzero* $\lambda \in \mathbb{R}^{\ell}$ *such that*

$$\sum_{j=1}^{\ell} \lambda_j G_j = 0.$$

Corollary 8.87. *Suppose* $G_1, \ldots, G_\ell$ *are rational expressions. If, for each* $n \in \mathbb{N}$ *and* $X \in (\mathbb{S}^{n\times n})^g$*, the set* $\{G_j(X)\colon j = 1, 2, \ldots, \ell\}$ *is linearly dependent, then the set* $\{G_j\colon j = 1, 2, \ldots, \ell\}$ *is linearly dependent.*

The point is that the λ_j are independent of X. Before proving Theorem 8.85 we shall introduce some terminology pursuant to our more general result.

Direct Sums

We present some definitions about direct sums and sets which respect direct sums, since they are important tools.

Definition 8.88. *Our definition of the direct sum is the usual one. Given pairs* (X_1, v_1) *and* (X_2, v_2)*, where* X_j *are* $n_j \times n_j$ *matrices and* $v_j \in \mathbb{R}^{n_j}$*,*

$$(X_1, v_1) \oplus (X_2, v_2) = (X_1 \oplus X_2, v_1 \oplus v_2),$$

where

$$X_1 \oplus X_2 := \begin{bmatrix} X_1 & 0 \\ 0 & X_2 \end{bmatrix}, \qquad v_1 \oplus v_2 := \begin{bmatrix} v_1 \\ v_2 \end{bmatrix}.$$

We extend this definition to μ *terms,* $(X_1, v_1), \ldots, (X_\mu, v_\mu)$ *in the expected way.*

In the definition below, we consider a set $\mathcal{B}$, which is the sequence

$$\mathcal{B} := (\mathcal{B}(n)),$$

where each $\mathcal{B}(n)$ is a set whose members are pairs (X,v), where X is in $(\mathbb{S}^{n\times n})^g$ and $v \in \mathbb{R}^n$.

Definition 8.89. *The set $\mathcal{B}$ is said to respect direct sums if (X^j, v^j) with $X^j \in (\mathbb{S}^{n_j\times n_j})^g$ and $v^j \in \mathbb{R}^{n_j}$ for $j = 1,\dots,\mu$ being contained in the set $\mathcal{B}(n_j)$ implies that the direct sum*

$$(X^1 \oplus \dots \oplus X^\mu, v^1 \oplus \dots \oplus v^\mu) = (\oplus_{j=1}^\mu X^j, \oplus_{j=1}^\mu v^j)$$

is also contained in $\mathcal{B}(\sum n_j)$.

Definition 8.90. *By a natural map G* on $\mathcal{B}$, we mean a sequence of functions $G(n) : \mathcal{B}(n) \to \mathbb{R}^n$, which *respects direct sums* in the sense that, if $(X^j, v^j) \in \mathcal{B}(n_j)$ for $j = 1, 2, \dots, \mu$, then

$$G\left(\sum_1^\mu n_j\right)(\oplus X^j, \oplus v^j) = \oplus_1^\mu G(n_j)(X^j, v^j).$$

Typically we omit the argument n, writing $G(X)$ instead of $G(n)(X)$.

Examples of sets which respect direct sums and of natural maps are provided by the following example.

Example 8.91. Let ρ be a rational expression.

(1) The set $\mathcal{B}^\rho = \{(X,v) : X \in \mathcal{D}^\rho \cap (\mathbb{S}^{n\times n})^g,\ v \in \mathbb{R}^n,\ n \in \mathbb{N}\}$ respects direct sums.

(2) If G is a matrix-valued noncommutative rational expression whose domain contains $\mathcal{D}^\rho$, then G determines a natural map on $\mathcal{B}(\rho)$ by $G(n)(X,v) = G(X)v$. In particular, every noncommutative polynomial determines a natural map on every noncommutative basic semialgebraic set $\mathcal{B}$. ∎

Main result on linear dependence

Theorem 8.92. *Suppose $\mathcal{B}$ is a set which respects direct sums and $G_1, \dots, G_\ell$ are natural maps on $\mathcal{B}$. If for each $(X,v) \in \mathcal{B}$ the set $\{G_1(X,v), \dots, G_\ell(X,v)\}$ is linearly dependent, then there exists a nonzero $\lambda \in \mathbb{R}^\ell$ so that*

$$0 = \sum_{j=1}^{\ell} \lambda_j G_j(X,v)$$

for every $(X,v) \in \mathcal{B}$. We emphasize that λ is independent of (X,v).

Before proving Theorem 8.92, we use it to prove an important earlier theorem.

***Proof of Theorem* 8.85.** Let $\mathcal{B}$ be given by

$$\mathcal{B}(n) = \{(X,v)\colon X \in \mathcal{D}^\rho \cap (\mathbb{S}^{n\times n})^g \text{ and } v \in \mathbb{R}^n\}.$$

Let G_j denote the natural maps, $G_j(X,v) = G_j(X)v$. Then $\mathcal{B}$ and $G_1,\dots,G_\ell$ satisfy the hypothesis of Theorem 8.92 and so the first conclusion of Theorem 8.85 follows.

The last conclusion follows because a noncommutative rational function r vanishing on a noncommutative basic open semialgebraic set is 0 on all $\mathrm{dom}(r)$ and hence is zero; cf. Exercise 8.33. □

Proof of Theorem 8.92

We start with a finitary version of Theorem 8.92.

Lemma 8.93. *Let $\mathcal{B}$ and G_i be as in Theorem 8.92. If $\mathcal{R}$ is a finite subset of $\mathcal{B}$, then there exists a nonzero $\lambda(\mathcal{R}) \in \mathbb{R}^\ell$ such that*

$$\sum_{j=1}^{\ell} \lambda(\mathcal{R})_j G_j(X)v = 0$$

for every $(X,v) \in \mathcal{R}$.

Proof. The proof relies on taking direct sums of matrices. Write the set $\mathcal{R}$ as

$$\mathcal{R} = \{(X^1,v^1),\dots,(X^\mu,v^\mu)\},$$

where each $(X^i,v^i) \in \mathcal{B}$. Since $\mathcal{B}$ respects direct sums,

$$(X,v) = (\oplus_{\nu=1}^{\mu} X^\nu, \oplus_{\nu=1}^{\mu} v^\nu) \in \mathcal{B}.$$

Hence, there exists a nonzero $\lambda(\mathcal{R}) \in \mathbb{R}^\ell$ such that

$$0 = \sum_{j=1}^{\ell} \lambda(\mathcal{R})_j G_j(X,v).$$

Since each G_j respects direct sums, the desired conclusion follows. □

***Proof of Theorem* 8.92.** The proof is essentially a compactness argument, based on Lemma 8.93. Let $\mathbb{B}$ denote the unit sphere in $\mathbb{R}^\ell$.

To $(X,v) \in \mathcal{B}$ associate the set

$$\Omega_{(X,v)} = \left\{ \lambda \in \mathbb{B} \colon \lambda \cdot G(X)v = \sum_j \lambda_j G_j(X,v) = 0 \right\}.$$

Since $(X,v) \in \mathcal{B}$, the hypothesis on $\mathcal{B}$ says $\Omega_{(X,v)}$ is nonempty. It is evident that $\Omega_{(X,v)}$ is a closed subset of $\mathbb{B}$ and is thus compact.

Let $\boldsymbol{\Omega} := \{\Omega_{(X,v)} \colon (X,v) \in \mathcal{B}\}$. Any finite subcollection from $\boldsymbol{\Omega}$ has the form $\{\Omega_{(X,v)} \colon (X,v) \in \mathcal{R}\}$ for some finite subset $\mathcal{R}$ of $\mathcal{B}$, and so by Lemma 8.93 has a nonempty intersection. In other words, $\boldsymbol{\Omega}$ has the finite intersection property. The compactness of $\mathbb{B}$ implies that there is a $\lambda \in \mathbb{B}$ which is in every $\Omega_{(X,v)}$. This is the desired conclusion of the theorem. □

8.5.3 Proof of the QuadratischePositivstellensatz

We are now ready to give the proof of Theorem 8.60. Accordingly, let $\mathcal{O}$ be a given basic open semialgebraic set. Suppose

$$q(x)[h] = V(x)[h]^\intercal\, Z(x)\, V(x)[h], \tag{8.22}$$

where V is the border vector and Z is the middle matrix; cf. (8.21). Clearly, if Z is matrix-positive on $\mathcal{O}$, then $q(X)[H]$ is positive semidefinite for each n, $X \in \mathcal{O}(n)$, and $H \in (\mathbb{S}^{n\times n})^g$.

The converse is less trivial and requires the CHSY lemma plus our main result on linear dependence of noncommutative rational functions. Let ℓ denote the degree of $q(x)[h]$ in the variable x. In particular, the border vector in the representation of $q(x)[h]$ itself has degree ℓ in x. Recall σ_ℓ from Exercise 8.28.

Suppose that for some s and g-tuple of symmetric matrices $\tilde{X} = (\tilde{X}_1, \dots, \tilde{X}_g) \in \mathcal{O}(s)$, the matrix $Z(\tilde{X})$ is not positive semidefinite. By Lemma 8.84 and Theorem 8.85, there is a t, a $Y \in \mathcal{O}(t)$, and a vector η so that $\{m(Y)\eta \colon m \in <x>_\ell\}$ is linearly independent. Let $X = \tilde{X} \oplus Y$ and $\gamma = 0 \oplus \eta \in \mathbb{R}^{s+t}$. Then $Z(X)$ is not positive semidefinite and $\{m(X)\gamma \colon m \in <x>_\ell\}$ is linearly independent.

Let $N = g^{\frac{\kappa(\kappa-1)}{2}} + 1$, where κ is given in Lemma 8.84, and let $n = (s+t)N$. Consider $W = X \otimes I_N = (X_1 \otimes I_N, \dots, X_g \otimes I_N)$ and vector $\omega = \gamma \otimes e$, for any nonzero vector $e \in \mathbb{R}^{N+1}$. The set $\{m(W)\omega \colon m \in <x>_\ell\}$ is linearly independent, and thus by Lemma 8.84, the codimension of $\mathcal{M} = \{V(W)[H]\omega \colon H \in (\mathbb{S}^{n\times n})^g\}$ is at most $N-1$. On the other hand, because $Z(X)$ has a negative eigenvalue, the matrix $Z(W)$ has an eigenspace $\mathcal{E}$, corresponding to a negative eigenvalue, of dimension at least N. It follows that $\mathcal{E} \cap \mathcal{M}$ is nonempty; i.e., there is an $H \in (\mathbb{S}^{n\times n})^g$ such that $V(W)[H]\omega \in \mathcal{E}$. In particular, this together with (8.22) implies

$$\langle q(W)[H]\omega, \omega\rangle = \langle Z(W)V(W)[H]\omega, V(W)\omega\rangle < 0,$$

and thus, $q(W)[H]$ is not positive semidefinite.

8.5.4 Exercises

Exercise 8.94. Prove Lemma 8.83.

Exercise 8.95. Let $A \in \mathbb{R}^{n\times n}$ be given. Show that if the rank of A is r, then the matrices $A, A^2, \dots, A^{r+1}$ are linearly dependent.

In the next exercise employ the Fock space (see Section 8.2.7) to prove a strengthening of Corollary 8.86 for noncommutative polynomials.

Exercise 8.96. Suppose $p_1, \dots, p_\ell \in \mathbb{R}<x>_k$ are noncommutative polynomials. Show that if the set of vectors

$$\{p_1(X)v, \dots, p_\ell(X)v\} \tag{8.23}$$

is linearly dependent for every $(X, v) \in (\mathbb{S}^{\sigma\times\sigma})^g \times \mathbb{R}^\sigma$, where $\sigma = \sigma(k) = \dim \mathbb{R}<x>_k$, then $\{p_1, \dots, p_\ell\}$ is linearly dependent.

Exercise 8.97. Redo Exercise 8.96 under the assumption that the vectors (8.23) are linearly dependent for all $(X, v) \in O \times \mathbb{R}^\sigma$, where $O \subseteq (\mathbb{S}^{\sigma\times\sigma})^g$ is a nonempty open set.

For a more algebraic view of the linear dependence of noncommutative polynomials we refer to [6].

Exercise 8.98. Prove that $f \in \mathbb{R}<x>$ is a sum of squares if and only if it has a positive semidefinite Gram matrix. Are then all of f's Gram matrices positive semidefinite?

8.6 Noncommutative Varieties with Positive Curvature Have Degree 2

This section looks at noncommutative varieties and their geometric properties. We see a very strong rigidity when they have positive curvature which generalizes what we have already seen about convex polynomials (their graph is a positively curved variety) having degree 2.

In the classical setting of a surface defined by the zero set

$$\nu(p) = \{x \in \mathbb{R}^g : p(x) = 0\}$$

of a polynomial $p = p(x_1, \dots, x_g)$ in g commuting variables, the second fundamental form at a smooth point x_0 of $\nu(p)$ is the quadratic form

$$h \mapsto -\langle (\operatorname{Hess} p)(x_0)h, h\rangle, \tag{8.24}$$

where $\operatorname{Hess} p$ is the Hessian of p, and $h \in \mathbb{R}^g$ is in the tangent space to the surface $\nu(p)$ at x_0; i.e., $\nabla p(x_0) \cdot h = 0$.[3]

We shall show that in the noncommutative setting the zero set $\mathcal{V}(p)$ of a noncommutative polynomial p (subject to appropriate irreducibility constraints) having positive curvature (even in a small neighborhood) implies that p is convex—and thus, p has degree at most two—and $\mathcal{V}(p)$ has positive curvature everywhere; see Theorem 8.103 for the precise statements.

In fact there is a natural notion of the signature $C_\pm(\mathcal{V}(p))$ of a variety $\mathcal{V}(p)$ and the bound

$$\deg(p) \le 2C_\pm(\mathcal{V}(p)) + 2$$

[3]The choice of the minus sign in (8.24) is somewhat arbitrary. Classically the sign of the second fundamental form is associated with the choice of a smoothly varying vector that is normal to $\nu(p)$. The zero set $\nu(p)$ has positive curvature at x_0 if the second fundamental form is either positive semidefinite or negative semidefinite at x_0. For example, if we define $\nu(p)$ using a concave function p, then the second fundamental form is negative semidefinite, while for the same set $\nu(-p)$ the second fundamental form is positive semidefinite.

on the degree of p in terms of the signature $C_\pm(\mathcal{V}(p))$ was obtained in [16]. The convention that $C_+(\mathcal{V}(p)) = 0$ corresponds to positive curvature, since in our examples, defining functions p are typically concave or quasiconcave. One could consider characterizing p for which $C_\pm(\mathcal{V}(p))$ satisfies a less restrictive hypothesis than being equal to zero, and this has been done to some extent in [14]; however, this higher level of generality is beyond our focus here. Since our goal is to present the basic ideas, we stick to positive curvature.

8.6.1 Noncommutative Varieties and Their Curvature

We next define a number of basic geometric objects associated to the noncommutative variety determined by a noncommutative polynomial p.

Varieties, tangent planes, and the second fundamental form

The *variety* (zero set) of a $p \in \mathbb{R}\langle x \rangle$ is

$$\mathcal{V}(p) := \bigcup_{n \geq 1} \mathcal{V}_n(p),$$

where

$$\mathcal{V}_n(p) := \left\{(X, v) \in (\mathbb{S}^{n\times n})^g \times \mathbb{R}^n \colon p(X)v = 0\right\}.$$

The *clamped tangent plane* to $\mathcal{V}(p)$ at $(X, v) \in \mathcal{V}_n(p)$ is

$$\mathcal{T}_p(X, v) := \{H \in (\mathbb{S}^{n\times n})^g \colon p'(X)[H]v = 0\}.$$

The *clamped second fundamental form* for $\mathcal{V}(p)$ at $(X, v) \in \mathcal{V}_n(p)$ is the quadratic form

$$\mathcal{T}_p(X, v) \to \mathbb{R}, \quad H \mapsto -\langle p''(X)[H]v, v\rangle.$$

Note that

$$\{X \in (\mathbb{S}^{n\times n})^g \colon (X, v) \in \mathcal{V}(p) \text{ for some } v \neq 0\} = \{X \in (\mathbb{S}^{n\times n})^g \colon \det(p(X)) = 0\}$$

is a variety in $(\mathbb{S}^{n\times n})^g$ and typically has a *true* (commutative) tangent plane at many points X, which of course has codimension one, whereas the clamped tangent plane at a typical point $(X, v) \in \mathcal{V}_n(p)$ has codimension on the order of n and is contained inside the true tangent plane.

Full rank points

The point $(X, v) \in \mathcal{V}(p)$ is a *full rank point* of p if the mapping

$$(\mathbb{S}^{n\times n})^g \to \mathbb{R}^n, \quad H \mapsto p'(X)[H]v$$

is onto. The full rank condition is a nonsingularity condition which amounts to a smoothness hypothesis. Such conditions play a major role in real algebraic geometry; see [5, Section 3.3].

As an example, consider the classical real algebraic geometry case of $n = 1$ (and thus $X \in \mathbb{R}^g$) with the commutative polynomial $\check{p}$ (which can be taken to be the *commutative collapse* of the polynomial p). In this case, a full rank point $(X, 1) \in \mathbb{R}^g \times \mathbb{R}$ is a point at which the gradient of $\check{p}$ does not vanish. Thus, X is a nonsingular point for the zero variety of $\check{p}$.

Some perspective for $n > 1$ is obtained by counting dimensions. If $(X, v) \in (\mathbb{S}^{n\times n})^g \times \mathbb{R}^n$, then $H \mapsto p'(X)[H]v$ is a linear map from the $g(n^2+n)/2$-dimensional space $(\mathbb{S}^{n\times n})^g$ into the n-dimensional space $\mathbb{R}^n$. Therefore, the codimension of the kernel of this map is no bigger than n. This codimension is n if and only if (X, v) is a full rank point, and in this case the clamped tangent plane has codimension n.

Positive curvature

As noted earlier, a notion of positive (really nonnegative) curvature can be defined in terms of the clamped second fundamental form.

The variety $\mathcal{V}(p)$ has *positive curvature* at $(X, v) \in \mathcal{V}(p)$ if the clamped second fundamental form is nonnegative at (X, v), i.e., if

$$-\langle p''(X)[H]v, v\rangle \geq 0 \quad \text{for every} \quad H \in \mathcal{T}_p(X, v)\,.$$

Irreducibility: The minimum degree defining polynomial condition

While there is no tradition of what is an effective notion of irreducibility for noncommutative polynomials, there is a notion of minimal degree noncommutative polynomial which is appropriate for the present context. In the commutative case the polynomial $\check{p}$ on $\mathbb{R}^g$ is a *minimal degree defining polynomial* for $\nu(\check{p})$ if there does not exist a polynomial q of lower degree such that $\nu(\check{p}) = \nu(q)$. This is a key feature of irreducible polynomials.

Definition 8.99. *A symmetric noncommutative polynomial p is a minimum degree defining polynomial for a nonempty set $\mathcal{D} \subseteq \mathcal{V}(p)$ if whenever $q \neq 0$ is another (not necessarily symmetric) noncommutative polynomial such that $q(X)v = 0$ for each $(X, v) \in \mathcal{D}$, then*

$$\deg(q) \geq \deg(p).$$

Note this contrasts with [15], where minimal degree meant a slightly weaker inequality holds.

The reader who is so inclined can simply choose $\mathcal{D} = \mathcal{V}(p)$ or $\mathcal{D}$ equal to the full rank points of $\mathcal{V}(p)$.

Now we give an example to illustrate these ideas.

8.6.2 A Very Simple Example

In the following example, the null space

$$\mathcal{T} = \mathcal{T}_p(X, v) = \{H \in (\mathbb{S}^{n\times n})^g : p'(X)[H]v = 0\}$$

is computed for certain choices of p, X, and v. Recall that if $p(X)v = 0$, then the subspace $\mathcal{T}$ is the *clamped tangent plane* introduced in Subsection 8.6.1.

Example 8.100. Let $X \in \mathbb{S}^{n\times n}$, $v \in \mathbb{R}^n$, $v \neq 0$, and let $p(x) = x^k$ for some integer $k \geq 1$. Suppose that $(X, v) \in \mathcal{V}(p)$, that is, $X^k v = 0$. Then, since

$$X^k v = 0 \Longleftrightarrow Xv = 0 \quad \text{when } X \in \mathbb{S}^{n\times n},$$

it follows that p is a minimum degree defining polynomial for $\mathcal{V}(p)$ if and only if $k = 1$.

It is readily checked that

$$(X, v) \in \mathcal{V}(p) \Longrightarrow p'(X)[H]v = X^{k-1}Hv$$

and hence that X is a full rank point for p if and only if X is invertible.

Now suppose $k \geq 2$. Then

$$\langle p''(X)[H]v, v\rangle = 2\langle HX^{k-2}Hv, v\rangle.$$

Therefore, if $k > 2$,

$$(X, v) \in \mathcal{V}(p) \quad \text{and} \quad p'(X)[H]v = 0 \quad \Longrightarrow \quad XHv = 0, \text{ and so}$$
$$\langle p''(X)[H]v, v\rangle = 0.$$

To count the dimension of $\mathcal{T}$ we can suppose without loss of generality that

$$X = \begin{bmatrix} 0 & 0 \\ 0 & Y \end{bmatrix} \quad \text{and} \quad v = \begin{bmatrix} 1 & 0 & \cdots & 0 \end{bmatrix}^\intercal,$$

where $Y \in \mathbb{S}^{(n-1)\times(n-1)}$ is invertible. Then, for the simple case under consideration,

$$\mathcal{T} = \{H \in \mathbb{S}^{n\times n} \colon h_{21}, \ldots, h_{n1} = 0\},$$

where h_{ij} denotes the ij entry of H. Thus,

$$\dim \mathcal{T} = \frac{n^2 + n}{2} - (n - 1),$$

i.e., $\operatorname{codim} \mathcal{T} = n - 1$. ∎

Remark 8.101. We remark that

$$X^k v = 0 \quad \text{and} \quad \langle p''(X)[H]v, v\rangle = 0 \Longrightarrow p'(X)[H]v = 0 \quad \text{if } k = 2t \geq 4,$$

as follows easily from the formula

$$\langle p''(X)[H]v, v\rangle = 2\langle X^{t-1}Hv, X^{t-1}Hv\rangle.$$

Exercise 8.102. Let $A \in \mathbb{S}^{n\times n}$ and let $\mathcal{U}$ be a maximal strictly negative subspace of $\mathbb{R}^n$ with respect to the quadratic form $\langle Au, u\rangle$. Prove that there exists a complementary subspace $\mathcal{V}$ of $\mathcal{U}$ in $\mathbb{R}^n$ such that $\langle Av, v\rangle \geq 0$ for every $v \in \mathcal{V}$.

8.6.3 Main Result: Positive Curvature and the Degree of p

Theorem 8.103. *Let p be a symmetric noncommutative polynomial in g symmetric variables, let $\mathcal{O}$ be a noncommutative basic open semialgebraic set, and let $\mathcal{R}$ denote the full rank points of p in $\mathcal{V}(p) \cap \mathcal{O}$. If*

1. *$\mathcal{R}$ is nonempty,*
2. *$\mathcal{V}(p)$ has positive curvature at each point of $\mathcal{R}$, and*
3. *p is a minimum degree defining polynomial for $\mathcal{R}$,*

then $\deg(p)$ *is at most* 2 *and p is concave.*

8.6.4 Ideas and Proofs

Our aim is to give the idea behind the proof of Theorem 8.103 under much stronger hypotheses. We saw earlier the positivity of a quadratic on a noncommutative basic open set $\mathcal{O}$ imparts positivity to its middle matrix there. The following shows this happens for thin sets (noncommutative varieties) too. Thus, the following theorem generalizes the QuadratischePositivstellensatz, Theorem 8.60.

Theorem 8.104. *Let $p, \mathcal{O}, \mathcal{R}$ be as in Theorem* 8.103. *Let $q(x)[h]$ be a polynomial which is quadratic in h having middle matrix representation $q = V^\intercal Z V$ for which* $\deg(V) \leq \deg(p)$. *If*

$$v^\intercal q(X)[H]v \geq 0 \quad \textit{for all} \quad (X, v) \in \mathcal{R} \textit{ and all } H, \tag{8.25}$$

then $Z(X)$ is positive semidefinite for all X with $(X, v) \in \mathcal{R}$.

Proof. The proof of this theorem follows the proof of the QuadratischePositivstellensatz, modified to take into account the set $\mathcal{R}$.

Suppose for each $(X, v) \in \mathcal{R}$ there is a linear combination $G_{(X,v)}(x)$ of the words $\{m(x)\colon \deg(m) < \deg(p)\}$ with $G_{(X,v)}(X)v = 0$ for all $(X, v) \in \mathcal{R}$. Then by Theorem 8.92 (note that $\mathcal{R}$ is closed under direct sums), there is a linear combination $G \in \mathbb{R}\langle x\rangle_{\deg(p)-1}$ with $G(X)v = 0$. However, this is absurd by the minimality of p. Hence there is a $(Y, v) \in \mathcal{R}$ such that $\{m(Y)v\colon \deg(m) < \deg(p)\}$ is linearly independent.

Assume for some g-tuple of symmetric matrices $\tilde{X} = (\tilde{X}_1, \ldots, \tilde{X}_g)$ there is a vector $\tilde{v}$ such that $(\tilde{X}, \tilde{v}) \in \mathcal{R}$, and the matrix $Z(\tilde{X})$ is not positive semidefinite. Let $X = \tilde{X} \oplus Y$ and $\gamma = \tilde{v} \oplus v$. Then $(X, \gamma) \in \mathcal{R}(\ell)$ for some ℓ; the matrix $Z(X)$ is not positive semidefinite; and $\{m(X)\gamma\colon \deg(m) < \deg(p)\}$ is linearly independent.

Let $N = g^{\frac{\kappa(\kappa-1)}{2}} + 1$, where κ is given in Lemma 8.84, and let $n = \ell N$. Consider $W = X \otimes I_N = (X_1 \otimes I_N, \ldots, X_g \otimes I_N)$ and vector $\omega = \gamma \otimes e$, where $e \in \mathbb{R}^N$ is the vector with each entry equal to 1. Then $(W, \omega) \in \mathcal{R}(n)$, and the set $\{m(W)\omega\colon m \in \langle x\rangle_\ell\}$ is linearly independent; thus by Lemma 8.84, the codimension of $\mathcal{M} = \{V(W)[H]\omega\colon H \in (\mathbb{S}^{n\times n})^g\}$ is at most $N - 1$. On the other hand, because $Z(X)$ has a negative eigenvalue, the matrix $Z(W)$ has an eigenspace $\mathcal{E}$, corresponding to a negative eigenvalue, of dimension at least N. It follows that $\mathcal{E} \cap \mathcal{M}$ is nonempty; i.e., there is an $H \in (\mathbb{S}^{n\times n})^g$ such that $V(W)[H]\omega \in \mathcal{E}$. In particular,

$$\langle q(W)[H]\omega, \omega\rangle = \langle Z(W)V(W)[H]\omega, V(W)\omega\rangle < 0,$$

and thus, $q(W)[H]$ is not positive semidefinite. □

The modified Hessian

Our main tool for analyzing the curvature of noncommutative varieties is a variant of the Hessian for symmetric noncommutative polynomials p. The curvature of $\mathcal{V}(p)$ is defined in terms of Hess (p) compressed to tangent planes, for each dimension n. This compression of the Hessian is awkward to work with directly, and so we associate to it a quadratic polynomial $q(x)[h]$ carrying all of the information of p'' compressed to the tangent plane, but having the key property (8.25). We shall call this q we construct the relaxed Hessian. The first step in constructing the relaxed Hessian is to consider the simpler *modified Hessian*

$$p''_{\lambda,0}(x)[h] := p''(x)[h] + \lambda\, p'(x)[h]^{\intercal} p'(x)[h],$$

which captures the conceptual idea. Suppose $X \in (\mathbb{S}^{n\times n})^g$ and $v \in \mathbb{R}^n$. We say that the *modified Hessian is negative* at (X, v) if there is a $\lambda_0 < 0$, so that for all $\lambda \leq \lambda_0$,

$$0 \leq -\langle p''_{\lambda,0}(X)[H]v, v\rangle$$

for all $H \in (\mathbb{S}^{n\times n})^g$. Given a subset $\mathcal{R} = (\mathcal{R}(n))_{n=1}^{\infty}$, with $\mathcal{R}(n) \subseteq (\mathbb{S}^{n\times n})^g \times \mathbb{R}^n$, we say that the *modified Hessian is negative on* $\mathcal{R}$ if it is negative at each $(X, v) \in S$.

Now we turn to motivation.

Example 8.105. *The classical* $n = 1$ *case.* Suppose that p is strictly smoothly quasi-concave, meaning that all superlevel sets of p are strictly convex with strictly positively curved smooth boundary. Suppose that the gradient ∇p (written as a row vector) never vanishes on $\mathbb{R}^g$. Then $G = \nabla p(\nabla p)^{\intercal}$ is strictly positive at each point X in $\mathbb{R}^g$. Fix such an X; the modified Hessian can be decomposed as a block matrix subordinate to the tangent plane to the level set at X, denoted T_X, and to its orthogonal complement (the gradient direction):

$$T_X \oplus \{\lambda\nabla p \colon \lambda \in \mathbb{R}\}.$$

In this decomposition the modified Hessian has the form

$$R = \begin{bmatrix} A & B \\ B^{\intercal} & D + \lambda G \end{bmatrix}.$$

Here, in the case of $\lambda = 0$, R is the Hessian and the second fundamental form is A or $-A$, depending on convention and the rather arbitrary choice of inward or outward normal to ν. If we select our normal direction to be ∇p, then $-A$ is the classical second fundamental form as is consistent with the choice of sign in our definition in Subsection 8.6.1. (All this concern with the sign is unimportant to the content of this chapter and can be ignored by the reader.)

Next, in view of the presumed strict positive curvature of each level set ν, the matrix A at each point of ν is negative definite but the Hessian could have a negative eigenvalue. However, by standard Schur complement arguments, R will be negative definite if

$$D + \lambda G - B^{\mathsf{T}} A^{-1} B \prec 0$$

on this region. Thus, strict convexity assumptions on the sublevel sets of p make the modified Hessian negative definite for negative enough λ. One can make this negative definiteness uniform in X in various neighborhoods under modest assumptions. ■

Very unfortunately in the noncommutative case, Remark 6.8 [17] implies that if n is large enough, then the second fundamental form will have a nonzero null space, thus strict negative definiteness of the A part of the modified Hessian is impossible.

Our trick for dealing with the likely reality that A is only positive semidefinite and obtaining a negative definite R is to add another negative term, say δI, with arbitrarily small $\delta < 0$. After adding such δ, the argument based on choosing $-\lambda$ large succeeds as before. This δ term plus the λ term produces the "relaxed Hessian," to be introduced next, and proper selection of these terms makes it negative definite.

The relaxed Hessian

Recall Let $V_k(x)[h]$ denotes the vector of polynomials with entries $h_j w(x)$, where $w \in {<}x{>}$ runs through the set of g^k words of length k, $j = 1, \dots, g$. Although the order of the entries is fixed in some of our earlier applications (see e.g. [16, (2.3)]) it is irrelevant for the moment. Thus, $V_k = V_k(x)[h]$ is a vector of height g^{k+1}, and the vectors

$$V(x)[h] = \operatorname{col}(V_0, \dots, V_{d-2}) \quad \text{and} \quad \widetilde{V}(x)[h] = \operatorname{col}(V_0, \dots, V_{d-1})$$

are vectors of height $g\sigma(d-2)$ and $g\sigma(d-1)$, respectively. Note that

$$\widetilde{V}(x)[h]^{\mathsf{T}} \widetilde{V}(x)[h] = \sum_{j=1}^{g} \sum_{\deg(w) \le d-1} w(x)^{\mathsf{T}} h_j^2 w(x).$$

The *relaxed Hessian* of the symmetric noncommutative polynomial p of degree d is defined to be

$$p''_{\lambda,\delta}(x)[h] := p''_{\lambda,0}(x)[h] + \delta\, \widetilde{V}(x)[h]^{\mathsf{T}} \widetilde{V}(x)[h] \in \mathbb{R}{<}x{>}[h].$$

Suppose $X \in (\mathbb{S}^{n\times n})^g$ and $v \in \mathbb{R}^n$. We say that the *relaxed Hessian is negative* at (X, v) if for each $\delta < 0$ there is a $\lambda_\delta < 0$, so that for all $\lambda \leq \lambda_\delta$,

$$0 \leq -\langle p''_{\lambda,\delta}(X)[H]v, v\rangle$$

for all $H \in (\mathbb{S}^{n\times n})^g$. Given an $\mathcal{R} = (\mathcal{R}(n))_{n=1}^{\infty}$, with $\mathcal{R}(n) \subseteq (\mathbb{S}^{n\times n})^g \times \mathbb{R}^n$, we say that the *relaxed Hessian is positive* (*respectively, negative*) *on* $\mathcal{R}$ if it is positive (respectively, negative) at each $(X, v) \in S$.

The following theorem provides a link between the signature of the clamped second fundamental form with that of the relaxed Hessian.

Theorem 8.106. *Suppose p is a symmetric noncommutative polynomial of degree d in g symmetric variables and $(X, v) \in (\mathbb{S}^{n\times n})^g \times \mathbb{R}^n$. If $\mathcal{V}(p)$ has positive curvature at $(X, v) \in \mathcal{V}_n(p)$, i.e., if*

$$\langle p''(X)[H]v, v\rangle \leq 0 \quad \textit{for every } H \in \mathcal{T}_p(X, v),$$

then for every $\delta < 0$ there exists a $\lambda_\delta < 0$ such that for all $\lambda \leq \lambda_\delta$,

$$\langle p''_{\lambda,\delta}(X)[H]v, v\rangle \leq 0 \quad \textit{for every } H \in (\mathbb{S}^{n\times n})^g;$$

i.e., the relaxed Hessian of p is negative at (X, v).

We leave the proof of Theorem 8.106 to the reader.

The basic idea of the proof of Theorem 8.103 is to obtain a negative relaxed Hessian q from Theorem 8.106 and then apply Theorem 8.104. We begin with the following lemma.

Lemma 8.107. *Suppose R and T are operators on a finite-dimensional Hilbert space $H = K \oplus L$. Suppose further that, with respect to this decomposition of H, the operator $R = CC^\intercal$ for*

$$C = \begin{bmatrix} r \\ c \end{bmatrix} : L \to K \oplus L \quad \textit{and} \quad T = \begin{bmatrix} T_0 & 0 \\ 0 & 0 \end{bmatrix}.$$

If c is invertible and if for every $\delta > 0$ there is a $\eta > 0$ such that for all $\lambda > \eta$,

$$T + \delta I + \lambda R \succeq 0,$$

then $T \succeq 0$.

Proof. Write

$$T + \delta I + \lambda R = \begin{bmatrix} T_0 + \delta I + \lambda r r^\intercal & \lambda r c^\intercal \\ \lambda c r^\intercal & \delta + \lambda c c^\intercal \end{bmatrix}.$$

From Schur complements it follows that

$$T_0 + \delta I + r(\lambda - \lambda^2 c^\intercal(\delta + \lambda c c^\intercal)^{-1} c) r^\intercal \succeq 0.$$

Now

$$\begin{aligned} r(\lambda - \lambda^2 c^\intercal(\delta + \lambda cc^\intercal)^{-1}c)r^\intercal &= \lambda rc^\intercal((cc^\intercal)^{-1} - \lambda(\delta + \lambda cc^\intercal)^{-1})cr^\intercal \\ &= \lambda rc^\intercal \delta(cc^\intercal)^{-1}(\delta + \lambda(cc^\intercal))^{-1}cr^\intercal \\ &\preceq \delta r(cc^\intercal)^{-1}r^\intercal. \end{aligned}$$

Hence,

$$T_0 + \delta I + \delta r(cc^\intercal)^{-1}r^\intercal \succeq 0.$$

Since the above inequality holds for all $\delta > 0$, it follows that $T_0 \succeq 0$. □

We now have enough machinery developed to prove Theorem 8.103.

***Proof of Theorem* 8.103.** Fix $\lambda, \delta > 0$ and consider $q(x)[h] = -p''_{\lambda,\delta}(x)[h]$. We are led to investigate the middle matrix $Z^{\lambda,\delta}$ of $q(x)[h]$, whose border vector $V(x)[h]$ includes all monomials of the form $h_j m$, where m is a word in x only of length at most $d-1$; here d is the degree of p. Indeed,

$$Z^{\lambda,\delta} = Z + \delta I + \lambda W,$$

where Z is the middle matrix for $-p''(x)[h]$ and W is the middle matrix for $p'(x)[h]^\intercal p'(x)[h]$. With an appropriate choice of ordering for the border vector V, we have $W = CC^\intercal$, where

$$C(x) = \begin{bmatrix} w(x) \\ c \end{bmatrix}$$

for a nonzero vector c, and at the same time,

$$Z(x) = \begin{bmatrix} Z^{0,0}(x) & 0 \\ 0 & 0 \end{bmatrix}.$$

By the curvature hypothesis at a given X with $(X, v) \in \mathcal{R}$, Theorem 8.106 implies for every $\delta > 0$ there is an $\eta > 0$ such that if $\lambda > \eta$

$$\langle q(X)[H]v, v\rangle \geq 0 \qquad \text{for all } (X, v) \in \mathcal{R} \text{ and all } H.$$

Hence, by Theorem 8.104, the middle matrix, $Z^{\lambda,\delta}(X)$ for $q(x)[h]$ is positive semidefinite. We are in the setting of Lemma 8.107, from which we obtain $Z^{0,0}(X) \succeq 0$. If this held for X in a noncommutative basic open semialgebraic set, then Theorem 8.71 forces p to have degree no greater than 2. The proof of that theorem applies easily here to finish this proof. □

8.6.5 Exercises

Exercise 8.108. Compute the BV-MM representation for the relaxed Hessian of x^3 and x^4.

8.7 Convex Semialgebraic Noncommutative Sets

In this section we will give a brief overview of convex semialgebraic noncommutative sets and positivity of noncommutative polynomials on them. We shall see that their structure is much more rigid than that of their commutative counterparts. For example, roughly speaking, each convex semialgebraic noncommutative set is a spectrahedron, i.e., a solution set of a linear matrix inequality (LMI) (cf. Section 8.7.1 below). Similarly, every noncommutative polynomial nonnegative on a spectrahedron admits a sum of squares representation with weights and optimal degree bounds (see Section 8.7.2 for details and precise statements).

8.7.1 Noncommutative Spectrahedra

Let L be an affine linear pencil. Then the solution set of the LMI $L(x) \succ 0$ is

$$\mathcal{D}_L = \bigcup_{n\in\mathbb{N}} \{X \in (\mathbb{S}^{n\times n})^g \colon L(X) \succ 0\}$$

and is called a *noncommutative spectrahedron*. The set $\mathcal{D}_L$ is convex in the sense that each

$$\mathcal{D}_L(n) := \{X \in (\mathbb{S}^{n\times n})^g \colon L(X) \succ 0\}$$

is convex. It is also a noncommutative basic open semialgebraic set as defined in Section 8.2.1 above. The main theorem of this section is the converse, a result which has implications for both semidefinite programming and systems engineering.

Most of the time we will focus on monic linear pencils. An affine linear pencil L is called *monic* if $L(0) = I$, i.e., $L(x) = I + A_1x_1 + \cdots + A_gx_g$. Since we are mostly interested in the set $\mathcal{D}_L$, there is no harm in reducing to this case whenever $\mathcal{D}_L \neq \emptyset$; see Exercise 8.111.

Let $p \in \mathbb{R}^{\delta\times\delta}\langle x\rangle$ be a given symmetric noncommutative $\delta \times \delta$ valued matrix polynomial. Assuming that $p(0) \succ 0$, the positivity set $\mathcal{D}_p(n)$ of a noncommutative symmetric polynomial p in dimension n is the component of 0 of the set

$$\{X \in (\mathbb{S}^{n\times n})^g \ : \ p(X) \succ 0\}.$$

The *positivity set*, $\mathcal{D}_p$, is the sequence of sets $(\mathcal{D}_p(n))_{n\in\mathbb{N}}$. The noncommutative set $\mathcal{D}_p$ is called *convex* if, for each n, $\mathcal{D}_p(n)$ is convex.

Theorem 8.109 (Helton–McCullough [38]). *Fix p, a $\delta \times \delta$ symmetric matrix of polynomials in noncommuting variables. Assume*

1. *$p(0)$ is positive definite;*

2. *$\mathcal{D}_p$ is bounded; and*

3. *$\mathcal{D}_p$ is convex.*

Then there is a monic linear pencil L such that

$$\mathcal{D}_L = \mathcal{D}_p.$$

Here we shall confine ourselves to a few words about the techniques involved in the proof, and refer the reader to [38] for the full proof. Since we are dealing with matrix convex sets, it is not surprising that the starting point for our analysis is the matricial version of the Hahn–Banach separation theorem of Effros and Winkler [20], which (itself a part of the theory of operator spaces and completely positive maps [4, 57, 58]) says that given a point x not inside a matrix convex set there is a (finite) LMI which separates x from the set. For a general matrix convex set $\mathcal{C}$, the conclusion is then that there is a collection, likely infinite, of LMIs which cut out $\mathcal{C}$.

In the case $\mathcal{C}$ is matrix convex and also semialgebraic, the challenge is to prove that there is actually a *finite* collection of LMIs which define $\mathcal{C}$. The techniques used to meet this challenge have little relation to the methods of noncommutative calculus and positivity in the previous sections. Indeed a basic tool (of independent interest) is a degree bounded type of free Zariski closure of a single point $(X, v) \in (\mathbb{S}^{n\times n})^g \times \mathbb{R}^n$,

$$Z_d(X,v) := \bigcup_m \{(Y,w) \in (\mathbb{S}^{m\times m})^g \times \mathbb{R}^m : q(Y)w = 0 \text{ if } q(X)v = 0,\ q \in \mathbb{R}\langle x\rangle_d\}.$$

Chief among a pleasant list of natural properties is the fact that there is an (X, v) with $X \in \partial\mathcal{D}_p$ and $p(X)v = 0$ for which $Z_d(X, v)$ contains all pairs (Y, w) such that $Y \in \partial\mathcal{D}_p$ and $p(Y)w = 0$. Combining this with the Effros–Winkler theorem and battling degeneracies is a bit tricky, but separation prevails in the end. See [38] for the details.

An unexpected consequence of Theorem 8.109 is that projections of noncommutative semialgebraic sets may not be semialgebraic; see Exercise 8.112. For perspective, in the commutative case of a basic open semialgebraic subset $\mathcal{C}$ of $\mathbb{R}^g$, there is a stringent condition, called the "*line test*" (see Chapter 6 for more details), which, in addition to convexity, is necessary for $\mathcal{C}$ to be a spectrahedron. In two dimensions the line test is necessary and sufficient [44], a result used by Lewis–Parrilo–Ramana [51] to settle a 1958 conjecture of Peter Lax on hyperbolic polynomials.

In summary, if a (commutative) bounded basic open semialgebraic convex set is a spectrahedron, then it must pass the highly restrictive line test; whereas a noncommutative basic open semialgebraic set is a spectrahedron if and only if it is convex.

8.7.2 Noncommutative Positivstellensätze under Convexity Assumptions

An algebraic certificate for positivity of a polynomial p on a semialgebraic set S is a Positivstellensatz. The familiar fact that a polynomial p in one variable which is positive on $\mathbb{R}$ is a sum of squares is an example.

The theory of Positivstellensätze—a pillar of the field of real algebraic geometry—underlies the main approach currently used for global optimization of polynomials. See [50] or Chapter 3 by Parrilo for a beautiful treatment of this, and other, applications of commutative real algebraic geometry. Further, because convexity of a polynomial p on a set S is equivalent to positivity of the Hessian of p on S, this theory also provides a link between convexity and semialgebraic geometry. Indeed, this link in the noncommutative setting ultimately leads to the conclusion that a matrix convex noncommutative polynomial has degree at most 2; cf. Subsection 8.4.4.

In this section we give a result of opposite type. We present a noncommutative Positivstellensatz for a polynomial to be nonnegative on a convex semialgebraic noncommutative set (i.e., on a spectrahedron). Again, this result is cleaner and more rigid than the commutative counterparts (cf. Theorem 8.10).

Theorem 8.110 ([33]). *Suppose L is a monic linear pencil. Then a noncommutative polynomial p is* positive semidefinite *on $\mathcal{D}_L$ if and only if it has a weighted sum of squares representation with optimal degree bounds. Namely,*

$$p = s^\intercal s + \sum_{j}^{\text{finite}} f_j^\intercal L f_j, \tag{8.26}$$

where s, f_j are vectors of noncommutative polynomials of degree no greater than $\frac{\deg(p)}{2}$.

The main ingredient of the proof is an analysis of rank preserving extensions of truncated noncommutative Hankel matrices; see [33] for details. We point out that with $L = 1$, Theorem 8.110 recovers Theorem 8.10.

Theorem 8.110 contrasts sharply with the commutative setting, where the degrees of s, f_j are vastly greater than $\deg(p)$ and assuming only p nonnegative yields a clean Positivstellensatz so seldom that the cases are noteworthy.

8.7.3 Exercises

Exercise 8.111. Suppose L is an affine linear pencil such that $0 \in \mathcal{D}_L(1)$. Show that there is a monic linear pencil $\check{L}$ with $\mathcal{D}_L = \mathcal{D}_{\check{L}}$.

Exercise 8.112. Chapters 5 and 6 discuss sets $D \subseteq \mathbb{R}^g$ which have a semidefinite representation as a strict generalization of a spectrahedron. For instance, consider the TV screen (cf. Section 8.2.1)

$$\text{ncTV}(1) = \{X \in \mathbb{R}^2 \colon 1 - X_1^4 - X_2^4 > 0\} \subseteq \mathbb{R}^2.$$

Given α a positive real number, choose $\gamma^4 = 1 + 2\alpha^2$ and let

$$L_0 = \begin{bmatrix} 1 & 0 & y_1 \\ 0 & 1 & y_2 \\ y_1 & y_2 & 1 - 2\alpha(y_1 + y_2) \end{bmatrix} \tag{8.27}$$

and

$$L_j = \begin{bmatrix} 1 & \gamma x_j \\ \gamma x_j & \alpha + y_j \end{bmatrix}, \quad j = 1, 2. \tag{8.28}$$

Note that the L_j are not monic, but because $L_j(0) \succ 0$, they can be normalized to be monic without altering the solution sets of $L_j(X) \succ 0$; cf. Exercise 8.111. Let $L = L_0 \oplus L_1 \oplus L_2$.

It is readily verified that ncTV(1) is the projection onto the first two (the x) coordinates of the set $\mathcal{D}_L(1)$; i.e.,

$$\text{ncTV}(1) = \{X \in \mathbb{R}^2 \colon \exists Y \in \mathbb{R}^2 \ L(X, Y) \succ 0\}.$$

1. Show that ncTV(1) is not a spectrahedron. (*Hint*: How often is $L_{\text{TV}}(tX, tY)$ for $t \in \mathbb{R}$ singular?)

2. Show that ncTV is not the projection of the noncommutative spectrahedron $\mathcal{D}_L$.

3. Show that ncTV is not the projection of *any* noncommutative spectrahedron.

4. Is ncTV(2) a projection of a spectrahedron? (Feel free to use the results about ncTV and LMI representable sets (spectrahedra), stated without proofs, from Sections 8.2.1 and 8.7.1.)

Exercise 8.113. If q is a symmetric concave matrix-valued polynomial with $q(0) = I$, then there exists a linear pencil L and a matrix-valued linear polynomial Λ such that

$$q = I - L - \Lambda^{\intercal}\Lambda.$$

Exercise 8.114. Consider the monic linear pencil

$$M(x) = \begin{bmatrix} 1 & x \\ x & 1 \end{bmatrix}.$$

1. Determine $\mathcal{D}_M$.

2. Show that $1 + x$ is positive semidefinite on $\mathcal{D}_M$.

3. Construct a representation for $1 + x$ of the form (8.26).

Exercise 8.115. Consider the univariate affine linear pencil

$$L(x) = \begin{bmatrix} 1 & x \\ x & 0 \end{bmatrix}.$$

1. Determine $\mathcal{D}_L$.

2. Show that x is positive semidefinite on $\mathcal{D}_L$.

3. Does x admit a representation of the form (8.26)?

Exercise 8.116. Let L be an affine linear pencil. Prove that

1. $\mathcal{D}_L$ is bounded if and only if $\mathcal{D}_L(1)$ is bounded;

2. $\mathcal{D}_L = \emptyset$ if and only if $\mathcal{D}_L(1) = \emptyset$.

Exercise 8.117. Let $L = I + A_1x_1 + \cdots + A_gx_g$ be a monic linear pencil and assume that $\mathcal{D}_L(1)$ is bounded. Show that $I, A_1, \ldots, A_g$ are linearly independent.

Exercise 8.118. Let

$$\Delta(x_1, x_2) = I + \begin{bmatrix} 0 & 1 & 0 \\ 1 & 0 & 0 \\ 0 & 0 & 0 \end{bmatrix} x_1 + \begin{bmatrix} 0 & 0 & 1 \\ 0 & 0 & 0 \\ 1 & 0 & 0 \end{bmatrix} x_2 = \begin{bmatrix} 1 & x_1 & x_2 \\ x_1 & 1 & 0 \\ x_2 & 0 & 1 \end{bmatrix}$$

and

$$\Gamma(x_1, x_2) = I + \begin{bmatrix} 1 & 0 \\ 0 & -1 \end{bmatrix} x_1 + \begin{bmatrix} 0 & 1 \\ 1 & 0 \end{bmatrix} x_2 = \begin{bmatrix} 1 + x_1 & x_2 \\ x_2 & 1 - x_1 \end{bmatrix}$$

be affine linear pencils. Show

1. $\mathcal{D}_\Delta(1) = \mathcal{D}_\Gamma(1)$.

2. $\mathcal{D}_\Gamma(2) \subsetneq \mathcal{D}_\Delta(2)$.

3. Is $\mathcal{D}_\Delta \subseteq \mathcal{D}_\Gamma$? What about $\mathcal{D}_\Gamma \subseteq \mathcal{D}_\Delta$?

Exercise 8.119. Let $L = A_1x_1 + \cdots + A_gx_g \in \mathbb{S}^{d\times d}\langle x \rangle$ be a (homogeneous) linear pencil. Then the following are equivalent:

(i) $\mathcal{D}_L(1) \neq \emptyset$;

(ii) If $u_1, \ldots, u_m \in \mathbb{R}^d$ with $\sum_{i=1}^m u_i^\intercal L(x) u_i = 0$, then $u_1 = \cdots = u_m = 0$.

8.8 From Free Real Algebraic Geometry to the Real World

Now that you have gone through the mathematics we return to its implications. In the linear systems engineering problems you have seen both in Section 8.1.1 and in Section 2.2.1, the conclusion was that the problem was equivalent to solving an LMI. Indeed this is what one sees throughout the literature. Thousands of engineering papers have a dimension free problem and it converts (often by serious cleverness) to an LMI in the best of cases, or more likely there is some approximate solution which is an LMI.

While engineers would be satisfied with convexity, what they actually do get is an LMI. One would hope that there is a rich world of convex situations not equivalent to an LMI. Then there would be a variety of methods waiting to be discovered for dealing with them. Alas what we have shown here is compelling evidence that any convex dimension free problem is equivalent to an LMI. Thus there is no rich world of convexity beyond what is already known and no armada of techniques beyond those for producing LMIs which we already see all around us.

Bibliography

[1] S. Balasubramanian and S. McCullough. Quasi-convex free polynomials. To appear in *Proc. Amer. Math. Soc.* http://arxiv.org/abs/1208.3582.

[2] G. M. Bergman. Rational relations and rational identities in division rings I. *J. Algebra*, 43:252–266, 1976.

[3] R. Bhatia. *Matrix Analysis*. Springer-Verlag, Berlin, 1997.

[4] D. P. Blecher and C. Le Merdy. *Operator Algebras and Their Modules—An Operator Space Approach*, Oxford Science Publications, Oxford, UK, 2004.

[5] J. Bochnak, M. Coste, and M. F. Roy. *Real Algebraic Geometry*. Springer-Verlag, Berlin, 1998.

[6] M. Brešar and I. Klep. A local-global principle for linear dependence of noncommutative polynomials. *Israel J. Math.*, to appear.

[7] K. Cafuta, I. Klep, and J. Povh. A note on the nonexistence of sum of squares certificates for the Bessis-Moussa-Villani conjecture. *J. Math. Phys.*, 51:083521, 2010.

[8] K. Cafuta, I. Klep, and J. Povh. `NCSOStools`: a computer algebra system for symbolic and numerical computation with noncommutative polynomials. *Optim. Methods Softw.*, 26:363–380, 2011.

[9] J. F. Camino, J. W. Helton, R. E. Skelton, and J. Ye. Matrix inequalities: A symbolic procedure to determine convexity automatically. *Integral Equations Operator Theory*, 46:399–454, 2003.

[10] P. M. Cohn. *Skew Fields. Theory of General Division Rings*. Cambridge University Press, Cambridge, UK, 1995.

[11] P. M. Cohn. *Free Ideal Rings and Localization in General Rings*. Cambridge University Press, Cambridge, UK, 2006.

[12] A. C. Doherty, Y.-C. Liang, B. Toner, and S. Wehner. The quantum moment problem and bounds on entangled multi-prover games. In *Twenty-Third Annual IEEE Conference on Computational Complexity*, 2008, pp. 199–210.

[13] M. de Oliviera, J. W. Helton, S. McCullough, and M. Putinar. Engineering systems and free semi-algebraic geometry. In *Emerging Applications of Algebraic Geometry*, IMA Vol. Math. Appl. 149. Springer-Verlag, Berlin, 2009, pp. 17–62.

[14] H. Dym, J. M. Greene, J. W. Helton, and S. McCullough. Classification of all noncommutative polynomials whose Hessian has negative signature one and a noncommutative second fundamental form. *J. Anal. Math.*, 108:19–59, 2009.

[15] H. Dym, J. W. Helton, and S. McCullough. Irreducible noncommutative defining polynomials for convex sets have degree four or less. *Indiana Univ. Math. J.*, 56:1189–1232, 2007.

[16] H. Dym, J. W. Helton, and S. McCullough. The Hessian of a non-commutative polynomial has numerous negative eigenvalues. *J. Anal. Math.*, 102:29–76, 2007.

[17] H. Dym, J. W. Helton, and S. McCullough. Noncommutative varieties with curvature having bounded signature, *Illinois J. Math.*, to appear.

[18] E. G. Effros. A matrix convexity approach to some celebrated quantum inequalities. *Proc. Natl. Acad. Sci. USA*, 106:1006–1008, 2009.

[19] A. Ebadiana, I. Nikoufarb, and M. E. Gordjic. Perspectives of matrix convex functions. *Proc. Natl. Acad. Sci. USA*, 108:7313–7314, 2011.

[20] E. G. Effros and S. Winkler. Matrix convexity: Operator analogues of the bipolar and Hahn-Banach theorems. *J. Funct. Anal.*, 144:117–152, 1997.

[21] W.S. Gray and Y. Li. Generating series for interconnected analytic nonlinear systems. *SIAM J. Control Optim.*, 44:646–672, 2005.

[22] W.S. Gray and M. Thitsa. A unified approach to generating series of mixed cascades of analytic nonlinear input-output systems. *Internat. J. Control*, 85:1737–1754, 2012.

[23] J. M. Greene, J. W. Helton, and V. Vinnikov. Noncommutative plurisubharmonic polynomials, Part I: Global assumptions. *J. Funct. Anal.*, 261:3390–3417, 2011.

[24] F. Hansen. Operator convex functions of several variables. *Publ. Res. Inst. Math. Sci.*, 33:443–463, 1997.

[25] F. Hansen and J. Tomiyama. Differential analysis of matrix convex functions. *Linear Algebra Appl.*, 420:102–116, 2007.

[26] D. M. Hay, J. W. Helton, A. Lim, and S. McCullough. Non-commutative partial matrix convexity. *Indiana Univ. Math. J.*, 57:2815–2842, 2008.

[27] J. W. Helton. "Positive" noncommutative polynomials are sums of squares. *Ann. of Math.* (2), 156:675–694, 2002.

[28] J. W. Helton, I. Klep, and S. McCullough. Analytic mappings between noncommutative pencil balls. *J. Math. Anal. Appl.*, 376:407–428, 2011.

[29] J. W. Helton, I. Klep, and S. McCullough. Proper analytic free maps. *J. Funct. Anal.*, 260:1476–1490, 2011.

[30] J. W. Helton, I. Klep, and S. McCullough. Relaxing LMI domination matricially. In *49th IEEE Conference on Decision and Control*, 2010, pp. 3331–3336.

[31] J. W. Helton, I. Klep, and S. McCullough. Convexity and semidefinite programming in dimension-free matrix unknowns. In M. Anjos and J. B. Lasserre, editors, *Handbook of Semidefinite, Cone and Polynomial Optimization.* Springer-Verlag, Berlin, 2012, pp. 377–405.

[32] J. W. Helton, I. Klep, and S. McCullough. The matricial relaxation of a linear matrix inequality. Preprint, http://arxiv.org/abs/1003.0908. To appear in *Math. Program.*

[33] J. W. Helton, I. Klep, and S. McCullough. The convex Positivstellensatz in a free algebra. *Adv. Math.*, 231:516–534, 2012.

[34] J. W. Helton, I. Klep, S. McCullough, and N. Slinglend. Noncommutative ball maps. *J. Funct. Anal.*, 257:47–87, 2009.

[35] J. W. Helton and O. Merino. Sufficient conditions for optimization of matrix functions. In *37th IEEE Conference on Decision and Control*, 1998, pp. 3361–3365.

[36] J. W. Helton and S. McCullough. Convex noncommutative polynomials have degree two or less. *SIAM J. Matrix Anal. Appl.*, 25:1124–1139, 2004.

[37] J. W. Helton and S. McCullough. A Positivstellensatz for noncommutative polynomials. *Trans. Amer. Math. Soc.*, 356:3721–3737, 2004.

[38] J. W. Helton and S. McCullough. Every free basic convex semialgebraic set has an LMI representation. *Ann. of Math.*, 176:979–1013, 2012.

[39] J. W. Helton, S. McCullough, and M. Putinar. A non-commutative Positivstellensatz on isometries. *J. Reine Angew. Math.*, 568:71–80, 2004.

[40] J. W. Helton, S. McCullough, M. Putinar, and V. Vinnikov. Convex matrix inequalities versus linear matrix inequalities. *IEEE Trans. Automat. Control*, 54:952–964, 2009.

[41] J. W. Helton, S. McCullough, and V. Vinnikov. Noncommutative convexity arises from linear matrix inequalities. *J. Funct. Anal.*, 240:105–191, 2006.

[42] J. W. Helton, M. de Oliveira, R. L. Miller, and M. Stankus. `NCAlgebra`: A Mathematica package for doing non commuting algebra, available from http://www.math.ucsd.edu/~ncalg/.

[43] J. W. Helton and M. Putinar. Positive polynomials in scalar and matrix variables, the spectral theorem and optimization, In *Operator Theory, Structured Matrices, and Dilations*, Theta Ser. Adv. Math. 7. American Mathematical Society, Providence, RI, 2007, pp. 229–306.

[44] J. W. Helton and V. Vinnikov. Linear matrix inequality representation of sets. *Comm. Pure Appl. Math.* 60:654–674, 2007.

[45] D. Kalyuzhnyi-Verbovetskiĭ and V. Vinnikov. Singularities of rational functions and minimal factorizations: The noncommutative and the commutative setting. *Linear Algebra Appl.*, 430:869–889, 2009.

[46] D. Kalyuzhnyi-Verbovetskiĭ and V. Vinnikov. Foundations of noncommutative function theory, in preparation.

[47] I. Klep and M. Schweighofer. Connes' embedding conjecture and sums of Hermitian squares. *Adv. Math.*, 217:1816–1837, 2008.

[48] I. Klep and M. Schweighofer. Sums of Hermitian squares and the BMV conjecture. *J. Stat. Phys.*, 133:739–760, 2008.

[49] F. Kraus. Über konvexe matrixfunktionen. *Math. Z.*, 41:18–42, 1936.

[50] J. B. Lasserre. *Moments, Positive Polynomials and Their Applications.* Imperial College Press, London, 2010.

[51] A. S. Lewis, P. A. Parrilo, and M. V. Ramana. The Lax conjecture is true. *Proc. Amer. Math. Soc.*, 133:2495–2499, 2005.

[52] T. Lyons, M. Caruana, and T. Lévy. Differential equations drive by rough paths. In *École d'Eté de Probabilités de Saint-Flour XXXIV*, Lecture Notes in Math. 1908, Springer-Verlag, Berlin, 2004.

[53] S. McCullough. Factorization of operator-valued polynomials in several noncommuting variables. *Linear Algebra Appl.*, 326:193–203, 2001.

[54] P. S. Muhly and B. Solel. Progress in noncommutative function theory. *Sci. China Ser. A*, 54:2275–2294, 2011.

[55] B. Sz.-Nagy, C. Foias, H. Bercovici, and L. Kerchy. *Harmonic Analysis of Operators on Hilbert Space.* Springer-Verlag, New York, 2010.

[56] H. Osaka, S. Silvestrov, and J. Tomiyama. Monotone operator functions, gaps and power moment problem. *Math. Scand.*, 100:161–183, 2007.

[57] V. Paulsen. *Completely Bounded Maps and Operator Algebras.* Cambridge University Press, Cambridge, UK, 2002.

[58] G. Pisier. *Introduction to Operator Space Theory.* Cambridge University Press, Cambridge, UK, 2003.

[59] S. Pironio, M. Navascués, and A. Acín. Convergent relaxations of polynomial optimization problems with noncommuting variables. *SIAM J. Optim.*, 20:2157–2180, 2010.

[60] G. Popescu. Free holomorphic functions on the unit ball of $\mathcal{B}(\mathcal{H})^n$. *J. Funct. Anal.*, 241:268–333, 2006.

[61] G. Popescu. Free holomorphic automorphisms of the unit ball of $B(H)^n$. *J. Reine Angew. Math.*, 638:119–168, 2010.

[62] K. Schmüdgen. A strict Positivstellensatz for the Weyl algebra. *Math. Ann.*, 331:779–794, 2005.

[63] K. Schmüdgen. Noncommutative real algebraic geometry—some basic concepts and first ideas. In *Emerging Applications of Algebraic Geometry*, IMA Vol. Math. Appl. 149. Springer-Verlag, Berlin, 2009, pp. 325–350.

[64] D. Shlyakhtenko and D.-V. Voiculescu. Free analysis workshop summary, American Institute of Mathematics, http://www.aimath.org/pastworkshops/freeanalysis.html.

[65] R. E. Skelton and T. Iwasaki. Eye on education. Increased roles of linear algebra in control education. *IEEE Control Syst. Mag.*, 15:76–90, 1995.

[66] R. E. Skelton, T. Iwasaki, and K. M. Grigoriadis. *A Unified Algebraic Approach to Linear Control Design.* Taylor & Francis, London, 1997.

[67] J. L. Taylor. Functions of several noncommuting variables. *Bull. Amer. Math. Soc.*, 79:1–34, 1973.

[68] M. Uchiyama. Operator monotone functions and operator inequalities. *Sugaku Expositions*, 18:39–52, 2005.

[69] D.-V. Voiculescu. Free analysis questions I: Duality transform for the coalgebra of $\partial_{X:B}$. *Int. Math. Res. Not.*, 16:793–822, 2004.

[70] D.-V. Voiculescu. Free analysis questions II: The Grassmannian completion and the series expansions at the origin. *J. Reine Angew. Math.*, 645:155–236, 2010.

[71] D.-V. Voiculescu, K. J. Dykema, and A. Nica. *Free Random Variables. A Noncommutative Probability Approach to Free Products with Applications to Random Matrices, Operator Algebras and Harmonic Analysis on Free Groups.* American Mathematical Society, Providence, RI, 1992.

Chapter 9

Sums of Hermitian Squares: Old and New

Mihai Putinar[†]

This final chapter marks a departure from the main framework of the book by putting emphasis on hermitian forms over the complex field rather than symmetric forms over the real field. The passage is both natural and necessary. To give a simple motivation: polynomial or rational functions with real coefficients, so much praised in the preceding chapters, may very well have complex roots or complex poles. Taking them into account greatly simplifies computations and conceptual thinking, as we all remember from elementary algebra. A second important observation goes back to the dictionary between elementary functions and matrices: by writing in complex coordinates a real valued polynomial (in any number of variables) $p(z, \overline{z}) = \sum c_{\alpha\beta} z^\alpha \overline{z}^\beta$ uniquely determines the hermitian matrix $(c_{\alpha\beta})$, while a similar decomposition $q(x) = \sum \gamma_{\alpha\beta} x^{\alpha+\beta}$, with real coefficients $\gamma_{\alpha\beta}$, so much needed for semidefinite programming, has a clear ambiguity. The appearance at this late stage of the book of imaginary "ghosts" related to the basic entities encountered so far should not discourage the truly real and very applied reader.

9.1 Introduction

A question arises from the very beginning: how much of the vast theory of hermitian forms (in a finite or infinite number of variables) should the student or practitioner in applied areas of real algebra, functional analysis, algebraic geometry, or optimization theory know? Due to the depth and wide ramifications of hermitian forms (over the complex field) versus forms over real fields, the answer is: quite a lot! The good news is that the material, old and new, either is well known, circulating in part as folklore, or is accessible, due to a century and a half of continuous development

†Mihai Putinar was supported by NSF grant DMS-1001071.

of hermitian forms, in all their impersonations. Without aiming at completeness, we touch below several basic aspects of the theory of hermitian forms. The historical and bibliographical notes, supplemented by the suggested problems will guide the reader though this field and will hopefully whet the appetite for a thorough study of some specific subtopics. A glimpse at the table of contents (of this chapter) will give an indication of what we aim at below: root separation of polynomials, the structure of stable polynomials, effective computation of bounds for analytic functions, Hilbert space realization of analytic functions, hermitian positivity in several complex variables, and a brief return to real algebra. The identification of a positive definite hermitian form with a Hilbert space structure cannot be underestimated, especially for the emerging domain of convex algebraic geometry whose frontiers are delimited in this book. In other words: it is not an accident that Hilbert spaces pop up unexpectedly in convex algebraic geometry.

It is important to state from the very beginning that a major source of the theory of hermitian forms is omitted by our survey: the study of linear integral equations as they appear in problems of mathematical physics, such as the stationary values of the energy functional in potential theory, vibrations of strings and membranes, elasticity theory, dissipation of heat, and so on. Major figures in this field were Riemann, Hilbert, and Poincaré and their contemporaries. Hilbert has collected six of his groundbreaking articles on integral equations in a booklet [21]. The modern reader can find them actual, accessible, and full of ideas. In particular, Hilbert regards the whole area of integral equations as a chapter of the theory of hermitian forms of infinitely many variables. His point of view has persisted through the first half of the twentieth century, as one can also see from the German Mathematical Encyclopedia article by Hellinger and Toeplitz [19]. Even today (quantum) mathematical physicists prefer to work with *hermitian forms* rather than with *linear unbounded operators*, and the distinction is not only cosmetic.

9.2 Hermitian Forms and Sums of Squares

We start by recalling a few well-known facts about canonical forms of matrices and positive definite kernels. Let $\mathbb{C}$ be the complex field and denote by $M_d(\mathbb{C})$ the algebra of $d \times d$ matrices over $\mathbb{C}$, regarded as linear transforms of the space $\mathbb{C}^d$. We endow $\mathbb{C}^d$ with its *hermitian structure*, that is, the inner product

$$\langle z, w \rangle = z \cdot \overline{w} = z_1\overline{w_1} + \cdots + z_d\overline{w_d},$$

where $z = (z_1, \ldots, z_d)$, $w = (w_1, \ldots, w_d) \in \mathbb{C}^d$. We put as usual $\|z\|^2 = \langle z, z \rangle$. The adjoint of a linear transform $A \in L(\mathbb{C}^d)$ is defined by the identity

$$\langle Az, w \rangle = \langle z, A^*w \rangle.$$

Let $e_1, \ldots, e_d$ denote the canonical orthonormal basis of $\mathbb{C}^d$. When representing $A = (a_{jk})_{j,k=1}^d$ and $z = z_1e_1 + \cdots + z_de_d$ as a column vector, we have

$$(Az)_j = \langle Az, e_j \rangle = \sum_{k=1}^d a_{jk}z_k,$$

whence A^* is represented by the transpose complex conjugate matrix $(\overline{a}_{kj})_{j,k=1}^d$. The linear transform A is called *self-adjoint* or *hermitian* if $A = A^*$. A linear transform $U \in L(\mathbb{C}^d)$ is called *unitary* if $UU^* = U^*U = I$, that is, U is isometric:

$$\langle Uz, Uw \rangle = \langle z, w \rangle, \quad z, w \in \mathbb{C}^d.$$

9.2.1 The Spectral Theorem

Theorem 9.1. *Let $A = A^*$ be a hermitian matrix. There exists a unitary matrix U and a diagonal matrix D with real entries, such that*

$$A = UDU^*.$$

The elements on the diagonal of D are determined by A, up to a permutation, as they coincide, multiplicity included, with the *eigenvalues* of A, that is the roots of the characteristic polynomial $\det(\lambda I - A)$. For proofs see Chapter IX in Gantmacher's monograph [15], or your favorite linear algebra textbook.

There are two other ways to look at the spectral theorem. One of them involves the *quadratic form* on $\mathbb{C}^d$:

$$q_A(z) = \langle Az, z \rangle, \quad z \in \mathbb{C}^d.$$

Note that $q_A(z)$ is a bihomogeneous polynomial of degree $(1,1)$ in the variables z, respectively, $\overline{z}$, where the latter denotes complex conjugation entry by entry. Conversely, we have the following lemma.

Lemma 9.2. *Every homogeneous polynomial $P(z, \overline{z})$ of bidegree $(1,1)$ which has real values for $z \in \mathbb{C}^d$ is of the form $q_A(z)$ for a unique self-adjoint matrix A.*

Proof. Write

$$P(z, \overline{z}) = \sum_{j,k=1}^{d} c_{jk} z_j \overline{z}_k = \langle Cz, z \rangle,$$

where C is the matrix of its coefficients. If $P(z, \overline{z}) \in \mathbb{R}$ for all $z \in \mathbb{C}^d$, we infer

$$\langle Cz, z \rangle = \langle z, Cz \rangle, \quad z \in \mathbb{C}^d.$$

But this identity can be polarized, that is,

$$\langle Cz, w \rangle = \langle z, Cw \rangle, \quad z, w \in \mathbb{C}^d,$$

which implies $C = C^*$. This operation also implies the uniqueness of the matrix C.

To explain the polarization operation it is sufficient to contemplate the identity

$$4\langle u, v \rangle = \langle u + v, u + v \rangle - \langle u - v, u - v \rangle + i\langle u + iv, u + iv \rangle - i\langle u - iv, u - iv \rangle,$$

where $u, v \in \mathbb{C}^d$ and $i = \sqrt{-1}$. $\square$

9.2.2 The Law of Inertia

The spectral theorem asserts that the quadratic form $q_A(z)$ can be written as a weighted sum of squares of complex linear forms:

$$q_A(z) = \sum_{j=1}^{d} \lambda_j |w_j|^2,$$

where

$$w_j = \sum_{k=1}^{d} u_{jk} z_k, \quad 1 \le j \le d,$$

is a new orthonormal system of coordinates in $\mathbb{C}^d$ and λ_j are the eigenvalues of A.

Now look at the level set

$$E = \{z \in \mathbb{C}^d;\ q_A(z) = 1\}.$$

In the new system of coordinates E has the equation

$$\lambda_1 |w_1|^2 + \lambda_2 |w_2|^2 + \cdots + \lambda_d |w_d|^2 = 1.$$

Thus the reciprocals of the eigenvalues, when nonzero, represent the semiaxes of this real quadratic hypersurface E in $\mathbb{R}^{2d} = \mathbb{C}^d$. The reader is invited to question what happens with E if one eigenvalue is zero.

In short, the quadratic form q_A can be written as

$$q_A(z) = \sum_{j=1}^{n} |P_j(z)|^2 - \sum_{j=n+1}^{r} |P_j(z)|^2, \tag{9.1}$$

where $P_j(z)$ are linear, homogeneous polynomials. Is this decomposition unique, or are at least the number of positive, respectively, negative squares unique? The answer to these important questions was given a long time ago by Jacobi and Sylvester. First observe that we should avoid obvious cancellations, such as $0 = |P(z)|^2 - |P(z)|^2$, or denoting by ζ a single complex variable $0 = |\zeta+1|^2 + |\zeta-1|^2 - |\sqrt{2}\zeta|^2 - |\sqrt{2}|^2$.

Theorem 9.3. *Let $q_A(z)$ be a hermitian form on $\mathbb{C}^d$. In any decomposition* (9.1) *with linearly independent complex linear forms $P_1, \ldots, P_r$, the number of positive or negative squares (n, respectively, $r - n$) is independent of the decomposition.*

For a proof and two classical methods (going back to Lagrange and Jacobi) of how to compute effectively the sums of hermitian squares decompositions, see Chapter X in [15]. To understand the intrinsic character of these numbers, simply note that r is the rank of the matrix A, while n is the maximal dimension of a vector subspace V of $\mathbb{C}^d$ on which $q_A(z), z \in V, z \neq 0$, has only positive values. The difference $n - (r - n)$ or sometimes the pair $(n, r - n)$ is called the *signature* of the hermitian form $q_A(z)$.

The quadratic form $q_A(z)$ is called *positive semidefinite*, respectively, *positive definite*, if $q_A(z) \ge 0$ for all z, respectively, $q_A(z) > 0$ for $z \neq 0$, in other terms the eigenvalues of the hermitian matrix A are nonnegative, respectively, positive. The terminology carries over to the matrix A.

9.2.3 Min-max Principle

Let $A = A^*$ be a hermitian matrix with associated quadratic form q_A. Since the eigenvalues of A are real, we can arrange them in decreasing order:

$$\lambda_1(A) \geq \lambda_2(A) \geq \cdots \geq \lambda_d(A).$$

The spectral decomposition and the interpretation of these numbers as reciprocals of the principal axes of the quadric $q_A(z) = 1$ lead to the following important variational principle, stated as below by Courant and Fischer; see, for instance, [22, Section 4.2].

Theorem 9.4. *The eigenvalues of the hermitian matrix A satisfy*

$$\lambda_k(A) = \min_{\dim V = d-k+1} \max_{z \in V \setminus \{0\}} \frac{q_A(z)}{\|z\|^2}, \quad 1 \leq k \leq d.$$

For this, and other, reasons, the numbers $\lambda_k(A)$ are also known as the *characteristic values* of the form $q_A(z)$.

9.2.4 Exercises

Exercise 9.5. A matrix A is called *symmetric* if it coincides with its transpose: $A = A^T$. Does the spectral theorem hold true for symmetric matrices over an arbitrary field? What about symmetric matrices over a real closed field?

Exercise 9.6. Let $A = A^* \in M_d(\mathbb{C})$ be a hermitian matrix and let $V \subset \mathbb{C}^d$ be a vector subspace of dimension $d - 1$. Prove, using the min-max principle, that the restriction to V of the quadratic form q_A has characteristic values interlaced with those of q_A.

Exercise 9.7. Let q_A, q_B be two hermitian forms. Try to define the relative characteristic numbers of A with respect to B via the Rayleigh quotient q_A/q_B. Relate these values to the zeros of the determinant of the linear pencil of matrices $A - \lambda B$.

9.3 Positive Definite Kernels

9.3.1 Hilbert Space Factorization

Let X be a set and let $K : X \times X \longrightarrow \mathbb{C}$ be a map. We call K a *positive semidefinite kernel* if, for every finite subset $I \subset X$, the matrix $(K(i,j))_{i,j \in I}$ is hermitian and positive semidefinite, or equivalently

$$K(i,j) = \overline{K(j,i)}$$

and

$$\sum_{i,j \in I} K(i,j) c_i \bar{c}_j \geq 0$$

for all complex numbers $c_i \in \mathbb{C}, i \in I$. The kernel K is *positive definite* if the matrix $(K(i,j))_{i,j \in I}$ is (strictly) positive definite for all finite subsets $I \subset X$. The following result, going back at least one century to Mercer and rediscovered by Aronsajn, respectively, Kolmogorov, gives a set theoretic analogue of the sums of squares decomposition of a Hermitian form; see, for instance, [24].

Theorem 9.8. *Let $K : X \times X \longrightarrow \mathbb{C}$ be a positive semidefinite kernel. Then there exists a complex Hilbert space H and a map $F : X \longrightarrow H$ such that*

$$K(x,y) = \langle F(x), F(y) \rangle, \quad x, y \in X.$$

Proof. Although tautological in its nature, the proof of this factorization theorem is quite important for its wide range of applications. We construct the Hilbert space as follows: let $\mathcal{F}(X)$ denote the set of all finitely supported functions $f, g : X \longrightarrow \mathbb{C}$, and define the inner product

$$\langle f, g \rangle = \sum_{x \in X} K(x,y) f(x) \overline{g(y)}.$$

Denote the vectors of zero norm by $N = \{f \in \mathcal{F}(X);\ \langle f, f \rangle = 0\}$. Note that by the classical Cauchy–Schwarz inequality we infer

$$|\langle f, g \rangle|^2 \leq \langle f, f \rangle \langle g, g \rangle.$$

Thus N is a vector subspace of $\mathcal{F}(X)$ and the quotient $\mathcal{F}(X)/N$ carries a non-degenerate inner product induced on equivalence classes by $\langle f, g \rangle$. The Hilbert space completion H then contains $\mathcal{F}(X)/N$ as a dense subspace and the map $F : X \longrightarrow H$ defined by the class of characteristic function $F(x)(y) = 0$ if $x \neq y$ and $F(x)(x) = 1$ induces then the factorization in the statement. $\square$

Note that in general the Hilbert space constructed in the proof is nonseparable. A uniqueness of the factorization can be immediately derived from the same proof.

Corollary 9.9. *Assume that the positive semidefinite kernel $L : X \times X \longrightarrow \mathbb{C}$ admits two factorizations $L(x,y) = \langle F(x), F(y) \rangle_H = \langle G(x), G(y) \rangle_K$, where H, K are Hilbert spaces and the maps $F : X \longrightarrow H, G : X \longrightarrow K$ both have dense ranges. Then there exists a unitary transformation $U : H \longrightarrow K$ with the property $U \circ F = G$.*

9.3.2 Positivity and Analyticity

In practice the positive definite kernel K satisfies some smoothness conditions on an appropriate supporting set X, and consequently the factorization takes place in a separable Hilbert space. We state only one possible result in this direction.

Proposition 9.10. *Let $\Omega \subset \mathbb{C}^d$ be an open set and let $K : \Omega \times \Omega \longrightarrow \mathbb{C}$ be a positive semidefinite kernel which is analytic in the first variable and antianalytic in the second. Then there exists a separable, complex Hilbert space H and an analytic map $F : \Omega \longrightarrow H$, such that*

$$K(z, w) = \langle F(z), F(w) \rangle, \qquad z, w \in \Omega.$$

Proof. By its very construction, the factorization proved in Theorem 9.8 has the property that the scalar function $z \mapsto \langle F(z), y \rangle$ is analytic for every vector y belonging to a dense subspace of H. By taking limits of sequences of the form $\langle F(z), y_n \rangle$ we find that the map $z \mapsto \langle F(z), u \rangle$ is analytic for *every* $u \in H$. Hence $F(z)$ is analytic, due to the equivalence between weak and strong analyticity of Hilbert space valued maps; see, for instance, [30].

To prove that the space H is separable, simply note that the vectors $F(\zeta)$, $\zeta \in G$, span H as soon as the countable set G is everywhere dense in Ω. □

When expanding in a Taylor, or Fourier, series we will encounter later the natural question of whether the matrix of coefficients of a kernel reflects its positivity as a map, as defined at the beginning of this section. For instance, take Ω to be a polydisk (that is, a product of disks) in $\mathbb{C}^d$ centered at $z = 0$, and assume that the map $K : \Omega \times \Omega \longrightarrow \mathbb{C}$ is analytic/antianalytic. Then a power series expansion

$$K(z, w) = \sum_{\alpha, \beta \in \mathbb{N}^d} c_{\alpha,\beta} z^\alpha \overline{w}^\beta \tag{9.2}$$

is convergent in $\Omega \times \Omega$. Under these conditions, we note the following simple but essential observation:

Proposition 9.11. *The kernel* (9.2) *is positive semidefinite in a polydisk of convergence if and only if the infinite matrix $(c_{\alpha,\beta})_{\alpha,\beta \in \mathbb{N}^d}$ is positive semidefinite.*

Proof. Remark that, for $\epsilon > 0$ sufficiently small,

$$c_{\alpha,\beta} = \int \cdots \int_{|z_j| = |w_k| = \epsilon} K(z, w) z^{-\alpha} \overline{w}^{-\beta} \prod_{j=1}^{d} \left(\frac{dz_j}{2\pi i z_j} \frac{dw_j}{2\pi i w_j} \right).$$

A Riemann sum approximation of the integral proves then that $(c_{\alpha,\beta})_{\alpha,\beta \in \mathbb{N}^d}$ is a positive semidefinite discrete kernel. Conversely, assuming that $(c_{\alpha,\beta})_{\alpha,\beta \in \mathbb{N}^d}$ is positive semidefninite, the convergence of the power series expansion implies the positivity of K. □

Exactly as in the case of hermitian forms, an analytic/antianalytic kernel $K(z, w)$ is determined by its values on the diagonal $K(z, z)$.

9.3.3 Hadamard's Product

Besides the natural operations which preserve positivity of kernels, their pointwise product stands out:

Theorem 9.12 (Schur). *Let $K_j : X \times X \longrightarrow \mathbb{C}$, $j = 1, 2$, be two positive semidefinite kernels. Then $K(x, y) = K_1(x, y)K_2(x, y)$, $x, y \in X$, is also positive semidefinite.*

For the proof see [15].

To give a single, illustrative application of Schur's theorem, consider an open set $\Omega \subset \mathbb{C}^d$ and a positive definite kernel $K : \Omega \times \Omega \longrightarrow \mathbb{C}$ which is analytic/antianalytic in the sense discussed above. Assume that there exists a positive constant M such that $K(z, z) < M$ for all $z \in \Omega$. Then the new kernel

$$\frac{1}{M^2 - K(z,w)}, \quad z, w \in \Omega,$$

has the same properties (i.e., analyticity and positivity). Indeed, by virtue of the Cauchy–Schwarz inequality, $|K(z, w)| < M$ for all $z, w \in \Omega$. Then Neumann series decomposition and Schur's theorem lead to the desired conclusion:

$$\frac{1}{M^2 - K(z,w)} = M^{-2}\left(1 + \frac{K(z,w)}{M^2} + \frac{K(z,w)^2}{M^4} + \dots\right).$$

9.3.4 Bergman's Kernel

A classical construction in the geometry of complex varieties relies on a positive definite kernel for constructing invariants to biholomorphisms. We briefly recall the construction in the particular case of a bounded domain Ω of $\mathbb{C}^d$. Let $A^2(\Omega)$ denote the *Bergman space* of all analytic functions $f : \Omega \longrightarrow \mathbb{C}$ which are square summable with respect to the Lebesgue volume measure $d\lambda_{2d}$:

$$\|f\|^2_{2,\Omega} = \int_\Omega |f(z)|^2 d\lambda_{2d}(z) < \infty.$$

It is easy to see that $A^2(\Omega)$ is complete with respect to this norm and that the evaluation functional $f \mapsto f(a)$ is continuous for every $a \in \Omega$ (for the proof use the mean value theorem on a polydisk centered at $z = a$ and fully contained in Ω). Thus, according to Riesz's representation theorem (see [30]), there exists a unique element $k_a \in A^2(\Omega)$ which represents this functional:

$$f(a) = \langle f, k_a \rangle, \quad f \in A^2(\Omega).$$

The positive definite kernel $K_\Omega(z, w) = \langle k_w, k_z \rangle$, also known as the *Bergman kernel of the domain* Ω, consequently satisfies the reproducing property

$$f(z) = \int_\Omega K_\Omega(z,w) f(w) d\lambda_{2d}(w), \quad z \in \Omega, \ f \in A^2(\Omega).$$

Moreover, this property characterizes K_Ω.

Assume that $\Phi : \Omega_1 \longrightarrow \Omega_2$ is a biholomorphic map between bounded domains of $\mathbb{C}^d$. Then the change of variables in the above identity and the uniqueness of the reproducing kernel yield

$$K_{\Omega_2}(\Phi(z), \Phi(w)) \frac{\partial \Phi}{\partial z}(z) \overline{\frac{\partial \Phi}{\partial w}(w)} = K_{\Omega_1}(z, w),$$

where $\frac{\partial \Phi}{\partial z}(z)$ denotes the complex Jacobian. Since $K_{\Omega_1}(z, z) > 0$ for all $z \in \Omega_1$ we infer that the differential form

$$\sum_{j,k=1}^{d} \frac{\partial \log K_{\Omega}(z, z)}{\partial z_j \partial \overline{z}_k} dz_j \wedge d\overline{z}_k$$

is invariant under biholomorphic mappings. See [33, 35] for details.

To put this invariant to work, let us consider the unit ball B and the unit polydisk Δ in $\mathbb{C}^d$. A power series argument leads to the closed forms

$$K_B(z, w) = \frac{1}{|B|(1 - \langle z, w \rangle)^{d+1}}, \quad K_\Delta(z, w) = \frac{1}{|\Delta|(1 - z_1\overline{w}_1)^2 \dots (1 - z_d\overline{w}_d)^2},$$

where $|A|$ denotes the volume of the set A. One can prove via these invariants that the ball and the polydisk are not biholomorphically equivalent as soon as $d \geq 2$; see [33].

9.3.5 Exercises

Exercise 9.13. Let (X, μ) be a compact space endowed with a Borel probability measure, and let

$$T_K : L^2(X, \mu) \longrightarrow L^2(X, \mu), \quad (T_K f)(x) = \int_X K(x, y) f(y) d\mu(y),$$

be a linear bounded integral operator with kernel $K : X \times X \longrightarrow \mathbb{C}$. Relate the positive definiteness of the kernel K to the positivity of T_K:

$$\langle T_K f, f \rangle_{2,\mu} \geq 0, \quad f \in L^2(X, \mu).$$

Exercise 9.14. Let $\Omega \subset \mathbb{C}^d$ be an open set, and let $H : \Omega \times \Omega \longrightarrow \mathbb{C}$ be an analytic/antianalytic function. Prove that $H(z, z) = 0, \quad z \in \Omega$ implies $H = 0$.

Exercise 9.15. Denote by $B \subset \mathbb{C}^d$ the unit ball. For which values of the parameter $\sigma \in \mathbb{R}$ is the kernel $(1 - \langle z, w \rangle)^\sigma$ positive definite in B?

Exercise 9.16. Let $\Omega \subset \mathbb{C}^d$ be a bounded domain, and let $(h_n)_{n=0}^\infty$ denote an orthonormal basis of Bergman's space $A^2(\Omega)$. Prove that

$$K_\Omega(z, w) = \sum_{n=0}^{\infty} h_n(z) \overline{h_n(w)},$$

where the series converges uniformly on compact subsets of $\Omega \times \Omega$.

9.4 Origins of Hermitian Forms

In an inspired and undeservedly forgotten work of his early career, Hermite has developed an algebraic method for counting the number of solutions of systems of polynomial equations which are contained in a prescribed basic semialgebraic subset of $\mathbb{R}^n$ or $\mathbb{C}^n$. He was aiming at bypassing, via purely algebraic methods, Cauchy's residue integral method for counting roots of complex polynomials, analogous with the widely circulated (at that time) algebraic algorithm developed by Sturm for counting real zeros of polynomials. For this very reason Hermite introduced and studied what we call today hermitian forms. For complete mathematical details and ample historical comments see the (also forgotten) little book by Krein and Naimark [27].

We illustrate below Hermite's ideas in a couple of typical examples. For simplicity we expose Hermite's idea in two variables, the transition to a larger number of variables being straightforward. Suppose that two polynomials $P_1, P_2 \in \mathbb{R}[x, y]$ of degrees n_1 (respectively, n_2) possess exactly $n = n_1 n_2$ common roots $V(P_1, P_2) = \{(a_j, b_j),\ 1 \leq j \leq n\}$, complex or real. Fix rational real functions $\chi, \psi_1, \ldots, \psi_n$ so that χ does not vanish on $V(P_1, P_2)$,

$$\det((\psi_j(a_k, b_k))_{j,k=1}^n) \neq 0,$$

and consider the hermitian form on $\mathbb{C}^n$:

$$H(z, \overline{z}) = \sum_{j=1}^{n} \chi(a_j, b_j)|z_1\psi_1(a_j, b_j) + z_2\psi_2(a_j, b_j) + \cdots + z_n\psi_n(a_j, b_j)|^2.$$

Since the sum is symmetric in the variables (a_j, b_j) the hermitian form H depends only on the coefficients of the polynomials P_1, P_2. Denote the number of roots in different sectors as follows:

$$N_c(P_1, P_2) = \#(V(P_1, P_2) \setminus \mathbb{R}^2),$$

$$N_+(P_1, P_2) = \#(V(P_1, P_2) \cap \{(x, y) \in \mathbb{R}^2;\ \chi(x, y) > 0\}),$$

$$N_-(P_1, P_2) = \#(V(P_1, P_2) \cap \{(x, y) \in \mathbb{R}^2;\ \chi(x, y) < 0\}).$$

By the inertia theorem we infer, following Hermite, that

$$n_-(H) = N_c(P_1, P_2) + N_-(P_1, P_2), \quad n_+(H) = N_c(P_1, P_2) + N_+(P_1, P_2),$$

where $(n_-(H), n_+(H))$ is the signature of the form H. Although it is difficult in general to eliminate the variables (a_j, b_j) in the form H, counting the number of real common zeros of some given polynomials contained in a rectangle leads to an elegant closed form, as pointed out by Hermite; see [27] for details.

We specialize the above ideas to polynomials of a single complex variable. For $p \in \mathbb{C}[\zeta]$ denote $p^*(\zeta) = \overline{p(\overline{\zeta})}$, that is, the polynomial obtained from p by conjugating the coefficients. Assume $n = \deg p$ and define the complex polynomial in two variables:

$$-i\frac{p(u)p^*(v) - p^*(u)p(v)}{u - v} = \sum_{k,l=1}^{n} c_{kl} u^{k-1} v^{l-1}.$$

By definition, the coefficients satisfy the reality condition $c_{kl} = \overline{c_{lk}}$, $1 \le k, l \le n$. Let $z = (z_1, \ldots, z_n) \in \mathbb{C}^n$ and define the hermitian form

$$H_p(z, \overline{z}) = \sum_{k,l=1}^{n} c_{kl} z_k \overline{z_l}. \tag{9.3}$$

Theorem 9.17 (Hermite). *Let H_p be the hermitian form (9.3) associated with a polynomial $p \in \mathbb{C}[\zeta]$ of degree n. Denote by $n_\pm(H_p)$ the number of negative, respectively, positive, squares in the decomposition of H_p. Then the polynomial p has $n_+(H_p)$ roots in the upper half-plane $\Im\zeta > 0$, $n_-(H_p)$ roots in the lower half-plane, and $n - n_-(H_p) - n_+(H_p)$ common roots between p and p^*, that is, real roots or complex conjugated roots.*

In particular we derive from here a stability criterion widely used in mechanics and engineering (compare with the similar criteria due to Routh and Hurwitz [15]).

Corollary 9.18. *Assume that the hermitian form H_p is positive definite. Then the polynomial p has all roots contained in the upper half-plane.*

The proof of Hermite's theorem relies on a product formula (well known today in the context of Bezoutian computations). The key identity is, assuming $p = p_1 p_2$:

$$\frac{p(u)p^*(v) - p^*(u)p(v)}{u - v} =$$

$$p_2(u)p_2^*(v)\frac{p_1(u)p_1^*(v) - p_1^*(u)p_1(v)}{u - v} + p_1^*(u)p_1(v)\frac{p_2(u)p_2^*(v) - p_2^*(u)p_2(v)}{u - v}.$$

A similar product rule is inherited by the form H_p, allowing us to use the inertia theorem and induction on the degree in order to prove Hermite's theorem.

9.4.1 Root Separation in the Unit Disk

The specific denominator $u - v$ and form of the conjugate p^* in the Hermite theorem are related to the Schwarz reflection with respect to the boundary of the domain of root separation. In the case of the upper half-plane the reflection is $\zeta \mapsto \overline{\zeta}$. When repeating the procedure for the unit disk, with Schwarz reflection $\zeta \mapsto \overline{\zeta}^{-1}$ one arrives at a similar conclusion. The computations were detailed by Schur (and independently by several other authors); see [27]. Specifically, let $p \in \mathbb{C}[\zeta]$ be a polynomial of degree n and define $p^\sharp(\zeta) = \zeta^n p^*(\zeta^{-1})$ as the polynomial with conjugated coefficients, arranged in reversed order. Consider the bivariate polynomial

$$\frac{p^\sharp(u)p^{\sharp *}(v) - p(u)p^*(v)}{1 - uv} = \sum_{k,l=1}^{n} a_{kl} u^{k-1} v^{l-1}.$$

Let S_p be the hermitian form with coefficients (a_{kl}). In complete analogy with Hermite's theorem we state the following well-known result.

Theorem 9.19 (Schur). *Let $p \in \mathbb{C}[\zeta]$ be a complex polynomial of degree n with associated form S_p and signature $n_\pm(S_p)$. Then p has $n_+(S_p)$ roots in the open unit disk, $n_-(S_p)$ roots in the complement of the closed unit disk, and $n-n_+(S_p)-n_-(S_p)$ roots of modulus one, or conjugated with respect to the unit circle.*

Again, a criterion for all roots to be in the unit disk is that the form S_p is positive definite. To complete the picture we remark that in the case

$$p(\zeta) = \prod_{j=1}^{n}(\zeta - a_j), \quad p^\sharp(\zeta) = \prod_{j=1}^{n}(1 - \overline{a_j}\zeta),$$

we obtain a *finite Blaschke* product as quotient,

$$m(\zeta) = \frac{p}{p^\sharp}(\zeta) = \prod_{j=1}^{n} \frac{\zeta - a_j}{1 - \overline{a_j}\zeta}, \tag{9.4}$$

and it is a rational n-fold covering of the disk onto the disk and its boundary onto the boundary.

9.4.2 Eigenvalue Separation

A too well charted and traveled area of control theory deals with stability criteria for linear systems of differential equations. In its turn, via a Laplace transform, this heavily relies on root separation criteria as presented above. We discuss below an instance of Hermite theory as transgressed and distilled by engineers.

Let A, B be complex $n \times n$ matrices, with $A = A^*$ self-adjoint. We consider the (spectrahedral) region in the complex plane

$$G = \{z \in \mathbb{C};\ A + zB + (zB)^* \prec 0\}.$$

We assume that G is nonempty and does not coincide with the full complex plane.

Theorem 9.20. *An $n \times n$ matrix M has all its eigenvalues in the region G if and only if there exists a positive definite matrix X such that*

$$A \otimes M + B \otimes (XM) + (XM)^* \otimes B^* \prec 0.$$

The most important examples are given by the following choices: the half-plane $n = 1, A = 0, B = 1$ and the disk centered at zero, of radius r, corresponding to $n = 2, A = \begin{pmatrix} -r & 0 \\ 0 & -r \end{pmatrix}$, $B = \begin{pmatrix} 0 & 1 \\ 0 & 0 \end{pmatrix}$.

For the proof of the theorem and more details see [7].

9.4.3 Exercises

Exercise 9.21. Let $p(\zeta) = (\zeta - \alpha)(\zeta - \beta)$ be a monic polynomial of degree 2. Compute the associated forms H_p, S_p and verify Hermite and Schur theorems, respectively.

Exercise 9.22. Let $p \in \mathbb{R}[\zeta]$ and fix an angle $\theta \in (0, \pi)$. Prove that the polynomial p has all roots in the wedge $-\theta < \arg \zeta < \theta$ if and only if the matrix of coefficients of the polynomial

$$-i\frac{f(e^{i\theta}u)f(e^{-i\theta}v) - f(e^{i\theta}v)f(e^{-i\theta}u)}{u-v}$$

is positive definite.

Exercise 9.23. Find a hermitian form whose positivity certifies that a polynomial has all roots contained in a given ellipse.

Exercise 9.24. Prove the eigenvalue separation theorem in the case of a disk or a half-plane.

9.5 Schur's Algorithm

Returning to Schur's theorem discussed in the last section, notice that the Blaschke product (9.4) produces a positive semidefinite kernel

$$K(u,v) = \frac{1 - m(u)\overline{m(v)}}{1 - u\overline{v}}.$$

Indeed, we have already seen that the kernel

$$p^\sharp(u)\overline{p^\sharp(v)}K(u,v) = \frac{p^\sharp(u)p^{\sharp *}(\overline{v}) - p(u)p^*(\overline{v})}{1 - u\overline{v}}$$

is positive semidefinite (as the polynomial p has all its roots contained in the unit disk), and in addition the function $p^\sharp$ does not vanish in the disk.

It was Schur who recognized in the above positivity a characterization of all power series

$$f(z) = a_0 + a_1 z + a_2 z^2 + \cdots \tag{9.5}$$

which map the disk into the disk. By different means, the same question was studied by Carathéodory, Féjer, and Toeplitz; again see [27] for more details. We focus below on Schur's approach, as it leads to a basic algorithmic way of verifying when $f(z)$ maps the disk into the disk. We call, in short, f *a contractive analytic function* in the disk.

Assume that the analytic function (9.5) satisfies $|f(z)| \leq 1$ whenever $|z| < 1$. In particular $|a_0| = |f(0)| \leq 1$. If $|a_0| = 1$, then the function $f(z) = a_0$ is a constant by the maximum principle. Assume that $|a_0| < 1$. Then a Möbius transform applied to f yields a new function from the disk to its closure, which in addition vanishes at $z = 0$, whence

$$zf_1(z) = \frac{f(z) - \gamma_0}{1 - \overline{\gamma_0}f(z)},$$

where $\gamma_0 = f(0)$ by definition. By virtue of Schwarz lemma, the factor $f_1(z)$ satisfies $|f_1(z)| \leq 1$ for all $|z| < 1$. By inverting the transform we find

$$f(z) = \frac{zf_1(z) + \gamma_0}{1 + \overline{\gamma_0}zf_1(z)}.$$

Let $\gamma_1 = f_1(0)$ and continue by induction. If $|\gamma_1| = 1$, stop: $f_1(z) = \gamma_1$ is a constant. If $|\gamma_1| < 1$, continue and define successively

$$f_{k-1}(z) = \frac{zf_k(z) + \gamma_{k-1}}{1 + \overline{\gamma_{k-1}} z f_k(z)}. \tag{9.6}$$

In this way we have associated with the finite section of the sequence of coefficients of $f(z)$ another sequence of the same length, called *the Schur parameters*:

$$(a_0, \ldots, a_n) \mapsto (\gamma_0, \gamma_1, \ldots, \gamma_n).$$

The transformation is real analytic, as one can easily prove by induction. The main result is:

Theorem 9.25 (Schur). *Let n be a positive integer and let $a_0, a_1, \ldots, a_n$ be complex numbers. There exists a power series*

$$f(z) = a_0 + a_1 z + a_2 z^2 + \cdots + a_n z^n + O(z^{n+1})$$

mapping the open disk into the closed disk if and only if the Schur parameters $\gamma_0, \gamma_1, \ldots, \gamma_n$ are of modulus less than or equal to one. If k is the first index with $|\gamma_k| = 1$, then there exists only one continuation of $a_0 + a_1 z + \cdots + a_k z^k$ into such a function f, and this is a Blaschke product of degree k.

Moreover, the recursion formula (9.6) labels all possible extensions of the polynomial $a_0 + a_1 z + a_2 z^2 + \cdots + a_n z^n$ to a contractive function as in the statement. The recursion formula and the representation of the function $f(z)$ by a chain of simple multiplication and division operations is a perfect analogue of the continued fraction algorithm in number theory.

One step further, Schur made the connection with the counting zeros form, by proving that $a_0 + a_1 z + a_2 z^2 + \cdots + a_n z^n$ can be continued to a function $f(z)$ which maps the disk into its closure if and only if the Toeplitz matrix

$$T = \begin{pmatrix} a_0 & a_1 & \ldots & a_n \\ 0 & a_0 & \ldots & a_{n-1} \\ \vdots & & \ddots & \vdots \\ 0 & 0 & \ldots & a_0 \end{pmatrix}$$

is contractive. In order to prove this fact we start with the observation that Schur's algorithm as presented above implies that every analytic function $f(z)$ mapping the disk into the disk can be uniformly approximated on compact subsets of the open disk by Blaschke products. Consequently, the kernel

$$\frac{1 - f(z)\overline{f(w)}}{1 - z\overline{w}}, \quad |z|, |w| < 1,$$

is positive semidefinite. We can dilate the argument of the function to $f(rz), r < 1$, and assume that f is analytic in a neighborhood of the closed unit disk.

Let $p(z) = p_0 + p_1 z + \cdots + p_n z^n$ be a polynomial of degree less than or equal to n. Computing by Cauchy's formula,

$$0 \leq \int_{|z|=1} \int_{|w|=1} \frac{1 - f(z)\overline{f(w)}}{1 - z\overline{w}} \overline{p(z)} p(w) \frac{dz}{2\pi i z} \frac{dw}{2\pi i w}$$

$$= \int_{|z|=1} |p(z)|^2 \frac{dz}{2\pi i z}$$

$$- \int_{|z|=1} \int_{|w|=1} (a_0 + a_1 z + \cdots + a_n z^n)(\overline{p_0} + \overline{p_1} z^{-1} + \cdots + \overline{p_n} z^{-n})$$

$$\cdot (\overline{a_0} + \overline{a_1} w^{-1} + \cdots + \overline{a_n} w^{-n})(p_0 + p_1 w + \cdots + p_n w^n)(1 + z\overline{w} + \cdots + z^n \overline{w}^n) \frac{dz}{2\pi i z} \frac{dw}{2\pi i w}$$

$$= \|v\|^2 - \|Tv\|^2,$$

where $v = (\overline{p_0}, \overline{p_1}, \ldots, \overline{p_n}) \in \mathbb{C}^n$ and T is the above Toeplitz matrix.

The reader can consult the monograph [13] for further details and many unexpected applications of the Schur parameters.

9.5.1 Exercises

Exercise 9.26. Prove that the only power series (9.5) associated with an extremal Schur parameter $|\gamma_n| = 1$ is a Blaschke product of degree n.

Exercise 9.27. Find all continuations to a contractive power series of a degree 2 polynomial $a_0 + a_1 z + a_2 z^2$. Describe explicitly the conditions on the coefficients a_0, a_1, a_2 that such a continuation exists, and if so, that it is unique.

Exercise 9.28. Let $f(z)$ be an analytic function mapping the disk into the disk. Prove that f can be approximated uniformly on compact subsets of the open disk by Blaschke products.

9.6 Riesz–Herglotz Theorem

The structure of contractive analytic functions in the disk revealed in the previous section can be related by a linear fractional transform to that of nonnegative harmonic functions in the disk. We briefly describe this new point of view.

Let $h(z), |h(z)| \leq 1$, be an analytic function in the disk $|z| < 1$. Leaving the case of a constant.function aside, we can assume that $|h(z)| < 1$ in the disk, and define the function $f(z) = \frac{1+h(z)}{1-h(z)}$, so that $\Re f(z) \geq 0$ for all $|z| < 1$. Let $f_r(z) = f(rz)$, $0 < r < 1$, so that the functions f_r are defined in a neighborhood of the closed disk and $\lim_{r \to 1} f_r = f$ uniformly on compact subsets of $\mathbb{D} = \{z \in \mathbb{C};\ |z| < 1\}$. A direct application of Cauchy's formula yields

$$f_r(w) = \int_{-\pi}^{\pi} \frac{e^{i\theta} + w}{e^{i\theta} - w} \frac{\Re f_r(e^{i\theta}) d\theta}{2\pi} + i\Im f(0).$$

Remark that the measures $\delta\mu_r = \frac{\Re f_r(e^{i\theta})d\theta}{2\pi}$ are nonnegative and of uniform mass equal to $\Re f(0)$; hence they form a compact set in the weak-$*$ topology of measures on the unit torus. By passing to a limit point we obtain a positive measure μ with the property

$$f(w) = \int_{-\pi}^{\pi} \frac{e^{i\theta}+w}{e^{i\theta}-w} d\mu(\theta) + i\Im f(0). \tag{9.7}$$

Since the trigonometric polynomials are dense in the space of continuous functions on the torus, we infer that the measure μ is unique with the above property.

Formula (9.7) is known as the *Riesz–Herglotz representation* of the nonnegative harmonic functions in the disk. Since $\mathbb{D}$ is simply connected, for any harmonic function $u : \mathbb{D} \longrightarrow \mathbb{R}$ there exists an analytic function $f : \mathbb{D} \longrightarrow \mathbb{C}$ such that $u = \Re f$. Putting together these observations we have proved the equivalence between the first two statements in the next theorem.

Theorem 9.29. *Let $f : \mathbb{D} \longrightarrow \mathbb{C}$ be an analytic function. The following assertions are equivalent:*

(a) $\Re f \geq 0$;

(b) *there exists a positive measure μ on $\partial\mathbb{D}$, such that* (9.7) *holds;*

(c) *the kernel $\frac{f(z)+\overline{f(w)}}{1-z\overline{w}}$ is positive semidefinite on $\mathbb{D}\times\mathbb{D}$.*

Proof. (a) $\Rightarrow$ (b) was proved before. If (b) holds true, then

$$\frac{f(z)+\overline{f(w)}}{1-z\overline{w}} = 2\int_{-\pi}^{\pi} \frac{d\mu(\theta)}{(e^{i\theta}-z)(e^{-i\theta}-\overline{w})},$$

whence (c) is true. Finally, (c) $\Rightarrow$ (a) because a positive semidefinite kernel has nonnegative values on the diagonal. $\square$

The above positivity result has a classical counterpart in the case f is a nonnegative polynomial on the boundary of the disk. Specifically, we have the following Riesz–Fejér theorem.

Theorem 9.30. *Let $p(z,\overline{z})$ be a polynomial with complex coefficients which is nonnegative on the unit torus $\mathbb{T}$. Then there exists a polynomial $q(z) \in \mathbb{C}[z]$ with the property*

$$p(z,\overline{z}) = |q(z)|^2, \quad z \in \mathbb{T}.$$

Proof. For $z \in \mathbb{T}$ write $z = e^{i\theta}$ and decompose

$$p(e^{i\theta}, e^{-i\theta}) = \sum_{-d}^{d} c_j e^{ij\theta}.$$

The assumption $p(e^{i\theta}, e^{-i\theta}) \geq 0$ for all $\theta \in [0, 2\pi]$ implies $c_{-j} = \overline{c_j}, \ \ 0 \leq j \leq d$. Consider the Laurent series $P(z) = \sum_{-d}^{d} c_j z^j$ and note that $P(z) = \overline{P(1/\overline{z})}$, first for $z \in \mathbb{T}$ and then for all $z \in \mathbb{C}$, by analytic continuation. Thus, the zeros and poles of the rational function P are symmetric with respect to the torus, whence

$$z^d P(z) = cz^\nu \prod_j (z - \lambda_j)^2 \prod_k (z - \mu_k)\left(z - \frac{1}{\overline{\mu_k}}\right),$$

where $c \neq 0$ is a constant, $|\lambda_j| = 1$, and $0 < |\mu_k| < 1$. By returning to the parametrization of the torus $z = e^{i\theta}$ we infer

$$p(e^{i\theta}, e^{-i\theta}) = |p(e^{i\theta}, e^{-i\theta})| = |e^{id\theta} P(e^{i\theta})|^2 = |c| \prod_j |e^{i\theta} - \lambda_j|^2 \prod_k \frac{|e^{i\theta} - \mu_k|^2}{|\mu_k|^2}. \qquad \square$$

9.6.1 Bounded Analytic Optimization

To remain in the spirit of this volume, and returning to Schur's theorem and the Riesz–Herglotz integral representation, we are in the position of stating the following direct optimization corollary of our computations.

Proposition 9.31. *Let $h(z)$ be a bounded analytic function in the disk. Then*

$$M_* = \|h\|_{\infty,\mathbb{D}} = \sup_{z\in\mathbb{D}} |h(z)|$$

is the smallest nonnegative number M with the property that the kernel

$$\frac{M^2 - h(z)\overline{h(w)}}{1 - z\overline{w}}$$

is positive semidefinite.

For the proof we simply substitute $f(z) = \frac{M+h(z)}{M-h(z)}$ in Theorem 9.29 and remark that

$$f(z) + \overline{f(w)} = 2\frac{M^2 - h(z)\overline{h(w)}}{(M - h(z))(M - \overline{h(w)})}.$$

9.6.2 Hilbert Space Realizations

Theorem 9.29 has a fourth equivalent statement which brings into focus very naturally Hilbert space representations of all bounded analytic functions in the disk. We start with the positive kernel appearing in Proposition 9.31. According to Proposition 9.10 there exists a Hilbert space H and an analytic function $F : \mathbb{D} \longrightarrow H$, such that

$$\frac{M^2 - h(z)\overline{h(w)}}{1 - z\overline{w}} = \langle F(z), F(w)\rangle, \quad z, w \in \mathbb{D},$$

or equivalently

$$M^2 + \langle zF(z), wF(w)\rangle = h(z)\overline{h(w)} + \langle F(z), F(w)\rangle.$$

Passing to the Hilbert space $\mathbb{C} \oplus H$, the latter identity becomes

$$\left\langle \begin{pmatrix} M \\ zF(z) \end{pmatrix}, \begin{pmatrix} M \\ wF(w) \end{pmatrix} \right\rangle = \left\langle \begin{pmatrix} h(z) \\ F(z) \end{pmatrix}, \begin{pmatrix} h(w) \\ F(w) \end{pmatrix} \right\rangle.$$

In other terms we have two Hilbert space factorizations of the same positive definite kernel. According to the uniqueness stated in Corollary 9.9, there exists a unitary map V between the linear span of the vectors $\binom{M}{zF(z)}$ into the linear span of the vectors $\binom{h(z)}{F(z)}$ such that $V\binom{M}{zF(z)} = \binom{h(z)}{F(z)}$. Extend V to the whole space $\mathbb{C} \oplus H$ to a linear contractive map

$$\hat{V} = \begin{pmatrix} d & \langle \cdot, b \rangle \\ c & A \end{pmatrix},$$

where $d \in \mathbb{C},\ b, c \in H,\ A : H \longrightarrow H$. In particular $|d| \leq 1, \|A\| \leq 1$. Therefore

$$dM + \langle zF(z), b \rangle = h(z), \quad Mc + zAF(z) = F(z)$$

for all $z \in \mathbb{D}$. By eliminating $F(z)$ we obtain

$$h(z) = M[d + z\langle (I - zA)^{-1}c, b \rangle], \quad z \in \mathbb{D}.$$

We have thus proved half of the following Hilbert space realization theorem, well known for its wide applications in control theory.

Theorem 9.32. *An analytic function h maps the disk into the disk if and only if it can be written as $h(z) = d + z\langle (I - zA)^{-1}c, b \rangle$, where*

$$\begin{pmatrix} d & \langle \cdot, b \rangle \\ c & A \end{pmatrix} : \begin{matrix} \mathbb{C} \\ \oplus \\ H \end{matrix} \longrightarrow \begin{matrix} \mathbb{C} \\ \oplus \\ H \end{matrix}$$

is a contractive linear operator and H is an auxiliary Hilbert space.

Proof. In order to prove the sufficiency of the above representation, define $F(z) = (I - zA)^{-1}c$, and remark that, due to the contractivity of the 2×2 block matrix

$$\left\| \begin{pmatrix} 1 \\ zF(z) \end{pmatrix} \right\| \geq \left\| \begin{pmatrix} h(z) \\ F(z) \end{pmatrix} \right\|, \quad z \in \mathbb{D}.$$

Thus

$$1 + |z|^2 \|F(z)\|^2 \geq |h(z)|^2 + \|F(z)\|^2 \geq |h(z)|^2 + |z|^2 \|F(z)\|^2,$$

which implies $|h(z)| \leq 1$ for all $z \in \mathbb{D}$. $\square$

Remember that originally V was a unitary map between two subspaces of $\mathbb{C} \oplus H$ and $\hat{V}$ was a linear contractive extension of it. With a little extra care one can adapt the construction of the realization so that $\hat{V}$ is unitary. Again, the monograph [13] is an invaluable source of information on these topics.

9.6.3 Exercises

Exercise 9.33. Assume that the analytic function f maps the disk into the right half-plane. Under which conditions has the kernel $\frac{f(z)+\overline{f(w)}}{1-z\overline{w}}$ finite rank?

Exercise 9.34. The set $\mathcal{H}$ of all functions f satisfying the conditions in Theorem 9.29 is a closed convex cone, as a subset of $\mathcal{O}(\mathbb{D})$, the Fréchet space of all analytic functions in the disk. Find the extremal rays of $\mathcal{H}$.

Exercise 9.35. Let Ω be a simply connected domain and let $\phi : \mathbb{D} \longrightarrow \Omega$ be a conformal mapping. Let $f \in \mathcal{O}(\Omega)$ be bounded. Find $\|f\|_{\infty,\Omega}$ via a positive definite optimization of a hermitian form involving f and ϕ.

Exercise 9.36. Derive a Hilbert space realization of all analytic functions mapping the disk into the right half-plane.

Exercise 9.37. [17] Let $A \in M_d(\mathbb{C})$ be a matrix with cyclic vector ξ and minimal monic polynomial $P_d(z)$. Prove that there are polynomials $P_k(z)$, of exact degree $\deg P_k = k, 0 \leq k < d$, such that

$$|P(z)|^2\|(A-z)^{-1}\xi\|^2 = |P_{d-1}(z)|^2 + |P_{d-2}(z)|^2 + \cdots + |P_0(z)|^2.$$

Conversely, every sum of hermitian squares of polynomials in exact decreasing order comes as above from a cyclic matrix.

9.7 von Neumann's Inequality

We are ready at this point to prove a new inequality involving functions and operators.

Theorem 9.38 (von Neumann). *Let $T : H \longrightarrow H$ be a linear contractive ($\|T\| \leq 1$) operator acting on a complex Hilbert space H and let f be an analytic function defined in a neighborhood of the closed unit disk. Then*

$$\|f(T)\| \leq \|f\|_{\infty,\mathbb{D}}. \tag{9.8}$$

By $f(T)$ we mean the analytic functional calculus obtained by replacing the variables in the power series expansion of f by the operator T.

Proof. The statement is equivalent to

$$(\text{for all } z \in \mathbb{D}, \Re f(z) \geq 0) \Rightarrow \Re f(T) \geq 0.$$

Indeed, if a linear operator S satisfies $\|S\| < 1$, then $\Re[(I+S)(I-S)^{-1}] > 0$, and vice versa, due to the identity

$$(I+S)(I-S)^{-1} + (I-S^*)^{-1}(I+S^*) = 2(I-S^*)^{-1}(I-S^*S)(I-S)^{-1}.$$

Let $f(z)$ be an analytic function with positive real part defined in a neighborhood of the closed unit disk. By the Riesz–Herglotz formula

$$f(z)+\overline{f(w)} = 2\int_{-\pi}^{\pi} \frac{1}{(e^{-i\theta}-\overline{w})}(1-\overline{w}z)\frac{1}{(e^{i\theta}-z)}d\mu(\theta),$$

where μ is a positive measure on the unit circle.

Expand everything into a series and replace z in the above identity by T and $\overline{w}$ by T^*, assuring that in the mixed terms contain T^* to the left of T. The result is

$$\Re f(T) = \int_{-\pi}^{\pi} (e^{-i\theta}-T^*)^{-1}(1-T^*T)(e^{i\theta}-T)^{-1}d\mu(\theta) \geq 0,$$

and the proof is complete. $\square$

Originally, von Neumann proved the above inequality using Schur's algorithm and a rational approximation. See [2] for further details and generalizations.

9.7.1 The Spectral Theorem

Among the many applications of von Neumann's inequality we sketch below the construction of the spectral measure of a unitary operator. The reader will easily adapt afterward the proof to the case of bounded self-adjoint operators.

Let H be a complex Hilbert space and let $U: H \longrightarrow H$ be a unitary operator, that is $U^*U = UU^* = I$. Then $zI - U$ is invertible for all $z \in \mathbb{C}, |z| \neq 1$. In other terms, the spectrum of U is contained in the unit circle $\partial\mathbb{D}$.

Let $p(z,\overline{z})$ be a polynomial satisfying $p(z,\overline{z}) \geq 0$ whenever $|z|=1$. By means of the identity $z\overline{z}=1$ along $\partial\mathbb{D}$ we can replace all mixed terms $z^m\overline{z}^n$ by a linear combination of pure terms z^k or $\overline{z}^\ell$, modulo $1-|z|^2$. Consequently we can write

$$p(z,\overline{z}) = p_1(z) + \overline{p_1(z)} + (1-|z|^2)p_2(z,\overline{z}),$$

where $p_1(z)$ is a polynomial which depends only on z. According to the von Neumann inequality we obtain

$$p(U,U^*) = 2\Re p_1(U) \geq 0,$$

since $\Re p_1(z) \geq 0$ on the circle.

Thus, the polynomial functional calculus $\phi : \mathbb{C}[z,\overline{z}] \longrightarrow L(H)$, defined by $\phi(p(z,\overline{z})) = p(U,U^*)$ is linear, multiplicative, unital, and positive:

$$(\text{for all } z \in \partial\mathbb{D}, p(z,\overline{z}) \geq 0) \Rightarrow p(U,U^*) \geq 0,$$

or equivalently,

$$\|p(U,U^*)\| \leq \|p\|_{\infty,\partial\mathbb{D}}, \quad \mathbb{C}[z,\overline{z}].$$

Since every continuous function on the circle is a uniform limit of trigonometric polynomials, we can extend ϕ by continuity to a continuous algebra homomorphism

$$\phi : C(\partial\mathbb{D}) \longrightarrow L(H), \quad \phi(z) = U,$$

which is in addition compatible with the involutions

$$\phi(\overline{f}) = \phi(f)^*, \quad f \in C(\partial\mathbb{D}),$$

and the order structures

$$f \geq 0 \ \Rightarrow \phi(f) \geq 0.$$

The positivity of the functional calculus ϕ allows a further extension defined on all bounded Borel functions on the circle, simply repeating the construction of the Lebesgue integral in this setting. The key observation being that $\lim_{n\to\infty} \phi(f_n)x$ exists for all $x \in H$ whenever f_n is a monotonic and uniformly bounded sequence of measurable functions; see, for instance, [30].

In conclusion we obtain the full spectral theorem for unitary operators.

Theorem 9.39. *Let $U \in L(H)$ be a unitary operator and denote by $\mathcal{B}(\partial\mathbb{D})$ the space of all bounded Borel measurable functions on the unit circle. There exists a positive, unital algebra homomorphism $\phi : \mathcal{B}(\partial\mathbb{D}) \longrightarrow L(H)$ with the properties*

(a) $\phi(z) = U$;

(b) $\|\phi(f)\| \leq \sup_{|z|=1} |f(z)|$;

(c) *for all $x \in H$ and every monotonic, pointwise convergent sequence $f_n \to f$ in $\mathcal{B}(\partial\mathbb{D})$, $\phi(f)x = \lim_{n\to\infty} \phi(f_n)x$.*

9.7.2 Exercises

Exercise 9.40. Let J_n be the (nilpotent) Jordan block of size $n \times n$. Prove that $\|J_n\| = 1$ and translate the matrix inequality $\|p(J_n)\| \leq \|p\|_{\infty,\partial\mathbb{D}}$ into numerical inequalities referring to an arbitrary polynomial $p(z)$.

Exercise 9.41. Let U_k, $1 \leq k \leq n$ be a finite system of commuting unitary matrices of size $d \times d$. Prove that there exists a unitary matrix U and polynomials $p_k(z)$ such that $U_k = p_k(U)$ for all $k, 1 \leq k \leq n$.

Exercise 9.42. Let $U \in L(H)$ be a unitary operator. Prove that the bicommutant $(U')'$ of U is equal to the range of the Borel functional calculus described in Theorem 9.39. The *commutant* of a set of operators $S \subset L(H)$ is $S' = \{T \in L(H);\ TX = XT,\ \ X \in S\}$.

9.8 Bounded Analytic Interpolation

One of the classical applications of the realization Theorem 9.32 has to do with the bounded analytic interpolation of discrete data in the disk. Contrary to the free polynomial interpolation, the data are in this case bound by a series of positivity conditions. The precise statement follows.

Theorem 9.43 (Nevanlinna–Pick). *Let $\{a_i\}, \{c_i\}$, $i \in I$, be subsets of $\mathbb{D}$, so that a_i does not have accumulation points, but the index set may be infinite. There* *exists an analytic function* $f : \mathbb{D} \longrightarrow \mathbb{D}$ *interpolating the data*

$$f(a_i) = c_i, \quad i \in I,$$

if and only if the kernel

$$\frac{1 - c_i\overline{c_j}}{1 - a_i\overline{a_j}}, \quad i, j \in I,$$

is positive semidefinite.

Proof. One implication follows from Theorem 9.29. In order to prove the converse, assume that the kernel in the statement is positive semidefinite. Then there exists a Hilbert space and a function $h : I \longrightarrow H$ with the property

$$\frac{1 - c_i\overline{c_j}}{1 - a_i\overline{a_j}} = \langle h(i), h(j) \rangle, \quad i, j \in I.$$

Starting from here we argue as in the proof of Theorem 9.32, namely,

$$1 + \langle a_i h(i), a_j h(j) \rangle = c_i\overline{c_j} + \langle h(i), h(j) \rangle$$

implies the existence of a contractive block matrix operator on $\mathbb{C} \oplus H$ satisfying

$$\begin{pmatrix} d & \langle \cdot, b \rangle \\ c & A \end{pmatrix} : \begin{pmatrix} 1 \\ a_i h(i) \end{pmatrix} = \begin{pmatrix} c_i \\ h(i) \end{pmatrix}.$$

From here we infer by eliminating $h(i)$:

$$c_i = d + \langle (I - a_i A)^{-1} c, b \rangle.$$

Hence the contractive analytic function

$$f(z) = d + \langle (I - zA)^{-1} c, b \rangle$$

interpolates the given data. □

A similar result is known for (higher multiplicity) *Hermite interpolation*, that is, by prescribing the values of finitely many derivatives of f at every point:

$$f^{(k)}(a_i) = c_i^{(k)}, \quad 0 \le k \le K(i), \ i \in I.$$

The reader can try to find and prove without too many additional complications the right statement.

Full details and complements on Nevanlinna–Pick interpolation can be found in [2, 13].

9.8.1 Exercises

Exercise 9.44. Prove and state a Nevanlinna–Pick theorem for matrix-valued functions.

Exercise 9.45. Let $a_1 = 0, a_2, a_3$ be three distinct points in the unit disk and choose $c_1 = 0, c_2, c_3$ also in the disk. Write the 3×3 conditions that there exists a contractive analytic function in the disk interpolating these data. Find when this function is unique.

Exercise 9.46. Prove that in the case of finitely many data (the set I in the statement is finite), there always exists a rational contractive interpolant. Estimate its degree.

9.9 Perturbations of Self-Adjoint Matrices

Riesz–Herglotz's theory on the unit circle has an obvious parallel on the line. A few details are worth a closer look, as they provide the background of perturbation theory of self-adjoint operators. We avoid below the complications related to unbounded symmetric operators or even general Hilbert space theory, focusing only on finite-dimensional computations. The reader can greatly benefit by filling these gaps by reading the relevant sections contained in the monograph by Gohberg and Krein [16].

Start with a self-adjoint matrix $A = A^* \in L(\mathbb{C}^d)$. We can arrange the eigenvalues in nondecreasing order:

$$\lambda_1(A) \leq \lambda_2(A) \leq \cdots \leq \lambda_d(A).$$

Consider a rank 1 self-adjoint operator $\xi\langle\cdot,\xi\rangle$ acting on $\mathbb{C}^d$, where $\xi \in \mathbb{C}^d$ is a vector. An immediate corollary of the min-max principle (see Exercises 9.2.4, exercise 2) shows that the perturbed matrix $B = A + \xi\langle\cdot,\xi\rangle$ has eigenvalues interlaced to those of A:

$$\lambda_1(A) \leq \lambda_1(B) \leq \lambda_2(A) \leq \cdots \leq \lambda_d(A) \leq \lambda_d(B).$$

Let $\phi = \sum_{j=1}^d \chi_{[\lambda_j(A),\lambda_j(B)]}$ be the characteristic function of the union of "spectral displacement" intervals between the two sets of eigenvalues. Then, for every $z \notin \mathbb{R}$ we obtain

$$\det[(B - zI)(A - zI)^{-1}] = \prod_{j=1}^d \frac{\lambda_j(B) - z}{\lambda_j(A) - z}$$

$$= \exp\left(\sum_{j=1}^d \log \frac{\lambda_j(B) - z}{\lambda_j(A) - z}\right) = \exp \int_{\mathbb{R}} \frac{\phi(t)dt}{t - z}.$$

On the other hand, by simply expanding the vector ξ in the orthonormal basis which diagonalizes A we find

$$\det[(B - zI)(A - zI)^{-1}] = \det[I + (A - zI)^{-1}\xi\langle\cdot,\xi\rangle]$$

$$= 1 + \langle (A - zI)^{-1}\xi, \xi\rangle = \sum_{j=1}^d \frac{c_j}{\lambda_j(A) - z},$$

where $c_j \geq 0$ for all j, $1 \leq j \leq d$.

One step further, we can put together the above observations in the form of equivalent representations of the same object.

Proposition 9.47. *The following classes are equivalent:*

(a) *rational functions $R(z) \in \mathbb{C}(z)$ satisfying $R(\infty) = 1$ and*

$$0 < \Im R(z)\Im(z) < \infty, \quad z \notin \mathbb{R};$$

(b) *finite atomic positive measures μ on the real line;*

(c) *characteristic functions $\phi(t)$ of bounded semialgebraic subsets of the real line;*

(d) *unitary equivalence classes of pairs (A, ξ), where $A = A^* \in L(\mathbb{C}^d)$ and ξ is a cyclic vector for A.*

The equivalence is given by the formulas

$$R(z) = 1 + \int_{\mathbb{R}} \frac{d\mu(t)}{t - z} = \exp \int_{\mathbb{R}} \frac{\phi(t)dt}{t - z}$$

$$= \det[(A + \xi\langle\cdot, \xi\rangle - zI)(A - zI)^{-1}] = 1 + \langle(A - zI)^{-1}\xi, \xi\rangle.$$

Proof. A bounded semialgebraic subset of the real line is simply a finite union of intervals. We prefer this fancy terminology due to higher-dimensional analogues; see [17]. Since in all formulas the operator A or its powers appear against the vector ξ, it is natural to assume that this vector is cyclic with respect to A; that is, $\xi, A\xi, \ldots, A^{d-1}\xi$ is a linear basis of $\mathbb{C}^d$.

To see that (a) $\Rightarrow$ (b) we remark that both zeros and poles of R must be real, interlaced (by the argument principle), and that the residues at every pole are positive. Then (b) $\Rightarrow$ (d) by considering the multiplier $A = M_t$ on the space $L^2(\mu)$, and (d) $\Rightarrow$ (c), (d) $\Rightarrow$ (b) by the computations preceding the statement. The implication (c) $\Rightarrow$ (b) follows by direct integration and exponentiation. Finally (d) $\Rightarrow$ (a) is a straightforward computation

$$\frac{\langle(A - zI)^{-1}\xi, \xi\rangle - \langle(A - \overline{z}I)^{-1}\xi, \xi\rangle}{z - \overline{z}} = \langle(A - zI)^{-1}(A - \overline{z}I)^{-1}\xi, \xi\rangle > 0$$

for all $z \notin \mathbb{R}$. □

To be in line with the theme of this chapter, we can add to the above equivalences the positivity (as a kernel) condition

(e) R is rational, $R(\infty) = 1$, and $\left[\frac{R(z) - \overline{R(z)}}{z - \overline{z}}\right] > 0, \quad z \notin \mathbb{R}.$

In this way the relation to the positivity theory in the disk exposed in the previous sections becomes more transparent.

The function $\phi_{A \to B} = \phi$ appearing in the statement is known as *the phase shift* or *the spectral shift* of the perturbation $A \to B = A + \xi\langle\cdot, \xi\rangle$. The name is justified by the following remarkable trace formula:

$$\text{Tr}(f(B) - f(A)) = \int_{\mathbf{R}} f'\phi_{A \to B}dt, \quad f \in \mathbb{C}[z].$$

Indeed,

$$\mathrm{Tr}(f(B) - f(A)) = \sum_{j=1}^{d}(f(\lambda_j(B)) - f(\lambda_j(A))) = \sum_{j=1}^{d}\int_{\lambda_j(A)}^{\lambda_j(B)} f'(t)dt.$$

By defining step by step via rank 1 additive perturbations to the spectral shift of a pair of self-adjoint matrices according to the rule

$$\phi_{A\to B} + \phi_{B\to C} = \phi_{A\to C},$$

we are led to the crucial observation

$$\int_{\mathbb{R}} |\phi_{A\to B}|dt \leq \mathrm{Tr}|A - B|.$$

This enables us to take limits and obtain the following well-known theorem.

Theorem 9.48 (Lifshitz–Krein). *Let $A, B \in L(H)$ be bounded self-adjoint operators acting on a complex Hilbert space H and assume that $A - B$ is trace-class. Then there exists a function $\phi \in L^1(\mathbb{R}, dt)$ with compact support, such that*

$$\mathrm{Tr}(f(B) - f(A)) = \int_{\mathbf{R}} f'\phi dt$$

for every function $f \in C_0^1(\mathbb{R})$, and

$$\int_{\mathbb{R}} |\phi|dt \leq \mathrm{Tr}|A - B|.$$

The reader can consult for details the original article [25] and the monograph [23].

9.9.1 Exercises

Exercise 9.49. Proposition 9.47 has a word-by-word counterpart for analytic functions mapping the upper half-plane into itself, with μ an arbitrary positive measure, compactly supported on the line and the function $\phi \in L^1(\mathbb{R}, dt)$ also of compact support. The analytically inclined reader will find pleasure in proving the complete equivalence between the four corresponding statements. Details can be found in [16].

9.10 Positive Forms in Several Complex Variables

Let $z \in \mathbb{C}^d$ denote the d-tuple of complex variables. We focus below on hermitian bihomogeneous forms

$$f(z, \overline{z}) = \sum_{|\alpha|=|\beta|=n} c_{\alpha\beta} z^\alpha \overline{z}^\beta,$$

where the standard multiindex notation is used: $z^\alpha = z_1^{\alpha_1} \cdots z_d^{\alpha_d}$. Note that the matrix of coefficients $(c_{\alpha\beta})$ is unambiguously determined by f. A diagonalization of this matrix yields a decomposition

$$f(z, \overline{z}) = \|F_1(z)\|^2 - \|F_2(z)\|^2,$$

where $F_j : C^d \longrightarrow \mathbb{C}^{n_j}$ are homogeneous (of degree n), vector-valued polynomial functions.

It is important to remark from the very beginning that, even if $f(z, \overline{z}) > 0$ for all $z \neq 0$, the form f may not be a sum of hermitian squares. The following simple example in two variables (z, w) singles out where the obstruction lies:

$$f_0 = |z|^4 + |w|^4 - c|zw|^2$$

is everywhere positive on $\mathbb{C}^2 \setminus \{0\}$ as soon as $c < 2$. Now define

$$f_N = (|z|^2 + |w|^2)^N (|z|^4 + |w|^4 - c|zw|^2).$$

The matrix associated with f_N is diagonal with entries containing binomial coefficients of the form

$$\binom{N}{p} + \binom{N}{p+2} - c\binom{N}{p+1}.$$

After elementary calculations, the condition that all these coefficients are positive is

$$N + 1 > \frac{2c}{2-c},$$

showing that N has to increase to infinity to ensure that f_N is a sum of squares, assuming that c tends to the value 2.

In analogy with Artin's positive solution to Hilbert's 17th problem (see [4, 28]), the following result casts well the same phenomenon in the complex domain.

Theorem 9.50 (Quillen). *Let $f(z, \overline{z})$ be a bihomogeneous form on $\mathbb{C}^d$. If $f(z, \overline{z}) > 0$ for all $z \neq 0$, then there exists a positive integer N, such that $\|z\|^{2N} f(z, \overline{z})$ is a sum of hermitian squares.*

In complex dimension one we know a much simpler factorization, offered by Riesz–Fejér Theorem 9.30. We will discuss three proofs of Quillen's theorem. The third one, which is purely algebraic, is the most accessible for the nonanalyst, but the other two have intrinsic value, as we shall see. For more details and examples the reader can consult the works of d'Angelo and his collaborators [5, 6, 9, 10, 11].

Proof 1. Quillen's original proof [31] is based on computations in the Bargmann–Fock space (a natural environment for quantum physicists). Specifically, we endow $\mathbb{C}^d$ with the Gaussian measure $dG = \pi^{-n} e^{-\|z\|^2} d\lambda_{2d}(z)$ and define $\mathcal{H}$ as the Hilbert space completion in $L^2(\mathbb{C}^d, dG)$ of the complex polynomials. One computes without difficulty

$$\langle z^\alpha, z^\beta \rangle_{\mathcal{H}} = \delta_{\alpha,\beta}\, \alpha!, \quad \alpha, \beta \in \mathbb{N}^d,$$

where $\delta_{\alpha,\beta}$ is Kronecker's symbol. Moreover, the Bargmann–Fock space carries the remarkable adjunction identity

$$\langle D^\alpha u, v\rangle_{\mathcal{H}} = \langle u, z^\alpha v\rangle_{\mathcal{H}}, \quad u, v \in \mathbb{C}[z],$$

where $D^\alpha = \frac{\partial^{|\alpha|}}{\partial z_1^{\alpha_1} \dots \partial z_d^{\alpha_d}}$. Fix a bihomogeneous form

$$f(z, \overline{z}) = \sum_{|\alpha|=|\beta|=n} a_{\alpha\beta} z^\alpha \overline{z}^\beta$$

and consider the integral operator

$$E_f : \mathbb{C}[z] \longrightarrow \mathbb{C}[z], \quad E_f(u)(z) = \int_{\mathbb{C}^d} f(z, \overline{w}) u(w) dG.$$

If $u(w) = \sum_{|\gamma|} u_\gamma w^\gamma$, then

$$E_f(u)(z) = \sum_{\alpha,\beta} a_{\alpha\beta} \beta! u_\beta z^\alpha$$

and

$$\langle E_f(u), v\rangle_{\mathcal{H}} = \sum_{\alpha,\beta} a_{\alpha\beta} \alpha! \beta! u_\alpha \overline{v_\beta},$$

where $v(z) = \sum_{|\beta|} v_\beta z^\beta$. If the matrix of coefficients $a_{\alpha\beta}$ is hermitian, then E_f turns out to be a symmetric operator acting on the space of homogeneous polynomials of degree n.

$$\langle E_f(u), v\rangle_{\mathcal{H}} = \langle u, E_f(v)\rangle_{\mathcal{H}}.$$

Note also that $E_f \succeq 0$ is a linear operator if and only if the matrix $(a_{\alpha,\beta})$ is positive semidefinite, if and only if the form f is a sum of hermitian squares.

Let N be a positive integer, and denote by $f_N(z, \overline{z}) = \frac{\|z\|^{2N}}{N!} f(z, \overline{z})$, a form of bidegree $(N + n, N + n)$. Consider two homogeneous polynomials u, v of degree $N + n$ each. Note that

$$\frac{\|z\|^{2N}}{N!} = \sum_{|\mu|=N} \frac{z^\mu \overline{z}^\mu}{\mu!}.$$

Then

$$\begin{aligned}
\langle E_{f_N} u, v\rangle_{\mathcal{H}} &= \sum_{\alpha,\beta,\mu} a_{\alpha\beta} \frac{(\alpha + \mu)!(\beta + \mu)!}{\mu!} u_{\alpha+\mu} \overline{v_{\beta+\mu}} \\
&= \sum_{\alpha,\beta} a_{\alpha\beta} \langle D^\alpha u, D^\beta v\rangle_{\mathcal{H}} \\
&= \left\langle \sum_{\alpha,\beta} a_{\alpha\beta} z^\beta D^\alpha u, v \right\rangle_{\mathcal{H}}.
\end{aligned}$$

Assume that $f(z, \overline{z}) > 0$ for all $z \neq 0$. From this point the proof becomes more technical and we merely mention the main idea: the differential operator $T_f = \sum_{|\alpha|=|\beta|} a_{\alpha\beta} z^\beta D^\alpha$ is elliptic and symmetric on the space of polynomials, with positive principal symbol equal to the form f (up to a constant). Hence, on a proper choice of Sobolev norms, T_f is Fredholm. In particular, when restricted on polynomials T_f possesses a finite number of negative eigenvalues. But T_f maps the space of homogeneous polynomials of degree $N+n$ into itself, and it coincides there with the operator with integral kernel f_N. Thus, for N sufficiently large, E_{f_N} is a positive operator; that is, the form f_N is a sum of hermitian squares.

Proof 2. The second proof of Quillen's theorem is due to Catlin and d'Angelo [5] and closely follows the same idea of compact perturbations of integral operators (only that it was published thirty years after Quillen). Start again with the (n, n)-homogeneous form $f(z, \overline{z}) = \sum_{|\alpha|=|\beta|=n} a_{\alpha\beta} z^\alpha \overline{z}^\beta$, and assume that it is positive on the unit sphere in $\mathbb{C}^d$. By homogeneity, this is the same condition as before: $f(z, \overline{z}) > 0$ for all $z \neq 0$. Let B denote the unit ball in $\mathbb{C}^d$ and consider the Bergman space $A^2(B)$ with reproducing kernel K_B (see [5, Section 3.4]). The operator

$$S_f : A^2(B) \longrightarrow A^2(B), \quad (S_f u)(z) = \int_B K_B(z, w) f(z, \overline{w}) u(w) d\lambda_{2d}(w)$$

is bounded, and it maps homogeneous polynomials of degree N into homogeneous polynomials of degree N, as one can easily see from the orthogonality of the monomials z^α, $\alpha \in \mathbb{N}^d$, and a power series expansion of the integral kernel:

$$K_B(z, w) f(z, \overline{w}) = \frac{\sum_{|\alpha|=|\beta|=n} a_{\alpha\beta} z^\alpha \overline{w}^\beta}{|B|(1 - z \cdot \overline{w})^{d+1}} = \sum_{N=0}^{\infty} c_N f_N(z, \overline{w}),$$

where $c_N > 0$ are constants and the forms $f_N(z, \overline{z}) = \frac{\|z\|^{2N}}{N!} f(z, \overline{z})$ are the same as in the previous proof.

Note that S_f is a self-adjoint operator which is a compact perturbation of the positive operator S'_f with integral kernel $K_B(z, w) f(w, \overline{w})$:

$$\begin{aligned} \langle S'_f u, u \rangle &= \int_B K_B(z, w) f(w, \overline{w}) u(w) \overline{u(z)} d\lambda_{2d}(z) d\lambda_{2d}(w) \\ &= \int_B f(w, \overline{w}) |u(w)|^2 d\lambda_{2d}(w) \geq 0. \end{aligned}$$

Hence S_f has only finitely many negative eigenvalues. Finally, we infer that for large enough N, the restriction of the operator S_f on the space of homogeneous polynomials of degree N is positive, that is, the form f_N is a sum of hermitian squares.

9.10.1 Pólya's Theorem

A well-known interplay between the complex coordinates in $\mathbb{C}^d$ and their moduli, seen as coordinates on $\mathbb{R}^d_+$, leads to a new proof of a classical theorem of Pólya referring to positive polynomials in an octant. To be more precise, let $f(x) \in \mathbb{R}[x]$ be a homogeneous polynomial in d variables $x = (x_1, \dots, x_d)$ so that the substitution $x_j = |z_j|^2$ produces a bihomogeneous form $F(z, \overline{z}) = f(|z_1|^2, \dots, |z_d|^2)$. Assume that

$$\left(|x|_1 := \sum x_j > 0\right) \Rightarrow f(x) > 0.$$

Then $F(z, \overline{z}) > 0$, $z \neq 0$, by homogeneity. Thus, in view of Theorem 9.50, there exists $N \geq 0$ such that $\|z\|^{2N} F(z, \overline{z})$ is a sum of hermitian squares. This observation can be carried back to $f(x)$ in the following form.

Assume $|x_1|^N f(x) = \sum_{|\alpha|} a_\alpha x^\alpha$. Then $\|z\|^{2N} F(z, \overline{z}) = \sum_{|\alpha|} a_\alpha z^\alpha \overline{z}^\alpha$; that is, the coefficient matrix associated with F is diagonal, and hence it has positive entries by Quillen's theorem. In conclusion we obtain the following theorem.

Theorem 9.51 (Pólya). *Let $f(x)$ be a homogeneous polynomial with real coefficients in d variables. If $f(x) > 0$ for all $x = (x_1, \dots, x_d) \in [0, \infty)^d \setminus \{0\}$, then there exists an integer $N \geq 0$ with the property that the form $(x_1 + \cdots + x_d)^N f(x)$ has positive coefficients.*

An important addition to Pólya's theorem is that one can estimate the degree N from the degree of f and its distance to zero on the standard simplex; see [28].

9.10.2 Exercises

Exercise 9.52. Prove, using the geometry of the zero set, that for every $N \geq 1$ the two complex variable form $(|z|^2 + |w|^2)^N (|z|^2 - |w|^2)^2$ is not a sum of squares of hermitian forms.

Exercise 9.53. Prove that the zero set of a sum of hermitian squares is a complex algebraic variety.

Exercise 9.54. Show that x^2 cannot be represented as a sum of hermitian squares in the variable $z = x + iy$.

9.11 Semirings of Hermitian Squares

The decomposition of a polynomial into a sum of hermitian squares can be treated with purely algebraic methods, in the spirit of the classical real algebraic geometry [4, 28]. The specific feature of the convex hull of hermitian squares is that it is closed under additions and multiplications, but it does not contain all real squares. We explain below this difference and indicate how one can include the hermitian squares into the existing theory.

To this aim we identify real affine space of even dimension $\mathbb{R}^{2d}$ with complex affine space $\mathbb{C}^d$, with coordinates $(x, y) = (x_1, \dots, x_d, y_1, \dots, y_d) \in \mathbb{R}^{2d}$ (respectively, $z = (z_1, \dots, z_d) \in \mathbb{C}^d$), so that $z_k = x_k + iy_k$ $(1 \leq k \leq d)$. The Euclidean norm is denoted as before: $\|z\|^2 = \sum_{k=1}^d |z_k|^2 = \sum_{k=1}^d (x_k^2 + y_k^2)$.

When we polarize the variables z and $\overline{z}$ we identify a real polynomial of $2d$ variables $f(x,y)$ with a complex polynomial of $2d$ variables:

$$\tilde{f}(x+iy, x-iy) = f(x,y), \quad x, y \in \mathbb{R}^d.$$

The $\mathbb{R}$-homomorphisms $\mathbb{R}[x,y] \longrightarrow \mathbb{R}$ correspond in this way either to points $(\xi,\eta) \in \mathbb{R}^{2d}$ or to couples $(\xi+i\eta, \xi-i\eta) \in \mathbb{C}^d$, with the associated evaluation map

$$(\xi,\eta) \mapsto f(\xi,\eta) = \tilde{f}(\xi+i\eta, \xi-i\eta).$$

We will carry this isomorphism throughout the section, without making it always explicit.

Let $I \subset \mathbb{R}[x,y]$ be an ideal, and let

$$X := V_{\mathbb{R}}(I) = \{\alpha = \xi + i\eta \in \mathbb{C}^d \colon \text{ for all } f \in I,\ f(\xi,\eta) = \tilde{f}(\alpha, \overline{\alpha}) = 0\}$$

be the real zero set of I. The elements of the quotient algebra $A = \mathbb{R}[x,y]/I$ can be considered as real polynomial functions on X. Let ΣA^2 denote the convex cone of sums of squares in A.

Let Σ_h denote the convex cone of sums of hermitian squares $|p(z)|^2$ in $\mathbb{R}[x,y]$, where $p \in \mathbb{C}[z]$. Clearly, Σ_h is contained in $\Sigma\mathbb{R}[x,y]^2$, and it is easy to see that this inclusion is proper. Given $A = \mathbb{R}[x,y]/I$ as above, we write $\Sigma_h A := (\Sigma_h + I)/I$ for the cone of all sums of hermitian squares restricted to X. There are nontrivial examples of ideals I for which $\Sigma_h A$ contains every function in A which is strictly positive on X. One of them is furnished by Quillen's theorem.

Proposition 9.55. *Let $p(x,y)$ be a positive polynomial on the unit sphere in $\mathbb{R}^{2d}$. Then there exists an integer $n \geq 1$ and polynomials $q_j \in \mathbb{C}[z]$, $1 \leq j \leq n$, $h \in \mathbb{R}[x,y]$, such that*

$$p(x,y) = \sum_{j=1}^{n} |q_j(z)|^2 + (1 - \|z\|^2)h(x,y).$$

Proof. Let $\tilde{p}(z,\overline{z}) = \sum_{\alpha,\beta} p_{\alpha\beta} z^\alpha \overline{z}^\beta$ be the polarization of $p(x,y)$ and assume that $\tilde{p}$ has degree n both in z and $\overline{z}$, but it is not necessarily homogeneous. One can assume that n is even by passing from $\tilde{p}$ to $\|z\|^2\tilde{p}$. Next we add a new complex variable $z_{d+1} = u + iv$, $z' = (z, z_{d+1})$ and homogenize $\tilde{p}$:

$$P(z', \overline{z'}) = \sum_{\alpha,\beta} p_{\alpha\beta} z^\alpha z_{d+1}^{n-|\alpha|} \overline{z}^\beta \overline{z}_{d+1}^{n-|\beta|}.$$

The bihomogeneous polynomial P is positive on the set $\{z' \in \mathbb{C}^{d+1};\ \|z\| = 1, |z_{d+1}| = 1\}$. (Prove!) Therefore, by homogeneity there exists a positive constant $C > 0$ with the property

$$(\|z\|^2 = |z_{d+1}|^2) \Rightarrow P(z', \overline{z'}) > C\|z'\|^{2m},$$

whence

$$P(z', \overline{z'}) + C(\|z\|^2 - |z_{d+1}|^2)^m > 0 \ \text{ for all } z' \neq 0.$$

By applying Theorem 9.50 we find a positive integer N such that $\|z'\|^{2N} P(z', \overline{z'}) \in \Sigma_h^2$.

Taking $z_{d+1} = 1$ we find polynomials $Q_k \in \mathbb{C}[z]$ such that

$$(\|z\| = 1) \Rightarrow \tilde{p}(z, \overline{z}) = \sum_k |Q_k(z)|^2,$$

and the proof is complete, noting that the ideal of the variety $\|z\|^2 = 1$ is radical. □

Let A be an $\mathbb{R}$-algebra, and let S be a subsemiring of A with $\mathbb{R}_+ \subset S$. Recall that S is said to be *Archimedean (in A)* if $\mathbb{R}+S = A$, that is, if for every $f \in A$ there exists a real number c such that $c \pm f \in S$. If A is generated by $x_1, \ldots, x_n$, then S is Archimedean if and only if there exist $c_i \in \mathbb{R}$ with $c_i \pm x_i \in S$ $(i = 1, \ldots, n)$. See [28, Definition 5.4.1], [32, references there].

Definition 9.56. *Let I be an ideal in $\mathbb{R}[x, y]$. We say that Σ_h is* Archimedean modulo I *if the semiring $\Sigma_h + I$ is Archimedean in $\mathbb{R}[x, y]$ or, equivalently, if the semiring $\overline{\Sigma}_h = (\Sigma_h + I)/I$ is Archimedean in $\mathbb{R}[x, y]/I$.*

By (a particular case of) the representation theorem [28, Theorem 5.4.4], we have the following theorem.

Theorem 9.57. *Let I be an ideal in $\mathbb{R}[x, y]$. The following conditions on I are equivalent:*

(i) *The set $V_{\mathbb{R}}(I)$ is compact and every $f \in \mathbb{R}[x, y]$ with $f > 0$ on $V_{\mathbb{R}}(I)$ lies in $\Sigma_h + I$;*

(ii) *Σ_h is Archimedean modulo I.*

(The representation theorem, in the version for semirings, asserts that (ii) implies (i). The opposite implication is obvious.)

We observe the following simple characterization of these ideals.

Proposition 9.58. *Let I be an ideal in $\mathbb{R}[x, y]$. Then Σ_h is Archimedean modulo I if and only if I contains a polynomial of the form*

$$f = c + ||z||^2 + \sum_{k=1}^{r} |q_k(z)|^2$$

with $c \in \mathbb{R}$ and $q_k(z) \in \mathbb{C}[z]$ $(k = 1, \ldots, r)$.

Proof. When Σ_h is Archimedean modulo I, there exists $c \in \mathbb{R}$ with $c - ||z||^2 \in \Sigma_h + I$, which implies the above condition. Conversely, if $-(c + ||z||^2) \in \Sigma_h + I$, then also

$$(1 - c) \pm 2x_j = |z_j \pm 1|^2 + \sum_{k \neq j} |z_k|^2 - (c + ||z||^2)$$

and

$$(1 - c) \pm 2y_j = |z_j \pm i|^2 + \sum_{k \neq j} |z_k|^2 - (c + ||z||^2)$$

lie in $\Sigma_h + I$, for $j = 1, \ldots, d$. This implies that Σ_h is Archimedean modulo I. □

This gives plenty of examples of ideals I such that every polynomial strictly positive on $V_{\mathbb{R}}(I)$ is a hermitian sum of squares modulo I. In particular we have obtained in this way an algebraic proof (the third one) and explanation of Quillen's phenomenon.

Proposition 9.59. *On a real hypersurface of $\mathbb{C}^d$ of equation*

$$\|z\|^2 + \sum_{k=1}^{r} |q_k(z)|^2 = M,$$

where $q_k \in \mathbb{C}[z]$ and $M > 0$, every positive polynomial is a sum of hermitian squares.

9.11.1 Exercises

Exercise 9.60. Let $F(z, \overline{z}) = \sum_k |q_k(z)|^2$ be a sum of hermitian squares. Prove that the polarization of F satisfies Cauchy–Schwarz inequality $|F(\alpha, \overline{\beta})|^2 \leq F(\alpha, \overline{\alpha})F(\beta, \overline{\beta})$.

Exercise 9.61. Let $P_1, \ldots, P_r \in \mathbb{C}[z]$ be polynomials in a single complex variable and let $a_1, \ldots, a_r$ be real numbers. Define the function

$$h(z, \overline{z}) = \sum_{j=1}^{r} |(z^2 - 1)P_j(z) + a_j|^2 - \sum_{j=1}^{r} a_j^2.$$

Prove that

$$h(1, 1) = h(1, -1) = h(-1, -1) = 0,$$

and deduce that Σ_h is not Archimedean modulo the ideal (h).

9.12 Multivariable Miscellanea

We comment below on a few recent advances pertaining to the theory of hermitian forms of several variables.

9.12.1 The Schur–Agler Class

Among all aspects of the theory of bounded analytic functions, results of Nevanlinna–Pick interpolation type have received by far the most attention. A pioneer on these topics is Jim Agler. His book with McCarthy [1] well illustrates the intricate nature of interpolation and realization theories in higher dimensions.

One of the starting points is the observation that an analytic function $f(z)$ is fit for Nevanlinna–Pick interpolation in the unit ball B only if the kernel

$$\frac{1 - f(z)\overline{f(w)}}{1 - \langle z, w \rangle}, \quad z, w \in B,$$

is positive semidefinite. Then an operator realization as in Theorem 9.32 holds true. These functions form the *Schur–Agler class*. Also, it is within the same class of functions that a von Neumann inequality remains valid. The main line of attack for all proofs is the interpretation of the positivity of the above kernel as the boundedness of the multiplier by the function f on the space of analytic functions in the disk with reproducing kernel $\frac{1}{1-\langle z,w\rangle}$, the so-called *Drury space* of the ball.

The polydisk in two dimensions is exceptional in this context, due to the following observation.

Theorem 9.62 (Ando). *Let (T_1, T_2) be two commuting contractive operators on a Hilbert space. Then for every polynomial $p(z_1, z_2)$, von Neumann's inequality*

$$\|p(T_1, T_2)\| \leq \|p\|_{\infty,\mathbb{D}^2}$$

is true.

A celebrated example of Varopoulos shows that such an inequality fails for three commuting contractions; see [2]. One preferred way of avoiding the complications related to the difference between the Schur–Agler class and all contractive analytic functions is to turn to functions of free, noncommuting variables.

9.12.2 Quotients of Sums of Hermitian Squares

Having as an example Artin's solution to Hilbert 17th problem, there were a few recent attempts to characterize quotients of sums of hermitian squares. The definitive result is due to Varolin [34], but before stating it we consider a few low-dimensional cases and examples.

Proposition 9.63 (d'Angelo). *A nontrivial real valued polynomial of a single complex variable $P(z, \overline{z})$ can be represented as*

$$P(z, \overline{z}) = \frac{\sum_j |p_j(z)|^2}{\sum_k |q_k(z)|^2},$$

with finitely many $p_j, q_k \in \mathbb{C}[z]$ if and only if there are complex numbers a_ℓ, positive or negative integers n_ℓ, and a polynomial $Q(z, \overline{z})$, such that

$$P(z, \overline{z}) = \prod_\ell |z - a_\ell|^{2n_\ell} Q(z, \overline{z}), \quad z \in \mathbb{C},$$

$$Q(z, \overline{z}) > 0, \quad z \in \mathbb{C},$$

and $2 \deg_z Q = \deg Q$.

The proof [9] is accessible as an exercise to the reader, with the only indication that if a quotient of sums of hermitian squares vanishes at the point $z = a$, then its Taylor series in $z - a$ and $\overline{z - a}$ has the lower degree term of the form $|z - a|^{2m}$.

A second obstruction for a polynomial $P(z,\overline{z})$ to be a quotient of sums of squares is that its zero set be finite.

Using these observations one can analyze the polynomial

$$P(z,\overline{z}) = 1 + bz^2 + \overline{b}\overline{z}^2 c|z|^2 + |z|^4$$

and obtain the following conclusions [9, 10]:

The hermitian form associated with P is positive semidefinite if and only if $c \geq 2|b| - 2$,

P is a quotient of sums of squares if and only if $c > 2|b| - 2$, or $b = 0, c > -2$, or $|b| = 1$ and $c = 0$.

The main result, proved even in a more general context than stated below (for sections of a holomorphic vector bundle on a projective manifold, see [34]) is putting the above observations in their natural higher-dimensional context.

Theorem 9.64 (Varolin). *Let $P(z,\overline{z})$ be a hermitian, bihomogeneous polynomial depending on $z \in \mathbb{C}^d$. Let $P(z,\overline{z}) = \sum_{j=1}^n |p_j(z)|^2 - \sum_{j+1}^N |p_j(z)|^2$ be a decomposition into squares, with linearly independent entries p_j. Let $V(P) = \{z \in \mathbb{C}^d;\ P(z,\overline{z}) = 0\}$ be the zero set of P.*

Then P is a quotient of square norms if and only if

$$\sup_{z \notin V(P)} \frac{\sum_{j=1}^n |p_j(z)|^2 + \sum_{j+1}^N |p_j(z)|^2}{P(z,\overline{z})} < \infty.$$

The proof uses algebraic geometry techniques and a refined estimate of the Bergman kernel of a carefully chosen metric in the ambient space.

9.12.3 Geometry of Proper Analytic Maps

Spectacular applications of sums of hermitian squares were recently obtained in the study of proper analytic maps between balls of unequal dimensions; see [5]. We confine ourselves to touch a couple of elementary aspects of this area.

The first observation can be immediately derived from Quillen's theorem.

Theorem 9.65 (Catlin–d'Angelo). *Let $P : \mathbb{C}^d \longrightarrow \mathbb{C}^n$ be a homogeneous polynomial map. If $\|P(z)\| < 1$ for $\|z\| = 1$, then there exists a polynomial map $Q : \mathbb{C}^d \longrightarrow \mathbb{C}^m$, such that $P \oplus Q$ is a proper analytic map between the unit balls of $\mathbb{C}^d$ and $\mathbb{C}^{m+n}$.*

Proof. According to Proposition 9.55 applied to the bihomogeneous polynomial $\|z\|^{2N} - \|P(z)\|^2$, there exists Q as in the statement, so that

$$\|z\|^2 = \|P(z)\|^2 + \|Q(z)\|^2.$$

Since the map $P \oplus Q$ is nonconstant, the maximum principle implies that it is proper (that is, by definition, it pulls back compact subsets of the open unit ball in $\mathbb{C}^{m+n}$ into compact subsets of the open unit ball in $\mathbb{C}^d$). □

To illustrate the complexity of the classification of proper analytic maps between balls, we reproduce below from the work of d'Angelo (see, for instance, [10]) a low-degree and low-dimensional analysis.

The main point is the following question: *given N, is there a polynomial or rational function g from $\mathbf{C}^2$ to $\mathbf{C}^N$ such that $|g(z)|^2 = 1 - |\zeta z_1 z_2|^2$ on the unit sphere?* Here is the result.

(a) *If $|\zeta|^2 \geq 4$, then for all N, the answer is no.*

(b) *If $N = 1$, then the answer is yes only when $\zeta = 0$.*

(c) *If $N = 2$, the answer is yes precisely when one of the following holds: $\zeta = 0$, $|\zeta|^2 = 1$, $|\zeta|^2 = 2$, $|\zeta|^2 = 3$.*

(d) *For each ζ with $|\zeta|^2 < 4$, there is a smallest N_ζ for which the answer is yes. The limit as $|\zeta|$ tends to 2 of N_ζ is infinity.*

Idea of proof. We are seeking a holomorphic polynomial mapping g such that

$$|g_1(z)|^2 + \cdots + |g_N(z)|^2 + |\zeta|^2 |z_1 z_2|^2 = 1$$

on the unit sphere.

The components of g and the additional term $\zeta z_1 z_2$ define a holomorphic mapping from the n ball to the $N + 1$ ball which maps the sphere to the sphere. Such a map is either constant or proper. The maximum of $|\zeta z_1 z_2|^2$ on the sphere is 1 when $|z_1|^2 = |z_2|^2 = \frac{1}{2}$. Hence $|\zeta|^2 \leq 4$ must hold if the question has a positive answer. We claim that $|\zeta|^2 = 4$ cannot hold either. Suppose $|\zeta|^2 = 4$ and g exists. Then we would have

$$|g(z)|^2 + 4|z_1|^2|z_2|^2 = 1 = (|z_1|^2 + |z_2|^2)^2$$

on the sphere, and hence

$$|g(z)|^2 = (|z_1|^2 - |z_2|^2)^2$$

on the sphere. No such g exists.

The only proper mappings from the 2-ball to itself are automorphisms and hence linear fractional transformations. Therefore the term $\zeta z_1 z_2$ can arise only if $\zeta = 0$. When $\zeta = 0$ we may of course choose $g(z)$ to be (z_1, z_2).

The next statement follows from Faran's classification of the proper holomorphic rational mappings from B_2 to B_3 [12]. We say that two maps g and h are *spherically equivalent* if there are automorphisms u, v of the domain and target balls such that $h = vgu$. If g existed, then there would be a proper polynomial mapping h from B_2 to B_3 with the monomial $\zeta z_1 z_2$ as a component. It follows from Faran's classification that h would have to be spherically equivalent to one of the four mappings

$$h(z_1, z_2) = (z_1, z_2, 0),$$

$$h(z_1, z_2) = (z_1, z_1 z_2, z_2^2),$$

$$h(z_1, z_2) = (z_1^2, \sqrt{2} z_1 z_2, z_2^2),$$

$$h(z_1, z_2) = (z_1^3, \sqrt{3} z_1 z_2, z_2^3).$$

These four mappings provide the four possible values for $|\zeta|$.

In the higher number of squares, if we allow one larger target dimension, then one can obtain a one-parameter family of maps:

$$f(z) = (z_1, z_2^2, \cos(t) z_1 z_2, \sin(t) z_1 z_2^2, \sin(t) z_1^2 z_2).$$

From this formula we see that we can recover all values of $|\zeta|$ up to unity, but not beyond.

If $N = 4$, for example, the answer is yes for $0 \leq |\zeta|^2 \leq 2$ and the following additional values for $|\zeta|^2$:

$$\frac{7}{2}, \frac{10}{3}, \frac{8}{3}, \frac{5}{2}.$$

Explicit maps where the constants $\sqrt{\frac{7}{2}}$ and $\sqrt{\frac{10}{3}}$ arise as coefficients of $z_1 z_2$ are

$$f(z) = \left(z_1^7, z_2^7, \sqrt{\frac{7}{2}} z_1 z_2, \sqrt{\frac{7}{2}} z_1^5 z_2, \sqrt{\frac{7}{2}} z_1 z_2^5 \right),$$

$$f(z) = \left(z_1^5, z_2^5, \sqrt{\frac{10}{3}} z_1 z_2, \sqrt{\frac{5}{3}} z_1^4 z_2 \sqrt{\frac{5}{3}} z_1 z_2^4 \right).$$

9.12.4 Exercises

Exercise 9.66. Find the Hilbert space realization of functions in the Schur–Agler class.

Exercise 9.67. Show that the polynomial $(|zw|^2 - |u|^2)^2 + |z|^8$ is not a quotient of sums of squares.

Exercise 9.68. The polynomial $1 + \alpha|z|^2 + |z|^4$ is a quotient of sums of squares for $\alpha > -2$, but for $\alpha = -2$ it is not.

Exercise 9.69. The polynomial $z^2 + \overline{z}^2 + 2|z|^2$ is not a quotient of sums of squares.

Open problem. A classification of polynomial or rational proper maps between balls is still unknown, even for maps defined on B_3.

9.13 Hermitian Squares in the Free $*$-Algebra

Among the many possible ramifications of the positivity of hermitian forms discussed in the preceding section, the case of so-called hereditary polynomials in a free $*$-algebra stands aside. First for its simplicity, and second for the applications to optimization problems outlined in other chapters of the present book. We confine ourselves to report a couple of significant results in this direction, recently proved in [20].

Let $\mathcal{A}$ denote the free $\mathbb{R}$-algebra with generators $\{x_1, \ldots, x_d, x_1^*, \ldots, x_d^*\}$ and $\mathbb{R}$-linear involution satisfying

$$(fg)^* = g^* f^*, \quad (x_k)^* = x_k^*, \ (x_k^*)^* = x_k, \quad 1 \le k \le d, \quad f, g \in \mathcal{A}.$$

An element $f \in \mathcal{A}$ is called *analytic* if it belongs to the algebra generated by $x_1, \ldots, x_d$, and it is called *hereditary* if all monomials in the decomposition of f have x_k^* to the left of x_j for all j, k. For instance $x_1^* x_3 + x_2^* x_2$ is hereditary, while $x_1^* x_3 + x_2 x_2^*$ is not.

We will state a generic Positivstellensatz and Nullstellensatz, quite different and simpler than the results we have seen in the commutative case. To this aim, let $p_1, \ldots, p_m$ be analytic elements of the free $*$-algebra and let

$$(p) = \{r_1 p_1 + \cdots + r_d p_d; \ r_1, \ldots, r_d \in \mathcal{A}\}$$

denote the left ideal generated by them. Also, let

$$\operatorname{sym}(p) = \left\{ \sum (r_j^* q_j + q_j^* r_j); \quad r_j \in \mathcal{A}, q_j \in (p) \right\}$$

be the associated symmetrized ideal.

The following result holds.

Theorem 9.70. *Let $p_1, \ldots, p_m \in \mathcal{A}$ be analytic elements. If a symmetric hereditary $q \in \mathcal{A}$ satisfies*

$$\langle q(X)v, v \rangle \ge 0,$$

for all pairs (X, v) of finite matrices and vectors satisfying $p_j(X)v = 0, \ 1 \le j \le d$, then

$$q = \sum_{k=1}^{n} f_k^* f_k + g,$$

where $g \in \operatorname{sym(p)}$ and every f_k is analytic.

If instead, $\langle Q(X)v, v \rangle = 0$ for all pairs (X, v) satisfying $p_j(X)v = 0$, for all j, then $q \in \operatorname{sym}(p)$.

The proof consists of a standard separation of convex sets argument and a Gelfand–Naimark–Segal construction. For details see [20]. A heuristic explanation of why such a strong result holds being that the free $*$-algebra representations on finite matrices and vectors better separate points and directions than the mere point evaluations and derivations of the commutative polynomial algebra.

9.14 Further Reading

The selection of topics related to hermitian positivity included in the present chapter is far from complete. While we have tried to make the text self-contained and illustrative for many theoretical ramifications, we did not touch the vast array of applications, classical and modern. We indicate below a few links to applied areas with the hope that the interested reader will pursue some of these threads.

To start with the most recent publications, one can consult the monograph [3], where the essential role played by hermitian sums of squares in signal processing, the prediction theory of stochastic processes and quantum information, is well explained. Then, for matrix completion problems, a subject of high interest nowadays, having its origin in the Schur parameter analysis, see the monograph [8]. Matrix completion problems are frequently invoked nowadays in image analysis, remote sensing, information theory, codification, and on and on. The monograph by Foiaş and Frazho [13] contains an interesting application of completion problems and Schur parameters to the study of the wave propagation in layered media.

The early discoveries of the shift of spectral lines is at the origin of theory of the perturbation theory of hermitian forms. Together with scattering theory, another foundation theme of quantum mechanics, perturbation of spectra remains a hot theme of research, with recent spectacular applications to solid state physics and submolecular chemistry. The old writings of the founders, such as Friedrichs [14] and Krein [25, 26] remain actual and inspiring. We must remark here on the imperative appearance of complex numbers and hermitian forms in the mathematical formulations of quantum mechanics. The textbook [30] and its three additional volumes are filled with hermitian forms formalism, it is true, in infinitely many variables.

The stability of motion of classical mechanical systems (for instance, oscillations of an elastic medium or fluid flows) naturally leads to the problem of enclosing the spectrum of a hermitian or dissipative operator (aka generator of a semigroup or hamiltonian) into a prescribed region of the complex plane. The classical root separation results presented in the first part of this chapter have immediate consequences to the stability of dynamical systems; see, for instance, [15, 26].

Finally, Hilbert space realization of contractive analytic functions and bounded analytic interpolation theorems are at the heart of moment problems and the control theory of systems of differential equations. Each subject is a big enterprise in itself. See again the most recent publications [2, 3] and track their bibliographies back to century-old sources.

Bibliography

[1] J. Agler and J. McCarthy. *Pick Interpolation and Hilbert Function Spaces*, Grad. Stud. Math. American Mathematical Society, Providence, RI, 2002.

[2] K. J. Astrom and R. M. Murray. *Feedback Systems.* Princeton University Press, Princeton, NJ, 2008.

[3] M. Bakonyi and H. Woerdeman. *Matrix Completion, Moments, and Sums of Hermitian Squares.* Princeton University Press, Princeton, NJ, 2011.

[4] J. Bochnak, M. Coste, and M.-F. Roy. *Real Algebraic Geometry,* Ergeb. Math. Grenzgeb. (3) 36. Springer, Berlin, 1998.

[5] D. W. Catlin and J. P. d'Angelo. A stabilization theorem for Hermitian forms and applications to holomorphic mappings. *Math. Res. Lett.,* 3:149–166, 1996.

[6] D. W. Catlin and J. P. d'Angelo. An isometric imbedding theorem for holomorphic bundles. *Math. Res. Lett.,* 6:43–60, 1999.

[7] M. Chilali, P. Gahinet, and P. Apkarian. Robust pole placement in LMI regions. *IEEE Trans. Automat. Control,* 44:2257–2270, 1999.

[8] T. Constantinescu. *Schur Parameters, Factorization and Dilation Problems.* Oper. Theory Adv. Appl. 82. Birkhäuser, Basel, 1996.

[9] J. P. d'Angelo. Complex variables analogues of Hilbert's seventeenth problem. *Int. J. Math.,* 16:609–627, 2005.

[10] J. P. d'Angelo and M. Putinar. Polynomial optimization on odd-dimensional spheres. In *Emerging Applications of Algebraic Geometry,* IMA Vol. Math. Appl. 149. Springer, New York, 2009, pp. 1–15.

[11] J. P. d'Angelo and D. Varolin. Positivity conditions for Hermitian symmetric functions. *Asian J. Math.,* 7:1–18, 2003.

[12] J. Faran. Maps from the two-ball to the three-ball. *Invent. Math.,* 68:441–475, 1982.

[13] C. Foiaş and A. Frazho. *The Commutant Lifting Approach to Interpolation Problems.* Birkhäuser, Basel, 1989.

[14] K. O. Friedrichs. *Perturbation of Spectra in Hilbert Space.* American Mathematical Society, Providence, RI, 1965.

[15] F. R. Gantmacher. *The Theory of Matrices,* Chelsea, New York, 1959.

[16] I. C. Gohberg and M. G. Krein. *Introduction to the Theory of Linear Nonselfadjoint Operators in Hilbert Space,* Transl. Math. Monogr. 18, American Mathematical Society, Providence, RI, 1969.

[17] B. Gustafsson and M. Putinar. Linear analysis of quadrature domains. II. *Israel J. Math.,* 119:187–216, 2000.

[18] P. R. Halmos. Normal dilations and extensions of operators. *Summa Bras. Math.,* 2:125–134, 1950.

[19] E. Hellinger and O. Toeplitz. *Integralgleichungen und Gleichungen mit unendlichvielen Unbekannten.* Reprint from the Encyclopaedia Math. Sci. Chelsea, New York, 1953.

[20] J. W. Helton, S. A. McCullough, and M. Putinar. Non-negative hereditary polynomials in a free $*$-algebra. *Math. Z.*, 250:515–522, 2005.

[21] D. Hilbert. *Grundzüge einer Allgemeinen Theorie der Linearen Integralgleichungen.* Chelsea, New York, 1953.

[22] R. A. Horn and C. R. Johnson. *Matrix Analysis.* Cambridge University Press, Cambridge, UK, 1985.

[23] T. Kato. *Perturbation Theory of Linear Operators.* Springer, Berlin, 1966.

[24] A. N. Kolmogorov. *Selected Works of A. N. Kolmogorov: Vol.* 2, *Probability Theory and Mathematical Statistics.* Kluwer, Norwell, MA, 1991.

[25] M. G. Krein. On the trace formula in perturbation theory (in Russian). *Mat. Sb.*, 33:597–626, 1953.

[26] M. G. Krein. *Topics in Differential Equations and Integral Equations and Operator Theory*, Oper. Theory Adv. Appl. 7. Birkhäuser, Basel, 1983.

[27] M. G. Krein and M. A. Naimark. The method of symmetric and Hermitian forms in the theory of the separation of the roots of algebraic equations. *Linear Multilinear Algebra*, 10:265–308, 1981.

[28] A. Prestel and Ch. N. Delzell. *Positive Polynomials.* Springer, Berlin, 2001.

[29] M. Putinar. Sur la complexification du problème des moments, *C. R. Acad. Sci. Paris Sér.* I *Math.*, 314:743–745, 1992.

[30] M. Reed and B. Simon. *Methods of Mathematics Physics Vol.* 1*: Functional Analysis* Academic Press, San Diego, 1972.

[31] D. G. Quillen. On the representation of hermitian forms as sums of squares. *Invent. Math.*, 5:237–242, 1968.

[32] C. Scheiderer. Positivity and sums of squares: A guide to recent results. In *Emerging Applications of Algebraic Geometry*, IMA Vol. Math. Appl. 149 Springer, New York, 2009, pp. 271–324.

[33] B. V. Shabat. *Introduction to Complex Analysis, Part* II*: Functions of Several Variables*, Transl. Math. Monogr. 110. American Mathematical Society, Providence, RI, 1991.

[34] D. Varolin. Geometry of hermitian algebraic functions. Quotients of squared norms. *Amer. J. Math.*, 130:291–315, 2008.

[35] A. Weil. *Introduction à L'Étude des Variétés Kähleriennes*, Hermann et Cie, Paris, 1958.

Appendix A

Background Material

Grigoriy Blekherman, Pablo A. Parrilo, and Rekha R. Thomas

The appendix consists of four parts: matrices and quadratic forms, convex optimization, convex geometry, and algebraic geometry. The material in this appendix is mostly standard and as such is presented for the convenience of the reader in a compact form.

A.1 Matrices and Quadratic Forms

We present here a few basic facts about linear algebra, symmetric matrices, and quadratic forms. There are many excellent references on the topic, including [11] and [15], among others.

A matrix $A \in \mathbb{R}^{n\times n}$ is *symmetric* if $a_{ij} = a_{ji}$ for $i, j = 1, \dots, n$. The set of symmetric matrices is denoted as $\mathcal{S}^n$ and is a real vector space of dimension $\binom{n+1}{2} = \frac{1}{2}(n+1)n$. Real quadratic forms can always be represented in terms of symmetric matrices, i.e., $q(x) = \sum_{i=1}^{n}\sum_{j=1}^{n} a_{ij}x_i x_j = x^T A x$, where $a_{ij} = a_{ji}$. We often identify a symmetric matrix with the corresponding quadratic form.

The *characteristic polynomial* of a matrix $A \in \mathcal{S}^n$ is $p_A(\lambda) := \det(\lambda I - A) = \lambda^n + \sum_{k=0}^{n-1} p_k \lambda^k = \prod_{k=1}^{n}(\lambda - \lambda_k)$, where λ_k are the *eigenvalues* of A. Given a subset $S \subseteq \{1, \dots, n\}$, let A_S be the submatrix of A whose rows and columns are indexed by S. The *principal minor* of A corresponding to the subset S is the determinant of A_S. If S has the form $\{1, 2, \dots, k\}$, then the corresponding minor is called a *leading principal minor*. It can be shown that the coefficient p_k of the characteristic polynomial is equal (up to sign) to the sum of all the principal minors of size $n - k$, i.e., $p_k = (-1)^{n-k}\sum_{S:|S|=n-k} \det A_S$. Notice that, in particular, $p_{n-1} = -\operatorname{Tr} A$ and $p_0 = (-1)^n \det A$.

A.1.1 Positive Semidefinite Matrices

If the quadratic form $x^T Ax$ takes only nonnegative values, we say that the matrix A is *positive semidefinite*. Similarly, if it takes only positive values (except at the origin, where it necessarily vanishes), then A is *positive definite*. There are several equivalent conditions for a matrix to be positive (semi)definite:

Proposition A.1. *Let $A \in \mathcal{S}^n$ be a symmetric matrix. The following statements are equivalent:*

1. *The matrix A is positive semidefinite ($A \succeq 0$).*
2. *For all $x \in \mathbb{R}^n$, $x^T Ax \geq 0$.*
3. *All eigenvalues of A are nonnegative.*
4. *All $2^n - 1$ principal minors of A are nonnegative.*
5. *The coefficients of $p_A(\lambda)$ weakly alternate in sign, i.e., $(-1)^{n-k} p_k \geq 0$ for $k = 0, \ldots, n-1$.*
6. *There exists a factorization $A = BB^T$, where $B \in \mathbb{R}^{n \times r}$ and r is the rank of A.*

For the definite case, there are similar characterizations:

Proposition A.2. *Let $A \in \mathcal{S}^n$ be a symmetric matrix. The following statements are equivalent:*

1. *The matrix A is positive definite ($A \succ 0$).*
2. *For all nonzero $x \in \mathbb{R}^n$, $x^T Ax > 0$.*
3. *All eigenvalues of A are strictly positive.*
4. *All n leading principal minors of A are strictly positive.*
5. *The coefficients of $p_A(\lambda)$ alternate in sign, i.e., $(-1)^{n-k} p_k > 0$ for $k = 0, \ldots, n-1$.*
6. *There exists a factorization $A = BB^T$, with B square and nonsingular.*

The set of positive semidefinite matrices is denoted as $\mathcal{S}^n_+$, and its interior (the set of positive definite matrices) as $\mathcal{S}^n_{++}$. The set $\mathcal{S}^n_+$ is invariant under nonsingular *congruence transformations*; i.e., if T is nonsingular, $A \succeq 0 \Leftrightarrow T^T AT \succeq 0$. The same statement holds for its interior, i.e., $A \succ 0 \Leftrightarrow T^T AT \succ 0$. For additional facts about the geometry of the set of positive semidefinite matrices, see Section A.3.5.

A.1.2 Matrix Factorizations

For a symmetric matrix A, there are several *matrix factorizations* that can be used to determine or certify properties of A; see, e.g., [11] for theoretical background and [9] for computational aspects. Among the most important matrix factorizations, we have the following.

Eigenvalue decomposition. Since A is symmetric, the eigenspaces corresponding to distinct eigenvalues are mutually orthogonal, and thus one can choose an orthonormal basis of eigenvectors. As a consequence, the matrix A is diagonalizable and there is always a decomposition

$$A = V \Lambda V^T, \qquad \Lambda = \operatorname{diag}(\lambda_1, \dots, \lambda_n),$$

where the matrix V is orthogonal ($VV^T = V^T V = I$). If A is positive semidefinite, we have $\lambda_i \geq 0$ for $i = 1, \dots, n$.

Cholesky decomposition. A positive semidefinite matrix A can be decomposed as

$$A = LL^T,$$

where L is a lower triangular matrix (i.e., $L_{ij} = 0$ for $j > i$). This is known as the *Cholesky decomposition* of the matrix A and can be obtained by solving the identity above column by column (or row by row). The Cholesky decomposition can be computed in $O(n^3)$ operations (in the dense case, faster if the matrix is sparse). Notice that, as opposed to eigenvalue methods, no iterative methods are required. This decomposition plays an important role in numerical algorithms for semidefinite programming.

LDL^T decomposition. This is a decomposition of the form

$$A = LDL^T,$$

where the matrix D is diagonal with nonnegative entries, and L is lower triangular with normalized diagonal entries $L_{ii} = 1$. It should be clear that this can be directly obtained from the Cholesky decomposition, by suitably normalizing its diagonal entries. The importance of the LDL^T decomposition is that, in contrast to the other two factorizations discussed above, it is a *rational* decomposition; i.e., if the matrix A is rational then all numbers that appear in the decomposition are rational (and also, polynomially sized).

Two distinct factorizations of the same positive semidefinite matrix can always be related through a suitable orthogonal transformation. The following result makes this precise.

Theorem A.3. *Let $A \in \mathcal{S}^n$ be a positive semidefinite symmetric matrix, with $A = FF^T$ and $A = GG^T$, where $F, G \in \mathbb{R}^{n \times p}$. Then, there exists a matrix $U \in \mathbb{R}^{p \times p}$ such that $F = GU$ and U is orthogonal (i.e., such that $UU^T = I$ and $U^T U = I$).*

A.1.3 Inertia and Signature

Definition A.4. *Consider a symmetric matrix A. The* inertia *of A, denoted $\mathcal{I}(A)$, is the integer triple (n_-, n_0, n_+), where n_-, n_0, n_+ are the number of negative, zero, and positive eigenvalues, respectively. The* signature *of A is equal to the number* *of positive eigenvalues minus the number of negative eigenvalues, i.e., the integer* $n_+ - n_-$.

Notice that, with the notation above, the rank of A is equal to $n_+ + n_-$. A symmetric positive definite $n \times n$ matrix has inertia $(0, 0, n)$, while a positive semidefinite one has $(0, k, n-k)$ for some $k \geq 0$.

The inertia is an important invariant of a quadratic form, since it holds that $\mathcal{I}(A) = \mathcal{I}(TAT^T)$, where T is nonsingular. This invariance of the inertia of a matrix under congruence transformations is known as *Sylvester's law of inertia*; see, for instance, [11]. This invariance makes it possible to efficiently compute the inertia of a matrix from its LDL^T decomposition, since in this case $\mathcal{I}(A) = \mathcal{I}(D)$, and the inertia of a diagonal matrix is trivial to compute.

A.1.4 Schur Complements

Given a block-partitioned matrix

$$\begin{bmatrix} A & B \\ B^T & C \end{bmatrix},$$

where A is square and invertible, the *Schur complement* of A is the matrix $C - B^T A^{-1} B$. Similarly, if C is square and invertible, its Schur complement is the matrix $A - BC^{-1}B^T$. Schur complements appear in many areas, including among others convex optimization (partial minimization of quadratic functions), probability and statistics (conditioning and marginalization of multivariate Gaussians), and algorithms (block matrix inversion). For several applications and generalizations, see, for instance, the classical survey [6].

Many of the properties of the Schur complement follow quite directly from the two factorizations:

$$\begin{aligned} \begin{bmatrix} A & B \\ B^T & C \end{bmatrix} &= \begin{bmatrix} I & 0 \\ B^T A^{-1} & I \end{bmatrix} \begin{bmatrix} A & 0 \\ 0 & C - B^T A^{-1} B \end{bmatrix} \begin{bmatrix} I & A^{-1}B \\ 0 & I \end{bmatrix} \\ &= \begin{bmatrix} I & BC^{-1} \\ 0 & I \end{bmatrix} \begin{bmatrix} A - BC^{-1}B^T & 0 \\ 0 & C \end{bmatrix} \begin{bmatrix} I & 0 \\ C^{-1}B^T & I \end{bmatrix}. \end{aligned}$$

Since the factorizations above are congruence transformations, this implies that the following conditions are equivalent:

$$\begin{bmatrix} A & B \\ B^T & C \end{bmatrix} \succ 0 \quad \Leftrightarrow \quad \begin{cases} A \succ 0, \\ C - B^T A^{-1} B \succ 0 \end{cases} \quad \Leftrightarrow \quad \begin{cases} C \succ 0, \\ A - BC^{-1}B^T \succ 0. \end{cases}$$

A.2 Convex Optimization

In this section we describe the basic elements of optimization theory, with an emphasis on convexity. For additional background, complete statements, and proofs, we refer the reader to the works [2, 3, 5].

A.2.1 Convexity and Hessians

A set $S \subset \mathbb{R}^n$ is a *convex set* if $x, y \in S$ implies $\lambda x + (1-\lambda)y \in S$ for all $0 \leq \lambda \leq 1$. A function $f : \mathbb{R}^n \to \mathbb{R}$ is a *convex function* if $f(\lambda x + (1-\lambda)y) \leq \lambda f(x) + (1-\lambda)f(y)$ for all $0 \leq \lambda \leq 1$ and $x, y \in \mathbb{R}^n$. A function f is convex if and only if its *epigraph* $\{(x,t) \in \mathbb{R}^n \times \mathbb{R} : f(x) \leq t\}$ is a convex set. A function f is *concave* if $-f$ is convex. When a function is differentiable there are several equivalent characterizations of convexity, in terms of the gradient $\nabla f(x)$ or the Hessian $\nabla^2 f(x)$:

Lemma A.5. *Let $f : \mathbb{R}^n \to \mathbb{R}$ be a twice differentiable function. The following propositions are equivalent:*

(i) *The function f is convex, i.e.,*

$$f(\lambda x + (1-\lambda)y) \leq \lambda f(x) + (1-\lambda)f(y) \quad \text{for all} \quad 0 \leq \lambda \leq 1, \quad x, y \in \mathbb{R}^n.$$

(ii) *The* first-order *convexity condition holds:*

$$f(y) \geq f(x) + (\nabla f(x))^T (y - x), \quad \text{for all} \quad x, y \in \mathbb{R}^n.$$

(iii) *The* second-order *convexity condition holds:*

$$\nabla^2 f(x) \succeq 0, \quad \text{for all} \quad x \in \mathbb{R}^n,$$

i.e., the Hessian is positive semidefinite everywhere.

A.2.2 Minimax Theorem

Given a function $f : S \times T \to \mathbb{R}$, the following inequality always hold:

$$\max_{t \in T} \min_{s \in S} f(s,t) \leq \min_{s \in S} \max_{t \in T} f(s,t). \tag{A.1}$$

If the maxima or minima in (A.1) are not attained, the inequality is still true by replacing "max" and "min" with "sup" and "inf," respectively.

It is of interest to understand situations under which (A.1) holds with equality. The following is a well-known condition for this.

Theorem A.6 (minimax theorem). *Let $S \subset \mathbb{R}^n$ and $T \subset \mathbb{R}^m$ be compact convex sets, and $f : S \times T \to \mathbb{R}$ be a continuous function that is convex in its first argument and concave in the second. Then*

$$\max_{t \in T} \min_{s \in S} f(s,t) = \min_{s \in S} \max_{t \in T} f(s,t).$$

A special case of this theorem, used in game theory to prove the existence of equilibria for zero-sum games, is when S and T are standard unit simplices and the function $f(s,t)$ is a bilinear form.

A.2.3 Lagrangian Duality

Consider a nonlinear optimization problem:

$$\begin{aligned} \underset{x\in\mathbb{R}^n}{\text{minimize}} \quad & f(x) \\ \text{subject to} \quad & g_i(x) \leq 0, \quad i = 1,\dots,m, \\ & h_j(x) = 0, \quad j = 1,\dots,p, \end{aligned} \tag{A.2}$$

and let $u^\star$ be its optimal value. Define the *Lagrangian* associated with the optimization problem (A.2) as

$$\begin{array}{rcl} L : \mathbb{R}^n \times \mathbb{R}^m_+ \times \mathbb{R}^p & \to & \mathbb{R}^n, \\ (x, \lambda, \mu) & \mapsto & f(x) + \sum_{i=1}^m \lambda_i g_i(x) + \sum_{j=1}^p \mu_j h_j(x). \end{array}$$

The *Lagrange dual function* is defined as

$$\phi(\lambda, \mu) := \min_{x\in\mathbb{R}^n} \quad L(x, \lambda, \mu),$$

Maximizing this function over the dual variables (μ, λ) yields

$$v^\star := \max_{\mu\in\mathbb{R}^p \text{ and } \lambda\geq 0} \phi(\lambda, \mu).$$

Applying the minimax inequality (A.1), we see that this is a lower bound on the value of the original optimization problem:

$$v^\star \leq \min_{x\in\mathbb{R}^n} \max_{\mu\in\mathbb{R}^p \text{ and } \lambda\geq 0} L(x, \lambda, \mu) = u^\star.$$

If the functions f, g_i are convex and h_i are affine, then the Lagrangian is convex in x and concave in (λ, μ). To ensure strong duality (i.e., equality in the expression above), compactness or other *constraint qualifications* are needed. An often used condition is the *Slater constraint qualification*: there exists a strictly feasible point, i.e., a point $z^\star \in \mathbb{R}^n$ such that $g_i(z^\star) < 0$ for all $i = 1,\dots,m$ and $h_j(z^\star) = 0$ for all $j = 1,\dots,p$. Under this condition, strong duality always holds.

Theorem A.7. *Consider the optimization problem* (A.2)*, where f, g_i are convex and h_i are affine. Assume Slater's constraint qualification holds. Then the optimal value of the primal is the same as the optimal value of the dual, i.e., $v^\star = u^\star$.*

A.2.4 KKT Optimality Conditions

Consider the nonlinear optimization problem in (A.2). The *Karush–Kuhn–Tucker* (KKT) optimality conditions are

$$\begin{array}{rrcl} \multicolumn{4}{l}{\nabla f\Big|_{x^\star} + \sum_{i=1}^m \lambda_i^\star \cdot \nabla g_i\Big|_{x^\star} + \sum_{j=1}^p \mu_j^\star \cdot \nabla h_j\Big|_{x^\star} = 0,} \\ \text{Primal feasibility:} & g_i(x^\star) & \leq 0 & \text{for } i = 1,\dots,m, \\ & h_j(x^\star) & = 0 & \text{for } j = 1,\dots,p, \\ \text{Dual feasibility:} & \lambda_i^\star & \geq 0 & \text{for } i = 1,\dots,m, \\ \text{Complementary slackness:} & \lambda_i^\star \cdot g_i(x^\star) & = 0 & \text{for } i = 1,\dots,m. \end{array} \tag{A.3}$$

Under certain constraint qualifications (e.g., the ones in the theorem below), the KKT conditions are necessary for local optimality.

Theorem A.8. *Assume any of the following* constraint qualifications *hold:*

- *The gradients of the constraints* $\{\nabla g_1(x^\star), \ldots, \nabla g_m(x^\star), \nabla h_1(x^\star), \ldots, \nabla h_p(x^\star)\}$ *are linearly independent.*
- *There exists a strictly feasible point (Slater constraint qualification), i.e., a point* $z^\star \in \mathbb{R}^n$ *such that* $g_i(z^\star) < 0$ *for all* $i = 1, \ldots, m$ *and* $h_j(z^\star) = 0$ *for all* $j = 1, \ldots, p$.
- *All constraints* $g_i(x)$, $h_i(x)$ *are affine functions.*

Then, at every local minimum $x^\star$ *of* (A.2) *the KKT conditions* (A.3) *hold.*

On the other hand, for *convex optimization problems*, i.e., if all functions f, g_i are convex and h_i are affine, then the KKT conditions are *sufficient* for local (and thus global) optimality:

Theorem A.9. *Let* (A.2) *be a convex optimization problem and* $x^\star$ *be a point that satisfies the KKT conditions* (A.3). *Then* $x^\star$ *is a global minimum.*

A.3 Convex Geometry

We give a summary of standard properties of convex sets and the cone of positive semidefinite matrices. We refer the reader to [2, 13, 14] for more background and proofs.

A.3.1 Basic Facts

Recall that a subset K of $\mathbb{R}^n$ is called convex if for all $x, y \in K$ we have $\lambda x + (1-\lambda) y \in K$ for all real $0 \leq \lambda \leq 1$.

For vectors $x_1, \ldots, x_k \in \mathbb{R}^n$ a linear combination $\lambda_1 x_1 + \cdots + \lambda_k x_k$ is called a *convex combination* if $\lambda_i \geq 0$ for $1 \leq i \leq k$ and $\lambda_1 + \cdots + \lambda_k = 1$. A linear combination is called a *conic combination* if we require only that $\lambda_i \geq 0$ for $1 \leq i \leq k$. Equivalently, a subset K of $\mathbb{R}^n$ is convex if it is closed under taking convex combinations, and K is *convex cone* if it is closed under taking conic combinations. Equivalently, a convex cone is a convex set that is also closed under multiplication by nonnegative scalars.

Let $S \subseteq \mathbb{R}^n$ be an arbitrary subset. The convex hull, $\operatorname{conv}(S)$, of S is the smallest convex set containing S. Equivalently $\operatorname{conv}(S)$ is the set of all convex combinations of points in S:

$$\operatorname{conv}(S) = \left\{ x \in \mathbb{R}^n \;\middle|\; \begin{array}{c} x = \lambda_1 y_1 + \cdots + \lambda_k y_k \quad \text{for some} \quad y_1, \ldots, y_k \in S \\ \lambda_i \geq 0, \quad \lambda_1 + \cdots + \lambda_k = 1 \end{array} \right\}.$$

The conic hull, cone(S), of S is the set of all conic combinations of the points in S:

$$\mathrm{cone}(S) = \left\{ x \in \mathbb{R}^n \;\middle|\; \begin{array}{c} x = \lambda_1 y_1 + \cdots + \lambda_k y_k \quad \text{for some} \quad y_1, \ldots, y_k \in S \\ \lambda_i \geq 0 \end{array} \right\}.$$

The set cone(S) is the smallest convex cone containing S.

A priori it is not clear how large, in terms of the number of points, the convex combinations of points in S need to be to write down a point in conv(S). Carathéodory's Theorem provides an upper bound.

Theorem A.10. *Let S be a subset of $\mathbb{R}^n$. Then any point in the convex hull of S can be written as a convex combination of at most $n+1$ points in S.*

A set defined by finitely many linear inequalities is called a *polyhedron*. The convex hull of finitely many points in $\mathbb{R}^n$ is called a *polytope*, and the conic hull of finitely many points in $\mathbb{R}^n$ is called a *polyhedral cone*. We then have the following theorem.

Theorem A.11. *A bounded polyhedron is a polytope.*

Convex sets possess a well-defined notion of dimension. Let $K \subseteq \mathbb{R}^n$ be a convex set and let $\mathrm{Aff}(K)$ be its *affine hull*, i.e., the affine linear subspace spanned by K. Then K has interior, as a subset of $\mathrm{Aff}(K)$ and $\dim K = \dim \mathrm{Aff}(K)$. The interior of K as a subset of $\mathrm{Aff}\, K$ is called the *relative interior* of K, and the boundary of K as a subset of $\mathrm{Aff}\, K$ is called the *relative boundary* of K. A compact convex set that is full dimensional in $\mathbb{R}^n$ is a called a *convex body*.

Let $K \subset \mathbb{R}^n$ be a closed convex set. A subset $F \subseteq K$ is called a *face* of K if for all $x, y \in K$ and any $0 \leq \lambda \leq 1$, if we have $\lambda x + (1-\lambda)y \in F$, then $x, y \in F$. A face F is called a *proper face* of K if F is a nonempty proper subset of K. It is easy to see that a proper face F does not contain any points in the relative interior of K and therefore it is a subset of the relative boundary of K. The intersection of any two faces of K is a face of K.

A face F of K is called *exposed* if there exists an affine hyperplane H in $\mathbb{R}^n$ such that $F = K \cap H$. The hyperplane H divides $\mathbb{R}^n$ into two half spaces, and it follows that $K \setminus F$ must lie in one of the two open half spaces. Equivalently F is an exposed face of K if there exists an affine linear functional $\ell : \mathbb{R}^n \to \mathbb{R}$ such that $\ell(x) \geq 0$ for all $x \in K$ and $\ell(s) = 0$ for all $s \in F$.

A point $x \in K$ is called an *extreme point* of K if x is a face of K; i.e., if $x = \lambda y_1 + (1-\lambda)y_2$ for some $y_1, y_2 \in K$ and $0 \leq \lambda \leq 1$, then $y_1 = y_2 = x$. A point x in a convex cone C is said to span an *extreme ray* of C if the ray spanned by x is a face of C; i.e., if $x = \lambda_1 y_1 + \lambda_2 y_2$ for some $y_1, y_2 \in K$ and $\lambda_1, \lambda_2 \geq 0$, then y_1 and y_2 lie on the ray spanned by x. The following is the finite-dimensional Krein–Milman theorem. It is also known as Minkowski's theorem.

Theorem A.12. *Let $K \subset \mathbb{R}^n$ be a compact convex set. Then K is the convex hull of its extreme points.*

Faces of a compact convex set K ordered by inclusion form a partially ordered set $\mathcal{F}(K)$. The poset $\mathcal{F}(K)$ is a *lattice*, where the meet operation is intersection: $F_1 \vee F_2 = F_1 \cap F_2$ and the join operation is the intersection of all faces containing F_1 and F_2: $F_1 \wedge F_2 = \bigcap_{F \supset F_1, F_2} F$.

It follows from the Krein–Milman theorem that the minimal proper faces in $\mathcal{F}(K)$ are the extreme points of K, and it will follow from separation theorems presented below that the maximal proper faces in $\mathcal{F}(K)$ are exposed. We note that the maximal proper faces of $\mathcal{F}(K)$ do not have to have the same dimension. In particular for $K \subset \mathbb{R}^n$, the dimension of all the maximal proper faces can be strictly smaller than $n-1$. This is the case for the cone of positive semidefinite matrices $\mathcal{S}_+^n$ as we will see below.

A.3.2 Cone Decomposition

Let K_1, K_2 be convex subsets of $\mathbb{R}^n$. Define $K_1 + K_2$ as the set of all sums of points from K_1 and K_2:

$$K_1 + K_2 = \{x \in \mathbb{R}^n \mid x = x_1 + x_2,\, x_1 \in K_1,\, x_2 \in K_2\}.$$

This operation is called *Minkowski addition*, and the set $K_1 + K_2$ is also convex.

A closed convex cone $C \subset \mathbb{R}^n$ is called *pointed* if C does not contain straight lines. A cone that is closed, full-dimensional in $\mathbb{R}^n$, and pointed is called a *proper* cone. The following theorem shows that a nonpointed cone can always be decomposed into a pointed cone and a subspace.

Theorem A.13. *Let $C \subset \mathbb{R}^n$ be a closed convex cone. Then C is the Minkowski sum of a pointed cone P and a linear subspace L:*

$$C = P + L.$$

This allows us to concentrate on pointed convex cones. Now we formulate the analogue of the Krein–Milman theorem for pointed cones.

Theorem A.14. *Let C be a closed pointed cone in $\mathbb{R}^n$. Then C is the conic hull of the points spanning its extreme rays.*

There is also decomposition theorem for polyhedra, called the Minkowski–Weyl theorem.

Theorem A.15. *Every polyhedron is a Minkowski sum of a polytope and a polyhedral cone.*

A.3.3 Separation Theorems

An important property of a convex set is that we can *certify* when a point is not in the set. This is usually done via a separation theorem. Let H be an affine hyperplane in $\mathbb{R}^n$. Then H divides $\mathbb{R}^n$ into two half spaces. We will use H_+ and H_- to denote the open half spaces and $\bar{H}_+$ and $\bar{H}_-$ to denote the closed half spaces.

We say that H *separates* two sets K_1 and K_2 if K_1 and K_2 belong to different closed half spaces $\bar{H}_+$ and $\bar{H}_-$. We say that H *strictly separates* K_1 and K_2 if they belong to different open subspaces H_+ and H_-.

Equivalently we can think of H as the zero set of an affine linear functional $\ell : \mathbb{R}^n \to R$. Then ℓ separates K_1 and K_2 if $\ell(x) \geq 0$ for all $x \in K_1$ and $\ell(x) \leq 0$ for all $x \in K_2$. Similarly ℓ strictly separates K_1 and K_2 if $\ell(x) > 0$ for all $x \in K_1$ and $\ell(x) < 0$ for all $x \in K_2$.

Now we state our most general separation theorem.

Theorem A.16. *Let K_1 and K_2 be convex subsets of $\mathbb{R}^n$ such that $K_1 \cap K_2 = \emptyset$. Then there exists an affine hyperplane H that separates K_1 and K_2.*

We observe that it follows from Theorem A.16 that every face of a convex set K is contained in an exposed face of K.

We will often be interested in strict separation, in which case we need to make further assumptions on K_1 and K_2.

Theorem A.17. *Let K_1 and K_2 be disjoint convex subsets of $\mathbb{R}^n$ and suppose that K_1 is compact and K_2 is closed. Then there exists an affine hyperplane H strictly separating K_1 and K_2.*

Theorem A.17 is often applied when K_1 is a single point. Separation theorems lead to certificates of not belonging to a convex set. Combined with notions of polarity explained below this leads to *theorems of the alternative.*

We need to adjust Theorems A.16 and A.17 to the setting of cones, since, for example, all cones contain the origin and are never disjoint. Also, any hyperplane separating two cones C_1 and C_2 must be linear. We will say that a linear functional $\ell : \mathbb{R}^n \to R$ separates C_1 and C_2 if $\ell(x) \geq 0$ for all $x \in C_1$ and $\ell(x) \leq 0$ for all $x \in C_2$. Similarly ℓ strictly separates C_1 and C_2 if $\ell(x) > 0$ for all nonzero $x \in C_1$ and $\ell(x) < 0$ for all nonzero $x \in C_2$. Then we have the following theorem.

Theorem A.18. *Let C_1 and C_2 be pointed closed convex cones in $\mathbb{R}^n$ such that $C_1 \cap C_2 = 0$. Then there exists a linear functional $\ell : \mathbb{R}^n \to \mathbb{R}$ strictly separating C_1 and C_2.*

A.3.4 Polarity and Duality

We can view a compact convex set K as the convex hull of its extreme points, but we can also view it as being cut out by linear inequalities. The set of affine linear inequalities defining K is a convex object itself, and this leads to very fruitful notions of polarity and duality in convex geometry.

Let $\langle \ , \ \rangle$ be an inner product on $\mathbb{R}^n$. Let $K \subset \mathbb{R}^n$ be a convex body with origin in its interior. Define the *polar* body K° as follows:

$$K^\circ = \{x \in \mathbb{R}^n \mid \langle x, y \rangle \leq 1 \quad \text{for all} \quad y \in K\}.$$

The polar body encodes all the affine linear defining inequalities of K. It is easy to see that K° is also a convex body with origin in its interior. Moreover $x \in K^\circ$ is on the boundary of K° if and only if $\langle x, y\rangle = 1$ for some $y \in K$. Polarity reverses inclusion: if K_1 and K_2 are convex bodies and

$$\text{if} \quad K_1 \subseteq K_2, \quad \text{then} \quad K_2^\circ \subseteq K_1^\circ.$$

First we observe that polarity is an involution on convex bodies with origin in the interior.

Theorem A.19 (biduality theorem). *Let K be a convex body with origin in its interior. Then*

$$(K^\circ)^\circ = K.$$

We now note that extreme points of K° define maximal proper faces of K (and vice versa): given $x \in K^\circ$ let F_x be the face of K defined by $F_x = \{y \in K \mid \langle x, y\rangle = 1\}$. More generally, if F is a face of K, we can define the corresponding exposed face F^Δ of the polar K° by $F^\Delta = \{y \in K^\circ \mid \langle x, y\rangle = 1 \text{ for all } x \in F\}$. We observe that $(F^\Delta)^\Delta$ is equal to F if and only if F is exposed.

We can similarly define the notion of a polar cone. Let $C \subset \mathbb{R}^n$ be a convex cone. We note that if for some $x \in \mathbb{R}^n$ we have $\langle x, y\rangle \leq 1$ for all $y \in C$, then it follows that $\langle x, y\rangle \leq 0$ for all $y \in C$. Accordingly we define the *polar cone* C° as

$$C^\circ = \{x \in \mathbb{R}^n \mid \langle x, y\rangle \leq 0 \quad \text{for all} \quad y \in C\}.$$

The dual cone C^* is defined as the negative of the polar cone:

$$C^* = \{x \in \mathbb{R}^n \mid \langle x, y\rangle \geq 0 \quad \text{for all} \quad y \in C\}.$$

We note that the dual cone is often defined as a subset of the dual space $(\mathbb{R}^n)^*$:

$$C^* = \{\ell \in (\mathbb{R}^n)^* \mid \ell(y) \geq 0 \quad \text{for all} \quad y \in C\}.$$

Here we used an explicit identification of the dual space $(\mathbb{R}^n)^*$ with $\mathbb{R}^n$ via the inner product $\langle\ ,\rangle$. We can similarly state a biduality theorem for cones.

Theorem A.20. *Let C be a closed convex cone in $\mathbb{R}^n$. Then*

$$(C^\circ)^\circ = C \quad \textit{and} \quad (C^*)^* = C.$$

A.3.5 Cone of Positive Semidefinite Matrices

Let $\mathcal{S}_+^n$ denote the cone of positive semidefinite $n \times n$ matrices. It is easy to show that $\mathcal{S}_+^n$ is a closed, pointed cone and it is full dimensional in $\mathcal{S}^n$. We define an inner product on $\mathcal{S}^n$ as follows: $\langle A, B\rangle = \mathrm{Tr}(AB)$. It is not hard to show that the cone $\mathcal{S}_+^n$ is self-dual.

Proposition A.21. *With the inner product $\langle A, B\rangle = \mathrm{Tr}(AB)$ for $A, B \in \mathcal{S}^n$ we have*

$$\left(\mathcal{S}_+^n\right)^* = \mathcal{S}_+^n.$$

From diagonalization of symmetric matrices we see that any positive semidefinite matrix $A \in \mathcal{S}_+^n$ can be written as a sum of rank 1 positive semidefinite matrices. Thus we see that the extreme rays of $\mathcal{S}_+^n$ are the rank 1 positive semidefinite matrices. Now let V be a linear subspace of $\mathbb{R}^n$. Let F_V be the set of all positive semidefinite matrices A such that $V \subseteq \ker A$. It is easy to show that F_V is a face of $\mathcal{S}_+^n$. It also happens that all faces of $\mathcal{S}_+^n$ have this form.

Theorem A.22. *Let F be a face of $\mathcal{S}_+^n$; then $F = F_V$ for some subspace V of $\mathbb{R}^n$.*

Using diagonalization of symmetric matrices again it follows that the face F_V is isomorphic to the cone of positive semidefinite matrices of dimension $n - \operatorname{codim} V$. Therefore faces of $\mathcal{S}_+^n$ have dimension $\binom{k+1}{2}$ for some $0 \le k \le n$. We note that if V and W are linear subspaces of $\mathbb{R}^n$ and $V \subseteq W$, then $F_W \subseteq F_V$. From this we obtain the following description of the face lattice $\mathcal{F}\left(\mathcal{S}_+^n\right)$.

Corollary A.23. *The face lattice $\mathcal{F}\left(\mathcal{S}_+^n\right)$ is isomorphic to the lattice of linear subspaces of $\mathbb{R}^n$ ordered by reverse inclusion.*

We also have the following important corollary.

Corollary A.24. *Let A, B be positive semidefinite matrices. Then $\langle A, B\rangle = 0$ if and only if $AB = 0$.*

Proof. Suppose that $AB = 0$. Then $\langle A, B\rangle = \operatorname{Tr}(AB) = 0$. Now suppose that $\langle A, B\rangle = 0$. We can write $B = \sum_{i=1}^k R_i$, where R_i are positive semidefinite rank 1 matrices. Then we have $\langle A, B\rangle = \sum_{i=1}^k \langle A, R_i\rangle = 0$. Since the cone $\mathcal{S}_+^n$ is self-dual, we know that $\langle A, R_i\rangle \ge 0$ and therefore $\langle A, R_i\rangle = 0$ for all i. Since matrices R_i have rank 1 we can write $R_i = v_i v_i^T$ for some vectors $v_i \in \mathbb{R}^n$. Therefore we see that $\langle A, R_i\rangle = v_i^T A v_i = 0$, and since $A \succeq 0$ we see that v_i is in the kernel of A. Therefore we see that $AR_i = Av_i v_i^T = 0$ for all i. Thus we have $AB = 0$. $\square$

A.3.6 Dimensional Inequalities

It is often of great interest to find a *low rank* positive semidefinite matrix given some linear conditions on the entries of a matrix. While existence of a positive semidefinite matrix subject to linear constraints can be solved via semidefinite programming, the existence of a solution of low rank is a nonconvex problem and thus quite challenging. It is therefore of interest to find some theoretical guarantees on the existence of low rank solutions, given that a positive semidefinite solution exists. We state the following bounds discovered and rediscovered independently by several authors [1].

Theorem A.25. *Let $\mathcal{A}$ be an affine subspace of $\mathcal{S}_+^n$ such that the intersection $\mathcal{A} \cap \mathcal{S}_+^n$ is nonempty and* $\operatorname{codim} \mathcal{A} \le \binom{r+2}{2} - 1$ *for some nonnegative integer r. Then there is a matrix $X \in \mathcal{S}_+^n \cap \mathcal{A}$ such that* $\operatorname{rank} X \le r$.

This bound is sharp in general, but it was improved by Barvinok in the case where the intersection $\mathcal{A} \cap \mathcal{S}_+^n$ is bounded [1].

Theorem A.26. *Suppose that $r \geq 0$ and $n \geq r+2$. Let A be an affine subspace of $\mathcal{S}_+^n$ such that* $\operatorname{codim} A = \binom{r+2}{2}$. *Suppose that the intersection $A \cap \mathcal{S}_+^n$ is nonempty and bounded. Then there is a matrix $X \in \mathcal{S}_+^n \cap A$ such that* $\operatorname{rank} X \leq r$.

A.4 Algebra of Polynomials and Ideals

There are excellent books for the basics of commutative algebra, algebraic geometry, and real algebraic geometry used in this book. For polynomials, ideals, Gröbner bases, and basic algebraic geometry we refer the reader to [7], an introduction to these topics at the undergraduate level. For basic real algebraic geometry concepts such as semialgebraic sets and the Tarski–Seidenberg quantifier elimination, see [12]. What we provide below is a brief tour through some of the algebraic themes that arise in this book with the goal of giving the absolute newcomer a quick grasp of the concepts. For a more serious appreciation of these topics, the reader is referred to the above-mentioned books.

A.4.1 Monomials, Polynomials, and the Polynomial Ring

A *monomial* in the n variables $x_1, \ldots, x_n$ (abbreviated as x) is a product $x^a := x_1^{a_1} x_2^{a_2} \cdots x_n^{a_n}$, where $a = (a_1, \ldots, a_n) \in \mathbb{N}^n$. A *polynomial* in $x_1, \ldots, x_n$ with coefficients in a *field* k is a finite linear combination of the form $f := \sum c_a x^a$, where $c_a \in k$. A monomial x^a is in the *support* of f if $c_a \neq 0$ in the expression $f = \sum c_a x^a$. The *degree* of $f = \sum c_a x^a$ is the maximum L_1-norm of the vectors a that appear as exponents of monomials in the support of f. The usual fields considered in this book are the set of real numbers denoted as $\mathbb{R}$ and the set of complex numbers denoted as $\mathbb{C}$. In what follows, we assume that the field k is either $\mathbb{C}$ or $\mathbb{R}$. The *polynomial ring* $k[x] := k[x_1, \ldots, x_n]$ is the set of all polynomials in $x_1, \ldots, x_n$ with coefficients in k. It is endowed with the two binary operations of addition and multiplication of pairs of polynomials.

Groups, rings, and fields are basic objects in abstract algebra that satisfy an increasing list of properties. See, for instance, [8] for definitions and examples. A binary operation $\star$ on a set S is

- *associative* if $(f \star g) \star h = f \star (g \star h)$ for all $f, g, h \in S$, and
- *commutative* if $f \star g = g \star f$ for all $f, g \in S$.

The pair $(S, \star)$

- has an *identity* if there exists an element $e \in S$ such that $f \star e = e \star f = f$ for all $f \in S$, and
- has *inverses* if for each $f \in S$, there exists an element $f^{-1} \in S$ such that $f \star f^{-1} = f^{-1} \star f = e$.

Definition A.27.

- *A set G with a binary operation $\star$ is a* group *if $\star$ is associative and $(G, \star)$ has an identity and inverses. If in addition, $\star$ is commutative in G, then G is called a* commutative group.

- *A set R with two binary operations $+$ (addition) and $\cdot$ (multiplication) is a* ring *if $(R,+)$ is a commutative group, and $(R,\cdot)$ is associative and $\cdot$ distributes over $+$ in the sense that $f\cdot(g+h)=f\cdot g+f\cdot h$ for all $f,g,h\in R$. If $(R,\cdot)$ has an identity and/or $(R,\cdot)$ is commutative, then R is a ring with identity and/or commutative.*

- *A* field *is a ring $(F,+,\cdot)$ in which $(F,\cdot)$ is also a commutative group with identity.*

The set of integers under addition and multiplication, $(\mathbb{Z},+,\cdot)$, forms a commutative ring with identity: $(\mathbb{Z},+)$ is a commutative group (with 0 as its additive identity and for each $z\in\mathbb{Z}$, $-z$ is the additive inverse of z), and 1 is the multiplicative identity in $(\mathbb{Z},+,\cdot)$. No element $z\in\mathbb{Z}, z\neq\pm1$ has a multiplicative inverse. On the other hand, $(\mathbb{R},+,\cdot)$ and $(\mathbb{C},+,\cdot)$ are fields. The set of $n\times n$ matrices under matrix addition and multiplication forms a noncommutative ring with identity.

The polynomial ring $k[x]$ is a commutative ring with identity under addition and multiplication of pairs of polynomials. The empty monomial $x_1^0\cdots x_n^0=1\in k[x]$ and hence k is a subset of $k[x]$ and is called the set of scalars in $k[x]$. It is customary to denote $f\cdot g$ as just fg when the multiplication operation is clear. The ring $k\langle x\rangle$ denotes the *free ring* where the variables $x_1,\ldots,x_n$ do not commute; i.e., the relation $x_ix_j=x_jx_i$ is not assumed. The free ring (and also $k[x]$) is an example of an *algebra* which is a ring that is also a vector space over its field of scalars. Hence it is often called the *free algebra* in n variables over k. This noncommutative ring plays a central role in Chapter 8.

A.4.2 Polynomial Ideals, Gröbner Bases, and Quotient Rings

Definition A.28.

1. *A subset $I\subset k[x]$ is an* ideal *if it satisfies the following properties:*
   - $0\in I$.
   - *If $f,g\in I$, then $f+g\in I$.*
   - *If $f\in I$ and $h\in k[x]$, then $hf\in I$.*

2. *The ideal generated by $f_1,\ldots,f_t\in k[x]$ is the set $I=\left\{\sum_{i=1}^t h_if_i\ :\ h_i\in k[x]\right\}$, denoted as $\langle f_1,\ldots,f_t\rangle$.*

Check that $\langle f_1,\ldots,f_t\rangle$ is an ideal in $k[x]$. A simple example of an ideal in the polynomial ring $\mathbb{R}[x_1,x_2]$ is the set of all polynomials that evaluate to 0 on the point $(0,0)$. This ideal consists of all polynomials of the form x_1f+x_2g, where $f,g\in k[x]$ and hence equals $\langle x_1,x_2\rangle$. An ideal $I\subset k[x]$ is *finitely generated* if it is generated by a finite set of polynomials in $k[x]$. A generating set of an ideal I is called a *basis* of I. An ideal can have bases of different cardinalities and, unlike a vector space basis, an ideal basis is just a generating set with no independence requirements.

Theorem A.29 (Hilbert's basis theorem). *If k is a field, then every ideal in $k[x]$ is finitely generated (has a finite basis).*

Here are two important types of ideals.

Definition A.30. *A polynomial $f = \sum c_a x^a$ is* homogeneous *if all monomials in its support have the same degree. An ideal $I \subset k[x]$ is* homogeneous *if it is generated by homogeneous polynomials.*

Definition A.31. *An ideal I is* principal *if it is generated by a single polynomial.*

Gröbner bases are special bases for a polynomial ideal. They enable many algorithms in computational algebraic geometry.

Definition A.32. *A* term order $\succ$ *on $k[x]$ is a total ordering on the monomials in $k[x]$ such that*

- $1 \prec x^a$ *for all $a \neq 0$, and*
- *if $x^a \succ x^b$ then $x^a x^c \succ x^b x^c$ for all monomials x^c.*

A common example of a term order is the *lexicographic/dictionary* order with $x_1 \succ x_2 \succ \cdots \succ x_n$ defined as $x^a \succ x^b$ if and only if the left most nonzero entry in $a - b$ is positive. Note that there are $n!$ lexicographic term orders on $k[x]$. A term order needed in Chapter 7 is the *total degree order* which first sorts monomials by degree and then breaks ties using a fixed term order such as the above lexicographic order. More precisely, $x^a \succ x^b$ if and only if either $\deg(x^a) > \deg(x^b)$ or $\deg(x^a) = \deg(x^b)$ and x^a is lexicographically larger than x^b.

Definition A.33. *Fix a term order $\succ$ on $k[x]$.*

- *The* initial term $\text{in}_\succ(f)$ *of a polynomial $f = \sum c_a x^a \in k[x]$ with respect to $\succ$ is that monomial $c_a x^a$ with $c_a \neq 0$ such that $x^a \succ x^b$ for all other monomials x^b in the support of f. The monomial x^a is called the* initial monomial *of f.*
- *The* initial ideal $\text{in}_\succ(I)$ *is the ideal generated by the initial monomials of all polynomials in I.*

By Hilbert's basis theorem, the initial ideal $\text{in}_\succ(I)$ is finitely generated. In fact, it has a unique set of minimal generators that are all monomials.

Definition A.34. *A* Gröbner basis $G_\succ$ *of a polynomial ideal $I \subset k[x]$ with respect to the term order $\succ$ is a finite set of polynomials $g_1, \ldots, g_t \in I$ such that $\langle \text{in}_\succ(g_1), \ldots, \text{in}_\succ(g_t) \rangle = \text{in}_\succ(I)$.*

Each term order $\succ$ gives rise to a *reduced* Gröbner basis of I which is unique. The Gröbner bases returned by a computer algebra package such as Macaulay2 are usually reduced. In the 1960s Buchberger provided an algorithm to find a Gröbner basis of an ideal given a term order. This algorithm underlies the Gröbner basis functionality in modern computer algebra packages such as Macaulay2, SINGULAR, Maple, Mathematica, etc.

Example A.35. An example of a reduced Gröbner basis with respect to the total degree ordering was given in Chapter 7. Consider the ideal

$$I = \langle x^4 - y^2 - z^2, x^4 + x^2 + y^2 - 1\rangle.$$

Using Macaulay2 [10] one can calculate a total degree reduced Gröbner basis of I as follows:

```
Macaulay2, version 1.3

i1 : R = QQ[x,y,z,Weights => {1,1,1}];
i2 : I = ideal(x^4-y^2-z^2, x^4+x^2+y^2-1);
i3 : G = gens gb I
o3 = | x2+2y2+z2-1 4y4+4y2z2+z4-5y2-3z2+1 |
```

which says that a total degree Gröbner basis consists of the two polynomials

$$x^2 + 2y^2 + z^2 - 1 \quad \text{and} \quad 4y^4 + 4y^2z^2 + z^4 - 5y^2 - 3z^2 + 1.$$

The reduced Gröbner basis of I would have the property that no initial term of an element is divisible by the initial term of another element and that all initial terms have unit coefficients. Hence the reduced Gröbner basis of I is

$$\left\{x^2 + 2y^2 + z^2 - 1,\ \ y^4 + y^2z^2 + \frac{1}{4}(z^4 - 5y^2 - 3z^2 + 1)\right\}.$$

In particular, the initial ideal of I with respect to this term order is $\langle x^2, y^4\rangle$. Check that both elements in the Gröbner basis lie in the ideal I. ∎

Gröbner bases enable a multitude of computations with ideals such as checking whether a polynomial lies in an ideal (*ideal membership*), checking whether an ideal equals the whole ring, finding all roots of a system of polynomial equations, finding the intersection of two ideals, etc. Ideal membership relies on a multivariate *division algorithm* that computes the remainder (called a *normal form*) of a polynomial f with respect to a Gröbner basis. A polynomial f lies in I if and only if the normal form of f (with respect to any reduced Gröbner basis of I) is zero. This in turn relies on the fact that the normal form of a polynomial with respect to a reduced Gröbner basis of I is unique.

Example A.36. The normal form of the monomial x^2y with respect to the Gröbner basis in Example A.35 is obtained by successively dividing out the initial monomial $\text{in}_\succ(g)$ of an element $g := \text{in}_\succ(g) - g'$ in the reduced Gröbner basis from x^2y and multiplying with g'. Let $g_1 := x^2 + 2y^2 + z^2 - 1$ and $g_2 := y^4 + y^2z^2 + \frac{1}{4}(z^4 - 5y^2 - 3z^2 + 1)$. Then x^2y can be divided by g_1 to give $-2y^3 - yz^2 + y$. The resulting initial term $-2y^3$ cannot be divided by either $in_\succ(g_1)$ or $\text{in}_\succ(g_2)$, which implies that the normal form of x^2y is $-2y^3 - yz^2 + y$. ∎

Given an ideal I in a polynomial ring $k[x]$, one can compute the *quotient ring* $k[x]/I$ which consists of all equivalence classes of polynomials in $k[x]$ mod the ideal I. Given two polynomials $f, g \in k[x]$, f is equivalent to g mod I if $f - g \in I$. This is denoted as $f \cong g$ mod I, and the equivalence class of f mod I is denoted as $f + I$. This notion is a generalization of the familiar modular arithmetic in the ring of integers, where we say that $z, z' \in \mathbb{Z}$ are equivalent mod a fixed integer p if $z - z'$ is an integer multiple of p. In this case the ideal I (in the ring of integers $\mathbb{Z}$) is the ideal generated by p, namely the set consisting of all integer multiples of p. If f' is the normal form of a polynomial f with respect to a reduced Gröbner basis of an ideal I in $k[x]$, then $f - f' \in I$ and hence $f \cong f'$ mod I. Since the normal form of a polynomial with respect to a reduced Gröbner basis is unique, if $f - g \in I$, then the normal form of $f - g$ is zero, which implies that both f and g have the same normal form. Hence every equivalence class of polynomials mod I can be represented by the unique normal form of all the elements in that class with respect to a fixed reduced Gröbner basis of I.

Example A.37. In Example A.35, the equivalence class of x^2y mod I consists of all polynomials $g \in \mathbb{Q}[x, y, z]$ such that $x^2y - g \in I$. In other words, $x^2y + I$ is the set of all $g \in \mathbb{Q}[x, y, z]$ with normal form $-2y^3 - yz^2 + y$ with respect to the reduced Gröbner basis

$$\left\{g_1 := x^2 + 2y^2 + z^2 - 1, g_2 := y^4 + y^2z^2 + \frac{1}{4}(z^4 - 5y^2 - 3z^2 + 1)\right\}. \qquad \blacksquare$$

The quotient ring $k[x]/I$ is a k-vector space. Addition in the ring is defined as $(f + I) + (g + I) = (f + g) + I$ and scalar multiplication as $\alpha(f + I) = \alpha f + I$ for all $\alpha \in k$. A primary use of Gröbner bases is that they provide a vector space basis for $k[x]/I$ in the following sense. Fix a term order $\succ$ on $k[x]$ and consider the initial ideal $\text{in}_\succ(I)$ of the ideal I. Recall that this initial ideal is generated by monomials. The monomials in $k[x]$ that do not lie in $\text{in}_\succ(I)$ are called the *standard monomials* of $\text{in}_\succ(I)$. The equivalence classes $m + I$ as m varies over the standard monomials of $\text{in}_\succ(I)$ form a vector space basis of $k[x]/I$. Buchberger's algorithm for Gröbner bases was motivated by the quest to find vector space bases for $k[x]/I$. It is easy to see why the equivalence classes of standard monomials provide a vector space basis for $k[x]/I$. We saw earlier that once a term order $\succ$ is fixed, every equivalence class $f + I$ has a unique representative $f' + I$, where f' is the normal form of f with respect to the reduced Gröbner basis $G_\succ$ of I corresponding to $\succ$. Note that f' cannot be divided by $\text{in}_\succ(g)$ for any $g \in G_\succ$ and hence all its monomials are standard with respect to $\text{in}_\succ(I)$. This shows that the elements $m + I$ span $k[x]/I$. If a collection of them are linearly dependent, then there exists standard monomials $m_1, \ldots, m_t$ and scalars $\alpha_1, \ldots, \alpha_t$ such that $\sum \alpha_i(m_i + I) = 0 + I$, or equivalently, $\sum \alpha_i m_i \in I$. However, if $\sum \alpha_i m_i \in I$, then its normal form with respect to $G_\succ$ is zero which implies that some m_i is divisible by some $\text{in}_\succ(g)$ for $g \in G_\succ$, which is a contradiction.

Example A.38. The vector space $\mathbb{Q}[x, y, z]/I$ for the ideal in Example A.35 has infinite dimension. The initial ideal of the total degree order $\succ$ used in this example is $\text{in}_\succ(I) = \langle x^2, y^4 \rangle$. Hence the standard monomials of this initial ideal are all monomials in x, y, z that are not divisible by x^2 and y^4. There are infinitely many such monomials since all powers of z are standard. Regardless, an infinite basis of $\mathbb{Q}[x, y, z]/I$ consists of $m+I$ as m varies over the standard monomials of $\text{in}_\succ(I)$. ∎

A.4.3 Algebraic Varieties

Definition A.39. *Given an ideal $I = \langle f_1, \ldots, f_t \rangle \subset k[x]$, its* affine variety *in k^n is the set $V_k(I) := \{p \in k^n : f(p) = 0$ for all $f \in I\}$.*

It can be checked that $V_k(I) = \{p \in k^n : f_1(p) = \cdots = f_t(p) = 0\}$ and is hence the set of solutions (zeros) in k^n of the system of polynomial equations $f_1(x) = \cdots = f_t(x) = 0$. The affine variety of a principal ideal $\langle f \rangle \subset k[x]$ is called a *hypersurface* in k^n and denoted simply as $V_k(f)$.

Example A.40. The affine variety of the ideal $\langle x_i^2 - x_i$ for all $i = 1, \ldots, n\rangle \subset \mathbb{R}[x_1, \ldots, x_n]$ is the set of all 0/1 vectors in $\mathbb{R}^n$. ∎

If f is a homogeneous polynomial in $k[x]$, then for every $p \in k^n$ such that $f(p) = 0$, we also have that $f(\lambda p) = 0$ for all $\lambda \in k$. Hence, solutions of I come in lines through the origin. Hence, it makes sense to declare all points on a line through the origin in k^n as being equivalent. This leads to *projective geometry*, where we replace k^n with the *projective space* $\mathbb{P}_k^{n-1}$ whose points are in bijection with the distinct lines through the origin in k^n. The *homogeneous coordinates* of the point in $\mathbb{P}_k^{n-1}$ corresponding to the line spanned by $(x_1, \ldots, x_n)$ is denoted as $(x_1 : \cdots : x_n)$ to denote that it is unique only up to scalar multiplication. For details on the construction of projective spaces, see Chapter 8 in [7]. If a polynomial is not homogeneous, then it is not true that $p(x) = 0$ implies $p(\lambda x) = 0$ for all $\lambda \neq 0$.

Definition A.41. *The* projective variety *of a homogeneous ideal $I \subset k[x]$ is $\{p \in \mathbb{P}_k^{n-1} : f(p) = 0$ for all $f \in I\}$.*

We do not introduce any notation for projective varieties here as we will not discuss them in this appendix. Chapter 8 in [7] gives an introduction to projective varieties and their relationship to affine varieties. As for affine varieties and their ideals, Gröbner bases play an important role in computations involving projective varieties and their (homogeneous) ideals.

Example A.42. The homogeneous principal ideal $I = \langle yz - x^2 \rangle \subset \mathbb{C}[x, y, z]$ contains all lines spanned by the points $(t, t^2, 1)$, $t \in \mathbb{C}$, in its affine variety in $\mathbb{C}^3$. Its projective variety is $\{(t : t^2 : 1) : t \in \mathbb{C}\} \cup \{(0 : u : 0) : u \in \mathbb{C}\} \subset \mathbb{P}_\mathbb{C}^2$. ∎

A field k is *algebraically closed* if every polynomial in $k[x]$ has all its roots in k^n. The field $\mathbb{C}$ is algebraically closed while $\mathbb{R}$ is not. Every ideal $I \subset k[x]$ has an affine variety $V_k(I) \subset k^n$, although different ideals can have the same affine variety. For instance, both $\langle x, y \rangle$ and $\langle x^2, y^2 \rangle$ in $\mathbb{C}[x, y]$ have the affine variety $\{(0, 0)\}$ in $\mathbb{C}^2$.

Given a variety $W \subset k^n$, its *vanishing ideal*,

$$I(W) := \{f \in k[x] \, : \, f(p) = 0 \text{ for all } p \in W\}$$

is the set of all polynomials in $k[x]$ that vanish on W. Check that $I \subseteq I(V_k(I))$ and that $V_k(I(V_k(I))) = V_k(I)$. The ideal $I(V_k(I))$ is the largest ideal with the affine variety $V_k(I)$. This vanishing ideal has the important property that if f^m belongs to it, then so does f since $f^m(p) = 0$ for all $p \in V_k(I)$ implies that $f(p) = 0$ for all $p \in V_k(I)$.

Definition A.43. *The* radical *of an ideal $I \subset k[x]$ is*

$$\sqrt{I} := \{f \in k[x] \, : \, f^m \in I \text{ for some positive integer } m\}.$$

An ideal I is radical if $I = \sqrt{I}$.

The radical $\sqrt{I}$ is an ideal and both I and $\sqrt{I}$ have the same affine variety. Further, the vanishing ideal $I(V_k(I))$ is a radical ideal. The following theorem shows that when k is an algebraically closed field, there is a bijection between radical ideals in $k[x]$ and affine varieties in k^n.

Theorem A.44 (Hilbert's strong Nullstellensatz). *If k is an algebraically closed field, then $I(V_k(I)) = \sqrt{I}$.*

The following example points out the importance of k being algebraically closed in the above Nullstellensatz.

Example A.45. The ideal $I = \langle x^2 + y^2 \rangle \subset \mathbb{C}[x, y]$ is radical. Its affine variety in $\mathbb{R}^2$ is $\{(0, 0)\}$, whose vanishing ideal is $J = \langle x, y \rangle$ and $J \neq I$. On the other hand, the affine variety of I in $\mathbb{C}^2$ consists of the two lines $x = \pm iy$ whose vanishing ideal is I. ∎

There is a strong Nullstellensatz for projective varieties as well that has the same statement. However, there is a *weak Nullstellensatz* that characterizes empty varieties whose statements are different for affine and projective varieties. We refer the reader to [7, Chapter 8] for details.

Theorem A.46 (Hilbert's weak Nullstellensatz). *Let k be an algebraically closed field.*

1. *If I is an ideal in $k[x]$, then its affine variety $V_k(I) \subseteq k^n$ is empty if and only if $I = k[x]$.*
2. *If I is a homogeneous ideal in $k[x]$, then its projective variety in $\mathbb{P}_k^{n-1}$ is empty if and only if for each $i = 1, \ldots, n$, there is a monomial $x_i^{m_i} \in I$ where m_i is some nonnegative integer.*

To end this subsection, we briefly discuss the notions of dimension, degree, and singular points of an algebraic variety. These notions are too subtle to be explained correctly here and we refer the reader to [7, Chapter 9]. Dimension and degree of a variety can be computed from an algebraic entity called the *Hilbert polynomial* of the vanishing ideal of the variety. A key feature of Gröbner basis theory is that an ideal I and all its initial ideals have the same Hilbert polynomial and the polynomial has a combinatorial expression that can be computed from the standard monomials of any of its initial ideals. Intuitively, the dimension of an ideal is the dimension of the largest component of its affine variety. For instance we expect a hypersurface in k^n to have dimension $n-1$ since it is constrained by a single polynomial. However, when the field is not algebraically closed, this intuition can be wrong. For instance, $V_{\mathbb{R}}(x^2+y^2) = \{(0,0)\}$ is a zero-dimensional variety in $\mathbb{R}^2$ while $V_{\mathbb{C}}(x^2+y^2)$ is a one-dimensional variety in $\mathbb{C}^2$.

The *degree* of a variety is also defined from the Hilbert polynomial of the vanishing ideal. Intuitively we expect that slicing an r-dimensional variety in k^n with a generic plane of dimension $n-r$ through the origin would create finitely many intersections. The number of intersection points should be constant if the plane is generic enough and is intuitively the degree of the variety. For instance, the parabola defined by $y - x^2$ has two points of intersection with a generic line through the origin saying that its degree is two, while the cubic curve $y = x^3$ cuts out a variety of degree three.

A *nonsingular* (also called *regular* or *smooth*) point p on a variety W is a point where the *tangent space* to W at p has the same dimension as the component of W containing p and hence serves as a reasonable linear approximation to this component near p. For a polynomial $f \in k[x]$, let $\nabla(f) := (\frac{\partial f}{\partial x_1}, \ldots, \frac{\partial f}{\partial x_n})$ be the *gradient* of f and $\nabla(f)(p) \in k^n$ be the evaluation of $\nabla(f)$ at $p \in k^n$. Since the structure of a variety is unchanged by translation, we may assume without loss of generality that $p = 0$. If $I(W) = \langle f_1, \ldots, f_s \rangle$, then the *tangent space* of W at p is the null space of the matrix $J(0)$ whose rows are $\nabla(f_1)(0), \ldots, \nabla(f_s)(0)$. The matrix J whose rows are the polynomials $\nabla(f_1), \ldots, \nabla(f_s)$ is called the *Jacobian matrix* of $f_1, \ldots, f_s$. Thus the rank of $J(0)$ determines whether 0 is singular on W or not. In particular, 0 is a singular point on a hypersurface $V_k(f)$ if and only if $\nabla(f)(u) = 0$.

A.4.4 Real Algebraic Geometry

A good deal of the algebraic geometry that appears in this book is over $\mathbb{R}$, which is not an algebraically closed field. As a result, many of the theorems that apply over $\mathbb{C}$ do not work in this setting making the study of real varieties and their ideals more tricky than their complex counterparts. A good introduction to the real algebraic geometry background needed in this book is [12]. We define a few of the key concepts and results.

Definition A.47. *A set* $S \subset \mathbb{R}^n$ *defined as* $S = \{x \in \mathbb{R}^n : f_i(x) \triangleright_i 0, i = 1, \ldots, t\}$, *where, for each* i, $\triangleright_i$ *is one of* $\geq, >, =, \neq$, *and* $f_i(x) \in \mathbb{R}[x]$, *is called a* basic semialgebraic set. *A* basic closed semialgebraic set *is a set of the form* $S = \{x \in \mathbb{R}^n : f_1(x) \geq 0, \ldots, f_t(x) \geq 0\}$.

Every basic semialgebraic set can be expressed with polynomial inequalities of the form $f(x) \geq 0$ and a single inequality $g \neq 0$. In this book we only encounter basic semialgebraic sets in which $\rhd$ is either $>, \geq$, or $=$. Note that every real algebraic variety is a basic closed semialgebraic set.

Definition A.48. *A finite union of basic semialgebraic sets in* $\mathbb{R}^n$ *is called a* semialgebraic set, *and a finite union of basic closed semialgebraic sets is a closed semialgebraic set.*

Semialgebraic sets are closed under finite unions, finite intersections, and complementation. The following theorem shows that semialgebraic sets are also closed under projections, a fact that is used several times in this book. For more details see [12] and [4].

Theorem A.49 (Tarski–Seidenberg theorem). *Let* $S \subset \mathbb{R}^{k+n}$ *be a semialgebraic set and* $\pi : \mathbb{R}^{k+n} \to \mathbb{R}^n$ *be the projection map that sends* $(y, x) \mapsto x$. *Then* $\pi(S)$ *is a semialgebraic set in* $\mathbb{R}^n$.

Recall that Σ denotes the set of sums of squares polynomials in $\mathbb{R}[x]$.

Definition A.50. *The* preorder *associated with a finite set of polynomials* $f_1, \ldots, f_t \in \mathbb{R}[x]$ *is the set*

$$\mathbf{preorder}(f_1, \ldots, f_t) := \left\{ \sum s_\sigma f_1^{\sigma_1} \cdots f_t^{\sigma_t} : \sigma = (\sigma_1, \ldots, \sigma_t) \in \{0,1\}^t \text{ and } s_\sigma \in \Sigma \right\}.$$

All the polynomials in $\mathbf{preorder}(f_1, \ldots, f_t)$ are nonnegative on the basic closed semialgebraic set $S = \{x \in \mathbb{R}^n : f_1(x) \geq 0, \ldots, f_t(x) \geq 0\}$.

Definition A.51. *The* real radical *of an ideal* $I = \langle f_1, \ldots, f_t \rangle$ *is the ideal*

$$\sqrt[\mathbb{R}]{I} := \{ f \in \mathbb{R}[x] : -f^{2m} \in \Sigma + I \text{ for some nonnegative integer } m \}.$$

The ideal I *is said to be* real radical *if* $I = \sqrt[\mathbb{R}]{I}$.

We conclude with the Positivstellensatz and the real Nullstellensatz that play the analogous role of Hilbert's Nullstellensatz for semialgebraic sets and real varieties.

Theorem A.52 (Positivstellensatz). *Let* $f_1, \ldots, f_t \in \mathbb{R}[x]$ *and* $S = \{x \in \mathbb{R}^n : f_1(x) \geq 0, \ldots, f_t(x) \geq 0\}$ *and* T *be the preorder associated to* $f_1, \ldots, f_t$. *For a polynomial* $f \in \mathbb{R}[x]$,

1. $f > 0$ *on* S *if and only if there exists* $p, q \in T$ *such that* $pf = 1 + q$;
2. $f \geq 0$ *on* S *if and only if there exists an integer* $m \geq 0$ *and* $p, q \in T$ *such that* $pf = f^{2m} + q$;

3. *$f = 0$ on S if and only if there exists an integer $m \geq 0$ such that $-f^{2m} \in T$;*

4. *$S = \emptyset$ if and only if $-1 \in T$.*

Corollary A.53 (Real Nullstellensatz). *If I is an ideal in $\mathbb{R}[x]$, then its real radical ideal $\sqrt[\mathbb{R}]{I}$ is the largest ideal that vanishes on $V_{\mathbb{R}}(I)$.*

Recall that $\sqrt{I} \subset \mathbb{R}[x]$, the radical ideal of I in $\mathbb{R}[x]$ is the largest ideal that vanishes on the complex variety $V_{\mathbb{C}}(I)$. Therefore, since $V_{\mathbb{R}}(I) \subseteq V_{\mathbb{C}}(I)$, we have that $I \subseteq \sqrt{I} \subseteq \sqrt[\mathbb{R}]{I}$.

The Positivstellensatz also gives a simple solution to Hilbert's 17th problem, which asked whether every nonnegative polynomial in $\mathbb{R}[x]$ can be written as a sum of squares of rational functions in x. This was answered in the affirmative by Artin in 1927. The two-variable case was shown by Hilbert in 1893.

Bibliography

[1] A. Barvinok. A remark on the rank of positive semidefinite matrices subject to affine constraints. *Discrete Comput. Geom.*, 25:23–31, 2001.

[2] A. Barvinok. *A Course in Convexity*, Grad. Stud. Math. 54. American Mathematical Society, Providence, RI, 2002.

[3] D. P. Bertsekas, A. Nedić, and A. E. Ozdaglar. *Convex Analysis and Optimization.* Athena Scientific, Belmont, MA, 2003.

[4] J. Bochnak, M. Coste, and M-F. Roy. *Real Algebraic Geometry.* Springer, Berlin, 1998.

[5] S. Boyd and L. Vandenberghe. *Convex Optimization.* Cambridge University Press, Cambridge, UK, 2004.

[6] R. W. Cottle. Manifestations of the Schur complement. *Linear Algebra Appl.*, 8:189–211, 1974.

[7] D. Cox, J. Little, and D. O'Shea. *Ideals, Varieties and Algorithms.* Springer-Verlag, New York, 1992.

[8] D. S. Dummit and R. M. Foote. *Abstract Algebra.* Prentice Hall Inc., Englewood Cliffs, NJ, 1991.

[9] G. H. Golub and C. F. Van Loan. *Matrix Computations*, 3rd edition. Johns Hopkins University Press, 1996.

[10] D. R. Grayson and M. E. Stillman. Macaulay 2, a software system for research in algebraic geometry. Available at http://www.math.uiuc.edu/Macaulay2/.

[11] R. A. Horn and C. R. Johnson. *Matrix Analysis.* Cambridge University Press, Cambridge, UK, 1995.

[12] M. Marshall. *Positive Polynomials and Sums of Squares.* Math. Surveys Monogr. 146, American Mathematical Society, Providence, RI, 2008.

[13] R. T. Rockafellar. *Convex Analysis.* Princeton University Press, Princeton, New Jersey, 1970.

[14] R. Schneider. *Convex Bodies: The Brunn-Minkowski Theory.* Cambridge University Press, Cambridge, UK, 1993.

[15] G. Strang. *Introduction to Linear Algebra*, 4th edition. Wellesley Cambridge Press, 2009.

Index